Lecture Notes in Artificial Inte

Edited by J. G. Carbonell and J. Siekmann

Subseries of Lecture Notes in Computer Science

Springer

Berlin
Heidelberg
New York
Hong Kong
London
Milan
Paris
Tokyo

Aldo de Moor Wilfried Lex
Bernhard Ganter (Eds.)

Conceptual Structures for Knowledge Creation and Communication

11th International Conference
on Conceptual Structures, ICCS 2003
Dresden, Germany, July 21-25, 2003
Proceedings

Springer

Series Editors

Jaime G. Carbonell, Carnegie Mellon University, Pittsburgh, PA, USA
Jörg Siekmann, University of Saarland, Saarbrücken, Germany

Volume Editors

Aldo de Moor
Tilburg University
Infolab, Dept. of Information Systems and Management
P.O.Box 90153, 5000 LE Tilburg, The Netherlands
E-mail: ademoor@uvt.nl

Wilfried Lex
Technische Universität Clausthal
Institut für Informatik
Julius-Albert-Str.4, 38678 Clausthal-Zellerfeld, Germany
E-mail: lex@informatik.tu-clausthal.de

Bernhard Ganter
Technische Universität Dresden
Institut für Algebra
01062 Dresden, Germany
E-mail: ganter@math.tu-dresden.de

Cataloging-in-Publication Data applied for

A catalog record for this book is available from the Library of Congress.

Bibliographic information published by Die Deutsche Bibliothek.
Die Deutsche Bibliothek lists this publication in the Deutsche Nationalbibliografie;
detailed bibliographic data is available in the Internet at <http://dnb.ddb.de>.

CR Subject Classification (1998): I.2, G.2.2, F.4.1, F.2.1, H.4

ISSN 0302-9743
ISBN 3-540-40576-3 Springer-Verlag Berlin Heidelberg New York

Springer-Verlag Berlin Heidelberg New York,
a member of BertelsmannSpringer Science+Business Media GmbH

http://www.springer.de

© Springer-Verlag Berlin Heidelberg 2003
Printed in Germany

Typesetting: Camera-ready by author, data conversion by Steingräber Satztechnik GmbH
Printed on acid-free paper   SPIN: 10929278   06/3142   5 4 3 2 1 0

Preface

This volume contains the proceedings of ICCS 2003, the 11th International Conference on Conceptual Structures. This conference series continues to be the main forum for the presentation and discussion of state-of-the-art research on conceptual structures. The theories, methodologies, and techniques presented here have grown considerably in scope in recent years. On the other hand, solid bridges spanning the boundaries between such diverse fields as Conceptual Graphs, Formal Concept Analysis, and others are increasingly being built in our community.

The theme of this year's conference was "Conceptual Structures for Knowledge Creation and Communication". In our increasingly (Inter)networked world, the potential support of information technology for the creation and communication of quality knowledge is almost boundless. However, in reality, many conceptual barriers prevent the use of this potential. The main problem is no longer in the technological infrastructure, but in how to navigate, use, and manage the wealth of available data resources. Thus, the question is: how to create and communicate from data the information and ultimately the knowledge required by an ever more complex and dynamic society? Conceptual structures research focuses on what is behind and between the data glut and the information overload that need to be overcome in answering this question. In this way, our field contributes important ideas on how to actually realize some of the many still ambitious visions.

All regular papers were reviewed in a thorough and open process by at least two reviewers and one editor. We thank the authors for their submissions, and express our gratitude to the members of the Editorial Board, the Program Committee and the additional reviewers for their active involvement in reviewing the papers and making constructive suggestions for improvement of the manuscripts. Their commitment has ensured yet again a high-quality conference proceedings.

In addition to the regular papers, we have included several invited papers by leading researchers in the field who present their own, personal views on conceptual structures. Their papers illustrate both the breadth and the depth of the ideas that give our field its distinctive flavour.

July 2003

Aldo de Moor
Wilfried Lex
Bernhard Ganter

Organization

The International Conference on Conceptual Structures (ICCS) is the annual conference and the principal research forum in the theory and practice of conceptual structures. Previous ICCS conferences were held at the Université Laval (Quebec City, 1993), at the University of Maryland (1994), at the University of California (Santa Cruz, 1995), in Sydney (1996), at the University of Washington (Seattle, 1997), in Montpellier (1998), at Virginia Tech (Blacksburg, 1999), at Darmstadt University of Technology (2000), at Stanford University (2001), and in Borovets, Bulgaria (2002).

General Chair

Bernhard Ganter   TU Dresden, Germany

Program Chairs

Aldo de Moor   Tilburg University, The Netherlands
Wilfried Lex   TU Clausthal, Germany

Editorial Board

Galia Angelova (Bulgaria)   Guy Mineau (Canada)
Michel Chein (France)   Bernard Moulin (Canada)
Dan Corbett (Australia)   Marie-Laure Mugnier (France)
Harry Delugach (USA)   Heather Pfeiffer (USA)
Peter Eklund (Australia)   Uta Priss (UK)
Bernhard Ganter (Germany)   John Sowa (USA)
Mary Keeler (USA)   Gerd Stumme (Germany)
Sergei Kuznetsov (Russia)   Bill Tepfenhart (USA)
Lotfi Lakhal (France)   Rudolf Wille (Germany)
Wilfried Lex (Germany)

Program Committee

Jean-François Baget (France)
David Benn (Australia)
Tru Cao (Vietnam)
Claudio Carpineto (Italy)
Frithjof Dau (Germany)
John Esch (USA)
Ad Feelders (The Netherlands)
David Genest (France)
Olivier Gerbé (Canada)
Robert Godin (Canada)
Ollivier Haemmerlé (France)
Roger Hartley (USA)
Cornelis Hoede (The Netherlands)
Adalbert Kerber (Germany)
Pavel Kocura (UK)
Wolfgang Lenski (Germany)
Carsten Lutz (Germany)

Philippe Martin (Australia)
Engelbert Mephu Nguifo (France)
Rokia Missaoui (Canada)
Peter Øhrstrøm (Denmark)
Silke Pollandt (Germany)
Simon Polovina (UK)
Anne-Marie Rassinoux (Switzerland)
Gary Richmond (USA)
Olivier Ridoux (France)
Dan Rochowiak (USA)
Eric Salvat (France)
Janos Sarbo (The Netherlands)
Henrik Scharfe (Denmark)
Finnegan Southey (Canada)
Thanwadee Thanitsukkarn (Thailand)
Petko Valtchev (Canada)
Karl Erich Wolff (Germany)

Further Reviewers

Udo Hebisch (Germany)
Jo Hereth (Germany)

Manfred Jeusfeld (The Netherlands)
Toru Tsujishita (Japan)

Table of Contents

The Many Facets of Conceptual Structures

Logical and Linguistic Aspects

Conceptual Representations of Time and Space

Deepening the Formal Theory

Applications of Conceptual Structures

Conceptual Contents as Information –
Basics for Contextual Judgment Logic

Rudolf Wille

Technische Universität Darmstadt, Fachbereich Mathematik,
Schloßgartenstr. 7, D–64289 Darmstadt
wille@mathematik.tu-darmstadt.de

Abstract. In Contextual Judgment Logic, Sowa's conceptual graphs (understood as graphically structured judgments) are made mathematically explicit as *concept graphs* which represent information formally based on a power context family and rhetorically structured by relational graphs. The *conceptual content* of a concept graph is viewed as the information directly represented by the graph together with the information deducible from the direct information by object and concept implications coded in the power context family. The main result of this paper is that the conceptual contents can be derived as extents of the so-called *conceptual information context* of the corresponding power context family. In short, the conceptual contents of judgments are formally derivable as concept extents.

1   Information in Contextual Judgment Logic

In this paper, *information* in the scope of Conceptual Knowledge Processing shall be understood in the same way as in Devlin's book "Infosense - Turning Information into Knowledge" [De99]. Devlin briefly summarizes his understanding of information and knowledge by the formulas:

Information = Data + Meaning
Knowledge = Internalized information + Ability to utilize the information

Through Devlin's understanding it becomes clear that *Formal Concept Analysis* [GW99] enables to support the representation and processing of information and knowledge as outlined in [Wi02a]. Since *Contextual Logic* [Wi00] with its semantics is based on Formal Concept Analysis, it is desirable to make also explicit why and how Contextual Logic may support the representation and processing of information and knowledge. In this paper, we concentrate on approaching this aim for *Contextual Judgment Logic* [DK03].

   In Contextual Judgment Logic, *judgments* - understood as asserting propositions - are formally mathematized by so-called *concept graphs* which have been semantically introduced in [Wi97] as mathematizations of Sowa's *conceptual graphs* [So84]. Since judgments are philosophically conceived as assertional combinations of concepts, their mathematizatzion is based on formal concepts of formal contexts (introduced in [Wi82]): The semantical base is given by a *power*

A. de Moor, W. Lex, and B. Ganter (Eds.): ICCS 2003, LNAI 2746, pp. 1–15, 2003.

context family defined as a sequence $\vec{\mathbb{K}} := (\mathbb{K}_0, \mathbb{K}_1, \mathbb{K}_2, \ldots)$ of formal contexts $\mathbb{K}_k := (G_k, M_k, I_k)$ with $G_k \subseteq (G_0)^k$ for $k = 1, 2, \ldots$; the formal concepts of $\mathbb{K}_k$ with $k = 1, 2, \ldots$ are called *relation concepts*, because their extents represent k-ary relations on the object set G_0. As an example of a (small) power context family, Empedocles' doctrine of the four basic elements is formally represented in Fig.1 by contexts $\mathbb{K}_0$ and $\mathbb{K}_2$; these contexts are intensionally determined by the two oppositions *cold↔warm* and *dry↔moist* ($\mathbb{K}_1$ might be considered to be the empty context $(\emptyset, \emptyset, \emptyset)$); Fig.2 shows the concept lattices of the formal contexts in Fig.1.

	cold	moisture	dryness	heat	element
water	×	×			×
earth	×		×		×
air		×		×	×
fire			×	×	×

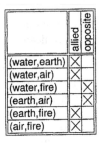

	allied	opposite
(water,earth)	×	
(water,air)	×	
(water,fire)		×
(earth,air)		×
(earth,fire)	×	
(air,fire)	×	

Fig. 1. A power context family $(\mathbb{K}_0, \mathbb{K}_2)$ representing Empedocles' doctrine of the four basic elements

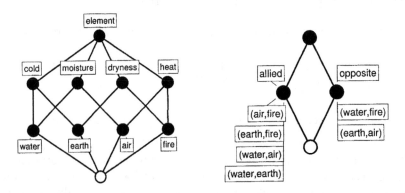

Fig. 2. The concept lattices of the formal contexts in Figure 1

The *affirmative information* represented by a power context family $\vec{\mathbb{K}}$ is given by all relationships of the form $gI_k m$ with $k = 0, 1, 2, \ldots$, $g \in G_k$, $m \in M_k$, and their related meaning. With respect to this underlying information, a formal concept with associated meaning can be understood as an *information unit* joining an extension and an intension (the formal structures of those information units are investigated in Contextual Concept Logic [DK03] which benefits strongly from more than twenty years of research in Formal Concept Analysis [GW99]). Information based on Contextual Judgment Logic is constituted by *elementary judgments* whose information, with respect to the power context family $\vec{\mathbb{K}}$, is formally represented by a pair $(g, \mathfrak{c})$ with $k = 0, 1, 2, \ldots, g \in G_k, \mathfrak{c} \in \mathfrak{B}(\mathbb{K}_k)$, and $g \in Ext(\mathfrak{c})$ (recall that $Ext(\mathfrak{c})$ denotes the extent of $\mathfrak{c}$).

More complex information combining a manifold of elementary judgments needs a *rhetoric structure* which makes the information intelligible. A general type of such rhetoric structures is associated with the *conceptual graphs* invented successfully by J. Sowa [So84]. In Contextual Judgment Logic, Sowa's rhetoric structures are made mathematically explicit as relational graphs which are defined as follows: A *relational graph* is a structure (V, E, ν) consisting of two disjoint sets V and E and a mapping $\nu : E \rightarrow \bigcup_{k=1,2,\ldots} V^k$; the elements of V and E are called *vertices* and *edges*, respectively, and $\nu(e) = (v_1, \ldots, v_k)$ is read: $v_1, \ldots, v_k$ are the *adjacent vertices* of the *k-ary edge* e ($|e| := k$ is the *arity* of e; the arity of a vertex v is defined by $|v| := 0$). Let $E^{(k)}$ be the set of all elements of $V \cup E$ of arity k ($k = 0, 1, 2, \ldots$) (for a modal extension of this definition see [SW03]).

Sowa's conceptual graphs combine rhetoric structures with information based on real world semantics. For making conceptual graphs mathematically explicit as concept graphs, the information is formally based on a power context family semantics and is rhetorically structured by relational graphs. This is elaborated in the following definition: A *concept graph* of a power context family $\vec{\mathbb{K}} := (\mathbb{K}_0, \mathbb{K}_1, \mathbb{K}_2, \ldots)$ with $\mathbb{K}_k := (G_k, M_k, I_k)$ for $k = 0, 1, 2, \ldots$ is a structure $\mathfrak{G} := (V, E, \nu, \kappa, \rho)$ for which

- (V, E, ν) is a relational graph,
- $\kappa : V \cup E \rightarrow \bigcup_{k=0,1,2,\ldots} \mathfrak{B}(\mathbb{K}_k)$ is a mapping such that $\kappa(u) \in \mathfrak{B}(\mathbb{K}_k)$ for all $u \in E^{(k)}$,
- $\rho : V \rightarrow \mathfrak{P}(G_0) \backslash \{\emptyset\}$ is a mapping such that $\rho(v) \subseteq Ext(\kappa(v))$ for all $v \in V$ and, furthermore, $\rho(v_1) \times \cdots \times \rho(v_k) \subseteq Ext(\kappa(e))$ for all $e \in E$ with $\nu(e) = (v_1, \ldots, v_k)$.

(for an extension of this definition to existential concept graphs with object variables see [Wi02b]). It is convenient to consider the mapping ρ not only on vertices but also on edges: For all $e \in E$ with $\nu(e) = (v_1, \ldots, v_k)$, let $\rho(e) := \rho(v_1) \times \cdots \times \rho(v_k)$.

Examples of concept graphs of the power context family represented in Fig.1 are depicted in Fig.3: The two concept graphs together inform that the element *water*, belonging to the concept *moisture*, is allied with the element *air*, belonging to the concept *heat*, and opposite to the element *fire*, belonging to the concept *heat* too.

2   Conceptual Contents of Concept Graphs

In the remaining paper, we discuss how the information transmitted by concept graphs might be mathematically represented; a further discussion of rhetoric

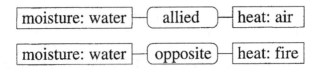

Fig. 3. Two concept graphs of the power context family in Fig.1

structures for formally represented information has to be postponed. Obviously, a concept graph $\mathfrak{G} := (V, E, \nu, \kappa, \rho)$ contains the *conceptual information units* $(g, \kappa(u))$ with $k = 0, 1, 2, \ldots$, $u \in E^{(k)}$, and $g \in \rho(u)$; for instance, the conceptual information units contained in the upper concept graph of Fig.3 are $(water, \mu_0(moisture))$, $(air, \mu_0(heat))$, $((water, air), \mu_2(allied))$.

An affirmative proposition might even transmit further information if *common preknowledge* is activated; for instance, the information "Gillian is a boy" with the conceptual information unit $(Gillian, boy)$ also transmits the conceptual information unit $(Gillian, male)$. To enable such inferences, J. Sowa assumes that the concepts in a conceptual graph belong to a corresponding concept hierarchy (type hierarchy). The philosopher R. B. Brandom assumes, in his widely discussed theory of discursive practice [Br94], even richer inference structures to derive the so-called *conceptual content* of an asserting proposition. For the foundation of Contextual Judgment Logic, such richer inferences can be rendered possible by assuming inference structures on object sets and on concept sets to capture common preknowledge; this opens rich possibilities for inferring new conceptual information units from given ones. For finally defining the conceptual content of concept graphs, it is helpful to show first that the assumed implication structures can be represented in power context families.

We start with the mathematically defined notions of an "implicational object structure" and an "implicational concept structure": An *implicational object structure* is a set structure $(G, \mathfrak{A})$ with $\mathfrak{A} \subseteq \mathfrak{P}(G)^2$; the elements of G and $\mathfrak{A}$ are called *objects* and *object implications*, respectively. An *implicational concept structure* is a set structure $(H, \mathfrak{B})$ with $\mathfrak{B} \subseteq \mathfrak{P}(H)^2$; the elements of H and $\mathfrak{B}$ are called *concepts*[1] and *concept implications*, respectively. If $(A, C) \in \mathfrak{A}$ resp. $(B, D) \in \mathfrak{B}$, we often write $A \to C$ resp. $B \to D$ and call A resp. B *premises* and C resp. D *conclusions*. The implications give rise to two closure systems defined by

$$\mathcal{C}(G, \mathfrak{A}) := \{E \subseteq G \mid \forall (A, C) \in \mathfrak{A} : A \not\subseteq E \text{ or } C \subseteq E\},$$
$$\mathcal{C}(H, \mathfrak{B}) := \{F \subseteq H \mid \forall (B, D) \in \mathfrak{B} : B \not\subseteq F \text{ or } D \subseteq F\}.$$

For each $A \subseteq G$ resp. $B \subseteq H$, there is a smallest superset $\overline{A}$ resp. $\overline{B}$ in the corresponding closure system, which is called the *implicational closure* of A resp. B. A consequence is that $(\mathcal{C}(G, \mathfrak{A}), \subseteq)$ and $(\mathcal{C}(H, \mathfrak{B}), \subseteq)$ are complete lattices.

Let us demonstrate the introduced notions by a small example about the size of celestial bodies with the following implicational object structure $(G, \mathfrak{A})$ and implicational concept structure $(H, \mathfrak{B})$:

---

[1] In this and the next paragraphs, "concept" is used as a primitive notion.

$(G, \mathfrak{A}) := (\{sun, earth, moon\}, \{(\{sun, moon\}, \{earth\})\}),$
$(H, \mathfrak{B}) := (\{large, small, not\, large, not\, small\}, \{(\{large\}, \{not\, small\}),$
    $(\{small\}, \{not\, large\}), (\{large, not\, large\}, H), (\{small, not\, small\}, H)\});$

the listed adjectives should all be understood as completed with the noun "celestial body" (c.b.), i.e., they name concepts of celestial bodies. The two closure structures $(\mathcal{C}(G, \mathfrak{A}), \subseteq)$ and $(\mathcal{C}(H, \mathfrak{B}), \subseteq)$ are depicted by lattice diagrams in Fig.4; in such a diagram, a circle represents the closure which consists of all elements whose names are attached to circles on pathes descending from that circle.

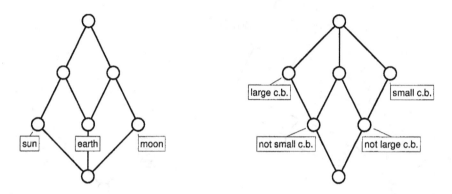

Fig. 4. Lattice diagrams of two closure systems concerning the size of celestial bodies

Next, we consider how conceptual information units may increase the preknowledge coded in an implicational object structure $(G, \mathfrak{A})$ and an implicational concept structure $(H, \mathfrak{B})$. Let us assume for our example that we have the conceptual information units $(sun, large\,c.b.)$ and $(moon, small\,c.b.)$ as further preknowledge. Then we can infer the conceptual information units $(sun, not\,small\,c.b.)$ and $(moon, not\,large\,c.b.)$. The preknowledge extended in this way is represented by the left diagram in Fig.5 in which the right diagram of Fig.4 is turned upside down (this makes it easier to read off the inferred information). The structure which combines the two implicational structures via conceptual information units shall be, in general, mathematized by the following notion: An *implicational lattice context* is defined as a formal context (K, L, R) formed by complete lattices K and L satisfying

(1) $a \leq_K c$ in K, $d \leq_L b$ in L, and $(c, d) \in R$ imply $(a, b) \in R$,
(2) $A \times B \subseteq R$ implies $(\bigvee_K A, \bigwedge_L B) \in R$.

In terms of the two closure systems, K can be understood as the complete lattice $(\mathcal{C}(G, \mathfrak{A}), \subseteq)$ and L as the dual of the complete lattice $(\mathcal{C}(H, \mathfrak{B}), \subseteq)$; R then yields the known conceptual information units including the ones inferred by the conditions (1) and (2) according to their closure-theoretic meaning. This can be described as follows:

(1): If the object a is in the closure of the object c and the concept b in the closure of the concept d where (c,d) is a known conceptual information unit then we can infer the conceptual information unit (a,b).

(2): If for an object set A and a concept set B the conceptual infomation unit (a,b) is known for all $a \in A$ and $b \in B$ then we can infer all conceptual information units $(\tilde{a}, \tilde{b})$ for which the object $\tilde{a}$ is in the closure of A and the concept $\tilde{b}$ is in the closure of B.

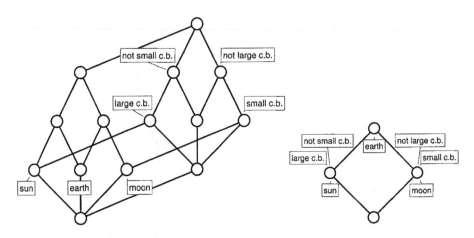

Fig. 5. Line diagrams of an implicational lattice context and its concept lattice

For our aim to represent the object and concept implications in one contextual structure, the following proposition yields an important step:

Proposition 1. *Let (K, L, R) be an implicational lattice context, let $K \dot{\cup} L$ be the disjoint union of K and L, and let $\leq$ be the transitive closure of the union of the relations $\leq_K$, R, and $\leq_L$ on $K \dot{\cup} L$ where $\leq_K$ and $\leq_L$ are the order relations of the complete lattices K and L, respectively. Then $(K \dot{\cup} L, \leq)$ is a complete lattice.*

Proof: Let $A \subseteq K \dot{\cup} L$ and let B be the set of all upper bounds of A in $(K \dot{\cup} L, \leq)$. Then, because of (2), we obtain $(\bigvee_K(A \cap K), \bigwedge_L(B \cap L)) \in R$. Since $A \cap K = A$ or $B \cap L = B$, we have to consider two cases: If $A \cap K = A$ then $\bigvee_K A \leq \bigwedge_L(B \cap L)$; hence $\bigvee_K A = \bigvee A$. If $B \cap L = B$ then $\bigwedge_L B \geq \bigvee_K(A \cap K)$; hence $\bigwedge_L B = \bigvee A$. Thus, $(K \dot{\cup} L, \leq)$ is a $\bigvee$-semilattice with 0_K as smallest element and therefore a complete lattice.   □

According to Proposition 1, the left diagram in Fig.5 is a line diagram of the lattice $(K \dot{\cup} L, \leq)$ representing the information of our celestial body example. Although the information is already represented in one lattice, the interrelationships between the object structure and the concept structure is not as tight as in a concept lattice. Thus, it might be better to consider the concept lattice of the implicational lattice context (K, L, R) (for our example, the concept lattice of the clarified context is depicted by the right diagram in Fig.5). To clarify the connection between $\underline{\mathfrak{B}}(K, L, R)$ and $(K \dot{\cup} L, \leq)$, we need some further definitions and results:

For $a \in K$, let $a^\uparrow := \bigwedge_L \{b \in L \mid (a, b) \in R\}$; then (2) implies $(a, a^\uparrow) \in R$. Dually, for $b \in L$, let $b^\downarrow := \bigvee_K \{a \in K \mid (a, b) \in R\}$; then (2) implies $(b^\downarrow, b) \in R$. The following consequences can be easily proved:

$$(\star) \quad a^\uparrow = a^{\uparrow\downarrow\uparrow}, \ b^\downarrow = b^{\downarrow\uparrow\downarrow}, \text{ and } (a, b) \in R \iff (a^{\uparrow\downarrow}, b^{\downarrow\uparrow}) \in R.$$

Let $R_\uparrow^\downarrow := \{(a, b) \in R \mid a^\uparrow = b \text{ and } b^\downarrow = a\}$.

Proposition 2. *The assignments $(a, b) \mapsto \gamma a (= \mu b)$ and $(A, B) \mapsto (\bigvee_K A, \bigwedge_L B)$ yield isomorphisms between $(R_\uparrow^\downarrow, \leq)$ and $\underline{\mathfrak{B}}(K, L, R)$ which are inverse to each other; in particular, $(R_\uparrow^\downarrow, \leq)$ is a complete lattice.*

Proof: We define $\varphi : (R_\uparrow^\downarrow, \leq) \to \underline{\mathfrak{B}}(K, L, R)$ by $\varphi(a, b) := \gamma a$ and $\psi : \underline{\mathfrak{B}}(K, L, R) \to (R_\uparrow^\downarrow, \leq)$ by $\psi(A, B) := (\bigvee_K A, \bigwedge_L B)$. Clearly, φ and ψ are order-preserving. Furthermore, $\psi\varphi(a, b) = \psi(([a]_K, [b]_L) = (a, b)$ and $\varphi\psi(A, B) = \varphi(\bigvee_K A, \bigwedge_L B) = (([\bigvee_K A]_K, [\bigwedge_L B]_L) = (A, B)$. Thus, φ and ψ are isomorphisms inverse to each other. $(R_\uparrow^\downarrow, \leq) \cong \underline{\mathfrak{B}}(K, L, R)$ obviously implies that $(R_\uparrow^\downarrow, \leq)$ is a complete lattice. $\square$

As Fig.5 shows, both lattices $(K \dot{\cup} L, \leq)$ and $\underline{\mathfrak{B}}(K, L, R)$ might have their advantages and disadvantages. Both lattices represent completely the given object and concept implications and the assumed and inferred conceptual information units, but $(K \dot{\cup} L, \leq)$ might miss interrelationships between the object structure and the concept structure, and $\underline{\mathfrak{B}}(K, L, R)$ might represent further object implications, concept implications, and conceptual information units. The disadvantages can be removed, in the case of $(K \dot{\cup} L, \leq)$, by appropriately assigning new objects and concepts to elements of $K \dot{\cup} L$ and, in the case of $\underline{\mathfrak{B}}(K, L, R)$, by appropriately assuming further conceptual information units in (K, L, R). In our celestial body example, the insertion of the two conceptual information units $(earth, not\, large\, c.b.)$ and $(earth, not\, small\, c.b.)$ yields an implicational lattice context $(K, L, \hat{R})$ for which the extent lattice is isomorphic to $(\mathcal{C}(G, \mathfrak{A}), \subseteq)$ and the concept lattice to $(\mathcal{C}(H, \mathfrak{B}), \subseteq)$; those isomorphic lattices can also be derived as an isomorphic copy of the quotient lattice $(K \dot{\cup} L, \leq)/(id_K \cup id_L \cup \hat{R}_\uparrow^\downarrow \cup (\hat{R}_\uparrow^\downarrow)^{-1})$ (cf. Fig.6). Thus, for representing object implications, concept implications, and conceptual information units by an implicational lattice context (K, L, R), we can assume in general $K = L$ and $R = \leq_L$ with the consequence that $L \cong \underline{\mathfrak{B}}(K, L, R)$.

Now, we can summarize that given object implications, concept implications, and conceptual information units may be appropriately represented by a formal context and its concept lattice. A power context family $\vec{\mathbb{K}} := (\mathbb{K}_0, \mathbb{K}_1, \mathbb{K}_2, \dots)$ can therefore be considered as a simultaneous representation of given k-ary object implications, concept implications, and conceptual information units by the

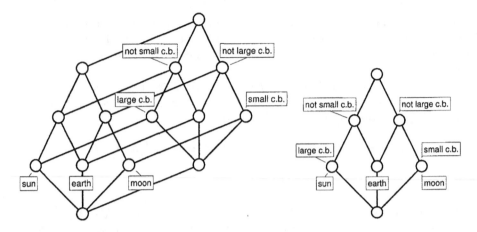

Fig. 6. Line diagrams of the improved implicational lattice context and its concept lattice

contexts $\mathbb{K}_k$ and their concept lattices $(k = 0, 1, 2, \ldots)$. The *material implications* coded in the power context family can be made formally explicit as follows (cf. [Wi02b]): Let $k = 0, 1, 2, \ldots$; for $A, C \subseteq G_k$, $\mathbb{K}_k$ satisfies $A \to C$ if $A^{I_k} \subseteq C^{I_k}$ and, for $\mathfrak{B}, \mathfrak{D} \subseteq \mathfrak{B}(\mathbb{K}_k)$, $\mathbb{K}_k$ satisfies $\mathfrak{B} \to \mathfrak{D}$ if $\bigwedge \mathfrak{B} \leq \bigwedge \mathfrak{D}$. The *formal implications* $A \to C$ and $\mathfrak{B} \to \mathfrak{D}$ give rise to a closure system $\mathcal{C}(\mathbb{K}_k)$ on $\mathbb{S}^{imp}(\mathbb{K}_k) := \{(g, \mathfrak{b}) \in G_k \times \mathfrak{B}(\mathbb{K}_k) \mid g \in Ext(\mathfrak{b})\}$ consisting of all subsets Y of $\mathbb{S}^{imp}(\mathbb{K}_k)$ which have the following property:

(P_k) If $A \times \mathfrak{B} \subseteq Y$ and if $\mathbb{K}_k$ satifies $A \to C$ and $\mathfrak{B} \to \mathfrak{D}$ then $C \times \mathfrak{D} \subseteq Y$.

Finally, we are able to define the conceptual content of concept graphs (understood as formalized judgments): The k-*ary conceptual content* $C_k(\mathfrak{G})$ (also called the $\mathbb{K}_k$-*conceptual content*) of a concept graph $\mathfrak{G} := (V, E, \nu, \kappa, \rho)$ of a power context family $\vec{\mathbb{K}}$ is defined as the closure of $\{(g, \kappa(u)) \mid u \in E^{(k)} \text{ and } g \in \rho(u)\}$ with respect to the closure system $\mathcal{C}(\mathbb{K}_k)$ $(k = 0, 1, 2, \ldots)$. Then

$$C(\mathfrak{G}) := C_0(\mathfrak{G}) \,\dot\cup\, C_1(\mathfrak{G}) \,\dot\cup\, C_2(\mathfrak{G}) \,\dot\cup\, \ldots$$

is called the $(\vec{\mathbb{K}}\text{-})$*conceptual content* of the concept graph $\mathfrak{G}$.

The conceptual contents give rise to an *information* (*quasi-*)*order* $\lesssim$ on the set of all concept graphs of a power context family: A concept graph $\mathfrak{G}_1 := (V_1, E_1, \nu_1, \kappa_1, \rho_1)$ is said to be *less informative* (*more general*) than a concept graph $\mathfrak{G}_2 := (V_2, E_2, \nu_2, \kappa_2, \rho_2)$ (in symbols: $\mathfrak{G}_1 \lesssim \mathfrak{G}_2$) if $C_k(\mathfrak{G}_1) \subseteq C_k(\mathfrak{G}_2)$ for $k = 0, 1, 2, \ldots$; $\mathfrak{G}_1$ and $\mathfrak{G}_2$ are called *equivalent* (in symbols: $\mathfrak{G}_1 \sim \mathfrak{G}_2$) if $\mathfrak{G}_1 \lesssim \mathfrak{G}_2$ and $\mathfrak{G}_2 \lesssim \mathfrak{G}_1$ (i.e., $C_k(\mathfrak{G}_1) = C_k(\mathfrak{G}_2)$ for $k = 0, 1, 2, \ldots$). The set of all equivalence classes of concept graphs of a power context family $\vec{\mathbb{K}}$ together with the order induced by the quasi-order $\lesssim$ is an ordered set denoted by $\vec{\Gamma}(\vec{\mathbb{K}})$. The ordered set $\vec{\Gamma}(\vec{\mathbb{K}})$ is even a complete lattice as shown in [Wi02b].

3   Conceptual Contents as Formal Concepts

The *conceptual contents* of the concept graphs of a power context family can be understood as the information representable by these concept graphs according to the object and concept implications coded in the power context family. For storing, activating, and processing such information, it is useful to find for it a suitable representation by *formal concepts* independent of concept graphs. That shall be elaborated in this section by first representing the k-ary conceptual contents as extents of formal concepts and then extending that result to conceptual contents in general.

We start with an implicational lattice context $(L, L, \leq_L)$ in which L can be understood as the concept lattice $\mathfrak{B}(\mathbb{K}_k)$ of all k-ary concepts of the given power context family and $\leq_L$ as the subconcept-superconcept-relation between the k-ary concepts. Now, the *implicational lattice structure* of such an implicational lattice context $(L, L, \leq_L)$ is defined as the ordered structure $\underline{\mathbb{S}}^{imp}(L) :=$ $(\mathbb{S}^{imp}(L), \leq, \check{\bigvee})$ with $\mathbb{S}^{imp}(L) := \{(a, b) \in L^2 \mid 0 <_L a \leq_L b\}$ where the order $\leq$ and the partial supremum $\check{\bigvee}$ are fixed by

(1) $(a, b) \leq (c, d) : \Longleftrightarrow a \leq_L c$ and $d \leq_L b$,

(2) $\check{\bigvee} A \times B := (\bigvee_L A, \bigwedge_L B)$ for $A \times B \subseteq \mathbb{S}^{imp}(L)$.

In this structure, the *implicational closure* $C^{imp}(X)$ of a subset $X \subseteq \mathbb{S}^{imp}(L)$ is the smallest order ideal of $(\mathbb{S}^{imp}(L), \leq)$ containing X closed under the partial supremum $\check{\bigvee}$. The set of all implicational closures of $\underline{\mathbb{S}}^{imp}(L)$ ordered by set-inclusion is denoted by $\underline{\mathcal{C}}(\underline{\mathbb{S}}^{imp}(L))$.

Lemma 1. *For* $(x, y) \in \mathbb{S}^{imp}(L)$, *the principal ideal* $((x, y)]$ *always equals the intersection* $((x, x)] \cap ((y, y)]$.

Proof: In general, we have $((z, z)] = \{(a, b) \in \mathbb{S}^{imp}(L) \mid a \leq_L z \leq_L b\}$. For $(x, y) \in \mathbb{S}^{imp}(L)$, this yields $((x, y)] = \{(a, b) \in \mathbb{S}^{imp}(L) \mid (a, b) \leq (x, y)\} = \{(a, b) \in \mathbb{S}^{imp}(L) \mid a \leq_L x$ and $y \leq_L b\} = ((x, x)] \cap ((y, y)]$. □

The assertion of Lemma 1 may be checked in the ordered set $(\mathbb{S}^{inf}(\underline{\mathfrak{B}}(\mathbb{K})), \leq)$ represented in Fig.7 for the left concept lattice in Fig.2.

Lemma 2. *Let* $\mathfrak{S}^{con}(L)$ *be the set of all convex subsets* S *of* $L \backslash \{0\}$ *for which* $S \cup \{0\}$ *is a complete sublattice of the interval* $[0, \bigvee S]$ *of* L. *There is an injective map* $\iota : \mathfrak{S}^{con}(L) \longrightarrow \underline{\mathcal{C}}(\underline{\mathbb{S}}^{imp}(L))$ *which is given by* $\iota(S) := \bigcup_{z \in S}((z, z)]$. *Conversely, for each implicational closure* C *in* $\underline{\mathbb{S}}^{imp}(L)$ *and each selection map* $\alpha : C \longrightarrow \bigcup_{(x,y) \in C}\{x, y\}$ *with* $\alpha(x, y) \in \{x, y\}$, $\alpha(C) := \{\alpha(x, y) \mid (x, y) \in C\}$ *is a convex subset of* $L \backslash \{0\}$ *and* $\alpha(C) \cup \{0\}$ *is a complete sublattice of the interval* $[0, \bigvee \alpha(C)]$ *of* L, *i.e.* $\alpha(C) \in \mathfrak{S}^{con}(L)$.

Proof: For $S \in \mathfrak{S}^{con}(L)$, $\iota(S) := \bigcup_{z \in S}((z, z)]$ is obviously an order ideal of $\underline{\mathbb{S}}^{imp}(L)$. For $\emptyset \neq A \times B \subseteq \iota(S)$ and for each $(a, b) \in A \times B$, there is a $z_{(a,b)} \in \iota(S)$ with $(a, b) \leq (z_{(a,b)}, z_{(a,b)})$ which is equivalent to $a \leq_L z_{(a,b)} \leq_L b$. It follows that $\bigvee A \leq_L \bigvee_{a \in A} z_{(a,b)} \leq_L b$ for all $b \in B$ and hence $\bigvee A \leq_L \bigwedge_{b \in B} \bigvee_{a \in A} z_{(a,b)} \leq_L$

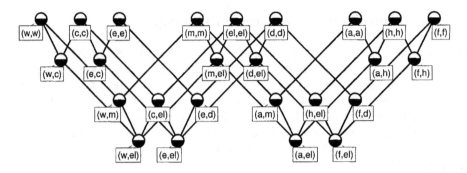

Fig. 7. Line diagrams of the implicational lattice structure of the left concept lattice in Fig.2

$\bigwedge B$. Since S is convex and $\bigvee A, \bigwedge B \in S$, we also obtain $\bigwedge_{b \in B} \bigvee_{a \in A} z_{(a,b)} \in S$ and therefore

$$\check{\bigvee}(A \times B) = (\bigvee A, \bigwedge B) \leq (\bigwedge_{b \in B} \bigvee_{a \in A} z_{(a,b)}, \bigwedge_{b \in B} \bigvee_{a \in A} z_{(a,b)}) \in \iota(S);$$

hence $\check{\bigvee}(A \times B) \in \iota(S)$ which completes the proof that $\iota(S)$ is an implicational closure in $\underline{\mathbb{S}}^{imp}(L)$. Since the pairs (z, z) with $z \in L \setminus \{0\}$ are exactly the maximal elements of $\underline{\mathbb{S}}^{imp}(L)$, the map ι is injective.

Conversely, let C be an implicational closure in $\underline{\mathbb{S}}^{imp}(L)$ and let $\alpha : C \longrightarrow \bigcup_{(x,y) \in C} \{x, y\}$ be a selection map with $\alpha(x, y) \in \{x, y\}$. Let $a, b \in \alpha(C)$ and $c \in L \setminus \{0\}$ with $a \leq c \leq b$. Since $(a, a), (b, b) \in C$ by Lemma 1 and therefore $(a, c), (c, b) \in C$, we have $\{a, c\} \times \{c, b\} \subseteq C$ and hence $\check{\bigvee}\{a, c\} \times \{c, b\} = (c, c) \in C$, i.e. $c \in \alpha(C)$. This proves that $\alpha(C)$ is convex. Now, let $\emptyset \neq A \subseteq \alpha(C)$. Since $(a, \bigvee A) \in ((a, a)]$ for $a \in A$, we obtain $\check{\bigvee}_{a \in A}(a, \bigvee A) = (\bigvee A, \bigvee A) \in C$ and hence $\bigvee A \in \alpha(C)$. If $\bigwedge A \neq 0$ then $(\bigwedge A, a) \in ((a, a)]$ for $a \in A$ and therefore $\check{\bigvee}_{a \in A}(\bigwedge A, a) = (\bigwedge A, \bigwedge A) \in C$; hence $\bigwedge A \in \alpha(C)$. Thus, we have proved that $\alpha(C) \cup \{0\}$ is a complete sublattice of the interval $[0, \bigvee \alpha(C)]$ of L. $\quad\square$

Lemma 3. *Each implicational closure in $\underline{\mathbb{S}}^{imp}(L)$ is the intersection of implicational closures of the form $\iota(S)$ with $S \in \mathfrak{S}^{con}(L)$.*

Proof: Let C be an implicational closure in $\underline{\mathbb{S}}^{imp}(L)$. With Lemma 1 we obtain

$$C = \bigcup_{(x,y) \in C} ((x, x)] \cap ((y, y)] = \bigcap_{\alpha \in \prod_{(x,y) \in C} \{x,y\}} \bigcup_{(x,y) \in C} ((\alpha(x, y), \alpha(x, y))].$$

Since $\iota(\alpha(C)) = \bigcup_{(x,y) \in C} ((\alpha(x, y), \alpha(x, y))]$, the obtained equalities together with Lemma 2 prove the assertion of the lemma. $\quad\square$

For an implicational lattice context $(L, L, \leq_L)$, the corresponding *conceptual information context* shall be defined as the formal context

$$\mathbb{K}^{inf}(L) := (\mathbb{S}^{imp}(L), \mathfrak{S}^{con}(L), \hat{\Delta})$$

with $(x, y)\hat{\Delta}S : \iff [x, y] \cap S \neq \emptyset$. Now, the Lemmas 1, 2, and 3 are summarized by the following proposition:

Proposition 3. *For an implicational lattice context* $(L, L, \leq_L)$, *the extents of the corresponding conceptual information context* $\mathbb{K}^{inf}(L)$ *are exactly the implicational closures of the implicational lattice structure* $\underline{\mathbb{S}}^{imp}(L)$.

Proof: In general, for $S \in \mathfrak{S}^{con}(L)$, we have the equivalences $(x, y) \in S^{\hat{\Delta}} \iff \exists z \in S : x \leq_L z \leq_L y \iff \exists z \in S : (x, y) \leq (z, z) \iff (x, y) \in \iota(S)$ which imply $S^{\hat{\Delta}} = \iota(S)$. By Lemma 2, this yields that each attribute extent $S^{\hat{\Delta}}$ is an implicational closure in $\underline{\mathbb{S}}^{imp}(L)$. An arbitrary extent X of $\mathbb{K}^{inf}(L)$ is the intersection of suitable attribute extents $S_t^{\hat{\Delta}}(= \iota(S_t))$ $(t \in T)$; hence X is also an implicational closure of $\underline{\mathbb{S}}^{imp}(L)$ and, conversely, each implicational closure is an extent of $\mathbb{K}^{inf}(L)$ by Lemma 3. $\qquad \square$

Proposition 3 can be easily generalized to arbitrary formal contexts $\mathbb{K} := (G, M, I)$. For doing this, we need the general definition of a *conceptual information context* corresponding to $\mathbb{K}$ as a formal context

$$\mathbb{K}^{inf}(\mathbb{K}) := (\mathbb{S}^{imp}(\mathbb{K}), \mathfrak{S}^{con}(\underline{\mathfrak{B}}(\mathbb{K})), \bar{\Delta})$$

with $\mathbb{S}^{imp}(\mathbb{K}) := \{(g, \mathfrak{b}) \in G \times \underline{\mathfrak{B}}(\mathbb{K}) \mid g \in Ext(\mathfrak{b})\}$ and $(g, \mathfrak{b})\bar{\Delta}S : \iff [\gamma g, \mathfrak{b}] \cap S \neq \emptyset$. The *implicational context structure* of a formal context $\mathbb{K} := (G, M, I)$ is then defined as the ordered structure $\underline{\mathbb{S}}^{imp}(\mathbb{K}) := (\mathbb{S}^{imp}(\mathbb{K}), \leq, \check{\bigvee})$ with $\mathbb{S}^{imp}(\mathbb{K}) := \{(g, \mathfrak{b}) \in G \times \underline{\mathfrak{B}}(\mathbb{K}) \mid g \in Ext(\mathfrak{b})\}$ where the order $\leq$ and the partial multi-operation $\check{\bigvee}$ are fixed by

    (1) $(g, \mathfrak{b}) \leq (h, \mathfrak{d}) : \iff \gamma g \leq \gamma h$ and $\mathfrak{d} \leq \mathfrak{b}$,
    (2) $\check{\bigvee} A \times \mathfrak{B} := \{(g, \bigwedge \mathfrak{B}) \mid g \in A''\}$ for $A \times \mathfrak{B} \subseteq \mathbb{S}^{imp}(\mathbb{K})$.

In this structure, the *implicational closure* $C^{imp}(X)$ of a subset $X \subseteq \mathbb{S}^{imp}(\mathbb{K})$ is the smallest order ideal of $(\mathbb{S}^{imp}(\mathbb{K}), \leq)$ containing X closed under the partial multi-operation $\check{\bigvee}$. The set of all implicational closures of $\underline{\mathbb{S}}^{imp}(\mathbb{K})$ ordered by set-inclusion is denoted by $\underline{\mathcal{C}}(\underline{\mathbb{S}}^{imp}(\mathbb{K}))$.

Now, we are able to deduce from Proposition 3 the main result about the *conceptual contents of formal contexts* $\mathbb{K}$ (which are understood as the conceptual contents of the concept graphs of a power context family having at the most $\mathbb{K}_0 := \mathbb{K}$ as a non-empty context):

Basic Theorem on $\mathbb{K}$-Conceptual Contents. *For a formal context $\mathbb{K}$ with $\emptyset^{II} = \emptyset$, the extents of the corresponding conceptual information context $\mathbb{K}^{inf}(\mathbb{K})$ are exactly the implicational closures of the implicational context structure $\underline{\mathbb{S}}^{imp}(\mathbb{K})$; those implicational closures are exactly the conceptual contents of $\mathbb{K}$.*

Proof: Let β be a map from the set $\mathfrak{U}(\mathbb{K}^{inf}(\mathbb{K}))$ of all extents of the conceptual information context $\mathbb{K}^{inf}(\mathbb{K})$ (corresponding to the formal context $\mathbb{K}$) into the power set of $\mathbb{S}^{imp}(\underline{\mathfrak{B}}(\mathbb{K}))$ and let δ be a map from the set $\mathfrak{U}(\mathbb{K}^{inf}(\underline{\mathfrak{B}}(\mathbb{K})))$ of all extents of the conceptual information context $\mathbb{K}^{inf}(\underline{\mathfrak{B}}(\mathbb{K}))$ (corresponding to the implicational lattice context $(\mathfrak{B}(\mathbb{K}), \mathfrak{B}(\mathbb{K}), \leq))$ into the power set of $\mathbb{S}^{imp}(\mathbb{K})$ defined by

$$\beta(U) := \{(\mathfrak{a}, \mathfrak{b}) \in \mathbb{S}^{imp}(\mathfrak{B}(\mathbb{K})) \mid \forall g \in Ext(\mathfrak{a}) : (g, \mathfrak{b}) \in U\},$$
$$\delta(V) := \{(g, \mathfrak{b}) \in \mathbb{S}^{imp}(\mathbb{K}) \mid \exists (\mathfrak{a}, \mathfrak{b}) \in V : g \in Ext(\mathfrak{a})\}.$$

For each extent U of $\mathbb{K}^{inf}(\mathbb{K})$ and each extent V of $\mathbb{K}^{inf}(\mathfrak{B}(\mathbb{K}))$, we obtain the equalities $U^{\hat{\Delta}} = \beta(U)^{\hat{\Delta}}$ and $V^{\hat{\Delta}} = \delta(V)^{\bar{\Delta}}$ by the following equivalences and their dual analogues, respectively:

$$S \in \beta(U)^{\hat{\Delta}} \iff \forall (\mathfrak{a}, \mathfrak{b}) \in \beta(U) \exists \mathfrak{c} \in S : \mathfrak{a} \leq \mathfrak{c} \leq \mathfrak{b}$$
$$\iff \forall (\mathfrak{a}, \mathfrak{b}) \in \beta(U) \forall g \in Ext(\mathfrak{a}) \exists \mathfrak{c}_g \in S : \gamma g \leq \mathfrak{c}_g \leq \mathfrak{b}$$
$$\iff \forall (g, \mathfrak{b}) \in U \exists \mathfrak{c}_g \in S : \gamma g \leq \mathfrak{c}_g \leq \mathfrak{b}$$
$$\iff S \in U^{\bar{\Delta}}.$$

For $(\mathfrak{a}, \mathfrak{b}) \in \beta(U)^{\hat{\Delta}\hat{\Delta}}$ and $S \in \beta(U)^{\hat{\Delta}} = U^{\bar{\Delta}}$, there is always some $\mathfrak{c} \in S$ with $\mathfrak{a} \leq \mathfrak{c} \leq \mathfrak{b}$ and hence $\gamma g \leq \mathfrak{c} \leq \mathfrak{b}$ for all $g \in Ext(\mathfrak{a})$; because of $U^{\bar{\Delta}\bar{\Delta}} = U$, it follows that $(g, \mathfrak{b}) \in U$ for all $g \in Ext(\mathfrak{a})$ and therefore $(\mathfrak{a}, \mathfrak{b}) \in \beta(U)$, which proves $\beta(U)^{\hat{\Delta}\hat{\Delta}} = \beta(U)$. Thus, β maps $\mathfrak{U}(\mathbb{K}^{inf}(\mathbb{K}))$ into $\mathfrak{U}(\mathbb{K}^{inf}(\mathfrak{B}(\mathbb{K})))$. For $(g, \mathfrak{b}) \in \delta(V)^{\bar{\Delta}\bar{\Delta}}$ and $S \in \delta(V)^{\bar{\Delta}} = V^{\hat{\Delta}}$, there is always some $\mathfrak{c} \in S$ with $\gamma g \leq \mathfrak{c} \leq \mathfrak{b}$; because of $V^{\hat{\Delta}\hat{\Delta}} = V$, it follows that $(g, \mathfrak{b}) \in \delta(V)$, which proves $\delta(V)^{\bar{\Delta}\bar{\Delta}} = \delta(V)$. Thus, δ maps $\mathfrak{U}(\mathbb{K}^{inf}(\mathfrak{B}(\mathbb{K})))$ into $\mathfrak{U}(\mathbb{K}^{inf}(\mathbb{K}))$. With $U^{\bar{\Delta}} = \beta(U)^{\hat{\Delta}}$ and $V^{\hat{\Delta}} = \delta(V)^{\bar{\Delta}}$, it can be easily checked that $\delta(\beta(U)) = U$ and $\beta(\delta(V)) = V$; therefore β and δ are bijective maps which are inverse to each other.

By Proposition 3, the extents of $\mathbb{K}^{inf}(\mathfrak{B}(\mathbb{K}))$ are exactly the implicational closures of $\underline{\mathbb{S}}^{imp}(\mathfrak{B}(\mathbb{K}))$. Next we prove that δ is a bijection from the set of all implicational closures of $\underline{\mathbb{S}}^{imp}(\mathfrak{B}(\mathbb{K}))$ onto the set of all implicational closures of $\underline{\mathbb{S}}^{imp}(\mathbb{K})$, which finally implies the assertion that the extents of $\mathbb{K}^{inf}(\mathbb{K})$ are exactly the implicational closures of $\underline{\mathbb{S}}^{imp}(\mathbb{K})$ because of $\delta^{-1} = \beta$:

For $(h, \mathfrak{d}) \leq (g, \mathfrak{b})$ in $\underline{\mathbb{S}}^{imp}(\mathbb{K})$, $(g, \mathfrak{b}) \in \delta(V)$ implies $\exists (\mathfrak{a}, \mathfrak{b}) \in V : g \in Ext(\mathfrak{a})$, hence $\exists (\mathfrak{a}, \mathfrak{d}) \in V : h \in Ext(\mathfrak{a})$, and therefore $(h, \mathfrak{d}) \in \delta(V)$; thus, $\delta(V)$ is an order ideal in $\underline{\mathbb{S}}^{imp}(\mathbb{K})$. $A \times \mathfrak{B} \subseteq \delta(V)$ implies $\forall g \in A \forall \mathfrak{b} \in \mathfrak{B} \exists (\mathfrak{a}_g^{\mathfrak{b}}, \mathfrak{b}) \in V : g \in Ext(\mathfrak{a}_g^{\mathfrak{b}})$, hence $(\bigvee_{g \in A, \mathfrak{b} \in \mathfrak{B}} \mathfrak{a}_g^{\mathfrak{b}}, \bigwedge \mathfrak{B}) \in V$ and $A \subseteq Ext(\bigvee_{g \in A, \mathfrak{b} \in \mathfrak{B}} \mathfrak{a}_g^{\mathfrak{b}})$, and therefore $\{(g, \bigwedge \mathfrak{B}) \mid g \in A''\} \subseteq \delta(V)$; thus, we obtain $\check{\bigvee} A \times \mathfrak{B} \subseteq \delta(V)$ which concludes the proof that $\delta(V)$ is an implicational closure of $\underline{\mathbb{S}}^{imp}(\mathbb{K})$.

Conversely, let C be an implicational closure of $\underline{\mathbb{S}}^{imp}(\mathbb{K})$. Then $(\mathfrak{a}, \mathfrak{b}) \in \beta(C)$ and $(\mathfrak{c}, \mathfrak{d}) \leq (\mathfrak{a}, \mathfrak{b})$ imply $\forall g \in Ext(\mathfrak{a}) : (g, \mathfrak{b}) \in C$, hence $\forall g \in Ext(\mathfrak{c}) : (g, \mathfrak{d}) \in C$, and therefore $(\mathfrak{c}, \mathfrak{d}) \in \beta(C)$; thus, $\beta(C)$ is an order ideal in $(\mathbb{S}^{imp}(\mathfrak{B}(\mathbb{K})), \leq)$. $(\mathfrak{A}, \mathfrak{B}) \subseteq \beta(C)$ implies $\forall (\mathfrak{a}, \mathfrak{b}) \in \mathfrak{A} \times \mathfrak{B} \forall g \in Ext(\mathfrak{a}) : (g, \mathfrak{b}) \in C$, hence $\forall \mathfrak{b} \in \mathfrak{B} \forall g \in \bigcup_{\mathfrak{a} \in \mathfrak{A}} Ext(\mathfrak{a}) : (g, \mathfrak{b}) \in C$, and therefore $\check{\bigvee}(\bigcup_{\mathfrak{a} \in \mathfrak{A}} Ext(\mathfrak{a})) \times \mathfrak{B} \subseteq C$ thus, we obtain $\{(g, \bigwedge \mathfrak{B}) \mid g \in Ext(\bigvee \mathfrak{A})\} \subseteq C$ and so $(\bigvee \mathfrak{A}, \bigwedge \mathfrak{B}) \in \beta(C)$ which concludes the proof that $\beta(C)$ is an implicational closure of $\underline{\mathbb{S}}^{imp}(\mathfrak{B}(\mathbb{K}))$.

It remains to prove that the implicational closures of $\underline{\mathbb{S}}^{imp}(\mathbb{K}))$ are exactly the conceptual contents of $\mathbb{K} := (G, M, I)$. Let U be such an implicational closure, let $\mathbb{K}$ satisfy $A \to C$ with $A, C \subseteq G$ and $\mathfrak{B} \to \mathfrak{D}$ with $\mathfrak{B}, \mathfrak{D} \subseteq \mathfrak{B}(\mathbb{K})$, and let $A \times \mathfrak{B} \subseteq U$. It follows that $(C'', C') \leq (A'', A') \leq \bigwedge \mathfrak{B} \leq \bigwedge \mathfrak{D}$ and hence $C \times \mathfrak{D} \subseteq$

$\mathbb{S}^{imp}(\mathbb{K})$; for $(g, \bigwedge \mathfrak{D}) \in C \times \mathfrak{D}$, we obtain $(g, \bigwedge \mathfrak{D}) \leq (g, \bigwedge \mathfrak{B}) \in \check{\bigvee} A \times \mathfrak{B} \subseteq U$ and therefore $(g, \mathfrak{d}) \in U$ for all $g \in C$ and $\mathfrak{d} \in \mathfrak{D}$, i.e. $C \times \mathfrak{D} \subseteq U$ and $U \in \mathcal{C}(\mathbb{K})$. Obviously, $(U, \emptyset, \emptyset, \kappa, \rho)$ with $\kappa(g, \mathfrak{b}) := \mathfrak{b}$ and $\rho(g, \mathfrak{b}) := \{g\}$ is a concept graph of $\mathbb{K}$ with U as conceptual content. Conversely, let U be a conceptual content of a concept graph of $\mathbb{K}$, i.e. $U \in \mathcal{C}(\mathbb{K})$. If $(g, \mathfrak{b}) \in U$ and $(h, \mathfrak{d}) \leq (g, \mathfrak{b})$ then $\mathbb{K}$ satisfies $\{g\} \rightarrow \{h\}$ and $\{\mathfrak{b}\} \rightarrow \{\mathfrak{d}\}$ and therefore $(h, \mathfrak{d}) \in U$; thus, U is an order ideal of $\mathbb{S}^{imp}(\mathbb{K})$. Finally, let $A \times \mathfrak{B} \subseteq U$. Since $\mathbb{K}$ satisfies $A \rightarrow A''$ and $\mathfrak{B} \rightarrow \{\bigwedge \mathfrak{B}\}$, we obtain $\check{\bigvee} A \times \mathfrak{B} = \{(g, \bigwedge \mathfrak{B}) \mid g \in A''\} \subseteq U$ which finishes the proof that the conceptual content U is an implicational closure of $\underline{\mathbb{S}}^{imp}(\mathbb{K})$. $\square$

For our example, Fig.8 shows, according to the Basic Theorem on $\mathbb{K}$-Conceptual Contents, the representation of the conceptual contents as implicational closures and also as extents of the depicted concept lattice.

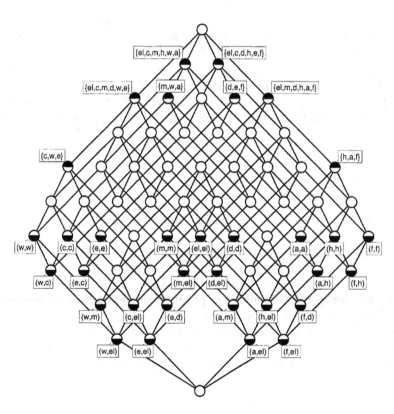

Fig. 8. Line diagram of the lattice of all implicational closures of the implicational lattice structure in Fig.7 and also of the concept lattice of the corresponding conceptual information context having as extents the conceptual contents of the formal context $\mathbb{K}_0$ in Fig.1

Finally, the Basic Theorem on $\mathbb{K}$-Conceptual Contents shall be extended to the general case of (limited) power context families. Let $\vec{\mathbb{K}} := (\mathbb{K}_0, \mathbb{K}_1, \ldots, \mathbb{K}_n)$ be a power context family with $\mathbb{K}_k := (G_k, M_k, I_k)$ $(k = 0, 1, \ldots, n)$. The *conceptual information context* corresponding to $\vec{\mathbb{K}}$ is defined as the formal context

$$\mathbb{K}^{inf}(\vec{\mathbb{K}}) := \underline{\mathbb{K}}^{inf}(\mathbb{K}_0) + \underline{\mathbb{K}}^{inf}(\mathbb{K}_1) + \cdots + \underline{\mathbb{K}}^{inf}(\mathbb{K}_n).$$

An extent U of $\mathbb{K}^{inf}(\vec{\mathbb{K}})$ is said to be *rooted* if $((g_1, \ldots, g_k), \mathfrak{b}_k) \in U$ always implies $(g_j, \top_0) \in U$ for $j = 1 \ldots, k$ and $\top_0 := (G_0, G_0^{I_0})$. Now we are able to formulate the desired theorem:

Basic Theorem on $\vec{\mathbb{K}}$-Conceptual Contents. *For a (limited) power context family $\vec{\mathbb{K}}$ with $\emptyset^{I_k I_k} = \emptyset$ for $k = 0, 1, 2, \ldots$, the conceptual contents of the concept graphs of $\vec{\mathbb{K}}$ are exactly the rooted extents of the corresponding conceptual information context $\mathbb{K}^{inf}(\vec{\mathbb{K}})$.*

Proof: The conceptual content of a concept graph $\mathfrak{G}$ of $\vec{\mathbb{K}} := (\mathbb{K}_0, \mathbb{K}_1, \ldots, K_n)$ is the disjoint uninion $C(\mathfrak{G}) := C_0(\mathfrak{G}) \dot\cap C_1(\mathfrak{G}) \dot\cap \cdots \dot\cap C_n(\mathfrak{G})$ where $C_k(\mathfrak{G})$ is an extent of $\mathbb{K}^{inf}(\mathbb{K}_k)$ $(k = 0, 1, \ldots, n)$ by the Basic Theorem on $\mathbb{K}$-Conceptual Contents. Therefore $C(\mathfrak{G})$ is an extent of $\mathbb{K}^{inf}(\vec{\mathbb{K}})$; this extent is rooted as a direct consequence of the definition of concept graphs. Conversely, let U be a rooted extent of $\mathbb{K}^{inf}(\vec{\mathbb{K}})$. Then $U_k := U \cap G_k$ is an extent of $\mathbb{K}_k := (G_k, M_k, I_k)$ for each $k = 0, 1, \ldots, n$ and hence an implicational closure of $\underline{\mathbb{S}}^{imp}(\mathbb{K}_k))$ by the Basic Theorem on $\mathbb{K}$-Conceptual Contents. This suggests the following construction of a concept graph $\mathfrak{G} := (V, E, \nu, \kappa, \rho)$ of $\vec{\mathbb{K}}$:

$$V := U_0, \; E := \bigcup_{k=1,\ldots,n} U_k, \; \nu((g_1, \ldots, g_k), \mathfrak{b}_k) := ((g_1, \top_0), \ldots, (g_n, \top_0)),$$
$$\kappa(g, \mathfrak{b}_0) := \mathfrak{b}_0 \text{ and } \kappa((g_1, \ldots, g_k), \mathfrak{b}_k) := \mathfrak{b}_k, \; \rho(g, \mathfrak{b}_0) := \{g\},$$
$$\text{and consequently } \rho((g_1, \ldots, g_k), \mathfrak{b}_k) := \{(g_1, \ldots, g_k)\}.$$

Clearly, this concept graph has U as its conceptual content. Thus the main theorem is proved. $\qquad\square$

Now, it would be desirable to extensively elaborate how the basic theorems on conceptual contents could be used to support the representation and processing of information and knowledge. Since that has to be postponed because of the restrictions given to this conference paper, it should only be mentioned that the representation of the conceptual contents of concept graphs by extents of suitable contexts opens the door to develop TOSCANA-systems for concept graphs of power context families (cf. [EW00]); in particular, the new TOSCANAJ 1.0 software (http://toscanaj.sourceforge.net) would strongly support such a development.

References

[Br94]   R. B. Brandom: *Making it explicit. Reasoning, representing, and discursive commitment.* Havard University Press, Cambridge 1994.

[DK03]   F. Dau, J. Klinger: From Formal Concept Analysis to Contextual Logic. FB4-Preprint, TU Darmstadt 2003.

[De99]   K. Devlin: *Infosense - turning information into knowledge.* Freeman, New York 1999.

[EW00]   P. Eklund, B. Groh, G. Stumme, R. Wille: A contextual-logic extension of TOSCANA. In: B. Ganter, G. Mineau (eds.): *Conceptual structures: logical, linguistic, and computational issues.* LNAI **1867**. Springer, Heidelberg 2000, 453–467.

[GW99]   B. Ganter, R. Wille: *Formal Concept Analysis: mathematical foundations.* Springer, Heidelberg 1999; German version: Springer, Heidelberg 1996.

[SW03]   L. Schoolmann, R. Wille: Concept graphs with subdivision: a semantic approach. This volume.

[So84]   J. F. Sowa: *Conceptual structures: information processing in mind and machine.* Adison-Wesley, Reading 1984.

[Wi82]   R. Wille: Restructuring lattice theory: an approach based on hierarchies of concepts. In: I. Rival (ed.): *Ordered sets.* Reidel, Dordrecht-Boston 1982, 445–470.

[Wi97]   R. Wille: Conceptual Graphs and Formal Concept Analysis. In: D. Lukose, H. Delugach, M. Keeler, L. Searle, J. F. Sowa (eds.): *Conceptual structures: fulfilling Peirce's dream.* LNAI **1257**. Springer, Heidelberg 1997, 290–303.

[Wi00]   R. Wille: Contextual Logic summary. In: G. Stumme (ed.): *Working with Conceptual Structures. Contributions to ICCS 2000.* Shaker, Aachen 2000, 265–276.

[Wi02a]   R. Wille: Why can concept lattices support knowledge discovery in databases? *J. Expt. Theor. Artif. Intell.* 14 (2002), 81–92.

[Wi02b]   R. Wille: Existential concept graphs of power context families. In: U. Priss, D. Corbett, G. Angelova (eds.): *Conceptual structures: integration and interfaces.* LNAI **2393**. Springer, Heidelberg 2002, 382–395.

Analogical Reasoning

John F. Sowa and Arun K. Majumdar

VivoMind LLC

Abstract. Logical and analogical reasoning are sometimes viewed as mutually exclusive alternatives, but formal logic is actually a highly constrained and stylized method of using analogies. Before any subject can be formalized to the stage where logic can be applied to it, analogies must be used to derive an abstract representation from a mass of irrelevant detail. After the formalization is complete, every logical step – of deduction, induction, or abduction – involves the application of some version of analogy. This paper analyzes the relationships between logical and analogical reasoning, and describes a highly efficient analogy engine that uses conceptual graphs as the knowledge representation. The same operations used to process analogies can be combined with Peirce's rules of inference to support an inference engine. Those operations, called the *canonical formation rules* for conceptual graphs, are widely used in CG systems for language understanding and scene recognition as well as analogy finding and theorem proving. The same algorithms used to optimize analogy finding can be used to speed up all the methods of reasoning based on the canonical formation rules.

1 Analogy and Perception

Before discussing the use of analogy in reasoning, it is important to analyze the concept of analogy and its relationship to other cognitive processes. General-purpose dictionaries are usually a good starting point for conceptual analysis, but they seldom go into sufficient depth to resolve subtle distinctions. A typical dictionary lists synonyms for the word *analogy*, such as *similarity, resemblance,* and *correspondence.* Then it adds more specialized word senses, such as *a similarity in some respects of things that are otherwise dissimilar, a comparison that determines the degree of similarity,* or *an inference based on resemblance or correspondence.* In AI, analogy-finding programs have been written since the 1960s, but they often use definitions of analogy that are specialized to a particular application.

The VivoMind Analogy Engine (VAE), which is described in Section 3, is general enough to be used in any application domain. Therefore, VAE leads to fundamental questions about the nature of analogy that have been debated in the literature of cognitive science. One three-party debate has addressed many of those issues:

1. *Thesis*: For the Structure Mapping Engine (SME), Falkenheimer, Forbus, and Gentner (1989) defined analogy as the recognition that "one thing is

A. de Moor, W. Lex, and B. Ganter (Eds.): ICCS 2003, LNAI 2746, pp. 16–36, 2003.

like another" if there is a mapping from a conceptual structure that describes the first one to a conceptual structure that describes the second. Their implementation in SME has been applied to a wide variety of practical applications and to psychological studies that compare the SME approach to the way people address the same problems.

2. *Antithesis*: In their critique of SME, Chalmers, French, and Hofstadter (1992) consider analogy to be an aspect of a more general cognitive function called *high-level perception* (HLP), by which an organism constructs a conceptual representation of a situation. They "argue that perceptual processes cannot be separated from other cognitive processes even in principle, and therefore that traditional artificial-intelligence models cannot be defended by supposing the existence of a 'representation module' that supplies representations ready-made." They criticize the "hand-coded rigid representations" of SME and insist that "content-dependent, easily adaptable representations" must be "an essential part of any accurate model of cognition."

3. *Synthesis*: In summarizing the debate, Morrison and Dietrich (1995) observed that the two positions represent different perspectives on related, but different aspects of cognition: SME employs structure mapping as "a general mechanism for all kinds of possible comparison domains" while "HLP views analogy as a process from the bottom up; as a representation-building process based on low-level perceptual processes interacting with high-level concepts." In their response to the critics, Forbus et al. (1998) admitted that a greater integration with perceptual mechanisms is desirable, but they repeated their claim that psychological evidence is "overwhelmingly" in favor of structure mapping "as a model of human analogical processing."

The VAE approach supports Point #3: a comprehensive theory of cognition must integrate the structure-building processes of perception with the structure-mapping processes of analogy. For finding analogies, VAE uses a high-speed implementation of structure mapping, but its algorithms are based on low-level operations that are also used to build the structures. In the first implementation, the conceptual structures were built during the process of parsing and interpreting natural language. More recently, the same low-level operations have been used to build conceptual structures from sensory data and from percept-like patterns used in scene recognition. VAE demonstrates that perception, language understanding, and structure mapping can be based on the same kinds of operations.

This paper discusses the interrelationships between logical and analogical reasoning, analyzes the underlying cognitive processes in terms of Peirce's semiotics and his classification of reasoning, and shows how those processes are supported by VAE. The same graph operations that support analogical reasoning can also be used to support formal reasoning. Instead of being mutually exclusive, logical reasoning is just a more cultivated variety of analogical reasoning. For many purposes, especially in language understanding, the analogical processes provide greater flexibility than the more constrained and less adaptable variety used in logic. But since logical and analogical reasoning share a common basis, they can be effectively used in combination.

2   Logical and Analogical Reasoning

In developing formal logic, Aristotle took Greek mathematics as his model. Like his predecessors Socrates and Plato, Aristotle was impressed with the rigor and precision of geometrical proofs. His goal was to formalize and generalize those proof procedures and apply them to philosophy, science, and all other branches of knowledge. Yet not all subjects are equally amenable to formalization. Greek mathematics achieved its greatest successes in astronomy, where Ptolemy's calculations remained the standard of precision for centuries. But other subjects, such as medicine and law, depend more on deep experience than on brilliant mathematical calculations. Significantly, two of the most penetrating criticisms of logic were written by the physician Sextus Empiricus in the second century AD and by the legal scholar Ibn Taymiyya in the fourteenth century.

Sextus Empiricus, as his nickname suggests, was an empiricist. By profession, he was a physician; philosophically, he was an adherent of the school known as the Skeptics. Sextus maintained that all knowledge must come from experience. As an example, he cited the following syllogism:

> Every human is an animal.
> Socrates is human.
> Therefore, Socrates is an animal.

Sextus admitted that this syllogism represents a valid inference pattern, but he questioned the source of evidence for the major premise *Every human is an animal.* A universal proposition that purports to cover every instance of some category must be derived by induction from particulars. If the induction is incomplete, then the universal proposition is not certain, and there might be some human who is not an animal. But if the induction is complete, then the particular instance Socrates must have been examimed already, and the syllogism is redundant or circular. Since every one of Aristotle's valid forms of syllogisms contains at least one universal affirmative or universal negative premise, the same criticisms apply to all of them: the conclusion must be either uncertain or circular.

The Aristotelians answered Sextus by claiming that universal propositions may be true by definition: since the type Human is defined as rational animal, the essence of human includes animal; therefore, no instance of human that was not an animal could exist. This line of defense was attacked by the Islamic jurist and legal scholar Taqi al-Din Ibn Taymiyya. Like Sextus, Ibn Taymiyya agreed that the form of a syllogism is valid, but he did not accept Aristotle's distinction between essence and accident (Hallaq 1993). According to Aristotle, the essence of human includes both rational and animal. Other attributes, such as laughing or being a featherless biped, might be unique to humans, but they are *accidental* attributes that could be different without changing the essence. Ibn Taymiyya, however, maintained that the distinction between essence and accident was arbitrary. Human might just as well be defined as laughing animal, with rational as an accidental attribute.

Denouncing logic would be pointless if no other method of reasoning were possible. But Ibn Taymiyya had an alternative: the legal practice of reasoning by cases and analogy. In Islamic law, a new case is *assimilated* to one or more previous cases that serve as precedents. The mechanism of assimilation is analogy, but the analogy must be guided by a cause that is common to the new case as well as the earlier cases. If the same cause is present in all the cases, then the earlier judgment can be transferred to the new case. As an example, it is written in the Koran that grape wine is prohibited, but nothing is said about date wine. The judgment for date wine would be derived in four steps:

1. Given case: Grape wine is prohibited.
2. New case: Is date wine prohibited?
3. Cause: Grape wine is prohibited because it is intoxicating; date wine is also intoxicating.
4. Judgment: Date wine is also prohibited.

In practice, the reasoning may be more complex. Several previous cases may have a common cause but different judgments. Then the analysis must determine whether there are mitigating circumstances that affect the operation of the cause. But the principles remain the same: analogy guided by rules of evidence and relevance determines the common cause, the effect of the mitigating circumstances, and the judgment.

Besides arguing in favor of analogy, Ibn Taymiyya also replied to the logicians who claimed that syllogistic reasoning is certain, but analogy is merely probable. He admitted that logical deduction is certain when applied to purely mental constructions in mathematics. But in any reasoning about the real world, universal propositions can only be derived by induction, and induction must be guided by the same principles of evidence and relevance used in analogy. Figure 1 illustrates Ibn Taymiyya's argument: Deduction proceeds from a *theory* containing universal propositions. But those propositions must have earlier been derived by induction using the methods of analogy. The only difference is that induction produces a theory as intermediate result, which is then used in a subsequent process of deduction. By using analogy directly, legal reasoning dispenses with the intermediate theory and goes straight from cases to conclusion. If the theory and the analogy are based on the same evidence, they must lead to the same conclusions.

The question in Figure 1 represents some known aspects of a new case, which has unknown aspects to be determined. In deduction, the known aspects are compared (by a version of structure mapping called *unification*) with the premises of some implication. Then the unknown aspects, which answer the question, are derived from the conclusion of the implication. In analogy, the known aspects of the new case are compared with the corresponding aspects of the older cases. The case that gives the best match may be assumed as the best source of evidence

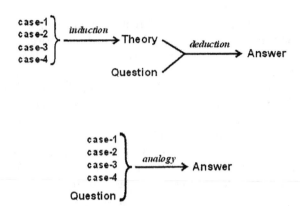

Fig. 1. Comparison of logical and analogical reasoning

for estimating the unknown aspects of the new case. The other cases show alternative possibilities for those unknown aspects; the closer the agreement among the alternatives, the stronger the evidence for the conclusion.

Both Sextus Empiricus and Ibn Taymiyya admitted that logical reasoning is valid, but they doubted the source of evidence for universal propositions about the real world. What they overlooked was the pragmatic value of a good theory: a small group of scientists can derive a theory by induction, and anyone else can apply it without redoing the exhaustive analysis of cases. The two-step process of induction followed by deduction has proved to be most successful in the physical sciences, which include physics, chemistry, molecular biology, and the engineering practices they support. The one-step process of case-based reasoning, however, is more successful in fields outside the so-called "hard" sciences, such as business, law, medicine, and psychology. Even in the "soft" sciences, which are rife with exceptions, a theory that is successful most of the time can still be useful. Many cases in law or medicine can be settled by the direct application of some general principle, and only the exceptions require an appeal to a long history of cases. And even in physics, the hardest of the hard sciences, the theories may be well established, but the question of which theory to apply to a given problem usually requires an application of analogy. In both science and daily life, there is no sharp dichotomy between subjects amenable to strict logic and those that require analogical reasoning.

The informal arguments illustrated in Figure 1 are supported by an analysis of the algorithms used for logical reasoning. Following is Peirce's classification of the three kinds of logical reasoning and the way that the structure-mapping operations of analogy are used in each of them:

- **Deduction.** A typical rule used in deduction is *modus ponens*: given an assertion p and an axiom of the form p implies q, deduce the conclusion q. In most applications, the assertion p is not identical to the p in the axiom, and structure mapping is necessary to *unify* the two ps before the rule can be applied. The most time-consuming task is not the application of a single rule, but the repeated use of analogies for finding patterns that may lead to successful rule applications.

- **Induction.** When every instance of p is followed by an instance of q, induction is peformed by assuming that p implies q. Since the ps and qs are rarely identical in every occurrence, a form of analogy called *generalization* is used to derive the most general implication that subsumes all the instances.
- **Abduction.** The operation of guessing or forming an initial hypothesis is what Peirce called abduction. Given an assertion q and an axiom of the form p implies q, the guess that p is a likely cause or explanation for q is an act of abduction. The operation of guessing p uses the least constrained version of analogy, in which some parts of the matching graphs may be more generalized while other parts are more specialized.

As this discussion indicates, analogy is a prerequisite for logical reasoning, which is a highly disciplined method of using repeated analogies. In both human reasoning and computer implementations, the same underlying operations can be used to support both.

3 Analogy Engine

The VivoMind Analogy Engine (VAE), which was developed by Majumdar, is a high-performance analogy finder that uses conceptual graphs for the knowledge representation. Like SME, structure mapping is used to find analogies. Unlike SME, the VAE algorithms can find analogies in time proportional to $(N \log N)$, where N is the number of nodes in the current knowledge base or context. SME, however, requires time proportional to N^3 (Forbus et al. 1995). A later version called MAC/FAC reduced the time by using a search engine to extract the most likely data before using SME to find analogies (Forbus et al. 2002). With its greater speed, VAE can find analogies in the entire WordNet knowledge base in just a few seconds, even though WordNet contains over 10^5 nodes. For that size, one second with an $(N \log N)$ algorithm would correspond to 30 years with an N^3 algorithm.

VAE can process CGs from any source: natural languages, programming languages, and any kind of information that can be represented in graphs, such as organic molecules or electric-power grids. In an application to distributed interacting agents, VAE processes both English messages and signals from sensors that monitor the environment. To determine an agent's actions, VAE searches for analogies to what humans did in response to similar patterns of messages and signals. To find analogies, VAE uses three methods of comparison, which can be used separately or in combination:

1. **Matching type labels.** Method #1 compares nodes that have identical labels, labels that are related as subtype and supertype such as Cat and Animal, or labels that have a common supertype such as Cat and Dog.
2. **Matching subgraphs.** Method #2 compares subgraphs with possibly different labels. This match succeeds when two graphs are isomorphic (inde-

pedent of the labels) or when they can be made isomorphic by combining adjacent nodes.

3. **Matching transformations.** If the first two methods fail, Method #3 searches for transformations that can relate subgraphs of one graph to subgraphs of the other.

These three methods of matching graphs were inspired by Peirce's categories of Firstness, Secondness, and Thirdness (Sowa 2000). The first compares two nodes by what they contain in themselves independent of any other nodes; the second compares nodes by their relationships to other nodes; and the third compares the mediating transformations that may be necessary to make the graphs comparable. To illustrate the first two methods, the following table shows an analogy found by VAE when comparing the background knowledge in WordNet for the concept types `Cat` and `Car`:

Analogy of Cat to Car	
Cat	Car
head	hood
eye	headlight
cornea	glass plate
mouth	fuel cap
stomach	fuel tank
bowel	combustion chamber
anus	exhaust pipe
skeleton	chasis
heart	engine
paw	wheel
fur	paint

Fig. 2. An analogy discovered by VAE

As Figure 2 illustrates, there is an enormous amount of background knowledge stored in lexical resources such as WordNet. It is not organized in a form that is precise enough for deduction, but it is adequate for the more primitive method of analogy.

Since there are many possible paths through all the definitions and examples of WordNet, most comparisons generate multiple analogies. To evaluate the evidence for any particular mapping, a *weight of evidence* is computed by using heuristics that estimate the closeness of the match. For Method #1 of matching type labels, the closest match results from identical labels. If the labels are not identical, the weight of evidence decreases with the distance between the labels in the type hierarchy:

1. Identical type labels, such as `Cat` to `Cat`.
2. Subtype to supertype, such as `Cat` to `Animal`.
3. Siblings of the same supertype, such as `Cat` to `Dog`.
4. More distant cousins, such as `Cat` to `Tree`.

For Method #2 of matching subgraphs, the closest match results from finding that both graphs, in their entirety, are isomorphic. The weight of evidence decreases as their common subgraphs become smaller or if the graphs have to be modified in order to force a match:

1. Match isomorphic graphs.
2. Match two graphs that have isomorphic subgraphs (the larger the subgraphs, the stronger the evidence for the match).
3. Combine adjacent nodes to make the subgraphs isomorphic.

The analogy shown in Figure 2 received a high weight of evidence because VAE found many matching labels and large matching subgraphs in corresponding parts of a cat and parts of a car:

- Some of the corresponding parts have similar functions: fur and paint are outer coverings; heart and engine are internal parts that have a regular beat; skeleton and chasis are structures to which other parts are attached; paw and wheel perform a similar function, and there are four of each.
- The longest matching subgraph is the path from mouth to stomach to bowel to anus of a cat, which matches the path from fuel cap to fuel tank to combustion chamber to exhaust pipe of a car. The stomach of a cat and the fuel tank of a car are analogous because they are both subtypes of Container. The bowel and the combustion chamber perform analogous functions. The mouth and the fuel cap are considered input orifices, and the anus and the exhaust pipe are outputs. The weight of evidence is somewhat reduced because adjustments must be made to ignore nodes that do not match: the esophagus of a cat does not match anything in WordNet's description of a car, and the muffler of a car does not match anything in its description of a cat.
- A shorter subgraph is the path from head to eyes to cornea of a cat, which matches the path from hood to headlights to glass plate of a car. The head and the hood are both in the front. The eyes are analogous to the headlights because there are two of each and they are related to light, even though the relationships are different. The cornea and the glass plate are in the front, and they are both transparent.

Each matching label and each structural correspondence contributes to the weight of evidence for the analogy, depending on the closeness of the match and the exactness of the correspondence.

As the cat-car comparison illustrates, analogy is a versatile method for using informal, unstructured background knowledge. But analogies are also valuable for comparing the highly formalized knowledge of one axiomatized theory to another. In the process of theory revision, Niels Bohr used an analogy between gravitational force and electrical force to derive a theory of the hydrogen atom as analogous to the earth revolving around the sun. Method #3 of analogy, which finds matching transformations, can also be used to determine the precise mappings required for transforming one theory or representation into another.

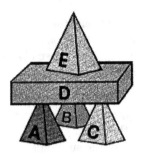

Fig. 3. A physical structure to be represented by data

As an example, Figure 3 shows a physical structure that could be represented by many different data structures.

Programmers who use different tools, databases, or programming languages often use different, but analogous representations for the same kinds of information. LISP programmers, for example, prefer to use lists, while FORTRAN programmers prefer vectors. Conceptual graphs are a highly general representation, which can represent any kind of data stored in a digital computer, but the types of concepts and relations usually reflect the choices made by the original programmer, which in turn reflect the options available in the original programming tools. Figure 4 shows a representation for Figure 3 that illustrates the typical choices used with relational databases.

Objects			Supports	
ID	Shape	Color	Supporter	Supportee
A	pyramid	red	A	D
B	pyramid	green	B	D
C	pyramid	yellow	C	D
D	block	blue	D	E
E	pyramid	orange	F	G
F	block	blue	H	G
G	block	orange		
H	block	blue		

Fig. 4. Two structures represented in a relational database

On the left of Figure 4 are two structures: a copy of Figure 3 and an arch constructed from three blocks. On the right are two tables: the one labeled Objects lists the identifiers of all the objects in both tables with their shapes and colors; the one labeled Supports lists each object that supports (labeled Supporter) and the object supported (labeled Supportee). As Figure 4 illustrates, a relational database typically scatters the information about a single object or structure of objects into multiple tables. For the structure of pyramids and blocks, each object is listed once in the `Objects` table, and one or more times in either or both columns of the `Supports` table. Furthermore, information about the two disconnected structures shown on the left is intermixed in both tables. When all the information about the structure at the top left is extracted from both tables of Figure 4, it can be mapped to the conceptual graph of Figure 5.

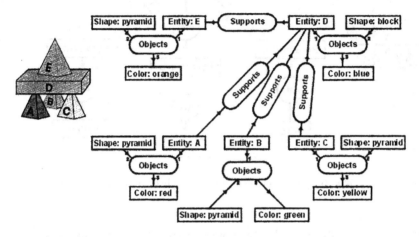

Fig. 5. A CG derived from the relational DB

In Figure 5, each row of the table labeled `Objects` is represented by a conceptual relation labeled `Objects`, and each row of the table labeled `Supports` is represented by a conceptual relation labeled `Supports`. The type labels of the concepts are mostly derived from the labels on the columns of the two tables in Figure 4. The only exception is the label `Entity`, which is used instead of `ID`. The reason for that exception is that `ID` is a metalevel term about the representation language; it is not a term that is derived from the entities in the domain of discourse. The concept [`Entity: E`], for example, says that E is an instance of type `Entity`. The concept [`ID: ''E''`], however, would say that the character string ''E'' is an instance of type `ID`. The use of the label `Entity` instead of `ID` avoids mixing the metalevel with the object level. Such mixing of levels is common in most programs, since the computer ignores any meaning that might be associated with the labels. In logic, however, the fine distinctions are important, and CGs mark them consistently.

When natural languages are translated to CGs, the distinctions must be enforced by the semantic interpreter. Figure 6 shows a CG that represents the English sentence, *A red pyramid A, a green pyramid B, and a yellow pyramid C support a blue block D, which supports an orange pyramid E.* The conceptual relations labeled `Thme` and `Inst` represent the case relations theme and instrument. The relations labeled `Attr` represent the attribute relation between a concept of some entity and a concept of some attribute of that entity. The type labels

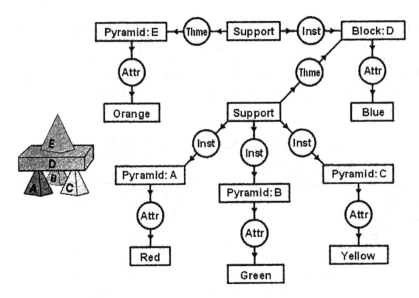

Fig. 6. A CG derived from an English sentence

of concepts are usually derived from nouns, verbs, adjectives, and adverbs in English.

Although the two conceptual graphs represent equivalent information, they look very different. In Figure 5, the CG derived from the relational database has 15 concept nodes and 9 relation nodes. In Figure 6, the CG derived from English has 12 concept nodes and 11 relation nodes. Furthermore, no type label on any node in Figure 5 is identical to any type label on any node in Figure 6. Even though some character strings are similar, their positions in the graphs cause them to be treated as distinct. In Figure 5, `orange` is the name of an instance of type `Color`; and in Figure 6, `Orange` is the label of a concept type. In Figure 5, `Supports` is the label of a relation type; and in Figure 6, `Support` is not only the label of a concept type, it also lacks the final S.

Because of these differences, the strict method of unification cannot show that the graphs are identical or even related. Even the more relaxed methods of matching labels or matching subgraphs are unable to show that the two graphs are analogous. Method #3 of analogy, however, can find matching transformations that can translate Figure 5 into Figure 6 or vice-versa. When VAE was asked to compare those two graphs, it found the two transformations shown in Figure 7. Each transformation determines a mapping between a type of subgraph in Figure 5 and another type of subgraph in Figure 6.

The two transformations shown in Figure 7 define a version of *graph grammar* for parsing one kind of graph and mapping it to the other. The transformation at the top of Figure 7 can be applied to the five subgraphs containing the relations of type `Objects` in Figure 5 and relate them to the five subgraphs containing the relations of type `Attr` in Figure 6. That same transformation could be applied in

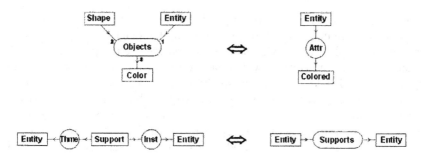

Fig. 7. Two transformations discovered by VAE

reverse to relate the five subgraphs of Figure 6 to the five subgraphs of Figure 5. The transformation at the bottom of Figure 7 could be applied from right to left in order to map Figure 6 to Figure 5. When applied in that direction, it would map three different subgraphs, which happen to contain three common nodes: the subgraph extending from [Pyramid: A] to [Block: D]; the one from [Pyramid: B] to [Block: D]; and the one from [Pyramid: C] to [Block: D]. When applied in the reverse direction, it would map three subgraphs of Figure 5 that contained only one common node.

The transformations shown in Figure 7 have a high weight of evidence because they are used repeatedly in exactly the same way. A single transformation of one subgraph to another subgraph with no matching labels would not contribute anything to the weight of evidence. But if the same transformation is applied twice, then its likelihood is greatly increased. Transformations that can be applied three times or five times to relate all the nodes of one graph to all the nodes of another graph have a likelihood that comes close to being a certainty.

Of the three methods of analogy used in VAE, the first two – matching labels and matching subgraphs – are also used in SME. Method #3 of matching transformations, which only VAE is capable of performing, is more complex because it depends on analogies of analogies. Unlike the first two methods, which VAE can perform in $(N \log N)$ time, Method #3 takes polynomial time, and it can only be applied to much smaller amounts of data. In practice, Method #3 is usually applied to small parts of an analogy in which most of the mapping is done by the first two methods and only a small residue of unmatched nodes remains to be mapped. In such cases, the number N is small, and the mapping can be done quickly. Even for mapping Figure 5 (with $N = 9$) to Figure 6 (with $N = 11$), the Method #3 took a few seconds, whereas the time for Methods #1 and #2 on graphs of such size would be less than a millisecond.

Each of the three methods of analogy determines a mapping of one CG to another. The first two methods determine a node-by-node mapping of CGs, where some or all of the nodes of the first CG may have different type labels from the corresponding nodes of the other. Method #3 determines a more complex mapping, which comprises multiple mappings of subgraphs of one CG to subgraphs of the other. These methods can be applied to CGs derived from any source, including natural languages, logic, or programming languages.

In one major application, VAE was used to analyze the programs and documentation of a large corporation, which had systems in daily use that were up to forty years old (LeClerc & Majumdar 2002). Although the documentation specified how the programs were supposed to work, nobody knew what errors, discrepancies, and obsolete business procedures might be buried in the code. The task required an analysis of 100 megabytes of English, 1.5 million lines of COBOL programs, and several hundred control-language scripts, which called the programs and specified the data files and formats. Over time, the English terminology, computer formats, and file names had changed. Some of the format changes were caused by new computer systems and business practices, and others were required by different versions of federal regulations. In three weeks of computation on a 750 MHz Pentium III, VAE combined with the Intellitex parser was able to analyze the documentation and programs, translate all statements that referred to files, data, or processes in any of the three languages (English, COBOL, and JCL) to conceptual graphs, and use the CGs to generate an English glossary of all processes and data, to define the specifications for a data dictionary, to create dataflow diagrams of all processes, and to detect inconsistencies between the documentation and the implementation.

4   Inference Engine

Most theorem provers use a tightly constrained version of structure mapping called *unification*, which forces two structures to become identical. Relaxing constaints in one direction converts unification to generalization, and relaxing them in another direction leads to specialization. With arbitrary combinations of generalization and specialization, there is a looser kind of similarity, which, if there is no limit on the extent, could map any graph to any other. When Peirce's rules of inference are redefined in terms of generalization and specialization, they support an inference procedure that can use exactly the same algorithms and data structures designed for the VivoMind Analogy Engine. The primary difference between the analogy engine and the inference engine is in the strategy that schedules the algorithms and determines which constraints to enforce.

When Peirce invented the implication operator for Boolean algebra, he observed that the truth value of the antecedent is always less than or equal to the truth value of the consequent. Therefore, the symbol $\leq$ may be used to represent implication: $p \leq q$ means that the truth value of p is less than or equal to the truth value of q. That same symbol may be used for generalization: if a graph or formula p is true in fewer cases than another graph or formula q, then p is more specialized and q is more generalized. Figure 8 shows a generalization hierarchy in which the most general CG is at the top. Each dark line in Figure 8 represents the $\leq$ operator: the CG above is a generalization, and the CG below is a specialization.

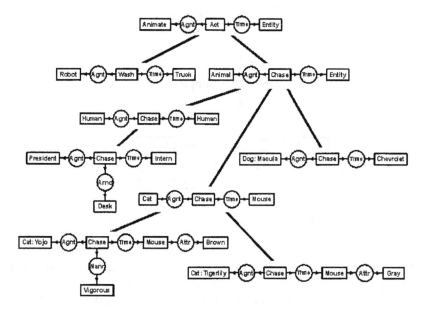

Fig. 8. A generalization hierarchy of CGs

The top CG says that an animate being is the agent of some act that has an entity as the theme of the act. Below it are two specializations: a CG for a robot washing a truck, and a CG for an animal chasing an entity. The CG for an animal chasing an entity has three specializations: a human chasing a human, a cat chasing a mouse, and the dog Macula chasing a Chevrolet. The two graphs at the bottom represent the most specialized sentences: *The cat Yojo is vigorously chasing a brown mouse*, and *the cat Tigerlily is chasing a gray mouse*.

The operations on conceptual graphs are based on combinations of six *canonical formation rules*, which perform the structure-building operations of perception and the structure-mapping operations of analogy. Logically, each rule has one of three possible effects on a CG: the rule can make it more specialized, more generalized, or logically equivalent but with a modified shape. Each rule has an inverse rule that restores a CG to its original form. The inverse of specialization is generalization, the inverse of generalization is specialization, and the inverse of equivalence is another equivalence.

All the graphs in Figure 8 belong to the *existential-conjunctive* subset of logic, whose only operators are the existential quantifier ∃ and the conjunction ∧. For this subset, the canonical formation rules take the forms illustrated in Figures 5, 6, and 7. These rules are fundamentally graphical: they are easier to show than to describe. Sowa (2000) presented the formal definitions, which specify the details of how the nodes and arcs are affected by each rule.

Figure 9 shows the first two rules: *copy* and *simplify*. At the top is a CG for the sentence "The cat Yojo is chasing a mouse." The down arrow represents two applications of the copy rule. The first copies the Agnt relation, and the second

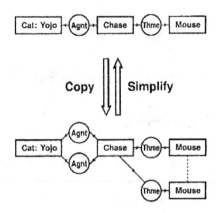

Fig. 9. Copy and simplify rules

copies the subgraph $\rightarrow$ (Thme) $\rightarrow$ [Mouse]. The two copies of the concept [Mouse] at the bottom of Figure 9 are connected by a dotted line called a *coreference link*; that link, which corresponds to an equal sign = in predicate calculus, indicates that both concepts must refer to the same individual. Since the new copies do not add any information, they may be erased without losing information. The up arrow represents the simplify rule, which performs the inverse operation of erasing redundant copies. The copy and simplify rules are called *equivalence rules* because any two CGs that can be transformed from one to the other by any combination of copy and simplify rules are logically equivalent. The two formulas in predicate calculus that are derived from Figure 9 are also logically equivalent. The top CG maps to the following formula:

$$(\exists x:Cat)(\exists y:Chase)(\exists z:Mouse)(name(x,'Yojo') \land agnt(y,x) \land thme(y,z)),$$

In the formula that corresponds to the bottom CG, the equality $z{=}w$ represents the coreference link that connects the two copies of [Mouse]:

$$(\exists x:Cat)(\exists y:Chase)(\exists z:Mouse)(\exists w:Mouse)(name(x,'Yojo') \land agnt(y,x)$$
$$\land\ agnt(y,x) \land thme(y,z) \land thme(y,w) \land z{=}w).$$

By the inference rules of predicate calculus, either of these two formulas can be derived from the other.

Figure 10 illustrates the *restrict* and *unrestrict* rules. At the top is a CG for the sentence "A cat is chasing an animal." Two applications of the restrict rule transform it to the CG for "The cat Yojo is chasing a mouse." The first step is a *restriction by referent* of the concept [Cat], which represents some indefinite cat, to the more specific concept [Cat: Yojo], which represents a particular cat named Yojo. The second step is a *restriction by type* of the concept [Animal] to a concept of the subtype [Mouse]. Two applications of the unrestrict rule perform the inverse transformation of the bottom graph to the top graph. The restrict rule is a *specialization rule*, and the unrestrict rule is a *generalization rule*. The

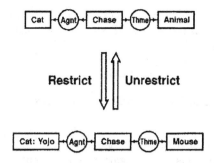

Fig. 10. Restrict and unrestrict rules

more specialized graph implies the more general one: if the cat Yojo is chasing a mouse, it follows that a cat is chasing an animal. The same implication holds for the corresponding formulas in predicate calculus. The more general formula

$$(\exists\, x{:}\text{Cat})(\exists\, y{:}\text{Chase})(\exists\, z{:}\text{Animal})(\text{agnt}(y,x) \wedge \text{thme}(y,z))$$

is implied by the more specialized formula

$$(\exists\, x{:}\text{Cat})(\exists\, y{:}\text{Chase})(\exists\, z{:}\text{Mouse})(\text{name}(x,\text{'Yojo'}) \wedge \text{agnt}(y,x) \wedge \text{thme}(y,z)).$$

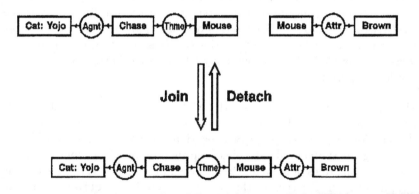

Fig. 11. Join and detach rules

Figure 11 illustrates the *join* and *detach* rules. At the top are two CGs for the sentences "Yojo is chasing a mouse" and "A mouse is brown." The join rule overlays the two identical copies of the concept [Mouse], to form a single CG for the sentence "Yojo is chasing a brown mouse." The detach rule performs the inverse operation. The result of join is a more specialized graph that implies the one derived by detach. The same implication holds for the corresponding formulas in predicate calculus. The conjunction of the formulas for the top two CGs

$(\exists\,x{:}Cat)(\exists\,y{:}Chase)(\exists\,z{:}Mouse)(name(x,'Yojo') \wedge agnt(y,x) \wedge thme(y,z))$
$\wedge\ (\exists\,w{:}Mouse)(\exists\,v{:}Brown)attr(w,v)\ (\exists\,x{:}Cat)(\exists\,y{:}Chase)(\exists\,z{:}Mouse)(\exists$
$v{:}Brown)(name(x,'Yojo')$

is implied by the formula for the bottom CG

$(\exists\,x{:}Cat)(\exists\,y{:}Chase)(\exists\,z{:}Mouse)(\exists\,v{:}Brown)(name(x,'Yojo') \wedge agnt(y,x)$
$\wedge\ thme(y,z) \wedge attr(z,v)).$

These rules can be applied to full first-order logic by specifying how they interact with negation. In CGs, each negation is represented by a context that has an attached relation of type Neg or its abbreviation by the symbol ¬ or ∼. A *positive context* is nested in an even number of negations (possibly zero). A *negative context* is nested in an odd number of negations. The following four principles determine how negations affect the rules:

1. *Equivalence rules.* An equivalence rule remains an equivalence rule in any context, positive or negative.
2. *Specialization rules.* In a negative context, a specialization rule becomes a generalization rule; but in a positive context, it remains a specialization rule.
3. *Generalization rules.* In a negative context, a generalization rule becomes a specialization rule; but in a positive context, it remains a generalization rule.
4. *Double negation.* A *double negation* is a nest of two negations in which no concept or relation node occurs between the inner and the outer negation. (It is permissibe for an arc of a relation or a coreference link to cross the space between the two negations, but only if one endpoint is inside the inner negation and the other endpoint is outside the outer negation.) Then drawing or erasing a double negation around any CG or any subgraph of a CG is an equivalence operation.

In short, a single negation reverses the effect of generalization and specialization rules, but it has no effect on equivalence rules. Since drawing or erasing a double negation adds or subtracts two negations, it has no effect on any rule.

By handling the syntactic details of conceptual graphs, the canonical formation rules enable the rules of inference to be stated in a form that is independent of the graph notation. For each of the six rules, there is an equivalent rule for predicate calculus or any other notation for classical FOL. To derive equivalent rules for other notations, start by showing the effect of each rule on the existential-conjunctive subset (no operators other than $\exists$ and $\wedge$). To handle negation, add one to the negation count for each subgraph or subformula that is governed by a ∼ symbol. For other operators ($\forall$, $\supset$, and $\vee$), count the number of negations in their definitions. For example, $p \supset q$ is defined as $\sim(p \wedge \sim q)$; therefore, the subformula p is nested inside one additional negation, and the subformula q is nested inside two additional negations.

When the CG rules are applied to other notations, some extensions may be necessary. For example, the *blank* or empty graph is a well-formed EG or CG, which is always true. In predicate calculus, the blank may be represented by a constant formula T, which is defined to be true. The operation of erasing a graph would correspond to replacing a formula by T. When formulas are erased or inserted, an accompanying conjunction symbol must also be erased or inserted in some notations. Other notations, such as the Knowledge Interchange Format (KIF), are closer to CGs because they only require one conjunction symbol for an arbitrarily long list of conjuncts. In KIF, the formula (and), which is an empty list of conjuncts, may be used as a synonym for the blank graph or T. Discourse representation structures (DRSs) are even closer to EGs and CGs because they do not use any symbol for conjunction; therefore, the blank may be considered a DRS that is always true.

Peirce's rules, which he stated in terms of existential graphs, form a sound and complete system of inference for first-order logic with equality. If the word *graph* is considered a synonym for *formula* or *statement*, the following adaptation of Peirce's rules can be applied to any notation for FOL, including EGs, CGs, DRS, KIF, or the many variations of predicate calculus. These rules can also be applied to subsets of FOL, such as description logics and Horn-clause rules.

- *Erasure.* In a positive context, any graph or subgraph u may be replaced by a generalization of u; in particular, u may be erased (i.e. replaced by the blank, which is a generalization of every graph).
- *Insertion.* In a negative context, any graph or subgraph u may be replaced by a specialization of u; in particular, any graph may be inserted (i.e. it may replace the blank).
- *Iteration.* If a graph or subgraph u occurs in a context C, another copy of u may be inserted in the same context C or in any context nested in C.
- *Deiteration.* Any graph or subgraph u that could have been derived by iteration may be erased.
- *Equivalence.* Any equivalence rule (copy, simplify, or double negation) may be performed on any graph or subgraph in any context.

These rules, which Peirce formulated in several equivalent variants from 1897 to 1909, form an elegant and powerful generalization of the rules of *natural deduction* by Gentzen (1935). Like Gentzen's version, the only axiom is the blank. What makes Peirce's rules more powerful is the option of applying them in any context nested arbitrarily deep. That option shortens many proofs, and it eliminates Gentzen's bookkeeping for making and discharging assumptions. For further discussion and comparison, see MS 514 (Peirce 1909) and the commentary that shows how other rules of inference can be derived from Peirce's rules.

Unlike most proof procedures, which are tightly bound to a particular syntax, this version of Peirce's rules is stated in notation-independent terms of generalization and specialization. In this form, they can even be applied to natural languages. For any language, the first step is to show how each syntax rule affects generalization, specialization, and equivalence. In counting the negation depth for natural languages, it is important to recognize the large number of negation words, such as *not, never, none, nothing, nobody,* or *nowhere*. But many other words also contain implicit negations, which affect any context governed by those words. Verbs like *prevent* or *deny*, for example, introduce a negation into any clause or phrase in their complement. Many adjectives also have implicit negations: a stuffed bear, for example, lacks essential properties of a bear, such as being alive. After the effects of these features on generalization and specialization have been taken into account in the syntactic definition of the language, Peirce's rules can be applied to a natural language as easily as to a formal language.

5   A Semeiotic Foundation for Cognition

Peirce developed his theory of signs or *semeiotic* as a species-independent theory of cognition. He considered it true of any "scientific intelligence," by which he meant any intelligence that is "capable of learning from experience." Peirce was familiar with Babbage's mechanical computer; he was the first person to suggest that such machines should be based on electical circuits rather than mechanical linkages; and in 1887, he published an article on "logical machines" in the *American Journal of Psychology*, which even today would be a respectable commentary on the possibilities and difficulties of artificial intelligence. His definition of *sign* is independent of any implementation in proteins or silicon:

> I define a sign as something, A, which brings something, B, its interpretant, into the same sort of correspondence with something, C, its object, as that in which itself stands to C. In this definition I make no more reference to anything like the human mind than I do when I define a line as the place within which a particle lies during a lapse of time. (1902, p. 235)

As a close friend of William James, Peirce was familiar with the experimental psychology of his day, which he considered a valuable study of what possibilities are realized in any particular species. But he considered semeiotic to be a more fundamental, implementation-independent characterization of cognition.

Peirce defined logic as "the study of the formal laws of signs" (1902, p. 235), which implies that it is based on the same kinds of semeiotic operations as all other cognitive processes. He reserved the word *analogy* for what is called "analogical reasoning" in this paper. For the kinds of structure mapping performed by SME and VAE, Peirce used the term *diagrammatic reasoning*, which he described as follows:

> The first things I found out were that all mathematical reasoning is diagrammatic and that all necessary reasoning is mathematical reasoning, no matter how simple it may be. By diagrammatic reasoning, I mean reasoning which constructs a diagram according to a precept expressed in general terms, performs experiments upon this diagram, notes their results, assures itself that similar experiments performed upon any diagram constructed according to the same precept would have the same results, and expresses this in general terms. This was a discovery of no little importance, showing, as it does, that all knowledge without exception comes from observation. (1902, pp. 91-92)

This short paragraph summarizes many themes that Peirce developed in more detail in his other works. In fact, it summarizes the major themes of this article:

1. By a diagram, Peirce meant any abstract pattern of signs. He included the patterns of algebra and Aristotle's syllogisms as diagrammatic, but he also said that his existential graphs were "more diagrammatic".

2. His "experiments" upon a diagram correspond to various AI procedures, such as generate and test, backtracking, breadth-first parallel search, and path following algorithms, all of which are performed on data structures that correspond to Peirce's notion of "diagram".

3. The first sentence of the paragraph applies the term "mathematical" to any kind of deduction, including Aristotle's syllogisms and any informal logic that may be used in a casual conversation.

4. The last sentence, which relates observation to diagrammatic reasoning, echoes the themes of the first section of this paper, which emphasized the need for an integration of of perception with the mechanisms of analogy. Peirce stated that point even more forcefully: "Nothing unknown can ever become known except through its analogy with other things known" (1902, p. 287).

In short, the operations of diagrammatic reasoning or structure mapping form the bridge from perception to all forms of reasoning, ranging from the most casual to the most advanced. Forbus et al. (2002) applied the term *reasoning from first principles* to logic, not to analogical reasoning. But Peirce, who invented two of the most widely used notations for logic, recognized that the underlying semeiotic mechanisms were more fundamental. The VivoMind implementation confirms Peirce's intuitions.

For natural language understanding, the constrained operations of unification and generalization are important, but exceptions, metaphors, ellipses, novel word senses, and the inevitable errors require less constrained analogies. When the VAE algorithms are used in the semantic interpreter, there are no exceptions: analogies are used at every step, and the only difference between unification, generalization, and looser similarities is the nature of the constraints on the analogy.

References

1. Chalmers, D. J., R. M. French, & D. R. Hofstadter (1992) "High-level perception, representation, and analogy: A critique of artificial intelligence methodology," *Journal of Experimental & Theoretical Artificial Intelligence* 4, 185-211.

2. Falkenhainer, B., Kenneth D. Forbus, Dedre Gentner (1989) "The Structure mapping engine: algorithm and examples," *Artificial Intelligence* 41, 1-63.

3. Forbus, Kenneth D., Dedre Gentner, & K. Law (1995) "MAC/FAC: A Model of Similarity-Based Retrieval," *Cognitive Science* 19:2, 141-205.

4. Forbus, Kenneth D., Dedre Gentner, Arthur B. Markman, & Ronald W. Fergu-son (1998) "Analogy just looks like high level perception: Why a domain-general approach to analogical mapping is right," *Journal of Experimental & Theoretical Artificial Intelligence* **10:2**, 231-257.
5. Forbus, Kenneth D., T. Mostek, & R. Ferguson (2002) "An analogy ontology for integrating analogical processing and first-principles reasoning," *Proc. IAAI-02* pp. 878-885.
6. Gentzen, Gerhard (1935) "Untersuchungen über das logische Schließen," trans-lated as "Investigations into logical deduction" in *The Collected Papers of Ger-hard Gentzen*, ed. and translated by M. E. Szabo, North-Holland Publishing Co., Amsterdam, 1969, pp. 68-131.
7. Hallaq, Wael B. (1993) *Ibn Taymiyya Against the Greek Logicians*, Clarendon Press, Oxford.
8. LeClerc, André, & Arun Majumdar (2002) "Legacy revaluation and the making of LegacyWorks," *Distributed Enterprise Architecture* **5:9**, Cutter Consortium, Ar-lington, MA.
9. Morrison, Clayton T., & Eric Dietrich (1995) "Structure-mapping vs. high-level perception: the mistaken fight over the explanation of Analogy," *Proc. 17th Annual Conference of the Cognitive Science Society*, pp. 678-682. Available at http://babs.cs.umass.edu/ clayton/CogSci95/SM-v-HLP.html
10. Peirce, Charles Sanders (1887) "Logical machines," *American Journal of Psychol-ogy*, vol. 1, Nov. 1887, pp. 165-170.
11. Peirce, Charles S. (1902) *Logic, Considered as Semeiotic*, MS L75, edited by Joseph Ransdell,
    http://members.door.net/arisbe/menu/LIBRARY/bycsp/L75/ver1/l75v1-01.htm
12. Peirce, Charles Sanders (1909) Manuscript 514, with commentary by J. F. Sowa, available at http://www.jfsowa.com/peirce/ms514.htm
13. Sowa, John F. (2000) *Knowledge Representation: Logical, Philosophical, and Com-putational Foundations*, Brooks/Cole Publishing Co., Pacific Grove, CA.

Hegel in a Strange Costume

Reconsidering Normative Science
for Conceptual Structures Research

Mary Keeler

Center for Advanced Research and Technology in the Arts and Humanities
University of Washington, Seattle
mkeeler@u.washington.edu

Abstract. Charles Peirce was influenced by German philosophy to conceive logic as one of a carefully related set of theoretical disciplines called the *Normative Sciences*, strategically located in his Classification of the Sciences between *Phenomenology* and *Metaphysics*. Barely enough evidence is available from his archived manuscripts to indicate how he might have developed that part of his philosophy in the three disciplines of esthetics, ethics, and logic-as-semiotic (I will use his preferred spelling, "semeotic"), which he says support his purpose for pragmatism in a theory of inquiry. I have investigated that evidence to find how his philosophy treats some neglected issues in modern philosophy (nominalism, intuition, automation) which limit the advancement of conceptual structures research, and suggest that his normative science will be required for effective knowledge creation and communication.

1 Overview of the Evidence

Early in his study of logic, Charles Peirce (1839-1914) was strongly influenced by German philosophy, especially the work of Kant (1724-1804) and Hegel (1770-1831). As he became involved in developing the logic of relatives to increase the power of logical analysis, he realized how the regime of traditional logic had fundamentally limited philosophical inquiry, and therefore all scientific inquiry. The effects of those limitations linger to this day, and continue to block advancements in theory and method, especially with regard to the phenomena of creating and communicating knowledge. I will examine those limitations to a theory of inquiry, considering them as the issues of *nominalism*, *intuition*, and *automation*, indicating how Peirce's pragmatism was conceived to address them methodologically, and then suggest what evidence we have to conclude that he eventually decided to develop the normative sciences—esthetics, ethics (or practics), and logic—to account for that methodology as required by his general theory of signs, or semeotic (I use the spelling he preferred [see *CP* 8.377]).

His writings (at least those few to which we have convenient access) do not provide much evidence for what he took to be the relationship between his pragmatism and his view of normative science. In fact, these two aspects of his work are rarely mentioned in the same context (and almost never explicitly), which may be simply because he realized the significance of the normative sciences very late in his

A. de Moor, W. Lex, and B. Ganter (Eds.): ICCS 2003, LNAI 2746, pp. 37–53, 2003.
© Springer-Verlag Berlin Heidelberg 2003

life, perhaps in the last decade (1902-1912). One particularly significant remark appears in a letter to William James, late in 1902: "These three normative sciences correspond to my three categories, which in their psychological aspect, appear as Feeling, Reaction, Thought. I have advanced my understanding of these categories much since Cambridge days; and can now put them in a much clearer light and more convincingly. The true nature of pragmatism cannot be understood without them. It does not, as I seem to have thought at first, take Reaction as the be-all, but it takes the end-all as the be-all, and the End is something that gives its sanction to action. It is of the third category. Only one must not take a nominalistic view of Thought as if it were something that a man had in his consciousness. Consciousness may mean any one of the three categories. But if it is to mean Thought it is more without us than within. It is we that are in it, rather than it in any of us" [CP 8.256].

Most of the evidence available is from his Lectures on Pragmatism of 1903. In fact, the first of these lectures is entitled "Pragmatism: The Normative Sciences." And between the fourth lecture, "The Reality of Thirdness," and the sixth lecture, "Three Types of Reasoning," the fifth is "Three Kinds of Goodness," which he begins: "Now I am going to make a series of assertions which will sound wild; for I cannot stop to argue them, although I cannot omit them if I am to set the supports of pragmatism in their true light" [CP 5.120]. In the final of these lectures, he concludes: "If you carefully consider the question of pragmatism you will see that it is nothing else than the question of the logic of abduction. That is, pragmatism proposes a certain maxim which, if sound, must render needless any further rule as to the admissibility of hypotheses to rank as hypotheses, that is to say, as explanations of phenomena held as hopeful suggestions; and, furthermore, this is all that the maxim of pragmatism really pretends to do, at least so far as it is confined to logic, and is not understood as a proposition in psychology.... For the maxim of pragmatism is that a conception can have no logical effect or import differing from that of a second conception except so far as, taken in connection with other conceptions and intentions, it *might conceivably* modify our practical conduct differently from that second conception" [CP 5.196; my emphasis].

We have even less evidence available to discern how he intended semeotic to relate to pragmatism and to normative science. From his voluminous manuscript, "Minute Logic" (1902-03), only one remark on the subject appears in print, in the *Collected Papers of Charles Sanders Peirce* [CP]: "With Speculative Rhetoric, Logic, in the sense of Normative Semeotic, is brought to a close" [CP 2.111]. We may conclude from that indication and only half a dozen others that, by this point, "Semeotic" was for him "logic," and should appear in his classification among the other normative sciences. I found it helpful to study his "Classification of the Sciences," once it had been constructed in outline from his textual accounts (see below, in abbreviated form; see [1] and [2] for more detail). That construction is only a first step in solving the puzzle of his many writings, archived in manuscript, concerning which Peirce himself remarks: "All that you can find in print of my work on logic are simply scattered outcroppings here and there of a rich vein which remains unpublished. Most of it I suppose has been written down; but no human being could ever put together the fragments. I could not myself do so.—1903" [printed at the beginning of Volume 2 of the CP, *Elements of Logic.*]

The Structure of Philosophy in a Classification of the Sciences of Discovery
represents philosophy as a positive science, in the sense of discovering what really is true but
limited to "so much of truth as can be inferred from common experience." The special sciences
are principally occupied with the accumulation of new facts inferred from specific human and
natural inquiries [see *CP*: 1.183-238]. The structure indicates that each science draws upon the
principles of the studies above it on the list, mathematics being the simplest and most abstract
science.

I. Mathematics (the Conditional or Hypothetical Science, studies what is and what is not
logically possible, without making itself responsible for any actual existence [see *CP*, 5.40])
II. Philosophy (the Theoretical Science in Three Branches: Phenomenology ascertains and
studies the kinds of elements universally present at any time to the mind in any way; Normative
science distinguishes what ought to be from what ought not to be; Metaphysics seeks to give an
account of the universe of mind and matter. [see *CP*: 1.186])

    A. Phenomenology (the science of experience in terms of Category Theory: Firstness,
    Secondness, Thirdness; "describing all the features that are common to whatever is
    experienced or might conceivably be experienced or become an object of study in any
    way direct or indirect"[*CP*: 4.390, 2.84, 5.37])

    B. The Normative Sciences ("the science of the laws of conformity of things [as
    phenomena] to ends or ideals, that is, perhaps, to Truth, Right, and Beauty" [*CP* 5.121])

        1. Esthetics (the science of ideals; considers those things whose ends are to
        embody qualities of feeling)

        2. Ethics (the theory of self-controlled or deliberate conduct; considers those
        things whose ends lie in action; should perhaps be called "Practics")

        3. Logic (Formal Semeotic, or the science of self-controlled or deliberate
        thought; considers those things whose end is to represent something)

            a. Philosophical Grammar (Speculative Grammar, or the theory of
            meaning)

            b. Critical Logic (the theory of inference)

            c. Philosophical Rhetoric (Speculative Rhetoric, i.e., theory of method,
            especially pragmatism)

    C. Metaphysics (the science of Reality. Reality consists in regularity.[see *CP*: 5.121-
    129; see also V 6, B 1])

        1. Ontology (general metaphysics)

        2. Psychical metaphysics (religion)

        3. Physical Metaphysics or Cosmology (natural laws)

III. The Special Sciences: these are now represented by the various disciplines in a college of
arts and sciences, apart from philosophy and mathematics. Peirce divides these into what we
now think of as the humanities along with other human studies and the natural sciences, which
would correspond roughly to the *Geisteswissenschaften* and the *Naturwissenschaften*) [3].

The "Introduction" to Peirce's "Bibliography" in volume 8 of the *CP* includes a
brief list (see below) to account for those of Peirce's manuscripts that arrived at the
Harvard University archive. The introduction says that his manuscript collection "is
divided according to Peirce's classification of the sciences as follows, with the
number of boxes and bundles listed after each division." Judging from the number of
boxes and bundles listed for each domain of his work, we should expect to find most
of his work devoted to normative science. An editor's note in the *CP* (Volume 2)
explains: "Peirce came to recognize the nature of the Normative Sciences at a very
late date (c.1903). He apparently wrote practically nothing on esthetics (see 2.197)
and linked most of his discussions of practics and ethics with those on pragmatism

and logic. Logic, the third of the Normative Sciences, being the subject on which Peirce spent about sixty years of intensive study and on which he left the most manuscripts." But his later writings (during his last decade) are not yet at all well represented in any print edition, because they become more difficult to decipher as his handwriting deteriorated with age. Because his manuscripts are now in very delicate condition, on high acid paper, we may never know whether he had the opportunity to pursue further development of his normative sciences.

Listing of Peirce's Manuscript Collection at Harvard University

I. Science of Discovery	
A. Mathematics	8
B. Philosophy	
1. Pragmatism and the Categories	2
2. Normative Sciences (Logic)	12
3. Metaphysics	2
C. Idioscopy	8
II. Science of Review: Classification of the Sciences	1
III. Practical Science and Miscellaneous	3
IV. Book Reviews	2
V. Life and Letters	
Unclassified: V{a}, V{b},..., V{z}	6
A. Biography	2
B. Correspondence	
1. Personal	6
2. Professional	5
3. Business	2
4. Official: Coast Survey	1
5. Applications	1

Based on the limited evidence available in the *CP*, I conjecture that Peirce's conception of the normative sciences, which gives us a new view of logic and its role in inquiry, might serve conceptual structures work in effectively building theory and constructing methods of research. I will present evidence from his writings for that conjecture, in considering how he contends with those issues that limit adequate account of knowledge creation and communication in pursuing a theory of inquiry on Hegel's triadic model. As Peirce remarks, in unidentified fragments of a manuscript (dated 1892), "My philosophy resuscitates Hegel, though in a strange costume" [*CP* 1.42].

By the time he writes the third draft of his second lecture for the 1903 series on pragmatism, "The Universal Categories," he has explicitly distinguished his view from those of other modern philosophers: "Philosophy, as I understand the word, is a positive theoretical science, and a science in an early stage of development.... The followers of Haeckel are completely in accord with the followers of Hegel in holding that what they call philosophy is a practical science and the best of guides in the formation of what they take to be Religious Beliefs. I simply note the divergence, and pass on to an unquestionable fact; namely, the fact that all modern philosophy is built upon Ockhamism; by which I mean that it is all nominalistic and that it adopts nominalism because of Ockham's razor. And there is no form of modern philosophy of which this is more essentially true than the philosophy of Hegel. But it is not modern philosophers only who are nominalists. The nominalistic Weltanschauung has become incorporated into what I will venture to call the very flesh and blood of the average modern mind" [*CP* 5.61].

2 Modern Philosophy's Costume: Nominalism, Intuition, and Automation

Peirce's aim of a scientific philosophy required him to examine and explicate in intricate detail the assumptions of modern philosophy *as phenomenal objects* of inquiry themselves. His "Minute Logic" attempts to discover exactly how each attempt has failed in the inquiry of its proper phenomena, and to respond to those deficiencies effectively in his own philosophy. Anyone who studies his work in depth will find much evidence of the affinities he locates in nearly every modern system, which his analysis approves as part of his "strange costume." He emphasizes many times that nothing in his philosophy is original, that only its placement among other ideas in a comprehensive perspective gives it a new role in his design. Those who have considered his work in piecemeal (which is how it appears in the *Collected Papers,* and in any other available source—since the chronological edition of his work [4] is still far from complete) often conclude that pragmatism is a new system of philosophy, or simply a new form of some particular old one. Instead, we can better appreciate it as a maxim for a method by which he constructs an ensemble from worthy features of other systems.

According to Peirce, modern philosophy has suffered a "tidal wave of nominalism," beginning with Descartes. Kant was a nominalist, whose philosophy "would have been rendered compacter, more consistent, and stronger if its author had taken up realism," he says, and Hegel was a nominalist "of realistic yearnings" [*CP*1.19]. In a review of a book on the works of Berkeley, in 1871, Peirce comments on how the issue of nominalism affects us: "But though the question of realism and nominalism has its roots in the technicalities of logic, its branches reach about our life. The question whether the genus homo has any existence except as individuals, is the question whether there is anything of any more dignity, worth, and importance than individual happiness, individual aspirations, and individual life. Whether men really have anything in common, so that the community is to be considered as an end in itself, and if so, what the relative value of the two factors is, is the most fundamental practical question in regard to every public institution the constitution of which we have it in our power to influence" [*CP* 8.38].

Peirce considers nominalism to be a naive sort of metaphysics ("the simplest possible of all Logico-Metaphysical theories, if it can be sustained" [*CP* 2.166]), which our modern minds seem predisposed to assume. He persistently argued against it, and was convinced that only with adequate logical theory and method could anyone hope to examine this mode of thinking effectively enough to change the habit of blindly accepting it. We can find evidence that he identified common ground between nominalism and realism in terms of how both conceive *reality*, such as in an untitled manuscript of 1873: "I do not think that the two views [realism and nominalism] are absolutely irreconcilable, although they are taken from very widely separated stand-points. The realistic view emphasizes particularly the permanence and fixity of reality; the nominalistic view emphasizes its externality. But the realists need not, and should not, deny that the reality exists externally to the mind; nor have they historically done so, as a general thing. That is external to the mind, which is what it is, whatever our thoughts may be on any subject; just as that is real which is what it is, whatever our thoughts may be concerning that particular thing" [CP 7.339].

Also significant is that his argument for pragmatism, based on the theory that every thought is a sign, was the doctrine of Leibniz, Berkeley, and all the other extreme nominalists beginning in the sixteenth century. Just as they did, he insists, "Every realist must, as such, admit that a general is a term and therefore a sign," but warns, "If, in addition, he holds that it is an absolute exemplar, this Platonism passes quite beyond the question of nominalism and realism; and indeed the doctrine of Platonic ideas has been held by the extremest nominalists" [CP 5.470]. Ockham, the leading extreme nominalist, proclaimed: "It is to be maintained, therefore, that every universal is one singular thing, and therefore there is no universal except by signification, that is, by its being the sign of many." Peirce responds: "Ockham always thinks of a mental conception as a logical term, which, instead of existing on paper, or in the voice, is in the mind, but is of the same general nature, namely, a sign. The conception and the word differ in two respects: first, a word is arbitrarily imposed, while a conception is a natural sign; second, a word signifies whatever it signifies only indirectly, through the conception which signifies the same thing directly" [CP 8.20 (1871)].

At the end of a manuscript for a proposed book on the history of science, in 1896, Peirce contends that all modern nominalist philosophers recognize only one mode of being, the being of an individual thing or fact, which he says, "consists in the object's crowding out a place for itself in the universe, so to speak, and reacting by brute force of fact, against all other things. I call that existence" [CP 1.21]. To this he responds: "My view is that there are three modes of being. I hold that we can directly observe them in elements of whatever is at any time before the mind in any way. They are the being of positive qualitative possibility, the being of actual fact, and the being of law that will govern facts in the future" [CP 1.23].   These are the three Universal Categories of his phenomenology: Firstness, Secondness, and Thirdness, which he uses to explain the nominalist's problem:

> Now for Thirdness. Five minutes of our waking life will hardly pass without our making some kind of prediction; and in the majority of cases these predictions are fulfilled in the event. Yet a prediction is essentially of a general nature, and cannot ever be completely fulfilled. To say that a prediction has a decided tendency to be fulfilled, is to say that the future events are in a measure really governed by a law ... "Oh," but say the nominalists, "this general rule is nothing but a mere word or couple of words!" I reply, "Nobody ever dreamed of denying that what is general is of the nature of a general sign; but the question is whether future events will conform to it or not. If they will, your adjective 'mere' seems to be ill-placed." A rule to which future events have a tendency to conform is *ipso facto* an important thing, an important element in the happening of those events. This mode of being which consists, mind my word if you please, the mode of being which consists in the fact that future facts of Secondness will take on a determinate general character, I call a Thirdness. [CP1.26]

In his second lecture of a series in 1898, "Detached Ideas on Vitally Important Topics," Peirce sketches a useful summary of the two opposing views: "Roughly speaking, the nominalists conceived the general element of cognition to be merely a convenience for understanding this and that fact and to amount to nothing except for cognition, while the realists, still more roughly speaking, looked upon the general, not only as the end and aim of knowledge, but also as the most important element of being.... But as for the average nominalist whom you meet in the streets, he reminds me of the blind spot on the retina, so wonderfully does he unconsciously smooth over his field of vision and omit facts that stare him in the face, while seeing all round them without perceiving any gap in his view of the world" [CP 4.1].

The crucial question between the two views, as Peirce puts it, is "whether all properties, laws of nature, and predicates of more than an actually existent subject are, without exception, mere figments or not." He offers an example in a footnote: "Anybody may happen to opine that 'the' is a real English word; but that will not constitute him a realist. But if he thinks that, whether the word 'hard' itself be real or not, the property, the character, the predicate, hardness, is not invented by men, as the word is, but is really and truly in the hard things and is one in them all, as a description of habit, disposition, or behavior, then he is a realist" [CP 1.27; fn] To make the distinction, as Kant did and all nominalists do, between the true conception of something and the thing itself, insists Peirce, "is only to regard one and the same thing from two different points of view"; a true conception would be the reality, which is what gives us reason to pursue any inquiry with the hope that is has real direction, in which "the mind of man is, on the whole and in the long run, tending" [CP 8.12].

Two significant remarks indicate how firm was his judgment of nominalism. In his Lowell Lectures of 1903, he reveals that since his 1871 review of the book on Berkeley's works he has "very carefully and thoroughly revised" his philosophical opinions, and has "modified them more or less on most topics," but has "never been able to think differently on that question of nominalism and realism" [CP 1.20]. And in an untitled manuscript that was apparently intended as part of a lecture, "Fallibilism, Continuity, and Evolution" (c. 1897), he says: "It is one of the peculiarities of nominalism that it is continually supposing things to be absolutely inexplicable. That blocks the road of inquiry" [CP 1.170].

Peirce demonstrates how detailed analysis of the confused state of modern philosophy can "unblock that road," when he considers the issue of *intuition,* in two essays published in the Journal of Speculative Philosophy (1868). The first essay, "Questions Concerning Certain Faculties Claimed for Man," takes on the fundamental nominalist claims and, in particular, Kant's legacy of *intuition as a form of uninferred or immediate knowledge* (such as our knowledge of space and time). He begins by formally stating that definition: "Throughout this paper, the term intuition will be taken as signifying a cognition not determined by a previous cognition of the same object, and therefore so determined by something out of the consciousness." He questions whether we have this faculty by asking how we could know we do if we did? The essay ends with its denial: "We have no power of Intuition (as defined), but every cognition is determined" [CP 5.213-63].

In his second essay (in the same journal, 1868), he says that this proposition cannot be regarded as certain and must be traced out to its consequences. [see CP 5.265]. Here he makes the case that we could not know for sure that anything is unknowable, or inexplicable, for we could not conceive of it as apart from what we think. We are only aware of or can identify feelings (what we call intuitions, for example) by *predicating* them (or relating them to something, as when we say a computer interface is intuitive). Therefore, only whatever is *incomparable* with anything else is inexplicable, he stresses, and whatever is incomparable would be unthinkable and we could have no concept of it *as* unknowable. From this result, he makes a logical case for meaning as the continuity of mind, or as having a sort of being that is *relational*:

> So that nothing which we can truly predicate of feelings is left inexplicable ... no present actual thought (which is a mere feeling) has any meaning, any intellectual value; for this lies not in what is actually thought, but in what this thought may be connected with in representation by subsequent thoughts; so that the meaning of a thought is altogether something virtual. It may be objected, that if no thought has any meaning, all thought is without meaning. But this is a fallacy similar to saying that, if in no one of the successive spaces which a body fills there is room for motion, there is no room for motion throughout the whole. At no one instant in my state of mind is there cognition or representation, but in the relation of my states of mind at different instants there is. [In a footnote, Peirce says: "Accordingly, just as we say that a body is in motion, and not that motion is in a body, we ought to say that we are in thought and not that thoughts are in us."] [*CP* 5.289]

Of course, much of what goes on in the flow of thought we are not explicitly aware of as inference; in fact, we could say that most of our opinions come to us almost *automatically*, unconsciously derived from previous thought. But if we can consider beliefs to be *habits of mind*, as Peirce contends, then the difference between instinctive habits and beliefs is the capability of self-control: "a deliberate, or self-controlled habit, is precisely a belief" [*CP* 5.480]. Furthermore, he says, "The feeling of believing is a more or less sure indication of there being established in our nature some habit which will determine our actions. Doubt never has such an effect" [*CP* 5.371]; doubt disturbs our belief-habits to which "we cling tenaciously, not merely to believing, but to believing just what we do believe" [*CP* 5.372]. He concludes, "all doubt is a state of hesitancy about an imagined state of things.—1893" [*CP* 5.373fn], but stresses: "genuine doubt always has an external origin, usually from surprise" [*CP* 5.443]. Indubitable beliefs, says Peirce, indicate "a somewhat primitive mode of life." With the introduction of any degrees of self-control, "occasions of action arise in relation to which the original beliefs, if stretched to cover them, have no sufficient authority. In other words, we outgrow the applicability of instinct—not altogether, by any manner of means, but in our highest activities" [*CP* 5.511].

The capability to regulate our lives by belief-habits that respond to doubt is a form of virtual adaptation and so of evolution that is not available to less capable minds. Belief in imagined states of things: which we hypothetically maintain, experiment with by prediction, and then regulate our actions according to the results we experience (in pursuing inquiry), is what distinguishes our conduct from automatic regulation [see *CP* 5.473]. Peirce reminds us how subtle—and crucial—this distinction is: "There is no evidence that we have this faculty, except that we seem to feel that we have it. But the weight of that testimony depends entirely on our being supposed to have the power of distinguishing in this feeling whether the feeling be the result of education, old associations, etc., or whether it is an intuitive cognition; or, in other words, it depends on presupposing the very matter testified to" [*CP* 2.214

(1868)]. He eventually rescues intuition from its nominalist fate with a new definition: "Intuition is the regarding of the abstract in a concrete form, by the realistic hypostatization of relations [or hypothetically regarding thoughts and things]; that is the one sole method of valuable thought. Very shallow is the prevalent notion that this is something to be avoided. You might as well say at once that reasoning is to be avoided because it has led to so much error; quite in the same philistine line of thought would that be; and so well in accord with the spirit of nominalism that I wonder some one does not put it forward. The true precept is not to abstain from hypostatization, but to do it intelligently" [*CP* 5.383 (1877)]. And finally he advises: "Everybody ought to be a nominalist at first, and to continue in that opinion until he is driven out of it by the *force majeure* of irreconcilable facts.   Still he ought to be all the time on the lookout for these facts, considering how many other powerful minds have found themselves compelled to come over to realism" [CP 4.1 (1898)].

In his unpublished "Grand Logic" of 1893, he admonishes: "Many men are so cocksure that necessity governs everything that they deny that there is anything substantially contingent. But it will be shown in the course of this treatise that they are unwarrantably confident, that wanting omniscience we ought to presume there may be things substantially contingent, and further that there is overwhelming evidence that such things are. ... The question of realism and nominalism, which means the question how far real facts are analogous to logical relations, and why, is a very serious one, which has to be carefully and deliberately studied, and not decided offhand, and not decided on the ground that one or another answer to it is 'inconceivable'" [*CP* 4.67-8].   This question challenges conceptual structures research to take account of the human capability of virtual adaptation by self-critical control, theoretically and methodologically, if knowledge representation tools are truly to augment human inquiry.   Peirce's scientific philosophy offers logical principles that are hypotheses in normative theory, which bases it on abduction, like any other scientific theory [see 5].

3 Peirce's Strange Costume:
Semeotic, Pragmatism, and Normative Science

In the forms which we have access to examine it, Peirce's "costume" appears even stranger than he could imagine. Scholarship based on his writings has been so demanding (especially before the production of the electronic *CP* [6]) that most Peirce scholars have been entirely unaware of how he relates semeotic to pragmatism and places them within his normative sciences. The relationships among these three realms are never explicated in the evidence compiled from his manuscripts in the *CP,* which is not surprising when we remember that especially his later work—and particularly that on semeotic—is not well represented there. Only six occurrences of the term (in its various spellings) appear in the *CP*'s eight volumes of Peirce's texts, and one footnote reference, in the title of an unpublished work: "Logic viewed as Semeiotics," c.1904 (the only case where that spelling occurs, and that manuscript is not included in the *CP*). Otherwise the term occurs, by date, in these forms: 1893 (semiotic), 1896 (semeiotic), 1897 (semiotic), 1898 (semeiotic), 1902 (semeotic), 1908 (semeotic), and again in 1908 (semeiotic). His references to normative science are only twice as frequent; in only fifteen *CP* entries does the term specifically occur (and only between 1901 and 1908). Most of the entries where it appears were written in 1903, but (perhaps) curiously no entries of that year contain "semeotic" (in any spelling). In fact, most of his works on semeotic in the *CP* are dated before 1900. Even the *CP*'s bibliography contains only the one item title mentioned above, "Logic viewed as Semeiotics." Joseph Esposito explains that Peirce's study of signs and development of a theory "grew out of his use of sign concepts to solve *specific* philosophic questions," many years prior to his realization of the need to develop that theory further, as semeotic [7: 19].

Nevertheless, we know from the entries where the term appears that he clearly considered it to apply to an expanded form of logic. The fragmentary manuscript of 1896 begins to make the core connections: "The term 'logic' is unscientifically by me employed in two distinct senses. In its narrower sense, it is the science of the necessary conditions of the attainment of truth. In its broader sense, it is the science of the necessary laws of thought, or, still better (thought always taking place by means of signs), it is general semeiotic, treating not merely of truth, but also of the general conditions of signs being signs (which Duns Scotus called *grammatica speculativa*), also of the laws of the evolution of thought, which since it coincides with the study of the necessary conditions of the transmission of meaning by signs from mind to mind, and from one state of mind to another, ought, for the sake of taking advantage of an old association of terms, be called *rhetorica speculativa*, but which I content myself with inaccurately calling objective logic, because that conveys the correct idea that it is like Hegel's logic" [*CP* 1.444].

More explicitly, in his second lecture of the 1898 series on logic ("Detached Ideas on Vitally Important Topics"), he explains:

The highest kind of symbol is one which signifies a growth, or self-development, of thought, and it is of that alone that a moving representation is possible; and accordingly, the central problem of logic is to say whether one given thought is truly, i.e., is adapted to be, a development of a given other or not. In other words, it is the critic of arguments. Accordingly, in my early papers I limited logic to the study of this problem. But since then, I have formed the opinion that the proper sphere of any science in a given stage of development of science is the study of such questions as one social group of men can properly devote their lives to answering; and it seems to me that in the present state of our knowledge of signs, the whole doctrine of the classification of signs and of what is essential to a given kind of sign, must be studied by one group of investigators. Therefore, I extend logic to embrace all the necessary principles of semeiotic, and I recognize a logic of icons, and a logic of indices, as well as a logic of symbols; and in this last I recognize three divisions: Stecheotic (or stoicheiology), which I formerly called Speculative Grammar; Critic, which I formerly called Logic; and Methodeutic, which I formerly called Speculative Rhetoric. [*CP* 4.9]

Finally in his "Minute Logic" (1902), he makes the definitive connection to normative science, by name: "With Speculative Rhetoric, Logic, in the sense of Normative Semeotic, is brought to a close. But now we have to examine whether there be a doctrine of signs corresponding to Hegel's objective logic" [*CP* 2.111].

Most significantly (in a letter of 1908), he clearly defines a triadic-relation, "the decisive breakthrough," as Klaus Oehler calls it, which distinguishes Peirce's semeotic from the nominalist dyadic sign theories [8: 71].

It seems to me that one of the first useful steps toward a science of semeiotic ({sémeiötiké}), or the cenoscopic science of signs, must be the accurate definition, or logical analysis, of the concepts of the science. I define a Sign as anything which on the one hand is so determined by an Object and on the other hand so determines an idea in a person's mind, that this latter determination, which I term the Interpretant of the sign, is thereby mediately determined by that Object. A sign, therefore, has a triadic relation to its Object and to its Interpretant. But it is necessary to distinguish the Immediate Object, or the Object as the Sign represents it, from the Dynamical Object, or really efficient but not immediately present Object. It is likewise requisite to distinguish the Immediate Interpretant, i.e. the Interpretant represented or signified in the Sign, from the Dynamic Interpretant, or effect actually produced on the mind by the Sign; and both of these from the Normal Interpretant, or effect that would be produced on the mind by the Sign after sufficient development of thought. On these considerations I base a recognition of ten respects in which Signs may be divided. I do not say that these divisions are enough. But since every one of them turns out to be a trichotomy, it follows that in order to decide what classes of signs result from them, I have 310 or 59049, difficult questions to carefully consider; and therefore I will not undertake to carry my systematical division of signs any further, but will leave that for future explorers. [CP 8.343]

In his later work on pragmatism (c.1906), we find evidence in a parenthetical comment recorded in the CP that his theory conceives a continuous process called "semiosis," or sign-action; he urges: "(It is important to understand what I mean by semiosis. All dynamical action, or action of brute force, physical or psychical, either takes place between two subjects [whether they react equally upon each other, or one is agent and the other patient, entirely or partially] or at any rate is a resultant of such actions between pairs. But by 'semiosis' I mean, on the contrary, an action, or influence, which is, or involves, a coöperation of three subjects, such as a sign, its object, and its interpretant, this tri-relative influence not being in any way resolvable into actions between pairs. {Sémeiösis} in Greek of the Roman period, as early as Cicero's time, if I remember rightly, meant the action of almost any kind of sign; and my definition confers on anything that so acts the title of a 'sign.')" [CP 5.484]. He generalizes semiosis beyond sign action of the mind: "The reader may well wonder why I do not simply confine my inquiry to psychical semiosis, since no other seems to be of much importance. My reason is that the too frequent practice, by those logicians who do not go to work [with] any method at all [or who follow] the method of basing propositions in the science of logic upon results of the science of psychology—as contradistinguished from common-sense observations concerning the workings of the mind, observations well-known even if little noticed ...—is unsound and insecure" [CP 5.485-7].

His semeotic theory investigates how this habit of semiosis is produced and what sort of habit it is, beginning with a general account of the dualistic conditions that precede its triadic sign action, which he identifies with self-control, as previously distinguished from automatic regulation [see CP 5.473].

Every sane person lives in a double world, the outer and the inner world, the world of percepts and the world of fancies. What chiefly keeps these from being mixed up together is (besides certain marks they bear) everybody's well knowing that fancies can be greatly modified by a certain non-muscular effort, while it is muscular effort alone ... that can to any noticeable degree modify percepts. A man can be durably affected by his percepts and by his fancies. The way in which they affect him will be apt to depend upon his personal inborn disposition and upon his habits. Habits differ from dispositions in having been acquired as consequences of the principle ... that multiple reiterated behaviour of the same kind, under similar combinations of percepts and fancies, produces a tendency— the habit—actually to behave in a similar way under similar circumstances in the future. Moreover—here is the point—every man exercises more or less control over himself by means of modifying his own habits; and the way in which he goes to work to bring this effect about in those cases in which circumstances will not permit him to practice reiterations of the desired kind of conduct in the outer world shows that he is virtually well-acquainted with the important principle that reiterations in the inner world—fancied reiterations—if well-intensified by direct effort, produce habits, just as do reiterations in the outer world; and these habits will have power to influence actual behaviour in the outer world; especially, if each reiteration be accompanied by a peculiar strong effort that is usually likened to issuing a command to one's future self. [CP 5.487]

After Peirce considers the possible causes of habit-change and concludes that no entirely new habit can be created by involuntary experiences, he contends that "[e]very concept, doubtless, first arises when upon a strong, but more or less vague, sense of need is superinduced some involuntary experience of a suggestive nature," which is similar to the "instinctive ideas of animals," for whom conditions are relatively unchanging, so that their ideas need not progress. But man's ideas first take the form of conjectures, although they may not be recognized as such; "[e]very concept, every general proposition of the great edifice of science, first came to us as a conjecture" [CP 5.480]. Inquiry anticipates answers, just as common sense makes intuitive guesses.

To the extent that we consider what we can conceive (however vaguely) as unknowable, or knowable only by intuition (and that to be unknowable), or knowable only automatically (as instinct), we can not fully inquire about our capability of inquiry and cannot effectively consider the consequences of our conduct. Nominalist modern philosophy and science only attempt to explain our sense of what apparently *is* observable before us, but not how we can imagine what might be, as we certainly *can do,* most obviously in science fiction and other forms of art. According to Peirce, nominalist assumptions eliminate any perspective on the *evolution* of knowledge: "The nominalists' difficulty ... is their habit of reducing the possible to the actual and of not distinguishing the actual [or our immediate sense of the existential] from the real [or our mediated experience of the actual as possible]" [CP: 1.422]. Especially in its empiricist form, beginning with Ockham, nominalism is theoretically blind to (and so cannot account for) the phenomena of communication and creativity, which Peirce proposes to do in his normative theory of inquiry as the evolution of thought. He clearly identifies its evolutionary nature with Hegel, but then comments: "I sometimes agree with the great idealist and sometimes diverge from his footsteps—for my own

method has resulted from a more deliberate examination of the exact theory of logic
... and consequently has a broader form" [*CP* 1.453].

Nominalist philosophy was conceived to study our sense of *apparent* reality, as
though what appears to be is simply all we can know, without even questioning why
we have a sense of what *might be*—of what we don't know yet but can imagine.
Peirce claims that, because of its underdeveloped logical analysis capabilities,
nominalism cannot treat general conceptions, such as *reality* (or *common sense,* for
that matter) *as objects in our experience*—that is, as phenomena to be scientifically
investigated.   For Peirce, such general conceptions *are* the very objects, or
phenomena, of philosophical inquiry.   In his view, these are the conceptions we
assume in our everyday instinctive or intuitive behavior based on unexamined
feelings, reaction, and beliefs, upon which all our judgments and rational conduct
depend.   Any theory of inquiry must begin with instinct as a primitive method, he
says, but no creature can have instincts for every possible circumstance [see *CP*:
2.178].   "All instinctive beliefs are vague.   The moment they are precinded, the
pragmatist will begin to doubt them" [*CP* 6.499 (c. 1906)].   Pragmatism is his
conjectured method for how to continue improving our methods of inquiry, which can
be explained in theoretical terms of semeosis, while normative science investigates
the conditions in experience which compel that habit of sign action to occur and to
progress.

In a 1904 letter to Dewey, who proposed (following what he takes to be Hegel's
suggestion) that normative science be considered as natural history, Peirce
distinguishes his view:

[I]t is one of the characteristics of all normative science that it does not concern
itself in the least with what actually takes place in the universe, barring always its
assumption that what is before the mind always has those characteristics that are
found there and which *Phänomenologie* is assumed to have made out. But as to
particular and variable facts, no normative science has any concern with them,
further than to remark that they form a constant constituent of the phenomenon.
Now nothing like the study the Comparative Anatomists are occupied with can be
made of mere possibilities. ... There is no anatomy of possibilities because one
can say in advance how pure possibilities vary and diverge from one another.
Namely, they do so in every possible way. What renders a Comparative Anatomy
possible is that certain conceivable forms do not occur. Only a minute proportion
of them occur.... Thus there is in the list of chemical elements just that
experiential diversity and absence of most possible forms that renders the kind of
study called anatomical possible. If then you have a "Natural History" (i.e. a
comparative anatomy) of thought,—it is not the merely possible thought that
Normative Science studies, but thought as it presents itself in an apparently
inexplicable and irrational experience. [*CP* 8.239]
      [*CP* 8.239].

4   Conceptual Structures Research in Peirce's Costume?

While Peirce clearly acknowledges his philosophy's affinities to ideas of Hegel, he
also emphatically criticizes: "Hegelians overlook the facts of volitional action and

reaction in the development of thought. I find myself in a world of forces which act upon me, and it is they and not the logical transformations of my thought which determine what I shall ultimately believe" [*CP* 8.45]. So normative science must begin with the pre-logical phenomena of experience, and explain how beliefs and judgments evolve under the influence of natural forces and feelings.   Further investigation of Peirce's unfinished project of normative science must rely on Vincent Potter's detailed scholarly examination of the evidence (primarily Peirce's "Minute Logic" of 1902-3, the seven Pragmatism Lectures of 1903, and a series of *The Monist* articles in 1905-6) to resolve the inconsistencies, paradoxes, and hesitations that Peirce's writings reveal, indicating his uncertainties about that part of his philosophy.

For many years, Peirce struggled to justify ethics and especially esthetics as normative sciences, against his suspicion that they were pre-normative and must be relegated to intuition or instinct (as pre-determined behavior on some bio-evolutionary grounds, making them no more controllable than automatic responses), which is how nominalism must treat them. But he eventually conceived all normative science to study *what ought to be* [see *CP* 1.281], to determine norms or rules which need not but ought to be followed [see *CP* 2.156]. The sense of ought frees mind from material determinism; we always have the possibility to act contrary to an "ought," which implies ideals, ends, purposes which attract and guide deliberate conduct [see *CP* 1.575]. Although Peirce's theory considers the human mind as only one manifestation of Mind which is everywhere in one form or another, ours may perhaps be distinguished in having the greatest capacity for self-control. That theoretical view gives his normative sciences their concern: "We know very well that mind, in some sense, acts on matter, and matter on mind: the question is how" [*CP* 6.101]. His normative sciences are to analyze or define what conditions are required for accomplishment of a purpose, or deliberately chosen conduct that responds to conditions represented as fact [see *CP* 1.575].

By 1903, in his Lectures on Pragmatism, Peirce refers to normative science as the science "which investigates the universal and necessary laws of the relation of phenomena to Ends" [*CP* 5.121]. Here, says Potter, he begins to develop the notion that these sciences all distinguish good from bad: "in the representation of truth, in the efforts of the will, and in objects regarded simply in their presentation, respectively (5.36)" [9: 39]. And he begins to consider the role of pragmatism: "For if, as pragmatism teaches us, what we think is to be interpreted in terms of what we are prepared to do, then surely *logic,* or the doctrine of what we ought to think, must be an application of the doctrine of what we deliberately choose to do, which is Ethics" [*CP* 5.35]. He explains pragmatism as a method that helps us to know what we think, and what we believe, the meaning of which is interpreted as our willingness to act on that thought—in terms of its conceived consequences. "Pragmatism is the principle that every theoretical judgment expressible in a sentence in the indicative mood is a confused form of thought whose only meaning, if it has any, lies in its tendency to enforce a corresponding practical maxim expressible as a conditional sentence having its apodosis in the imperative mood" [*CP* 5.18].

In his seven lectures of 1903, Peirce explains that the essence of logic is to criticize arguments, and to judge arguments as good or bad implies that they are subject to control, so that in the future we can choose good arguments and avoid bad ones, giving us the ability to control and to correct. "Reasoning essentially involves self-control; so that [it] is a species of morality" [*CP* 5.108]. Potter stresses, "This is the very heart of the matter ... of Peirce's logic and of his entire philosophical outlook.

To make a normative judgment is to criticize; to criticize is to attempt to correct; to attempt to correct supposes a measure of control over what is criticized in the first place. Any other conception of goodness and badness is idle. In this Peirce was directly opposed to almost all other schools of thought of his day" [9: 41]. Even though Peirce concludes, in the 1903 lectures, that "esthetics considers those things whose ends are to embody qualities of feeling, ethics those things whose ends lie in action, and logic those things whose end is to represent something" [*CP* 5.129], he still hesitates to be sure about esthetics, because esthetic qualities seem to be beyond control. But the importance of that matter for pragmatism is now obvious: "For if the meaning of a symbol consists in how it might cause us to act, it is plain that this 'how' cannot refer to the description of mechanical motions that it might cause, but must intend to refer to a description of the action as having this or that aim. In order to understand pragmatism, therefore, well enough to subject it to intelligent criticism, it is incumbent upon us to inquire what an ultimate aim, capable of being pursued in an indefinitely prolonged course of action, can be" [*CP* 5.135].

By 1905, Potter says Peirce knew a distinction was required "between esthetic qualities themselves, that is, in their own intrinsic reality, and the conscious adoption of them as ideals to be pursued." Then he could argue that the role of esthetics is "to seek out through reflective analysis (1.580) what end is ultimate (can be consistently pursued in any and all circumstances) and to use this as the norm in adopting any particular esthetic quality as an ideal. According to this account of esthetics there would be the necessary element of criticism and control even with respect to the ultimate ideal, not in the sense that the objective reality of that ideal would be affected, but in the sense that one would accept it and conform to it willingly and deliberately" [9: 49]. Potter is certain that Peirce had this in mind, but says it would require many more pages of analysis to interpret the evidence. We do know that in 1905, he wrote the statement: "Pragmatism consists in holding that the purport of any concept is its conceived bearing upon our conduct" [*CP*: 5.460]; meaning cannot be reduced to action, as nominalism must assume. Peirce argues against this nominalism in 1906: "It has been a great, but frequent, error of writers on ethics to confound an ideal of conduct with a motive to action. The truth is that these two objects belong to different categories. Every action has a motive; but an ideal only belongs to a line [of] conduct which is deliberate. To say that conduct is deliberate implies that each action, or each important action, is reviewed by the actor and that his judgment is passed upon it, as to whether he wishes his future conduct to be like that or not. His ideal is the kind of conduct which attracts him upon review. If conduct is to be thoroughly deliberate, the ideal must be a habit of feeling which has grown up under the influence of a course of self-criticisms and of hetero-criticisms; and the theory of the deliberate formation of such habits of feeling is what ought to be meant by esthetics" [*CP* 1.574].

On Potter's reading of all the evidence, Peirce concludes that the *summum bonum* (or the ultimate end or aim) is reasoned and reasonable conduct; that Ethics and logic are specifications of esthetics; and that Ethics proposes what goals may reasonably be chosen in various circumstances, while logic proposes what means are available to pursue those ends [see 9:34]. The pragmatic maxim itself is normative in function: "the meaning of a concept does not lie in any individual reactions at all, but in the manner in which those reactions contribute to [expectation about future experience]" [*CP* 5.3; MS 462 (1903)]; we judge the meaning of a concept by the contribution of the reactions it evokes toward the realization of thought's ultimate aim. And Peirce concludes: "An aim which cannot be adopted and consistently pursued is a bad aim. It cannot properly be called an ultimate aim at all. The only moral evil is not to have an ultimate aim" [*CP* 5.133]. In normative science, then, Peirce's strange costume keeps his word (of 1896): "I follow an order of evolution ... the possibility evolves the actuality. So does Hegel" [*CP* 1.453].

How might this "order of evolution," in Peirce's unfinished "strange costume," affect conceptual structures research? Generally, it tells us that nominalist philosophy limits us to the mechanistic—or *causal*—theories of empirical science that cannot help us learn what to expect of ourselves, nor why we should and how we might improve our conduct (which is, after all, the ultimate aim of inquiry and knowledge), and so cannot assist us in asking what we want our tools to do for us—in augmenting our conduct. Self-consciously conditional, normative theory would study the laws of belief-habit growth, or the evolution of self-control. The science of semeotic, as an expanded logic, could identify our uncriticized or automatic habits of thought and make explicit our self-controlled habits (or beliefs, as intellectual concepts), so that pragmatism could improve their evolution and clarify the aims of any inquiry. Most significantly, we would realize that conceptual structures represent what we more-or-less doubtfully *conjecture*, what we think *ought* to be true, not simply what we *know*. All of us want our activity to amount to something, and especially our inquiry not to become idle. To the extent that we thoughtlessly relegate our goals and aims to intuition or autonomous behavior, we lose control of how we evolve; we become *demoralized*. To the extent that our agency in the evolution of mind and matter is lost in nominalism, we become whatever our machines make of us. Peirce's theory invites conceptual structures research to do otherwise in pursuing normative science [see 10, for more specific indications encouraging future work].

References

General Note: "MS" references are to Peirce's manuscripts archived at the Houghton Library, Harvard; for *CP* references, *Collected Papers of Charles Sanders Peirce,* 8 vols., ed. Arthur W. Burks, Charles Hartshorne, and Paul Weiss (Cambridge: Harvard University Press, 1931-58).

1. Keeler, M. "Pragmatically Yours," In: Ganter, B. and Mineau, G. (eds): *Lecture Notes in Artificial Intelligence,* vol. 1867, Springer-Verlag, (2000) 82-99.
2. Vehkavarra, T. "Peirce's Classification of the Sciences. "<http://mtlserver.uta.fi/~attove/peirce_systems3.pdf>
3. Ransdell, J. M. "Arisbe, a philosophical website, providing coordinated access to the internet resources relevant to the life and work of, and continuing interest in the American philosopher, scientist, and humanist, Charles Sanders Peirce": www.door.net/arisbe.
4. See the Peirce Edition Project <http://www.iupui.edu/~peirce>
5. Meyers, R.G. "Peirce's Extension of Empiricism." *Transactions of the Charles S. Peirce Society,* Winter/Spring, 2002, Vol. XXXVIII, No. 1/2 (137-154).
6. See the *Past Masters* series at Intelex: <http://www.nlx.com>
7. Esposito, J. L. "On the Origins and Foundation of Peirce's Semiotic." *Studies in Peirce's Semiotic.* Lubbock, Texas: Institute for Studies in Pragmaticism, 1979.
8. Oehler, K. "Peirce's Foundation of a Semiotic Theory of Cognition." *Studies in Peirce's Semiotic.* Lubbock, Texas: Institute for Studies in Pragmaticism, 1979.
9. Potter, V. G. *Charles S. Peirce on Norms and Ideals.* New York: Fordham University Press, 1997.

10. Some recently published material hints at what Peirce's archive still holds toward further development of his normative science. Manuscript number 283, "The Basis of Pragmatism," in his Harvard collection (a very small part of which is printed in the *CP*, but all of which now appears in Volume 2 of the *Essential Peirce*, see reference below), is his sixth attempt to write his third Monist essay of the 1906 series. The editors of the *EP* have added to its title "in the Normative Sciences," because here he clarifies his meaning of "science" as *essentially* self-critical inquiry, and distinguishes scientific philosophy (or Cenoscopic science) from retrospective philosophy as "embracing all that positive science which rests on familiar experience" [*E:* 372]. And here, he explains that *duality*, the object of normative science investigation, consists in no quality of both or either of two things conceived, but consists only in a relation between them [see *EP:* 381]. Not action but *reciprocal action*—action and reaction—make duality; so that "[a]ll inhibition of action, or action upon action, involves duality." Since all self-control essentially involves inhibition, "[a]ll direction toward an end or good supposes self-control; and thus the normative sciences are infused with duality" [*E:* 385]. In this essay he poses the central question of normative science: What is a sign? Although we might all agree that we use signs to communicate ideas "from the mind of yesterday to the mind of tomorrow into which yesterday's has grown," Peirce cautions here: "This is a question of no ordinary difficulty, to which the answer must be sought by a well-considered method. To begin with, let us consider what the question means; and first, What is its general nature? We are not studying lexicography. ... We all have a ragged-outlined notion of what we call a sign. We wish to replace that by a well-defined concept, which may exclude some things ordinarily called signs, and will almost certainly include some things not ordinarily so called. So that our new concept may have the highest utility for the science of logic, which is the purpose of the investigation, the terms of the definition must be strictly relevant to logic. As far as this condition will allow, it is to express that which is most essential in the vulgar notion of a sign or representamen. Now a sign as ordinarily understood is an implement of intercommunication; and the essence of an implement lies in its function, that is in its purpose together with the general idea,—not however, the plan,—of the means of attaining that purpose" [*EP:* 389]. The rest of the essay sketches how normative science would investigate the definition of "sign," finally to give a satisfactory account of hypostasis, an inquiry most fundamental to conceptual structures research. [The Peirce Edition Project. *The Essential Peirce, Volume 2.* Bloomington, IN: Indiana University Press, 1998.]

Different Kinds of Comparisons between Fuzzy Conceptual Graphs

Rallou Thomopoulos[1], Patrice Buche[1], and Ollivier Haemmerlé[1,2]

[1] UMR INA P-G/INRA BIA,
16 rue Claude Bernard, F-75231 Paris Cedex 05, France
{Rallou.Thomopoulos,Patrice.Buche,Ollivier.Haemmerle}@inapg.fr
[2] LRI (UMR CNRS 8623 - Université Paris-Sud) / INRIA (Futurs),
Bâtiment 490, 91405 Orsay Cedex, France

Abstract. In the context of a microbiological application, our study proposes to extend the Conceptual Graph Model in order to allow one: (i) to represent imprecise data and queries that include preferences, by using fuzzy sets (from fuzzy set theory) in concept vertices, in order to describe either an imprecise concept type or an imprecise referent; (ii) to query a conceptual graph that may include imprecise data (factual graph) using a conceptual graph that may include preferences (query graph). This is performed in two steps: firstly by extending the projection operation to fuzzy concepts, secondly by defining a comparison operation characterised by two matching degrees: the possibility degree of matching and the necessity degree of matching between two graphs, and particularly between a query graph and a factual graph.

1 Introduction

Our research project is part of a national program that aims at building a tool for microbial risk analysis in food products. This tool is based on an information system which is composed of two parts: a relational database and a conceptual graph knowledge base. The latter is used to store information that was not expected when the schema of the database was designed, and is useful nevertheless. As modifying the schema of the database is quite an expensive operation, the conceptual graph model [11] appears as a flexible way of representing complementary information until the relational database is updated. More precisely the formalization of [9] is used in our application.

Both bases are queried simultaneously by a unified querying system. This part of our work has already been presented in [2]. Both have to deal with the following two specificities. (i) Some of the data are imprecise, like data whose precision is limited by the measuring techniques. For instance, when using a method able to detect bacteria beyond a given concentration threshold (e.g. 10^2 cells per gramme), not detecting any bacterium means that their concentration is below this threshold, which is an imprecise value denoted "$< 10^2$ cells/g". (ii) The bases are incomplete, as they will never contain information about all possible food products and all possible pathogenic germs. Those two characteristics led us

A. de Moor, W. Lex, and B. Ganter (Eds.): ICCS 2003, LNAI 2746, pp. 54–68, 2003.

to propose, firstly the handling of imprecise values, and secondly the expression of different levels of preferences in the user's selection criteria so as to allow flexible querying.

In the bibliography concerning databases, the fuzzy set framework has been shown to be a sound scientific way of modelling both flexible queries [1] and imprecise values by means of possibility distributions [10]. Besides, the introduction of the fuzzy set theory into the conceptual graph model has been initially studied by Morton [8] and then extended by several works such as [14, 4]. Morton distinguished different kinds of fuzziness. In particular, linguistic fuzziness concerns metric attribute concepts, which can have either a precise measure or a label that stands for a crisp or fuzzy subset of what is called the "universe of discourse". In [14], linguistic fuzziness is proposed for non-metric attributes, and fuzzy relation vertices are introduced, by associating a certainty degree with relations.

In these studies, the notion of fuzzy marker is not clearly introduced: what is fuzzy is the measure associated with a given concept, and the definition domain of this measure seems external to the support. The notion of fuzzy concept type is not handled. Our work introduces the notions of fuzzy marker and fuzzy type. In both cases, fuzziness is represented homogeneously, by means of a normalized fuzzy set. The "universe of discourse" is clearly defined as part of the support of the conceptual graph model, and metric and non-metric concepts are not distinguished as they are treated in the same way. We do not propose fuzzy relations, as in our context fuzziness concerns the data and not the way they are linked. However, our study about fuzzy types could be applied to any hierarchy and thus to the relation type set.

In [4], the notion of conjunctive fuzzy type is proposed, which is a conjunction of types associated with the same individual marker (with different fuzzy truth values). In our approach to fuzzy types, we do not question the unicity of an individual marker's type: a fuzzy type represents a disjunction of possible types (with different possibility degrees).

We presented preliminary studies of our approach in [3, 13, 12], where basic principles were formalized, including the treatment of numerical values and the introduction of fuzzy markers and fuzzy types. In this paper, we focus on different ways of querying imprecise data using fuzzy queries. A first way is the extension of the projection operation to fuzzy conceptual graphs, which is an all-or-nothing process. A second way is the use of comparison degrees of the fuzzy set theory, which allows one to perform fuzzy querying. In order to introduce these two comparisons, we went deeper into the notion of fuzzy type, by distinguishing a fuzzy type in intention (or simply, a fuzzy type) from its associated devel oped form called fuzzy type in extension.

In Section 2, we recall our choices for representing fuzzy values in the conceptual graph model, that is, fuzzy markers and fuzzy types, and we introduce the notion of fuzzy type in extension. In Section 3, we propose an extension of the projection operation to handle fuzzy values. In Section 4, we introduce a more flexible way of comparing fuzzy conceptual graphs, using two comparison degrees of the fuzzy set theory, the possibility degree of matchin g and the necessity degree of matching.

2   Fuzzy Values in the Conceptual Graph Model

Information of the application stored in conceptual graphs (factual graphs or query graphs) may be represented in two ways: (i) as individual markers, as it is the case for numerical values (30, 50, etc.) or for bacterial strains (E-3, A-86, etc.); (ii) as concept types, as it is the case for substrates (*Milk*, *Beef*, etc.). In both cases, we must be able to represent them as fuzzy information, as explained in Part 2.1. We thus have introduced the representation of fuzzy values conc erning both markers, presented in Part 2.2, and concept types, presented in Part 2.3. As mentioned in the introduction, in this paper we propose a more thorough definition of fuzzy types, compared to [3, 13, 12].

2.1   Preliminary Notions: Fuzzy Sets

In our application we need firstly to be able to represent imprecise data, secondly to express preferences in queries. To perform this we use the fuzzy set theory [15].

Definition 1. *a fuzzy set A on a domain X is defined by a membership function μ_A from X to [0, 1] that associates the degree to which x belongs to A with each element x of X .*

The domain X may be continuous or discrete. These two cases are illustrated by the examples given in Figure 1. The fuzzy set *MyMilkProductPreferences* is also denoted : *1/Whole milk + 0.5/Half skim milk*, which indicates the degree associated with each element.

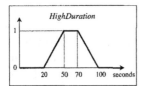 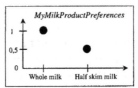

Fig. 1. Fuzzy sets *HighDuration* and *MyMilkProductPreferences*

A fuzzy set may be interpreted in two ways:

1. in a query, it expresses preferences on the domain of a selection criterion. For example, the fuzzy set *HighDuration* in Figure 1 may be interpreted as a preference concerning the required value of the criterion *Duration*: a duration between 50 and 70 seconds is fully satisfactory; values outside this interval may also be acceptable, but with smaller preference degrees;

2. in a datum, it describes an imprecise datum represented by a possibility distribution. For example, the fuzzy set *MyMilkProductPreferences* may be interpreted as an imprecise datum if the kind of milk that was used in the experiment is not clearly known: it is very likely to be whole milk, but half skim milk is not excluded.

In the following we recall the definitions of fuzzy markers and fuzzy types and we introduce the new notion of fuzzy type in extension. We refer to the conceptual graph model of [11], and more precisely to the formalisation of [9].

2.2   Fuzzy Markers

Definition 2. *A* **fuzzy marker** m_f *of concept type t is a fuzzy set defined on the set of individual markers I. It takes values between 0 and 1 for every individual marker that conforms[1] to t, and 0 elsewhere.*

A fuzzy marker represents a disjunction of individual markers that conform to type t, weighted by a coefficient between 0 and 1.

Remark 1. *A "classic" individual marker m of type t can be considered as a particular fuzzy marker: its membership function associates the value 1 with m, and the value 0 with the rest of I.*
*The generic marker * can be considered as a particular fuzzy marker of type t: its membership function associates the value 1 with any element that conforms to t, and 0 with the rest of I.*

Definition 3. *A* **concept with a fuzzy marker** *is a concept vertex whose label is a pair* (t, m_f), *where* m_f *is a fuzzy marker of concept type t.*

The conceptual graph represented in Figure 2 includes a concept with a fuzzy marker, of type *NumericalValue*.

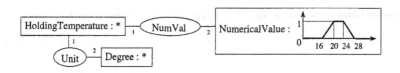

Fig. 2. An example of a concept with a fuzzy marker

---

[1] The set of individual markers that conform to t is $\{m \in I \mid \tau(m) \leq t\}$, where τ is the application from I to T_C, defined in the support, that associates a minimum concept type with each individual marker.

2.3 Fuzzy Types

Definition 4. *A* **fuzzy type** t_f *is a fuzzy set defined on a subset D_{t_f} of elements of the concept type set T_C.*

For example, the fuzzy set *MyMilkProductPreferences* represented in Figure 1 is a fuzzy type defined on a subset of the concept type set.

Remark 2. *A "classic" concept type t can be considered as a particular fuzzy type: its membership function is defined on one element $\{t\}$ and takes the value 1 for this element.*

We can note that no restriction has been imposed concerning the concept types that compose the definition domain of a fuzzy type. In particular, the user may associate a given degree d with a type t and another degree d' with a subtype t' of this type. $d' \leq d$ represents a semantic of restriction for t' compared to t, whereas $d' \geq d$ represents a semantic of reinforcement for t' compared to t. For instance, if the user is particularly interested in skim milk because he studies the properties of low fat products, but also wants to retrieve complementary information about other kinds of milk, he will be able to express his preferences using e.g. the following fuzzy type: *1/Skim milk + 0.5/Milk.*

However the information *1/Skim milk* and *0.5/Milk* may sound contradictory. Indeed, the criterion *0.5/Milk* seems to include skim milk, which is a kind of milk, though skim milk has the degree 1. In fact, a fuzzy type defined on a subset of T_C implicitly gives information about the rest of T_C. In our example, if the user's preferences are those expressed by the fuzzy type *1/Skim milk + 0.5/Milk,* we can deduce in addition that he is not interested in vegetable.

This observation led us to introduce the notion of **fuzzy type in extension**, which is the developed form of the fuzzy type defined in Definition 4, which we now call **fuzzy type in intention**. A fuzzy type in extension is defined on the whole concept type set T_C.

Definition 5. *Let t_{fi} be a fuzzy type in intention defined on a subset D_i of T_C. Its membership function is denoted μ_i. The* **fuzzy type in extension** t_{fe} *associated with t_{fi} is defined as follows.*

The membership function of t_{fe} is denoted μ_e. For each element $t \in T_C$:

- *if there exists, at least, one element $u \in D_i$ such that $u \geq t$, then we distinguish two cases:*
  - *if there exists one single smallest[2] element $u \in D_i$ such that $u \geq t$, then*
    $\mu_e(t) = \mu_i(u)$
    The interpretation is the following:
    * *if the fuzzy set expresses preferences, then being interested in a given type u with a degree d implies being interested in all the types t more specific than u with the same degree of preference.*

---

[2] with the meaning of the relation **a kind of** that partially orders T_C

* *if the fuzzy set expresses an imprecise datum, then declaring that a given type u has a degree of possibility d implies that, for all t which are more specific than u, their degree of possibility is d.*

• *otherwise, we call $u_1, u_2, \ldots, u_n$ the smallest elements of D_i such that $\forall k \in \{1, 2, \ldots, n\}$ $u_k \geq t$. These elements are not comparable[3]. We distinguish two cases:*

  * *if the fuzzy set expresses preferences, then the degree of preference associated with t must be at least equal to the degree of preference associated with each element of the list $u_1, u_2, \ldots, u_n$, and we define*
  $$\mu_e(t) = max_{k \in \{1,2,\ldots,n\}} \mu_i(u_k)$$
  * *if the fuzzy set expresses an imprecise datum, then t cannot have a degree of possibility greater than each of those associated with $u_1, u_2, \ldots, u_n$, and we define $\mu_e(t) = min_{k \in \{1,2,\ldots,n\}} \mu_i(u_k)$*

— *otherwise $\mu_e(t) = 0$*

Figure 3 shows the fuzzy type in extension associated with the fuzzy type in intention *0.8/Milk + 1/Whole milk + 0.3/Evaporated milk* on a part of the concept type set. We assume that the fuzzy set expresses an imprecise datum.

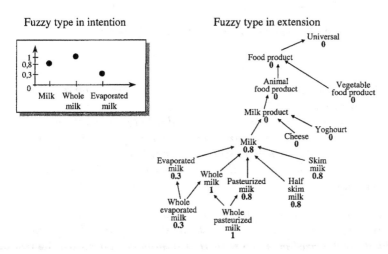

Fig. 3. An example of a fuzzy type in extension

In the fuzzy type in intention of Figure 3, the user has associated the degree 1 with *Whole milk* but only 0.3 with *Evaporated milk*. The minimum of these two degrees is thus associated with their common subtype *Whole evaporated milk* in the fuzzy type in extension, because we assume that the fuzzy type expresses an imprecise datum. If, on the contrary, it expressed preferences, the degree associated with *Whole evaporated milk* would be 1 instead of 0.3.

---

[3] using the partial order induced by the relation **a kind of**

On the other hand, there is no ambiguity about *Whole pasteurized milk* because nothing has been specified about *Pasteurized milk* in the fuzzy type in intention. Therefore, *Whole pasteurized milk* has the degree associated with *Whole milk*, that is 1, and not 0.8 which is associated with the more general type *Milk*.

Definition 6. *A* **concept with a fuzzy type** *is a concept vertex whose label is a pair* (t_f, m), *where* t_f *is a fuzzy type and* m *is the generic marker* *.

For instance, let us suppose that the user's preferences concerning the substrate are *MyMilkProductPreferences* represented in Figure 1. In conceptual graph terms, this substrate corresponds to the concept type *Whole milk* with the degree 1, or the concept type *Half skim milk* with the degree 0.5, which is represented by the concept with a fuzzy type of Figure 4.

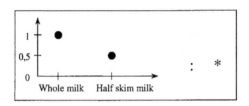

Fig. 4. An example of a concept with a fuzzy type

Definition 7. *In a concept with a fuzzy type* t_f, *the generic marker* * *has the following interpretation:*

Let t_{fe} *be the fuzzy type in extension associated with* t_f *and* μ_e *its membership function. The generic marker* * *is a fuzzy marker that associates the degree* $\mu_e(t)$ *with each individual marker* m *of* I, *where* $t = \tau(m)$ *and* τ *is the application from* I *to* T_C, *defined in the support, that attributes a minimum concept type to each individual marker.*

In Figure 4, the fuzzy type in extension t_{fe} associated with t_f is the fuzzy set that associates the degree 1 with *Whole milk* and its subtypes *Whole evaporated milk* and *Whole pasteurized milk*, 0.5 with *Half skim milk*, and 0 with the rest of T_C. The generic marker thus stands for the fuzzy set that associates the degree 1 with the markers of *Whole milk*, *Whole evaporated milk* and *Whole pasteurized milk*, 0.5 with the markers of *Half skim milk*, and 0 with the rest of I.

Remark 3. *The information brought by a generic marker, with the interpretation of Definition 7, is redundant with that given by its associated type* t_f: *the degree associated with a given individual marker in* * *is the same as the one associated with its type in* t_f.

Remark 4. *In the particular case where* t_f *is a "classic" type* t, t_{fe} *is the fuzzy type in extension that associates the value 1 with* t *and its subtypes, and the value*

0 on the rest of T_C. The associated generic marker is thus the fuzzy marker that associates the value 1 with the markers of t and its subtypes, that is, with the markers that conform to t, which is in conformity with the Remark 1.

3   The Specialization Relation for Fuzzy Conceptual Graphs

The specialization relation of the conceptual graph model allows one to perform comparisons of conceptual graphs. After having extended the model to represent fuzzy concepts (concepts with a fuzzy marker or with a fuzzy type), the next step is to preserve the specialization relation for conceptual graphs that include fuzzy concepts (called "fuzzy conceptual graphs"), in order to be able to compare fuzzy conceptual graphs, and particularly a fuzzy query graph with fuzzy factual graphs.

3.1   The Notion of Specialization for Fuzzy Sets

The notion of specialization for fuzzy sets is based on the inclusion relation: given two fuzzy sets A and B, A is a specialization of B if and only if A is included in B. An example is given in Figure 5 on a continuous domain. This definition applies to both discrete and continuous domains.

Definition 8. *Let A and B be two fuzzy sets defined on a domain X. A is included in B (denoted $A \subseteq B$) if and only if their membership functions μ_A and μ_B satisfy the condition:*

$$\forall x \in X, \mu_A(x) \leq \mu_B(x).$$

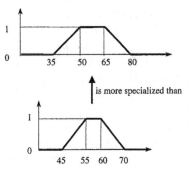

Fig. 5. Example of specialization for fuzzy sets

3.2 Extension of the Specialization Relation to Fuzzy Conceptual Graphs

Definition 9. *Let m and m' be two markers of types t and t'. m and m' are thus defined on the same domain I. m' is a specialization of m if and only if $m' \subseteq m$, where $\subseteq$ is the classic inclusion relation defined for fuzzy sets.*

An example of specialization involving fuzzy markers is given in Figure 6.

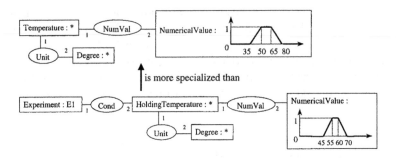

Fig. 6. An example of specialization involving fuzzy markers

Note that in Definition 9 there are four possible cases for m (resp. m'). m (resp. m') can be: an individual marker of a simple type; a fuzzy marker of a simple type; a generic marker of a simple type; a generic marker of a fuzzy type.

Remark 5. *In the "classic" conceptual graph model, there are three cases where m' is a specialization of m: (i) m and m' are individual and $m' = m$ (ii) $m = *$ and m' is individual (iii) $m = *$ and $m' = *$. In these cases, Definition 9 is in agreement with the classic specialization relation:*

If m and m' are two individual markers (of the simple types t and t', $t' \leq t$): m (resp. m') is represented by the fuzzy set that associates the value 1 with m (resp. m') and 0 with the rest of I, as specified in Remark 1. Then m' is a specialization of m iff $m' \subseteq m$, that is iff $m' = m$.

If m is the generic marker of a simple type t and m' an individual marker of a simple type t' ($t' \leq t$): m is represented by the fuzzy set that associates the value 1 with any individual marker that conforms to t (which includes m') and 0 with the rest of I, as specified in Remarks 1 and 4; m' is represented by the fuzzy set that associates the value 1 with m' and 0 with the rest of I. Then m' is a specialization of m because $m' \subseteq m$ is true.

If m is the generic marker of a simple type t and m' the generic marker of a simple type t' ($t' \leq t$): m is represented by the fuzzy set that associates the value 1 with any individual marker that conforms to t (which includes the individual markers that conform to t') and 0 with the rest of I; m' is represented by the fuzzy set that associates the value 1 with any individual marker that conforms to t' and 0 with the rest of I. Then m' is a specialization of m because $m' \subseteq m$ is true.

Definition 10. *Let t and t' be two types, t_{fe} and t'_{fe} their associated types in extension. t_{fe} and t'_{fe} are thus defined on the same domain T_C. t' is a specialization of t if and only if $t'_{fe} \subseteq t_{fe}$, where $\subseteq$ is the classic inclusion relation defined for fuzzy sets.*

An example of specialization involving fuzzy types is given in Figure 7.

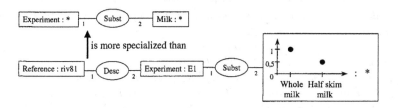

Fig. 7. An example of specialization involving fuzzy types

Remark 6. *In the "classic" case where t and t' are simple types, Definition 10 is in agreement with the classic specialization relation: t (resp. t') is represented by the fuzzy set that associates the value 1 with t (resp. t') and its subtypes, and 0 with the rest of T_C. Then t' is a specialization of t iff $t' \subseteq t$, that is iff the set of subtypes (including itself) of t' is included in the set of subtypes of t, that is iff $t' \leq t$.*

Definition 11. *Let $l = (t, m)$ and $l' = (t', m')$ be the labels of two concept vertices, where t and t' can be fuzzy types, m and m' can be fuzzy markers. Then l' is a specialization of l if and only if t' is a specialization of t and m' is a specialization of m.*

Definition 12. *The projection operation remains unchanged as a graph morphism that allows a restriction of the vertices labels [9], except that this restriction is now based on the specialization relation extended to fuzzy concepts as defined in Definitions 9 to 11.*

Using this extended projection operation, the comparison of two conceptual graphs leads to a binary result: a graph G can be projected into a graph G' or cannot, there is no intermediate solution. However, a more flexible comparison of fuzzy sets should allow one to evaluate the compatibility between two fuzzy conceptual graphs, and particularly between a fuzzy query graph and a fuzzy factual graph. Therefore, we propose to introduce a more flexible comparison that performs fuzzy querying.

4 A More Flexible Comparison of Fuzzy Conceptual Graphs

Two scalar measures are classically used in fuzzy set theory to evaluate the compatibility between a fuzzy selection criterion and a corresponding imprecise datum: (i) a possibility degree of matching [16]; (ii) a necessity degree of matching [5].

Definition 13. *Let m and m' be two markers of types t and t', with membership functions μ_m and $\mu_{m'}$. Then m' is compatible with m with the possibility degree $\Pi(m\ ;m')$ and the necessity degree $N(m\ ;m')$:*

- *$\Pi(m\ ;m')$, possibility degree of matching between m and m', is an "optimistic" degree of overlapping that measures the maximum compatibility between m and m', and is defined by:*
  $\Pi(m\ ;m') = sup_{x\in I}min(\mu_m(x),\mu_{m'}(x))$.
  sup denotes the supremum value of a function.
- *$N(m\ ;m')$, necessity degree of matching between m and m', is a "pessimistic" degree of inclusion that estimates the extent to which it is certain that m' is compatible with m. It is equal to the complement in 1 of the possibility degree of matching between the complement of m and m':*
  $N(m\ ;m') = 1 - sup_{x\in I}min(1 - \mu_m(x),\mu_{m'}(x))$.

Note that the measure of the possibility degree with which m' is compatible with m is symmetrical, whereas the necessity degree is not.

An example is given in Figure 8.

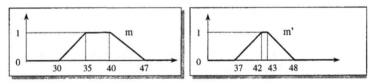

m' is compatible with m with the possibility degree Π(m ; m') and the necessity degree N(m ; m') obtained as follows :

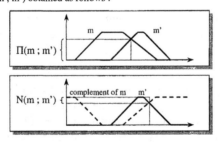

Fig. 8. Flexible comparison of two markers m and m'

Remark 7. *For the "classic" cases detailed in Remark 5, $\Pi(m\ ;m')$ and $N(m;m')$ take the value 1.*

Definition 14. *Let t and t' be two types, t_{fe} and t'_{fe} their associated types in extension with membership functions μ_e and μ'_e. Then t' is compatible with t with the possibility degree $\Pi(t\ ;t')$ and the necessity degree $N(t\ ;t')$:*

$\Pi(t\ ;t') = sup_{x\in T_C} min(\mu_e(x), \mu'_e(x))$
$N(t\ ;t') = 1 - sup_{x\in T_C} min(1 - \mu_e(x), \mu'_e(x))$.

Remark 8. *For two "classic" types t and t': in the "classic" specialization case where $t' \leq t$, $\Pi(t\ ;t')$ and $N(t\ ;t')$ take the value 1. $\Pi(t\ ;t')$ takes the value 1 iff t and t' have common subtypes (in the broad sense). $N(t\ ;t')$ takes the value 1 iff $t' \leq t$.*

Definition 15. *Let $l = (t, m)$ and $l' = (t', m')$ be the labels of two concepts c and c', where t and t' can be fuzzy types, m and m' can be fuzzy markers. Then c' is compatible with c with the possibility degree $\Pi(c\ ;c')$ and the necessity degree $N(c\ ;c')$, where $\Pi(c\ ;c')$ and $N(c\ ;c')$ are defined as follows:*
$\Pi(c\ ;c') = min(\Pi(t\ ;t'), \Pi(m\ ;m'))$.
$N(c\ ;c') = min(N(t\ ;t'), N(m\ ;m'))$.

The *min* operator is used for the conjunction of the compatibility degrees, as presented in [6].

For instance, for two concepts c and c':
c = [*Whole milk : 1/sample32 + 0.5/sample35*] and
c' = [*0.8/Whole milk + 1/Half skim milk : ***], we have:
$\Pi(t\ ;t') = 0.8$ (given by *Whole milk* which has the degree 1 in c and 0.8 in c'),
$\Pi(m\ ;m') = 0.8$ (both 'sample32' and 'sample35', which are individual markers of type *Whole milk*, have the degree 0.8 in c', where the generic marker $*$ stands for the fuzzy set that associates the degree 0.8 with every marker that conforms to *Whole milk* and 1 with every marker that conforms to *Half skim milk*; the result is due to 'sample32' which has the degree 1 in c),
$N(t\ ;t') = 0$ (*Half skim milk* is indeed fully possible in c' whereas it does not belong to the type of c),
$N(m\ ;m') = 0$ (for the same reason: the markers that conform to *Half skim milk* are fully possible in c' whereas they do not belong to the marker of c),
$\Pi(c\ ;c') = min(0.8, 0.8) = 0.8$,
$N(c\ ;c') = min(0, 0) = 0$.

Definition 16. *Let G and G' be two conceptual graphs that can possibly include fuzzy concepts. Then the graph G' is compatible with the graph G with the possibility degree $\Pi(G\ ;G')$ and the necessity degree $N(G\ ;G')$ if there is an ordered pair (f, g) of mappings, f (resp. g) from the set of relation types (resp. concept types) of G to the set of relation types (resp. concept types) of G', such that:*

– *the edges and their numbering are preserved;*
– *the relation vertex labels may be restricted.*

$\Pi(G\ ;G')$ *and* $N(G\ ;G')$ *are then determined as follows:*
Let C_G be the set of concept vertices of G. For each concept vertex $c \in C_G$, let $\Pi(c\ ;g(c))$ be the possibility degree and $N(c\ ;g(c))$ the necessity degree with which $g(c)$ is compatible with c. Then $\Pi(G\ ;G') = min_{c\in C_G} \Pi(c\ ;g(c))$ and $N(G\ ;G') = min_{c\in C_G} N(c\ ;g(c))$

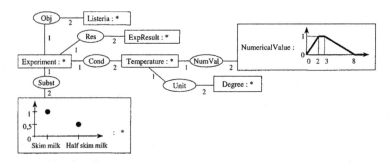

Fig. 9. An example of a query graph G

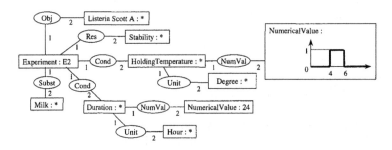

Fig. 10. An example of a factual graph G'

For example, let us consider the graph G given in Figure 9 and the graph G' given in Figure 10: G' is compatible with G with the possibility degree 0.8. This value corresponds to the possibility degree with which the concept with a fuzzy marker of G' (representing the numerical interval [4;6]) is compatible with the concept with a fuzzy marker of G (of type *NumericalValue*). All the other concept vertices of G' are compatible with the possibility degree 1 with their antecedent in G (including the concept vertex [*Milk* : *]). G' is compatible with G with the necessity degree 0, which is obtained for the [*Milk* : *] vertex.

Remark 9. *In the case of a "classic" projection operation (G and G' are not fuzzy, and G can be projected into G'), $\Pi(G ; G')$ and $N(G ; G')$ take the value 1. The opposite is not true. We can also note that the extended projection operation presented in Definition 12 is more restrictive than a graph compatibility with the possibility degree 1, but less restrictive than a graph compatibility with the necessity degree 1.*

5   Conclusion and Perspectives

In this paper, the conceptual graph model has been extended in order to be able to (i) represent imprecise data as well as preferences in queries (ii) propose the extension of the projection operation in order to take into account fuzzy values

(iii) perform a softer comparison using the possibility degree of matching and the necessity degree of matching between two graphs.

This work is part of a food risk control application which uses two databases: a relational database that has been well experimented, and the conceptual graph base mentioned in this article. The latter has been implemented using the CoG-ITaNT platform [7], including all the mechanisms presented in this paper and the unified querying system presented in [2]. It has been successfully presented to our microbiologist partners and is now operational. At the present time, it contains more than a hundred graphs and appears as an interesting and useful complement to the relational database. Conceptual graphs are drawn through the graphic interface of CoGITaNT, and registered as text files.

An important progress in our work has been the definition of the fuzzy type in extension, presented in this paper. This notion allowed us to deal with fuzzy types that are defined on the same domain for all of them and thus to be able to apply results, namely comparison degrees, from the fuzzy set theory.

Our next work will be focused on three different points: (i) the logical interpretation of the conceptual graph model extended to fuzzy values, (ii) the notion of equivalence classes for fuzzy types in intention that are associated with the same fuzzy type in extension, (iii) the study of the complexity of the model extended to fuzzy values, which has already been approached in [12].

References

1. P. Bosc, L. Lietard, and O. Pivert. Soft querying, a new feature for database management system. In *Proceedings DEXA'94 (Database and EXpert system Application), Lecture Notes in Computer Science*, volume 856, pages 631–640. Springer-Verlag, 1994.

2. P. Buche and O. Haemmerlé. Towards a unified querying system of both structured and semi-structured imprecise data using fuzzy views. In *Proceedings of the 8th International Conference on Conceptual Structures, Lecture Notes in Artificial Intelligence*, volume 1867, pages 207–220, Darmstadt, Germany, August 2000. Springer-Verlag.

3. P. Buche, O. Haemmerlé, and R. Thomopoulos. Representation of semi-structured imprecise data for fuzzy querying. In *Proceedings of the Joint 9th IFSA World Congress and 20th NAFIPS International Conference (IFSA-NAFIPS'2001)*, pages 2126–2131, Vancouver, Canada, July 2001.

4. T.H. Cao. *Foundations of Order-Sorted Fuzzy Set Logic Programming in Predicate Logic and Conceptual Graphs*. PhD thesis, University of Queensland, Australia, 1999.

5. D. Dubois and H. Prade. *Possibility Theory - An Approach to Computerized Processing of Uncertainty*. Plenum Press, New York, 1988.

6. D. Dubois and H. Prade. *Fuzziness in Database Management Systems, P. Bosc and J. Kacprzyk eds.*, chapter Tolerant fuzzy pattern matching : an introduction, pages 42–58. Heidelberg:Physica Verlag, 1995.

7. D. Genest and E. Salvat. A platform allowing typed nested graphs: How cogito became cogitant. In *Proceedings of the 6th International Conference on Conceptual Structures (ICCS'1998), Lecture Notes in Artificial Intelligence*, volume 1453, pages 154–161, Montpellier, France, August 1998. Springer.

8. S.K. Morton. *Conceptual graphs and fuzziness in artificial intelligence.* PhD thesis, University of Bristol, 1987.
9. M.L. Mugnier and M. Chein. Représenter des connaissances et raisonner avec des graphes. *Revue d'Intelligence Artificielle,* 10(1):7–56, 1996.
10. H. Prade. Lipski's approach to incomplete information data bases restated and generalized in the setting of zadeh's possibility theory. *Information Systems,* 9(1):27–42, 1984.
11. J.F. Sowa. *Conceptual structures - Information processing in Mind and Machine.* Addison-Welsey, 1984.
12. R. Thomopoulos, P. Buche, and O. Haemmerlé. Representation of weakly structured imprecise data for fuzzy querying. *To be published in Fuzzy Sets and Systems.*
13. R. Thomopoulos, P. Buche, and O. Haemmerlé. Extension du modèle des graphes conceptuels à la représentation de données floues. In *Actes du 13e Congrès Francophone AFRIF-AFIA de Reconnaissance des Formes et d'Intelligence Artificielle (RFIA'2002),* volume 1, pages 319–328, Angers, France, January 2002.
14. V. Wuwongse and M. Manzano. Fuzzy conceptual graphs. In *Proceedings of the First International Conference on Conceptual Structures, Lecture Notes in Artificial Intelligence,* volume 699, pages 430–449, Quebec City, Canada, August 1993. Springer-Verlag.
15. L.A. Zadeh. Fuzzy sets. *Information and Control,* 8:338–353, 1965.
16. L.A. Zadeh. Fuzzy sets as a basis for a theory of possibility. *Fuzzy Sets and Systems,* 1:3–28, 1978.

Intensional Formalization
of Conceptual Structures

Nikolaj Oldager

Informatics and Mathematical Modelling
Technical University of Denmark, DK-2800 Lyngby, Denmark
sno@imm.dtu.dk

Abstract. In this paper, I: (1) argue that the issue of intensionality is of great importance for formalization of conceptual structures, (2) show that this issue is underestimated in contemporary formalizations of conceptual structures, and (3)—as a remedy to this discrepancy—introduce an intensional language for formalization of conceptual structures. The language has a syntax similar to the description logic $\mathcal{ALC}$ with the exception that an additional equivalence relation is introduced. The purpose of this relation is to enable formalization of intensional equivalence of concepts. The intensional semantics is defined by a novel algebraic semantics which basically is an algebraic generalization of the well-known extensional semantics.

1 Introduction

The aims of this paper are: to argue that the issue of intensionality is of great importance for formalization of conceptual structures, to investigate whether the issue is underestimated in contemporary formalizations of conceptual structures, and to introduce an intensional language for formalization of conceptual structures.

To start with, we may compare concepts with sets. Whereas the notion of a set is well understood and precisely formalized, the notion of a concept is notoriously intricate, and no general formalization of it exists—at least not today. Nevertheless, formalization of concepts, as well as the similar properties, is a subject which has received a lot of attention in the literature, and most agree that in general concepts can be characterized by the following properties:

1. A concept has members, like a set has. In this paper, the members of a concept, that is, the individuals falling under the concept, are called the *extension* of the concept.
2. A concept is not defined by its extension in the sense that if two concepts have the same extension, we cannot in general deduce their identity. Concepts are, in contrast to sets, non-extensional.

Regrettably, this characterization leaves open a problem of fundamental importance, namely, how do we provide a condition for identifying concepts?

A. de Moor, W. Lex, and B. Ganter (Eds.): ICCS 2003, LNAI 2746, pp. 69–82, 2003.

Carnap's Method of Extension and Intension [11] proposes a solution (or explicatum) to this problem: concepts are defined by their *intension* in the sense that concepts with the same intension are identified.[1] A concept possesses therefore, in addition to its extension, a so-called intension. Moreover, the intension should determine the extension. A vast number of theories formalizing intensions (intensional logics in particular) exist, see for instance [21], and [12] and its successors. However, it seems to be the case that none of them can be considered adequate as a general intensional theory, cf. [1].

The lack of a general theory of intensions need not hinder our task, which is formalizing intensions of the concepts occurring in conceptual structures, because these concepts are rather elementary in that self-referential and higher-order concepts are omitted. And, as we shall see in Section 6, we can formalize an elementary theory of the intensions of these concepts. This enables us to define the semantics of an intensional language for formalization of conceptual structures which is introduced in the previous Section 4. At the end we show that the language is intensional, present three small examples, and argue that the language gives us a means for formalizing more well structured and fine-grained conceptual structures.

The issue of intensional representation of concepts and conceptual structures has been raised previously within AI [19, 8, 24, 27]. However, with the exception of [19], these papers are all rather sketchy from a formal point of view, and none of them presents a general purpose language for intensional formalization of conceptual structures.

2   On Extensionality and Intensionality

The subject of intensionality is of great importance for conceptual structures because concepts are non-extensional in general. As this is a comprehensive subject, we will only present the main results. The aim of this section is twofold. First, a well-known condition for determining whether a language is extensional or intensional is presented. Then we use it to investigate whether the contemporary formalizations of conceptual structures are extensional or intensional.

The common definition of intensionality is based on the definition of extensionality. A language is *extensional* if all co-extensional (equivalent) expressions may be substituted (replaced) for each other with preservation of truth. Moreover, it is *intensional* if at least one substitution of co-extensional expressions does not preserve truth.[2] In philosophy of language, see e.g. [2], this definition is applied to natural language sentences. For example,

the morning star is a planet

---

[1] The nomenclature is not uniform. In this paper, 'intension' is preferred over 'sense', 'conceptual content', or 'comprehension'. Note, we use 'intension' in a wider sense than the possible-world tradition.

[2] Intensionality is simply taken to be non-extensionality. It has been suggested, cf. [11], that an intensional language additionally should have the property that co-intensional expressions should be substitutable. We will return to this subject later.

is extensional. It is equivalent to, i.e. has the same truth value as,

the evening star is a planet,

which is the result of substituting the expression *morning star* with the co-extensional expression *evening star* (*morning star* and *evening star* both refer to (denote) the planet Venus). Whereas the true sentence

the ancients believed that the morning star is the morning star

is intensional, since substitution of the latter occurrence of *morning star* with *evening star* yields

the ancients believed that the morning star is the evening star,

which is false, meaning truth is not preserved under the substitution.

We may apply the definition to different logical languages. We get that first-order predicate logic is extensional because co-extensionality (for all individuals) implies substitutability, i.e.

$$((\forall x_1, x_2, \ldots)(\phi \leftrightarrow \psi)) \rightarrow (\chi \leftrightarrow \chi[\psi/\phi])$$

is valid for all formulae ϕ, ψ, χ and variables $x_1, x_2, \ldots$; where $\chi[\psi/\phi]$ denotes the result of substituting some but not necessarily all occurrences of ϕ with ψ in χ. In contrast, modal logic is intensional because co-extensionality does not imply substitutability, i.e.

$$(\phi \leftrightarrow \psi) \rightarrow (\Box\phi \leftrightarrow \Box\phi[\psi/\phi]),$$

that is,

$$(\phi \leftrightarrow \psi) \rightarrow (\Box\phi \leftrightarrow \Box\psi)$$

is *not* valid in a normal modal logic like **S5**.

We can now prove that the description logic $\mathcal{ALC}$ [23, 4] is extensional. It follows from the following proposition. (In description logics, implication is not a part of the syntax of concept descriptions, so the result is formulated by means of the notation for inference $\models$.)

Proposition 1. *Let C, D, ϕ be $\mathcal{ALC}$ concept descriptions, and let $\phi[D/C]$ denote the result of substituting some but not necessarily all of the occurrences of C with D in ϕ. Then*

$$C \equiv D \models \phi \equiv \phi[D/C].$$

Proof. Induction on the structure of the concept description ϕ.   □

This only shows that $\mathcal{ALC}$ is extensional, however, it should be clear that the result also holds for similar languages. Moreover, as conceptual graphs [24] and the Knowledge Interchange Format (KIF) [15, 16] have the same model-theoretic semantics as predicate logic [24], they are extensional too.[3] We may therefore conclude that intensionality is underestimated in contemporary formalizations of conceptual structures. As a remedy to the discrepancy between the intensional properties of concepts and their extensional formalizations, an intensional language for formalization of conceptual structures is introduced below.

3 An Informal Appetizer

One of the key ideas of the following language is that both of the properties of concepts which we listed in the introduction are fundamental in the process of formalizing conceptual knowledge. Therefore, a concept language should be comprised of:

- an extensional part in which the relations between the extensions of the concepts are formalized;
- an intensional part in which the relations between the intensions of the concepts are formalized.

We thus recognize two kinds of relations between concepts: extensional and intensional. Conceptual knowledge should in general be formalized by means of both—if we always employ the extensional relations, then the knowledge will not be genuine conceptual (at least not in accordance with the above-mentioned properties).

When two concepts c and d are co-extensional we shall write

$$c \equiv d.$$

When two concepts are intensionally equivalent, i.e. co-intensional, we shall write

$$c = d.$$

Since concepts are identified with their intension, as discussed in Section 1, $c = d$ means that c and d denote the same concept.[4]

Now we may capture an infamous example:

$$human \equiv featherless\ biped,$$
$$human = rational\ animal.$$

The former equation expresses that *human* and *featherless biped* are co-extensional. The latter expresses that *rational animal* and *human* are two names for the same concept, meaning human is defined as rational animal.

---

[3] It is a little more complicated than this. It *is* possible to formalize an intensional theory by means of conceptual graphs or KIF, just as it is possible to do so in predicate logic, however, this does not change the fact that these languages are extensional by their model-theoretic semantics, and that very few theories are intensional.

[4] One may compare our two notions of equivalence with Carnap's notion of equivalence and L-equivalence [11]. There is one important difference, though, our notion of equivalence is not dependent on extra-linguistic knowledge.

Similarly, we have an extensional subsumption relation, $c \sqsubseteq d$, which expresses that every individual which belongs to the extension of c also belongs to the extension of d. And an intensional subsumption relation, $c \leq d$, which expresses that c is intensionally included in d. We will give an example of the different kinds of subsumptions in Section 9.

4   Definition of the Conceptual Language

The syntax of the following language is similar to the syntax of T-boxes of the description logic $\mathcal{ALC}$ with the exception that two kinds of concept axioms exist.[5] From a given set of atomic concepts and atomic roles we inductively form a set of concept descriptions according to the following definition.[6]

Definition 1. *Let C be a set of atomic concepts and let R be a set of atomic roles. The set of concept descriptions $T(C)$ is defined as the least set satisfying*

1. *Every atomic concept, including the universal concept $\top$, is a concept description.*
2. *If c and d are concept descriptions then the* conjunction $c \sqcap d$ *is a concept description.*
3. *If d is a concept description then the* negation $\neg d$ *is a concept description.*
4. *If d is a concept description and r is an atomic role, then the* value restriction $\forall r.d$ *is a concept description.*

Then we may define different kinds of concept axioms.

Definition 2. *Let $c, d \in T(C)$ be concept descriptions, then*

1. *$c \equiv d$ is a* concept equivalence *between c and d;*
2. *$c = d$ is a* concept identity *between c and d;*
3. *$c \sqsubseteq d$ is an* extensional subsumption relation *between c and d;*
4. *$c \leq d$ is an* intensional subsumption relation *between c and d.*

Concept axiom *is used as common term for concept equivalence, concept identity, extensional subsumption relation, or intensional subsumption relation.*

5   Extensional Semantics

The following extensional semantics is a usual model-theoretic semantics of a first-order language.

---

[5] The syntaxes are not exactly the same, however, any $\mathcal{ALC}$ formula may be translated to one of our formulae through a few abbreviations, like $\bot$ is $\neg\top$, and back again.
[6] For those not familiar with description logics: roles are binary relations between extensions of concepts, like part-of or causal relations. See e.g. [3] for some examples.

Definition 3. *Let Γ be a set of concept axioms and let U be a universe of discourse. An* extensional model *of Γ is a mapping $\varepsilon : C \to 2^U$ (as well as a mapping $\varepsilon : R \to 2^{U \times U}$) which satisfies for all $c, d \in T(C)$ and $r \in R$:*

$$\varepsilon(\top) = U$$
$$\varepsilon(c \sqcap d) = \varepsilon(c) \cap \varepsilon(d)$$
$$\varepsilon(\neg c) = \complement \varepsilon(c) = U \backslash \varepsilon(c)$$
$$\varepsilon(\forall r.c) = \{x \in U \mid (\forall y \in U)\langle x, y \rangle \in \varepsilon(r) \Rightarrow y \in \varepsilon(c)\}$$

and which satisfies the axioms of Γ, that is, for all $c \equiv d \in \Gamma$, $\varepsilon(c) = \varepsilon(d)$; and for all $c = d \in \Gamma$, $\varepsilon(c) = \varepsilon(d)$; and for all $c \sqsubseteq d \in \Gamma$, $\varepsilon(c) \subseteq \varepsilon(d)$; and for all $c \leq d \in \Gamma$, $\varepsilon(c) \subseteq \varepsilon(d)$.

The extensional part of the language is equivalent to the description logic $\mathcal{ALC}$ in the sense that they have the same models. Note, under the extensional semantics, there is no difference between extensional concept axioms (including $\equiv, \sqsubseteq$) and intensional axioms (including $=, \leq$).

The following definition of entailment (logical consequence) gives us a means for inferring extensional knowledge from a given set of concept axioms.

Definition 4. *Let Γ be a set of concept axioms and let $c, d \in T(C)$. Then we say that Γ* entails $c \equiv d$ *and write*

$$\Gamma \models c \equiv d$$

if for all extensional models ε of Γ we have $\varepsilon(c) = \varepsilon(d)$. Moreover, we say that Γ entails $c \sqsubseteq d$ *and write*

$$\Gamma \models c \sqsubseteq d$$

if for all extensional models ε of Γ we have $\varepsilon(c) \subseteq \varepsilon(d)$.

6   Intensional Semantics

In order to understand the intensional semantics, it is useful to consider the extensional semantics from an algebraic point of view. The intensional semantics is namely an algebraic generalization of the extensional semantics. The fact that description logics may be given an algebraic semantics is not new, see [10, 9] and [7] which shows an algebraic approach to the related multi-modal logics.

The syntactic domain, i.e. the set $T(C)$ of concept descriptions, may be considered an algebra $(T(C), \sqcap, \neg, (\forall r.(\text{-}))_{r \in R}, \top)$. This algebra consists of a whole family $(\forall r.(\text{-}))_{r \in R}$ of operations, one for each atomic role. As this algebra is the most general algebra over the signature $(\sqcap, \neg, (\forall r.(\text{-}))_{r \in R}, \top)$ (it does not satisfy any equalities), it is the term algebra over C (for each r). The range of a given extensional model $\varepsilon : C \to 2^U$ is a Boolean algebra $(2^U, \cap, \complement, U)$. It is endowed with additional structure due to the value restrictions. A family of operations $(f_r^\cap)_{r \in R}$ exist, each corresponding to the interpretation of an atomic role (note, $\forall r.(\text{-})$ is a piece of syntax, $f_r^\cap$ is a corresponding operator). Each mapping $f_r^\cap : 2^U \to 2^U$ of this family is defined by

$$f_r^\cap(Z) = \{x \in U \mid (\forall y \in U)\langle x, y \rangle \in \varepsilon(r) \Rightarrow y \in Z\}$$

given an extensional model ε. The range of an extensional model is therefore an algebra $(2^U, \cap, \complement, (f_r^\cap)_{r \in R}, U)$, and such an algebra is called a Boolean algebra with operators [18], however, since we exclusively use them in the extensional semantics, we will call them *extensional algebras*.[7] An extensional model ε is a mapping from C to a Boolean algebra with operators:

$$\varepsilon : C \longrightarrow (2^U, \cap, \complement, (f_r^\cap)_{r \in R}, U).$$

The intensional semantics works the same way. An intensional model is also a mapping from C, however, there is one important difference, the range of the mapping is not a Boolean algebra (with operators). Instead, we are inspired by the property theory of G. Bealer, see [5,6] and the related [20,25].[8] The underlying idea is that the intensional semantics arises through interpretation over weaker structures. In our case, these structures are algebras, which we call *intensional algebras*. These are simply semilattices endowed with additional structure due to negations and value restrictions.

Definition 5. *An* intensional algebra *is a tuple* $(I, \times, \sim, (f_r^\times)_{r \in R}, 1)$ *consisting of a non-empty set* I*, a binary operation* $\times$ *on* I*, a unary operation* $\sim$ *on* I*, and an* R*-indexed family of mappings on* I *which satisfies*

$$x \times x = x$$
$$x \times y = y \times x$$
$$x \times (y \times z) = (x \times y) \times z$$
$$\sim(\sim x) = x$$
$$x \times 1 = x$$
$$f_r^\times(x \times y) = f_r^\times(x) \times f_r^\times(y)$$
$$f_r^\times(1) = 1$$

for all $x, y, z \in I$ *and* $r \in R$.

We may give the intensional algebras the following informal interpretation. Intensional algebras are formalizations of intensions of concepts, meaning the operations $\times, \sim, f_r^\times$, and 1 are operations on intensions. The axioms simply state that intensional conjunction $\times$ is idempotent, commutative, and associative, moreover, intensional negation $\sim$ is involution. For instance, this means that the conjunction of an intension with its negation does not (necessarily) yield a least

---

[7] Our operators are structure preserving with respect to $\cap$, meaning they are dual of the ones in [18].

[8] Our approach is, amongst other things, distinguished from Bealer's because we have both extensional and intensional interpretations.

intension (although this is the case in the extensional algebras, i.e. $x \cap \complement x = \emptyset$), because we are not guaranteed existence of a least intension. It has been philosophically justified in [25] that conjunction and negation have these algebraic properties. Moreover, as the roles preserve the structure of their arguments, they are strict endomorphisms on I. I find this intuitively acceptable, unfortunately, it is unjustified in the philosophical literature.

Before we present the intensional semantics the following relation is needed. Let $(I, \times, \sim, (f_r^\times)_{r \in R}, 1)$ be an intensional algebra. We can construct a partial order $\preceq$ on I by

$$x \preceq y \text{ if and only if } x = x \times y$$

It can be shown that $\preceq$ is an order-theoretic semilattice. Conversely, given a (meet) semilattice defined as a partial order $\preceq'$ on I we can construct a binary operation $\times'$ on I by $x \times' y$ is equal to the greatest lower bound of x and y. It can also be shown that $(I, \times')$ is a semilattice, and that the constructions are each other's inverses. The proofs can be found in [13].

Therefore, there is—in some sense—a one-to-one correspondence between the partial order and the binary operation of a semilattice. The partial order $\preceq$ is to formalize subsumption. So it is no coincidence that we have chosen semilattices as intensional algebras, since they are the most general algebras with this property. Note, with this definition of the ordering of the I elements, 1 becomes maximum.

Definition 6. *Let Γ be a set of concept axioms and let $(I, \times, \sim, (f_r^\times)_{r \in R}, 1)$ be an intensional algebra. An* intensional model *of Γ is a mapping $\iota : C \to I$ (as well as a mapping $\iota : R \to (I \to I)$) which satisfies*

$$\iota(\top) = 1$$
$$\iota(c \sqcap d) = \iota(c) \times \iota(d)$$
$$\iota(\neg c) = \sim \iota(c)$$
$$\iota(\forall r.c) = f_r^\times(\iota(c)) \text{ where } f_r^\times = \iota(r)$$

for all $c, d \in T(C)$ and $r \in R$, and which satisfies the intensional axioms of Γ, that is, for all $c = d \in \Gamma$, $\iota(c) = \iota(d)$; and for all $c \leq d \in \Gamma$, $\iota(c) \preceq \iota(d)$.

Definition 7. *Let Γ be a set of concept axioms and let $c, d \in T(C)$. Then we say that Γ entails $c = d$ and write*

$$\Gamma \models c = d$$

if for all intensional models ι of Γ we have $\iota(c) = \iota(d)$. Moreover, we say that Γ entails $c \leq d$ and write

$$\Gamma \models c \leq d$$

if for all intensional models ι of Γ we have $\iota(c) \preceq \iota(d)$.

One may object that our intensional algebras seem rather *ad hoc*, but this is not the case. We could have chosen another class of algebras instead of the intensional algebras (which would give rise to a different kind of intensionality), but the intensional algebras have a special status among these. Firstly, because there is the one-to-one correspondence between conjunctions and subsumptions, both of which are fundamental for formalization of conceptual structures. Secondly, they are—as mentioned above—philosophically justifiable.

7 Verification of Intensionality

The following proposition shows that the intensional semantics subsumes the extensional semantics.

Proposition 2. *Let $\varepsilon : C \to 2^U$ be an extensional model of a set Γ of concept axioms. Then ε is an intensional model of Γ.*

Proof. We verify that $(2^U, \cap, \complement, (f_r^{\cap})_{r \in R}, U)$ is an intensional algebra. Firstly, $\cap$ is idempotent, commutative, and associative. Secondly, $\complement$ is involutive. Thirdly, U is maximum, and fourthly, each operator $f_r^{\cap} : 2^U \to 2^U$ is structure preserving (a strict endomorphism):

$$
\begin{aligned}
f_r^{\cap}(W \cap Z) &= \{x \in U \mid (\forall y \in U)\langle x, y \rangle \in \varepsilon(r) \Rightarrow y \in (W \cap Z)\} \\
&= \{x \in U \mid (\forall y \in U)\langle x, y \rangle \in \varepsilon(r) \Rightarrow y \in W\} \cap \\
&\quad \{x \in U \mid (\forall y \in U)\langle x, y \rangle \in \varepsilon(r) \Rightarrow y \in Z\} \\
&= f_r^{\cap}(W) \cap f_r^{\cap}(Z)
\end{aligned}
$$

for all $W, Z \subseteq U$ and $r \in R$;

$$
f_r^{\cap}(U) = \{x \in U \mid (\forall y \in U)\langle x, y \rangle \in \varepsilon(r) \Rightarrow y \in U\} = U.
$$

Moreover, every concept identity and every intensional subsumption of Γ are satisfied because every concept axiom is satisfied. $\qquad\Box$

However, the converse does not hold, every intensional model is not an extensional model. For instance, existence of a least element in the intensional algebra is not guaranteed. If $c, d \in T(C)$, it is therefore the case that

$$
c \equiv d \not\models c = d,
$$

and since the right hand side of $\not\models$ is the result of $c = c[d/c]$, we have

$$
c \equiv d \not\models c = c[d/c].
$$

This shows that substitution of equivalent concept descriptions is not allowed in our language. The language is thus intensional.

We can also prove that co-intensional formulae may be freely substituted for each other. Let c, d, ϕ be concept descriptions, then we have the following theorem:

$$
c = d \models \phi = \phi[d/c].
$$

This can be shown by structural induction on the concept description ϕ. Our language follows therefore Carnap's criterion for intensionality which was mentioned in footnote 2.

8   Relations between the Extensional and Intensional Semantics

The important result from the previous section which showed that our language is intensional may be obtained in another way. This section shows how this can be done.[9]

Let Γ be a set of concept axioms, and let $F_E(C)$ denote the free algebra over the class of extensional algebras of Γ with free generating family C, and let $F_I(C)$ denote the free algebra over the class of intensional algebras of Γ with free generating family C. These algebras exist because the classes are equationally defined. As shown in proposition 2 every extensional model of Γ is an intensional model of Γ, so $F_E(C)$ belongs to the class of intensional algebras. This means that there exists a unique homomorphism $\tau : F_I(C) \rightarrow F_E(C)$ such that the following figure, where the remaining mappings are inclusions, commutes:

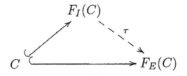

We can actually define τ:[10]

$$\tau(1) = U$$
$$\tau(x \times y) = \tau(x) \cap \tau(y)$$
$$\tau(\sim x) = \complement(\tau(x)) = U \backslash \tau(x)$$
$$\tau(f_r^\times(x)) = f_r^\cap(\tau(x))) = \{x \in U \mid (\forall y \in U)\langle x, y\rangle \in \tau(r) \Rightarrow y \in \tau(x)\}.$$

This characterization suggests the following definition.

Definition 8. *Let $\iota : C \rightarrow I$ be an intensional model of a set of concept axioms Γ. A homomorphism $\tau : I \rightarrow E$ such that $\tau \circ \iota$ is an extensional model of Γ is a translation mapping with respect to Γ and ι.*

A translation mapping gives an extensional (set-theoretic) interpretation of the intensional algebra. Now we have three kinds of mappings: extensional, intensional and translation. These are actually related as the following propositions show.

Proposition 3. *Let $\iota : C \rightarrow I$ be an intensional model of a set of concept axioms Γ and let $\tau : I \rightarrow E$ be a translation mapping w.r.t. Γ and ι. Then there exists a unique extensional model $\varepsilon : C \rightarrow E$ of Γ such that $\varepsilon = \tau \circ \iota$.*

Proof. Existence and uniqueness of ε follows from the definition $\varepsilon = \tau \circ \iota$. $\square$

---

[9] This section adopts notions from Universal Algebra [17].

[10] We have defined τ such that its range is a field of sets which is isomorphic to $F_E(C)$.

Proposition 4. *Let $\varepsilon : C \to E$ be an extensional model of a set of concept axioms Γ. Then there exists an intensional model $\iota : C \to I$ of Γ and a translation mapping $\tau : I \to E$ w.r.t. Γ and ι such that $\varepsilon = \tau \circ \iota$.*

Proof. We use the fact that every extensional model is an intensional (proposition 2). So we define ι to be ε, and τ to be the identity mapping id_E, and then we get $\varepsilon = \tau \circ \iota$. By definition, τ is a translation mapping w.r.t. Γ and ι.   □

In proposition 3, we can show existence and uniqueness, whereas we in 4 only can show existence. This may be taken as another formulation of the well-known dogma that intension (uniquely) determines extension.

Functional equalities like $\varepsilon = \tau \circ \iota$ are usually represented by commutative diagrams:

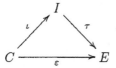

One may ask why we are interested in commutativity, i.e. mappings which satisfy $\varepsilon = \tau \circ \iota$? The answer is that, if commutativity holds, there is a coherence between intensions and what the intensions refer to on one hand, and extensions on the other hand.

Using another terminology: commutativity expresses coherence between content and reference of a term. Such coherence was noted by Frege [14] (although he used *sense* instead of *content* or *intension* as explicatum). A similar coherence has also been acknowledged in semiotics. It is usually presented as "Ogden's triangle" [22]:

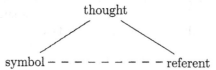

We may prove the main result. (Note, its converse does not hold.)

Proposition 5. $\Gamma \models c = d$ *implies* $\Gamma \models c \equiv d$.

Proof. Let ε be an extensional model of Γ such that $\varepsilon(c) \neq \varepsilon(d)$. Then by proposition 4 there is an intensional model ι of Γ and a translation mapping τ such that $\varepsilon = \tau \circ \iota$. If $\iota(c) = \iota(d)$ then $\tau(\iota(c)) = \tau(\iota(d))$, which gives us the contradiction $\varepsilon(c) = \varepsilon(d)$. Thus, $\iota(c) \neq \iota(d)$.   □

As a corollary we have that if $\Gamma \models c \leq d$ then $\Gamma \models c \sqsubseteq d$.

9   Applications

The expressivity of our language is higher than the description logic $\mathcal{ALC}$. This gives rise to a number of novel applications with respect to formalization of conceptual structures. This section describes three examples.

We know a priori that a bachelor is an unmarried man. This may be conceptually formalized by means of intensional equivalence:

$$bachelor = unmarried \sqcap man.$$

Moreover, we know that a bachelor has the property of being a featherless biped (in so far as every bachelor is a featherless biped). As this is based on a more empirical knowledge, it may be formalized differently, by means of extensional equivalence:

$$bachelor \equiv bachelor \sqcap featherless \sqcap biped.$$

The $\sqsubseteq$ relation may formalize extensional subsumption, and the $\leq$ relation may formalize intensional subsumption. For example, from the above equivalences we get

$$bachelor \leq unmarried$$
$$bachelor \sqsubseteq biped.$$

So *bachelor* is an intensional subconcept of *unmarried* and an extensional subconcept of *biped*.

In a series of papers, Guarino and Welty (see e.g. [26]), have introduced a methodology for construction of cleaner taxonomies. Basically, what they do, is to take a messy taxonomy from which they remove, add, or rearrange subsumption vertices (IS-A links) based on philosophically motivated meta-properties. As an example, they suggest that *apple* should only be a subconcept of *fruit* instead of a subconcept of both *fruit* and *food*. This means that the knowledge that an apple is a kind of food is removed from the taxonomy. However, for applications within e.g. nutrition, this knowledge is quite relevant. So it would be nice if we did not have to give it up. By means of our two kinds of subsumption, their proposal can be captured without giving up knowledge:

$$apple \leq fruit$$
$$apple \sqsubseteq food.$$

This approach does not mess up the taxonomy, because the two kinds of knowledge are clearly separated (in a graphical presentation of the taxonomy, extensional relations could for instance be removable). Our language thus supplies conceptual modellers with a tool for constructing more clean and more fine-grained taxonomies—both at the same time.

10   Conclusion

The important points of the paper reappear in the following list:

- Two fundamental properties of concepts have been identified.
- Based on these properties, two kinds of equivalence (and two kinds of subsumption), extensional and intensional, have been introduced into a language similar to the description logic $\mathcal{ALC}$.

- The extensional equivalences have been given a well-known model-theoretic semantics. The intensional equivalences have been given a novel algebraic semantics. The latter was accomplished through introduction of so-called intensional algebras, which basically are algebraic generalizations of the extensional, set-theoretic universes of discourse in model-theoretic semantics.
- Extensional equivalence was interpreted as identity in the extensional algebras, and intensional equivalence was interpreted as identity in the intensional algebras.
- It was justified that the language is intensional.
- Finally, we showed, by means of three all-too-brief examples, that this language supplies conceptual modellers with a tool for formalizing cleaner as well as more fine-grained taxonomies.

We may elaborate on the introduced intensional semantics. Firstly, we may ask whether this approach is limited to the presented language only, or whether it is applicable to more complex languages? I think, that with introduction of appropriate algebras, it should be possible to extend the semantical principles, although only future investigations can prove this. Secondly, we may ask what the nature of the intensional properties of this approach is. The short answer is that intensionality arises through generalization (or abstraction). For example, although every individual which is a bachelor has other properties besides being an unmarried man, like being featherless and biped (see Section 9), we are able to abstract from this in the definition of the concept *bachelor*. I am not aware of any intensional approach which is based on this principle. Therefore, I think that this approach contributes to the general theory of intensionality.

References

1. C. Anthony Anderson. General intensional logic. In D. Gabbay and F. Guenthner, editors, *Handbook of Philosophical Logic*, volume II, pages 355–385. D. Reidel Publishing Company, 1984.
2. R. Audi, editor. *The Cambridge Dictionary of Philosophy*. Cambridge University Press, second edition, 1999.
3. F. Baader, D. Calvanese, D. McGuinness, D. Nardi, and P. Patel-Schneider, editors. *The Description Logic Handbook*. Cambridge University Press, 2003.
4. Franz Baader and Werner Nutt. Basic description logics. In Baader et al. [3], pages 47–100.
5. George Bealer. *Quality and Concept*. Clarendon Press, 1982.
6. George Bealer and Uwe Mönnich. Property theories. In D. Gabbay and F. Guenthner, editors, *Handbook of Philosophical Logic*, volume IV, pages 133–251. D. Reidel Publishing Company, 1989.
7. P. Blackburn, M. de Rijke, and Y. Venema. *Modal Logic*. Cambridge University Press, 2001.
8. Magnus Boman, Jr Janis A. Bubenko, Paul Johannesson, and Benkt Wangler. *Conceptual Modelling*. Prentice Hall, 1997.
9. Chris Brink, Katarina Britz, and Renate A. Schmidt. Peirce algebras. *Formal Aspects of Computing*, 6:339–358, 1994.

10. Chris Brink and Renate Schmidt. Subsumption computed algebraically. *Comput. Math. Appl.*, 23:329–342, 1992. Special Issue on semantic networks in Artificial Intelligence.

11. Rudolf Carnap. *Meaning and Necessity*. The University of Chicago Press, second edition, 1956.

12. Alonzo Church. A formulation of the logic of sense and denotation. In Paul Henle, Horace M. Kallen, and Susanne K. Langer, editors, *Structure, Method, and Meaning: Essays in Honor of Henry M. Scheffer*. Liberal Arts Press, New York, 1951.

13. Brian A. Davey and Hilary A. Priestley. *Introduction to Lattices and Order*. Cambridge University Press, 1990.

14. Gottlob Frege. On sense and meaning. In Brian McGuinness, editor, *Gottlob Frege—Collected Papers on Mathematics, Logic, and Philosophy*, pages 157–177. Basil Blackwell Publisher, 1984. Originally published under the title "Über Sinn und Bedeutung" in Zeitschrift für Philosophie und philosophische Kritik 100 (1892), pp. 25–50. Translated by Max Black.

15. M. R. Genesereth and R. E. Fikes. Knowledge Interchange Format, Version 3.0 Reference Manual. Technical Report Logic-92-1, Computer Science Department, Stanford University, Stanford, CA, USA, June 1992.

16. Michael R. Genesereth. Knowledge interchange format. In James F. Allen, Richard Fikes, and Erik Sandewall, editors, *KR'91: Principles of Knowledge Representation and Reasoning*, pages 599–600. Morgan Kaufmann, 1991.

17. George Grätzer. *Universal Algebra*. Springer, New York, second edition, 1979. (First edition 1968).

18. Bjarni Jónsson and Alfred Tarski. Boolean algebras with operators, Part I. *Amer. J. Math.*, 73:891–939, 1951.

19. John McCarthy. First order theories of individual concepts and propositions. First published in Machine Intelligence 9 in 1979. Revised version from http://www-formal.stanford.edu/jmc/.

20. Christopher Menzel. *A Complete, Type-free "Second-order" Logic and Its Philosophical Foundations*. CSLI, 1986.

21. Richard Montague. *Formal Philosophy: Selected Papers of Richard Montague*. Yale University Press, New Haven, Connecticut, 1974. Edited, with an introduction, by Richmond H. Thomason.

22. C. K. Ogden and I. A. Richards. *The Meaning of Meaning*. Routledge & Kegan Paul, London, 1923.

23. Manfred Schmidt-Schauß; and Gert Smolka. Attributive concept descriptions with complements. *Artificial Intelligence*, 48(1):1–26, 1991.

24. John F. Sowa. *Knowledge Representation: logical, philosophical and computational foundations*. Brooks/Cole, 2000.

25. Chris Swoyer. Complex predicates and logics for properties and relations. *Journal of Philosophical Logic*, 27(3):295–325, 1998.

26. Christopher Welty and Nicola Guarino. Supporting ontological analysis of taxonomic relationships. *Data & Knowledge Engineering*, 39:51–74, 2001.

27. William A. Woods. Understanding subsumption and taxonomy: A framework for progress. In John F. Sowa, editor, *Principles of Semantics Networks*, pages 45–94. Morgan Kaufmann Publishers, 1991.

Hypotheses and Version Spaces

Bernhard Ganter[1] and Sergei O. Kuznetsov[2]

[1] Technische Universität Dresden
[2] VINITI Moscow

Abstract. We consider the relation of a learning model described in terms of Formal Concept Analysis in [3] with a standard model of Machine Learning called version spaces. A version space consists of all possible functions compatible with training data. The overlap and distinctions of these two models are discussed. We give an algorithm how to generate a version space for which the classifiers are closed under conjunction. As an application we discuss an example from predicitive toxicology. The classifiers here are chemical compounds and their substructures. The example also indicates how the methodology can be applied to Conceptual Graphs.

1 Hypotheses and Implications

Our paper deals with a standard model of Machine Learning called version spaces. But first, to make the paper self-contained, we introduce basic definitions of Formal Concept Analysis (FCA) [5]. We consider a set M of "structural attributes", a set G of objects (or observations) and a relation $I \subseteq G \times M$ such that $(g, m) \in I$ if and only if object g has the attribute m. Instead of $(g, m) \in I$ we also write gIm. Such a triple $\mathbb{K} := (G, M, I)$ is called a *formal context*. Using the *derivation operators*, defined for $A \subseteq G$, $B \subseteq M$ by

$$A' := \{m \in M \mid gIm \text{ for all } g \in A\},$$
$$B' := \{g \in G \mid gIm \text{ for all } m \in B\},$$

we can define a *formal concept* (of the context $\mathbb{K}$) to be a pair (A, B) satisfying $A \subseteq G$, $B \subseteq M$, $A' = B$, and $B' = A$. A is called the *extent* and B is called the *intent* of the concept (A, B). These concepts, ordered by

$$(A_1, B_1) \geq (A_2, B_2) \iff A_1 \supseteq A_2$$

form a complete lattice, called *the concept lattice* of $\mathbb{K} := (G, M, I)$.

Next, we use the FCA setting to describe a learning model from [2]. In addition to the structural attributes of M, we consider (as in [3]) a *goal attribute* $w \notin M$. This divides the set G of all objects into three subsets: The set G_+ of those objects that are known to have the property w (these are the *positive examples*), the set G_- of those objects of which it is known that they do not have w (the *negative examples*) and the set G_τ of *undetermined examples*, i.e.,

A. de Moor, W. Lex, and B. Ganter (Eds.): ICCS 2003, LNAI 2746, pp. 83–95, 2003.

of those objects, of which it is unknown if they have property w or not. This gives three subcontexts of $\mathbb{K} = (G, M, I)$:

$$\mathbb{K}_+ := (G_+, M, I_+), \quad \mathbb{K}_- := (G_-, M, I_-), \quad \text{and} \quad \mathbb{K}_\tau := (G_\tau, M, I_\tau),$$

where for $\varepsilon \in \{+, -, \tau\}$ we have $I_\varepsilon := I \cap (G_\varepsilon \times M)$.

Intents, as defined above, are attribute sets shared by some of the observed objects. In order to form hypotheses about structural causes of the goal attribute w, we are interested in sets of structural attributes that are common to some positive, but to no negative examples. Thus, a *positive hypothesis* for w is an intent of $\mathbb{K}_+$ that is not contained in the intent

$$g^- := \{m \in M \mid (g, m) \in I_-\}$$

of any negative example $g \in G_-$. *Negative hypotheses* are defined accordingly.

These hypotheses can be used as predictors for the undetermined examples. If the intent

$$g' := \{m \in M \mid (g, m) \in I\}$$

of an object $g \in G_\tau$ contains a positive, but no negative hypothesis, then g is *classified positively*. Negative classifications are defined similarly. If g' contains hypotheses of both kinds, or if g' contains no hypothesis at all, then the classification is *contradictory* or *undetermined*, respectively.

In [11] we argued that the consideration can be restricted to *minimal* (w.r.t. inclusion $\subseteq$) hypotheses, positive as well as negative, since an object intent obviously contains a positive hypothesis if and only if it contains a minimal positive hypothesis, etc.

2   Version Spaces

The term "version space" was proposed in [12], [13] and became the name of a certain class of Machine Learning models [14]. Version spaces can be defined in different ways, e.g., Mitchell defined them in terms of sets of maximal and minimal elements [12], [13]. In [20] they are defined in terms of minimal elements and sets of negative examples. They can also be defined in terms of some matching predicate. These representations are equivalent, however transformations from one into another are not always polynomially tractable. We will use the representation with matching predicates.

The next lines describe the basic notions of version spaces. We follow [12], [20], but with slightly modified terms, in order to avoid confusion when merging this approach with that of Formal Concept Analysis. We need

– An *example language* L_e by means of which the instances are described (elsewhere also called *instance language*). For our purposes it is not relevant to discuss how such a language is defined in detail. It suffices to assume that it describes a *set E of examples*.

- A *classifier language* L_c describing the possible classifiers (elsewhere called *concepts*). Again, for us it is not necessary to discuss a precise definition of L_c. We assume that it describes a set C of *classifiers* and a (partial) order $\sqsubseteq$ on C, called the *subsumption order*.
- A *matching predicate* $M(c, e)$ that defines if a classifier c does or does not *match* an example e: We have $M(c, e)$ iff e is an example of classifier c. The matching predicate must be compatible with the subsumption order in the following sense: for $c_1, c_2 \in L_c$,

$$c_1 \sqsupseteq c_2 \Rightarrow \forall_{e \in E} \ M(c_1, e) \rightarrow M(c_2, e).$$

- Sets E_+ and E_- of *positive* and *negative examples* of a *goal attribute* with $E_+ \cap E_- = \emptyset$. The goal attribute is not explicitly given.

On the basis of these data one defines a

- *consistency predicate* cons(c):
  cons(c) holds if for every $e \in E_+$ the matching predicate $M(c, e)$ holds and for every $e \in E_-$ the negation $\neg M(c, e)$ holds. The set of all consistent classifiers is called the *version space*

$$\text{VS}(L_c, L_e, M(c, e), E_+, E_-).$$

The learning problem is then defined as follows:

Given $L_c, L_e, M(c, e), E_+, E_-$.
Find the version space $\text{VS}(L_c, L_e, M(c, e), E_+, E_-)$.

In the sequel, we shall usually fix L_c, L_e, and $M(c, e)$, but work with varying sets of positive and negative examples. We therefore sometimes write

$$\text{VS}(E_+, E_-) \quad \text{or even just} \quad \text{VS}$$

for short.

Version spaces are often considered in terms of their *boundary sets*, as proposed in [13]. These can be defined if the language L_c is *admissible*, i.e., if every chain in the subsumption order has a minimal and a maximal element. In this case,

$$G(\text{VS}) := \text{MAX}(\text{VS}) := \{c \in \text{VS} \mid \neg \exists_{c_1 \in \text{VS}} \ c \leq c_1\},$$
$$S(\text{VS}) := \text{MIN}(\text{VS}) := \{c \in \text{VS} \mid \neg \exists_{c_1 \in \text{VS}} \ c_1 \leq c\}.$$

According to [14], the ideal result of learning a goal attribute is the case where the version space consists of a single element. Otherwise, the goal attribute is said to be *learned partially*.

It was said above that the goal attribute is not given explicitly. The elements of the version space can be used as potential classifiers for the goal attribute: A classifier $c \in \text{VS}$ *classifies* an example e positively if c matches e and negatively else. Then, all positive examples are classified positively, all negative examples are classified negatively, and undetermined examples may be classified either way. If it is assumed that E_+ and E_- carry sufficient information about the goal attribute, we may expect that an undetermined example is likely to have the goal attribute if it is classified positively by a large precentage of the version space (cf. [14]). We say that example e is *α-classified* if no less than $\alpha \cdot |\text{VS}|$ classifiers classify it positively.

3   Version Spaces in Terms of Galois Connections

We show that basic notions for version spaces can smoothly be expressed in the terminology of Formal Concept Analysis, more precisely, using the derivation operators. Let us remark that these form a Galois connection [5].

Consider the formal context (E, C, I), where E is the set of examples containing the disjoint sets of observed positive and negative examples: $E \supseteq E_+ \cup E_-$, $E_+ \cap E_- = \emptyset$, C is the set of classifiers and relation I corresponds to the matching predicate $M(c, e)$: for $c \in C$, $e \in E$ the relation eIc holds iff $M(c, e) =$ TRUE. The complementary relation, $\bar{I}$, corresponds to the negation: $e\bar{I}c$ holds iff $M(c, e) =$ FALSE.

Proposition 1.
$$\text{VS}(E_+, E_-) = E_+{}^I \cap E_-{}^{\bar{I}}.$$

Proof. The set of classifiers that match all positive examples is $E_+{}^I$. Since $\bar{I} = (E \times C) \setminus I$, the relation $\bar{I}$ corresponds to the predicate "classifier c does not match example e" and the set of classifiers that do not match any negative examples is $E_-{}^{\bar{I}}$. The version space is by definition the set of classifiers that match all positive examples and do not match any negative examples, i.e., is the intersection $E_+{}^I \cap E_-{}^{\bar{I}}$.   □

From this Proposition we immediately get a result from [7] about *merging version spaces*:

Corollary 1. *For fixed L_c, L_e, $M(c, e)$ and two sets E_{+1}, E_{-1} and E_{+2}, E_{-2} of positive and negative examples one has*

$$\text{VS}(E_{+1} \cup E_{+2}, E_{-1} \cup E_{-2}) = \text{VS}(E_{+1}, E_{-1}) \cap \text{VS}(E_{+2}, E_{-2}).$$

Proof. This follows from Proposition 1 and the relation $(A \cup B)' = A' \cap B'$, which holds for an arbitrary derivation operator $(\cdot)'$.   □

Proposition 2. *The set of all 100%-classified examples defined by the version space $\text{VS}(E_+, E_-)$ is given by*

$$(E_+{}^I \cap E_-{}^{\bar{I}})^I.$$

Proof. By the previous proposition, the version space given by the set of positive examples E_+ and the set of negative examples E_- is $E_+^I \cap E_-^{\bar{I}}$, therefore the set of all examples that are classified positively by all elements of the version space is $(E_+^I \cap E_-^{\bar{I}})^I$. $\qquad\square$

The next proposition explains what it means when the set E_+ of all positive examples is closed with respect to $(\cdot)^{II}$.

Proposition 3. *If $E_+{}^{II} = E_+$, then there cannot be any 100%-classified undetermined example, no matter what E_- is.*

Proof. By the previous propositions, $E_+{}^I$ is the set of all classifiers that match positive examples and $E_+{}^{II}$ is the set of all examples matched by these classifiers. If $E_+{}^{II} = E_+$, then, independent of any possible examples, these classifiers do not match any "new" examples (i.e., those outside E_+). $\qquad\square$

Proposition 4. *The set of examples that are classified positively by at least one element of the version space* $\mathrm{VS}(E_+, E_-)$ *is given by*

$$E \setminus (E_+{}^I \cap E_-{}^{\bar{I}})^{\bar{I}}.$$

Proof. By Proposition 1, $(E_+{}^I \cap E_-{}^{\bar{I}})$ is the version space given by examples E_+ and E_-. The set of all examples that are not matched by any classifier from this version space is $(E_+{}^I \cap E_-{}^{\bar{I}})^{\bar{I}}$. Therefore, the set of examples that are matched by at least one classifier is its complement w.r.t. the set of all possible examples, i.e., $E \setminus (E_+{}^I \cap E_-{}^{\bar{I}})^{\bar{I}}$. $\qquad\square$

4   Classifier Semilattices

In the preceding section we have described version spaces in the language of Formal Concept Analysis. But for many examples this description is too simple, because it does not assume any additional structure for the set of classifiers. We have mentioned that for version spaces the classifiers are given in terms of some language L_c, and it is therefore natural to consider the case when the ordered set $(C, \leq)$ of classifiers is a meet-semilattice. Let $\sqcap$ denote the meet operation of this semilattice.

This is not an inept assumption. Classifiers given as logical formulae form a meet semilattice when the set of these formulae is closed under conjunction. Classifiers given as attributes show the same effect if arbitrary attribute combinations are allowed as classifiers, too. This also covers the case of attributes with values. In the setting of [14], for example, each attribute takes one of possible values, either constant or "wildcard" $*$, the latter being the shortcut for universal quantification over constant values of this attribute. Examples are given by conjunctions of attribute values. A classifier c matches example e if all attribute values of c that do not coincide with the corresponding values of e are wildcards.

Formal Concept Analysis offers *scaling* techniques [5] that reduce the case of many-valued attributes to the case of plain "one-valued" ones. We then may assume that each classifier c corresponds to a subset S_c of a fixed set of attributes, and c matches an example e if and only if e has each of the attributes in S_c. The subsumption order then corresponds to the set inclusion:

$$c_1 \sqsubseteq c_2 \iff S_{c_1} \subseteq S_{c_2}.$$

If $\sqcap$ makes a complete semilattice (e.g., when the semilattice is finite), then the dual join operation $\sqcup$ can be (uniquely) defined. This operation on classifiers corresponds to the intersection of attribute sets: $c_1 \sqcup c_2$ matches exactly those examples that are matched by both c_1 and c_2.

Proposition 5. *If the classifiers, ordered by subsumption, form a complete $\sqcup$-semi"-lat"-tice, then the version space is a complete subsemilattice for any sets of examples E_+ and E_-.*

Proof. Recall that a classifier belongs to the version space iff it matches all positive and no negative examples. If c_1 and c_2 satisfy this condition, then $c_1 \sqcup c_2$, being more specific than c_1 and c_2, does not match any negative examples. On the other hand, the join $c_1 \sqcup c_2$ covers the intersection of sets of examples covered by c_1 and c_2. Since all positive examples are covered by both c_1 and c_2, their join $c_1 \sqcup c_2$ covers all positive examples too. $\qquad\square$

In [4] we have introduced a variant of Formal Concept Analysis based on so-called pattern structures. Here we use a slightly restricted form of them. First we assume that the set of all possible classifiers forms a complete semilattice $(C, \sqcap)$, which allows us to define a dual join operation $\sqcup$. A *pattern structure* is a tuple $(E, (C, \sqcap), \delta)$, where E is a set (of "examples"), δ is a mapping

$$\delta : E \to C,$$

$\delta(E) := \{\delta(e) \mid e \in E\}$. The subsumption order can be reconstructed from the semilattice operation:

$$c \sqsubseteq d \iff c \sqcap d = c.$$

For such a pattern structure $(E, (C, \sqcap), \delta)$ we can define derivation operators as we did for formal contexts[1] by

$$A^\diamond := \sqcap_{e \in A} \delta(e) \qquad \text{for } A \subseteq E$$

and

$$c^\diamond := \{e \in E \mid c \sqsubseteq \delta(e)\} \qquad \text{for } c \in C.$$

The pairs (A, c) satisfying

$$A \subseteq E, \ c \in C, \ A^\diamond = c, \ c^\diamond = A$$

---

[1] In fact, pattern structures can be represented as formal contexts.

are called the *pattern concepts* of $(E, (C, \sqcap), \delta)$, with extent A and *pattern intent* c. The pattern concepts form a complete lattice.

Not all classifieres are pattern intents, and it may be the case that many classifiers describe the same extent A. Only one of them, namely A°, then is a pattern intent. Let us call two classifiers c_1 and c_2 equivalent if $c_1^\diamond = c_2^\diamond$. We then observe that for every classifier c the equivalence class of c contains a unique pattern intent, $c^{\diamond\diamond}$. This is the *representing* pattern intent for c.

If, as above, E_+ and E_- are disjoint subsets of E representing positive and negative examples for some goal attribute, we can define a *positive hypothesis* h as a pattern intent of $(E_+, (C, \sqcap), \delta)$ that is not subsumed by any classifier from $\delta(E_-)$ (for short: not subsumed by any negative example). Formally, $h \in C$ is a positive hypothesis iff

$$h^\circ \cap E_- = \emptyset \quad \text{and} \quad \exists_{A \subseteq E_+} A^\circ = h.$$

A disappointing situation is when the version space is empty, i.e., when there are no classifiers at all separating all positive examples from all negative ones. This happens, for example, if there are *hopeless* positive examples, by which we mean elements $e_+ \in E_+$ that have a negative counterpart $e_- \in E_-$ such that every classifier which matches e_+ also matches e_-. An equivalent formulation of the hopelessness of e_+ is that $(e_+)^{\diamond\diamond} \cap E_- \neq \emptyset$.

Theorem 1. *Suppose that the classifiers, ordered by subsumption, form a complete meet-semilattice $(C, \sqcap)$, and let $(E, (C, \sqcap), \delta)$ denote the corresponding pattern structure. Then the following are equivalent:*

1. *The version space $\mathrm{VS}(E_+, E_-)$ is not empty.*
2. $(E_+)^{\diamond\diamond} \cap E_- = \emptyset$.
3. *There are no hopeless positive examples and there is a unique minimal positive hypothesis h_{min}.*

In this case, $h_{min} = (E_+)^\circ$ and the version space is represented by a convex set in the lattice of all pattern intents[2] with maximal element h_{min}.

Proof. The version space consists of precisely those classifiers c that satisfy $E_+ \subseteq c^\circ$ and $c^\circ \cap E_- = \emptyset$. The join m of all such classifiers has the same property; it therefore is the maximal element of the version space and the only element of $S(\mathrm{VS})$. From $E_+ \subseteq m^\circ$ we get that $(E_+)^{\diamond\diamond} \subseteq m^\circ$ and therefore $(E_+)^{\diamond\diamond} \cap E_- = \emptyset$.

Suppose $(E_+)^{\diamond\diamond} \cap E_- = \emptyset$, then $h := (E_+)^\circ$ is a positive hypothesis subsumed by every positive hypothesis A°, $A \subseteq E_+$. If $e_+ \in E_+$ the $(e_+)^{\diamond\diamond} \subseteq (E_+)^{\diamond\diamond}$. Thus, if $(E_+)^{\diamond\diamond} \cap E_- = \emptyset$, then there cannot be any hopeless positive examples.

If there is a unique minimal hypothesis, h_{min}, then $E_+ \subseteq A := h_{min}^\circ$. Suppose not, then there is some positive example $e \in E_+$, $e \notin A$. Since e is not hopeless, e° is a hypothesis, and this hypothesis is incomparable to h_{min}, a contradiction. Thus h_{min} is the maximal element of the version space. $\square$

---

[2] ordered by subsumption

The theorem gives access to an algorithm for generating the version space. To be more precise: our algorithm will not generate the version space, but the set of representing pattern intents. A well-known algorithm for generating all formal concepts of a formal context can be modified to only generate a convex set of the type described in Theorem 1. This can easily be described here, but without proof. For the latter we refer to [5].

Fix a linear order $\leq$ on the set G of all examples in such a way that all positive examples come first, then all negative ones, and the other elements come last. Let $n_{\max}$ denote the maximal negative example in this order. Also with respect to this order, we reformulate two elementary notations from [5]:

– For $A, B \subseteq G$ and $i \in G$ define

$$A <_i B : \iff i \in B, i \notin A, \text{ and } (j \in A \iff j \in B \text{ for all } j < i).$$

– For $A \subseteq G$ and $i \in G$, $i \notin A$, define

$$A \oplus i := (\{i\} \cup \{j \in A \mid j < i\})^{\circ\circ}.$$

Now the pattern intents representing the version space are recursively given as follows:

1. The first element is $h_{\min} = (E_+)^{\circ}$, provided that $(E_+)^{\circ\circ} \cap E_- = \emptyset$. Otherwise, the version space is empty.
2. If an element h of the version space has been computed, then the "next" element h_{next} is computed as follows: Let $A := h^{\circ}$, and let i be the largest element that is greater than $n_{\max}$ and that satisfies

$$A <_i A \oplus i.$$

Then $h_{\text{next}} = (A \oplus i)^{\circ}$. If no such i exists, the algorithm terminates.

Corollary 2. *If the classifiers, ordered by subsumption, form a finite meet-semi"-lat"-tice, then a system of representatives for the version space can be computed using the algorithm described above.*

5   Boundary Sets, Hypotheses, and Predictors

According to [5] a subset $A \subseteq M$ is a *proper premise of an attribute* $m \in M$ if $m \notin A$, $m \in A''$ and for any $A_1 \subset A$ one has $m \notin A_1''$. It would be useful to know the *proper premises of the goal attribute* w. But as long as there are undecided examples, i.e., as long as the set E_τ is not empty, our information does not suffice to determine these. We therefore generalize the notion to describe good candidates for such premises:

Definition 1 We call $d \in L_c$ a *positive proper predictor* with respect to example sets E_+, E_- if the following conditions are satisfied:

1. $d^\circ \cap E_- = \emptyset$,
2. $d^\circ \cap E_+ \neq \emptyset$,
3. if $d \sqsupseteq d_1$ and $d \neq d_1$ then $d_1^\circ \cap E_- \neq \emptyset$.

Thus, a positive proper predictor matches no negative example (due to Condition 1), at least one positive example (Condition 2), and is a most general classifier with these properties (Condition 3). The set of all such predictors for a pattern structure $\Pi = (E, (C, \sqcap), \delta)$ and example sets E_+ and E_- is denoted by $\mathrm{PP}_+(\Pi, E_+, E_-)$.

Our next proposition describes the interplay between the set of predictors of a pattern structure $\Pi := (E, (C, \sqcap), \delta)$ with example sets E_+ and E_-, the set $\mathrm{H}_+(\Pi, E_+, E_-)$ of its positive hypotheses and the boundary sets $G(\mathrm{VS})$ and $S(\mathrm{VS})$ of the corresponding version space. For a set $X \subseteq C$ of classifiers let $\min_{\sqsubseteq}(X)$ denote the minimal elements:

$$\min_{\sqsubseteq}(X) := \{x \in X \mid y \not\sqsubseteq x \text{ for all } y \neq x \text{ in } X\}.$$

Proposition 6. Let $\Pi = (E, (C, \sqcap), \delta)$ be a pattern structure and $E = E_+ \mathbin{\dot{\cup}} E_\tau \mathbin{\dot{\cup}} E_-$. Then

1. $\mathrm{PP}_+(\Pi, E_+, E_-) = \min_{\sqsubseteq}\left(\bigcup_{F_+ \subseteq E_+} G(\mathrm{VS}(\Pi, F_+, E_-))\right)$,
2. $\mathrm{H}_+(\Pi, E_+, E_-) = \bigcup_{F \subseteq E_+} S(\mathrm{VS}(\Pi, F_+, E_-))$.

Proof. 1. Consider a positive proper predictor p w.r.t. examples E_+ and E_-. Let $F_+ := \{g \in E_+ \mid p \sqsubseteq g^\circ\}$. By Condition 3 of Definition 1, no element of the version space $\mathrm{VS}(F_+, E_-)$ can be more general than p. Hence, by the definition of $G(\mathrm{VS})$, we have $p \in G(\mathrm{VS}(\Pi, F_+, E_-))$,

$$PP_+(\Pi, E_+, E_-) \subseteq \bigcup_{F_+ \subseteq E_+} G(\mathrm{VS}(\Pi, F_+, E_-)).$$

Moreover, p should be among the most general elements of

$$\bigcup_{F_+ \subseteq E_+} G(\mathrm{VS}(\Pi, F_+, E_-)),$$

since otherwise, there should have existed a positive classifier more general than p and p could not have been a positive proper predictor w.r.t. E_+, E_-.

In the other direction, let p be a maximal element of the set

$$\bigcup_{F_+ \subseteq E_+} G(\mathrm{VS}(\Pi, F_+, E_-)).$$

Then, by the definition of a version space, for some $F_+ \subseteq E_+$, we have $F_+ \subseteq p^\circ$, $E_- \not\subseteq p^\circ$, and hence $p^\circ \subseteq E_+ \cup E_\tau$, thus Condition 1 of Definition 1 is satisfied. Condition 2 is also satisfied, since for all $g \in F_+ \subseteq E_+$ one has $g \in p^\circ$. Condition 3 is satisfied, since otherwise, for some F_+ in the version space $VS(\Pi, F_+, E_-)$

there had existed a classifier $p_1 \sqsupseteq p$, which contradicts the assumption that p is maximal w.r.t. subsumption $\sqsubseteq$ in $\bigcup_{F_+ \subseteq E_+} G(\mathrm{VS}(\Pi, F_+, E_-))$. Therefore, p is a positive proper predictor w.r.t. examples E_+, E_-, and E_τ.

2. For each $F_+ \subseteq E_+$, if $\mathrm{VS}(\Pi, F_+, E_-)$ is not empty then, by Theorem 1, the unique element of the set $S(\mathrm{VS}(\Pi, F_+, E_-))$ is a minimal hypothesis for the pattern structure Π and sets of examples F_+, E_-. Thus,

$$H_+(\Pi, E_+, E_-) \supseteq \bigcup_{F \subseteq E_+} S(\mathrm{VS}(\Pi, F_+, E_-)).$$

In the other direction, each positive hypothesis H w.r.t. (Π, E_+, E_-) is a minimal positive hypothesis w.r.t. (Π, F_+, E_-) for $F_+ = H^\circ \cap E_+$. Hence, H is a unique element of $S(\mathrm{VS}(\Pi, F_+, E_-))$ and

$$H_+(\Pi, E_+, E_-) \subseteq \bigcup_{F \subseteq E_+} S(\mathrm{VS}(\Pi, F_+, E_-)).$$

$\square$

6   An Example

The 12th European Conference on Machine Learning was held jointly with the 5th European Conference on Principles of Knowledge Discovery in Databases. The program included a workshop on Predictive Toxicology Challenge [21], which consisted in a competition of machine learning programs for generation of hypothetical causes of toxicity from positive and negative examples of toxicity. The learning program presented in [1] turned out to be Pareto-optimal among all classification rules generated by learning models participating in the competition in terms of the relative number of false and true positive and negative classifications made by hypotheses generated by a learning program (see [21] for details). Their learning program is based on the JSM-method and generates minimal hypotheses along the lines of definitions in Section 1. It can be viewed as an instance for the methods of this article.

The "examples" in the toxicology challenge were chemical compounds; the goal attribute was toxicity. In the original representation the goal attribute was not binary, however obviously positive examples, i.e., those known to be certainly toxic and negative examples, i.e., those known to be certainly nontoxic, were isolated. In the toxocology challenge we used a standard representation by formal contexts. If we want to represent the chemical substances more properly, we can associate to each chemical compound $g \in G$ as a corresponding pattern $\delta(g)$ its molecular graph. As classifiers we use these graphs and their subgraphs and, more generally, subgraph combinations. The use of subgraph combinations leads to a semilattice of classifiers, as treated in Section 4, which is a subsemilattice of the lattice of order ideals of the ordered (w.r.t. subgraph isomorphism relation, see below) set of labeled graphs. This approach has already been used in [4].

It has to be specified precisely what we mean by a subgraph. A useful definition corresponds to that of the injective specialization relation [16] or injective morphism [15] used for Conceptual Graphs.

Let P be the set of all finite graphs with vertex and edge labels from L, up to isomorphism. A typical such graph is of the form $\Gamma := ((V, l_v), (E, l_e))$, with vertex set V, edge set E, and label assignments l_v, l_e for vertices and edges, respectively. We say that

$$\Gamma_2 := ((V_2, l_{v2}), (E_2, l_{e2}))$$

is **isomorphic to a subgraph** of

$$\Gamma_1 := ((V_1, l_{v1}), (E_1, l_{e1})),$$

or for short, Γ_2 is a **subgraph** of Γ_1, denoted by $\Gamma_2 \leq \Gamma_1$, if there exists a one-to-one mapping $\varphi : V_2 \to V_1$ that (for all $v, w \in V_2$)

- respects edges: $(v, w) \in E_2 \Rightarrow (\varphi(v), \varphi(w)) \in E_1$,
- fits under labels: $l_{v2}(v) = l_{v1}(\varphi(v))$, $l_{e2}(v, w) = l_{e1}(\varphi(v), \varphi(w))$.

Obviously, $\leq$ is an order relation. A more general definition, which takes into account ordering of labels (a label can be less or more "general" than another one) can be found in [9], [10], [4].

Having defined $\leq$ we define a lattice on graph sets along the lines of [9]. In terms of lattice theory, FCA, and pattern structures, this can be described as follows. First, the lattice of order ideals $LI(P, \leq)$ of the order $\leq$ is given by Birkhoff's theorem (see [5], Theorem 39). This gives us infimum and supremum operations $\sqcap$ and $\sqcup$. Then we define a pattern structure $(G, (LI(P, \leq), \sqcup), \delta)$, where δ takes each object $g \in G$ to an element of the lattice of order ideals on graphs described above. Using this pattern structure we generate hypotheses as described in Section 4 above.

The vertex labels in case of molecular graphs corresponding to atom types (e.g., C, N, Cl staying for carbon, nytrogene, chlorine, respectively) and edges are labeled by bond types (single bond, double bond, triple bond, aromatic bond, etc.). An aromatic bond in cyclic components of a molecular graph is usually denoted by a circle inside a cycle.

Consider the disconnected graph in Figure 1. It turns out that this is a minimal hypothesis for toxicity. The connected component to the right belongs to the version space, whereas the connected component to the left is not a premise of the goal attribute at all, since it is a subgraph of some graphs representing negative examples.

7   Discussion and Relation to Other Work

The major drawback of the version spaces, where classifiers (also called "concepts" in some publications on version spaces) are defined syntactically, is the very likely situation when - in case of too restrictive choice of the classifiers -

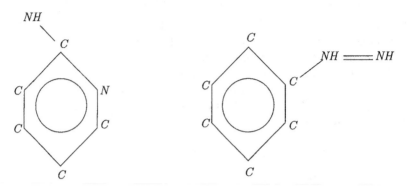

Fig. 1. Two connected components of a hypothesis

there is no classifier that matches all positive examples (so-called "collapse of the version space"). This can easily happen for example when classifiers are just conjunctions of attribute value assignments and "wildcards", a case mentioned above. In other words: The situation discussed in Theorem 1, which presupposes that there are classifiers that match all positive and no negative examples, is too narrow.

If the expressive power is increased syntactically, e.g., by introducing disjunction, then the version space tends to become trivial, while the most specific generalization of positive examples becomes "closer" to or just coincide with the set of positive examples.

As a remedy, we suggest to use sets of hypotheses as we defined them in terms of patterns structures. They offer in fact a sort of "context-restricted" disjunction: not all disjunctions are possible, but only those of minimal hypotheses.

Clearly, the relation of our work with that in Inductive Logic Programming ILP should be realized. In ILP (see, e.g., [17]) the definition of a version space is specified by taking the $\leq$ relation to be the inference relation in logic and the set of classifiers as a subset of the set of logical programming formulas.

As the origin of the $\leq$ order can be arbitrary, one can use the above constructions for learning of "generalized" descriptions from descriptions of positive and negative examples given in a description logic, by conceptual graphs, etc. One should just specify an initial "more general than or equal to" relation $\leq$.

References

1. V.G. Blinova, D.A. Dobrynin, V.K. Finn, S.O. Kuznetsov, and E.S. Pankratova, Toxicology Analysis by Means of the JSM-Method, *Bioinformatics*, 2003.
2. V. K. Finn, Plausible Reasoning in Systems of JSM Type, *Itogi Nauki i Tekhniki, Seriya Informatika*, **15**, 54-101, 1991 [in Russian].
3. B. Ganter and S. Kuznetsov, Formalizing Hypotheses with Concepts, *Proc. 8th Int. Conf. on Conceptual Structures, ICCS'00*, G. Mineau and B. Ganter, Eds., Lecture Notes in Artificial Intelligence, **1867**, 2000, pp. 342-356.

4. B. Ganter and S.O. Kuznetsov, Pattern Structures and Their Projections, *Proc. 9th Int. Conf. on Conceptual Structures, ICCS'01*, G. Stumme and H. Delugach, Eds., Lecture Notes in Artificial Intelligence, **2120**, 2001, pp. 129-142.

5. B. Ganter and R. Wille, *Formal Concept Analysis: Mathematical Foundations*, Springer, 1999.

6. C.A. Gunter, T.-H. Ngair, D. Subramanian, The Common Order-Theoretic Structure of Version Spaces and ATMSs, *Artificial Intelligence* **95**, 357-407, 1997.

7. H. Hirsh, Generalizing Version Spaces, *Machine Learning* **17**, 5-46, 1994.

8. H. Hirsh, N. Mishra, and L. Pitt, Version Spaces Without Boundary Sets, in *Proc. of the 14th National Conference on Artificial Intelligence (AAAI97)*, AAAI Press/MIT Press, 1997.

9. S.O. Kuznetsov, JSM-method as a machine learning method, *Itogi Nauki i Tekhniki, ser. Informatika*, **15**, pp.17-50, 1991 [in Russian].

10. S.O. Kuznetsov, Learning of Simple Conceptual Graphs from Positive and Negative Examples. In: J. Zytkow, J. Rauch (eds.), Proc. *Principles of Data Mining and Knowledge Discovery, Third European Conference, PKDD'99*, Lecture Notes in Artificial Intelligence, **1704**, pp. 384-392, 1999.

11. S.O. Kuznetsov and V.K. Finn, On a model of learning and classification based on similarity operation, *Obozrenie Prikladnoi i Promyshlennoi Matematiki* **3**, no. 1, 66-90, 1996 [in Russian].

12. T. Mitchell, Version Space: An Approach to Concept Learning, PhD thesis, Stanford University, 1978.

13. T. Mitchell, Generalization as Search, *Artificial Intelligence* **18**, no. 2, 1982.

14. T. Mitchell, Machine Learning, The McGraw-Hill Companies, 1997.

15. M.-L. Mugnier and M. Chein, Représenter des connaissances et raisonner avec des graphes, Revue d'Intelligence Artificielle, **10**(1), 1996, pp. 7-56.

16. M.-L. Mugnier, Knowledge Representation and Reasonings Based on Graph Homomorphisms, in *Proc. 8th Int. Conf. on Conceptual Structures, ICCS'2000*, G. Mineau and B. Ganter, Eds., Lecture Notes in Artificial Intelligence, **1867**, 2000, pp. 172-192.

17. S.-H. Nienhuys-Cheng and R. de Wolf, Foundations of Inductive Logic Programming, *Lecture Notes in Artificial Intelligence*, **1228**, 1997.

18. M. Sebag, Using Constraints to Building Version Spaces, in L. de Raedt and F. Bergadano, eds., *Proc. of the European Conference on Machine Learning (ECML-94)*, pp. 257-271, Springer, 1994.

19. M. Sebag, Delaying the Choice of Bias: A Disjunctive Version Space Approach, in L. Saitta ed., *Proc. of the 13th International Conference on Machine Learning*, pp. 444-452, Morgan Kaufmann, 1996.

20. E.N. Smirnov and P.J. Braspenning, Version Space Learning with Instance-Based Boundary Sets, in H. Prade, ed., *Proceedings of the 13th European Conference on Artificial Intelligence*, J. Wiley, Chichester, 460-464, 1998.

21. The web-site on Predictive Toxicology: http://www.predictive-toxicology.org/ptc/

Description Logics for Ontologies

Ulrike Sattler

TU Dresden, Germany

1 Introduction to Description Logics

Description logics (DLs) [6, 8, 21] are a family of logic-based knowledge representation formalisms designed to represent and reason about the knowledge of an application domain in a structured and well-understood way.

The basic notions in DLs are *concepts* (unary predicates) and *roles* (binary relations), and a specific DL is mainly characterised by the *constructors* it provides to form complex concepts and roles from atomic ones. Intuitively, the following concept describes "A cooler connected to a reactor which, in turn, has a part that is a stirrer and whose functionality is to stir or to cool (or both)":

$$\text{Cooler} \sqcap \exists \text{connectedTo.}(\text{Reactor} \sqcap (\exists \text{hasPart.Stirrer} \sqcap \tag{1}$$
$$\forall \text{functionality.}(\text{Cooling} \sqcup \text{Stirring})))$$

In addition to such a set of constructors, DLs are usually equipped with a terminological component, often called a *TBox*. In its simplest form, a TBox can be used to introduce names (abbreviations) for complex concepts. For example, we could introduce the abbreviation CooledStirringReactor for the concept in Concept 1 from above. More expressive TBox formalisms allow the statement of *general concepts inclusion axioms* (GCIs) such as

$$\exists \text{hasPart.Stirrer} \;\dot{\sqsubseteq}\; \text{Reactor} \sqcap \exists \text{functionality.Stirring,}$$

which says that only stirring reactors can have stirrers.

Description logic systems provide their users with various reasoning capabilities that deduce implicit knowledge from the one explicitly stated in the TBox. The *subsumption* algorithm determines subconcept-superconcept relationships: a concept C is subsumed by a concept D w.r.t. a TBox if, in each model of the TBox, each instance of C is also an instance of D. Such an algorithm can be used to compute the *taxonomy* of a TBox, i.e., the subsumption hierarchy of all those concepts introduced in the TBox. The *satisfiability* algorithm tests whether a given concept can ever be instantiated.

Unsurprisingly, the higher the expressive power of a DL is, the more complex are the subsumption and the satisfiability problem. To use a DL for a certain application, it has to provide enough expressive power to describe the relevant properties of the objects in this application. On the other hand, the system services should be "practical" in that they run in realistic time and space. Thus, we are confronted with the well-known trade-off between expressivity and complexity, as in many other areas of computer science.

A. de Moor, W. Lex, and B. Ganter (Eds.): ICCS 2003, LNAI 2746, pp. 96–116, 2003.

In the last decade, a lot of work was devoted to investigate DLs w.r.t. their expressive power and computational complexity. It turned out that the first DL systems were based on undecidable logics [74, 61]. As a reaction, the expressive power was restricted severely, thus yielding a DL with polynomial reasoning problems. Then, in parallel with the discovery of the close relation between description and modal logics [73, 23], PSPACE-complete DLs were specified [75] and a tableau-based reasoning algorithm was implemented for such a DL [7]. After certain optimisation, it turned out that this implementation behaves much better than the high worst-case complexity of the underlying reasoning problem suggests. As a reaction, tableau-based reasoning algorithms for EXPTIME-complete DLs were implemented [41, 37]. Again, these implementations proved to be amenable to optimisation and behave surprisingly well in practice. This fostered the design and investigation of other EXPTIME-complete DLs together with tableau-based, "practicable" reasoning algorithms. In parallel, the investigation of the complexity of description logics continued successfully such that, today, we have a good understanding of the effects of the combination of concept and role constructors on the computational complexity and the expressive power of DLs; see, e.g., [26, 18, 24, 21, 20, 84, 54, 57, 56].

Today, industrial strength DL systems are being developed for very expressive DLs with system services being based on highly optimised tableau algorithms and with applications like the Semantic Web or knowledge representation and integration in bio-informatics in mind.

1.1   Preliminaries

In this section, we define the basic description logic $\mathcal{ALC}$, TBox formalisms, and reasoning problems.

Definition 1. *Let* **C** *and* **R** *be disjoint sets of* concept *and* role name. *The set of* $\mathcal{ALC}$*-concepts is the smallest set such that each concept name* $A \in \mathbf{C}$ *is an* $\mathcal{ALC}$*-concept and, if* C *and* D *are* $\mathcal{ALC}$*-concepts and* r *is a role name, then*

$$\neg C, \ C \sqcap D, \ C \sqcup D, \ \exists r.C, \ and \ \forall r.C \ are \ also \ \mathcal{ALC}\text{-concepts.}$$

A general concept inclusion axiom (GCI) *is of the form* $C \stackrel{.}{\sqsubseteq} D$ *for* C, D $\mathcal{ALC}$*-concepts. A* TBox *is a finite set of GCIs.*

An interpretation $\mathcal{I} = (\Delta^{\mathcal{I}}, \cdot^{\mathcal{I}})$ *consists of a non-empty set* $\Delta^{\mathcal{I}}$*, the interpretation domain, and a mapping* $\cdot^{\mathcal{I}}$ *which associates, with each concept name* A*, a set* $A^{\mathcal{I}} \subseteq \Delta^{\mathcal{I}}$ *and, with each role name* r*, a binary relation* $r^{\mathcal{I}} \subseteq \Delta^{\mathcal{I}} \times \Delta^{\mathcal{I}}$*. The interpretation of complex concepts is defined as follows:*

$$(C \sqcap D)^{\mathcal{I}} = C^{\mathcal{I}} \cap D^{\mathcal{I}}, \quad (C \sqcup D)^{\mathcal{I}} = C^{\mathcal{I}} \cup D^{\mathcal{I}}, \quad \neg C^{\mathcal{I}} = \Delta^{\mathcal{I}} \setminus C^{\mathcal{I}},$$
$$(\exists r.C)^{\mathcal{I}} = \{d \in \Delta^{\mathcal{I}} \mid \textit{There exists an } e \in \Delta^{\mathcal{I}} \textit{ with } \langle d, e \rangle \in r^{\mathcal{I}} \textit{ and } e \in C^{\mathcal{I}}\},$$
$$(\forall r.C)^{\mathcal{I}} = \{d \in \Delta^{\mathcal{I}} \mid \textit{For all } e \in \Delta^{\mathcal{I}}, \textit{ if } \langle d, e \rangle \in r^{\mathcal{I}}, \textit{ then } e \in C^{\mathcal{I}}\}.$$

An interpretation $\mathcal{I}$ satisfies *a GCI* $C \stackrel{.}{\sqsubseteq} D$ *if* $C^{\mathcal{I}} \subseteq D^{\mathcal{I}}$*;* $\mathcal{I}$ *satisfies a TBox* $\mathcal{T}$ *if* $\mathcal{I}$ *satisfies all GCIs in* $\mathcal{T}$—*in this case,* $\mathcal{I}$ *is called a* model *of* $\mathcal{T}$*. An* *element* $d \in C^{\mathcal{I}}$ *is called an* instance *of* C *and, if* $\langle d, e \rangle \in r^{\mathcal{I}}$, *then* e *is called an* r-successor *of* d.

A concept C is satisfiable *w.r.t. a TBox* $\mathcal{T}$ *if there is a model* $\mathcal{I}$ *of* $\mathcal{T}$ *with* $C^{\mathcal{I}} \neq \emptyset$. *A concept* C *is* subsumed *by a concept* D *w.r.t.* $\mathcal{T}$ *(written* $C \sqsubseteq_{\mathcal{T}} D$*) if, for each model* $\mathcal{I}$ *of* $\mathcal{T}$*,* $C^{\mathcal{I}} \subseteq D^{\mathcal{I}}$. *Two concepts are* equivalent *if they mutually subsume each other.*

As usual, we use $\top$ as an abbreviation for $A \sqcup \neg A$, $\bot$ for $\neg\top$, $C \Rightarrow D$ for $\neg C \sqcup D$, and $C \Leftrightarrow D$ for $(C \Rightarrow D) \sqcap (D \Rightarrow C)$. Moreover, we use $C \doteq D$ as an abbreviation for $C \sqsubseteq D$ and $D \sqsubseteq C$.

Some remarks are in order here. Firstly, in $\mathcal{ALC}$, the two reasoning problems satisfiability and subsumption can be mutually reduced to each other: C is satisfiable w.r.t. $\mathcal{T}$ iff C is *not* subsumed by $\bot$ w.r.t. $\mathcal{T}$. And $C \sqsubseteq_{\mathcal{T}} D$ iff $C \sqcap \neg D$ is *not* satisfiable w.r.t. $\mathcal{T}$. Secondly, it can be shown that satisfiability (and thus subsumption) w.r.t. a general TBox is EXPTIME-complete [73], whereas these problems become PSPACE-complete when considered w.r.t. the empty TBox [75].

2   Description Logics as Ontology Languages

A well-known attempt to define what constitutes an ontology is due to Gruber [35]: *an ontology is an explicit specification of a conceptualisation*, where "a conceptualisation" means an abstract model of some aspect of the world. This was later elaborated to "a formal specification of a shared conceptualisation" [16]. In this abstract model, relevant concepts of the aspect in question are defined, including a description of the interesting properties of their instances.

In the last decade, ontologies became rather popular through applications like the Semantic Web [15], enterprise knowledge management systems [85], and medical terminology systems [80, 66, 79] and through the growing amount of data available electronically.

An ontology is *built*—possibly by a group of domain experts—and *evolves* over time in any application that changes over time. Moreover, it is advisable to *integrate* existing ontologies if a larger aspect of the world is to be covered—instead of building a new one from scratch. Finally, if an ontology is *deployed*, knowledge is shared using the concepts defined in the ontology, e.g., concrete objects are described using the vocabulary defined in an ontology. Each of these tasks is rather complex: e.g. building and evolution involves a huge amount of creativity, integration requires knowledge in a large aspect, and all three tasks involve co-operation, thus risking misunderstanding, redundancy, etc.

The increasing importance of ontologies and their processing in computers has led to the development of ontology editors [64, 22, 12]. Due to the above mentioned complexity of ontology engineering tasks, it is highly desirable that these editors support the user in the design, evolution, integration, and deployment of ontologies through corresponding, intelligent system services. Moreover, an *unambiguous* language, e.g., a logic-based language, not only decreases the risk of misunderstandings among the domain experts, but also enables the design of provably correct or optimal such services. Now *description logics* are such a class of logic-based knowledge representation languages that come with a knowledge base formalism which makes DLs good candidates for ontology languages: an ontology can be formalised in a TBox, which can be divided into the following two, disjoint parts.

Background Knowledge. GCIs of the form $C \sqsubseteq D$ for C and D complex concepts can be used to formalise background knowledge of the application domain and thus to constrain the set of models.

For example, we can express that two concepts A and B are disjoint by $A \sqsubseteq \neg B$ and that each individual having an r-successor which is an instance of B is an instance of A by $\exists r.B \sqsubseteq A$.

Definitiorial Part. For each concept relevant in the application domain, we can introduce a concept name A and a *concept definition* $A \sqsubseteq C$ or $A \doteq C$ for C a complex concept describing necessary or necessary and sufficient conditions for individuals to be an instance of A. We say that A is a *primitively defined* or a *defined* concept.

For example, we can primitively define connections as being devices having some input and some output, and then define a hose as a flexible connection:

$$\text{Connection} \sqsubseteq \text{Device} \sqcap \exists \text{hasComp.Output} \sqcap \exists \text{hasComp.Input}$$
$$\text{Hose} \doteq \text{Connection} \sqcap \text{Flexible}$$

System services provided by DL-based knowledge representation systems include

- a satisfiability test for each concept defined in a TBox.
- the computation of the *taxonomy*: for each pair A_1, A_2 of concepts defined in the definitorial part of the TBox $\mathcal{T}$, we test whether $A_1 \sqsubseteq_{\mathcal{T}} A_2$ and $A_2 \sqsubseteq_{\mathcal{T}} A_1$. A taxonomy is the partial order of the defined concepts w.r.t. $\sqsubseteq_{\mathcal{T}}$, and is often presented as the corresponding Hasse-diagram.

Clearly, unsatisfiable defined concepts and unintended or missing subsumption relationships are signs of modelling flaws, and thus these system services can be used to support the engineering of ontologies: in the design phase and when modifying or integrating an ontology, we can repeatedly use both system services to ensure that the TBox is consistent, that it reflects our intuition, and that it does not contain unintended redundancies, i.e., equivalent defined concepts. Unsurprisingly, it turned out that, in applications where the knowledge engineer is no description logic expert, ontology engineering requires more support [69, 58], e.g., the domain expert wants to see automatically generated suggestions for a new concept definition as a generalisation of a set of example instances. This observation lead to the investigation of *non-standard inferences* in description logics such as computing the least common subsumer of several concepts, matching a concept that contains concept variables to concept expressions, or computing the approximation of a concept expressed in a more expressive logic in a less expressive logic [52, 5, 9, 17].

State-of-the-art DL-based systems such as FaCT or Racer [41, 37] provide the above *standard* system services such as deciding the satisfiability and computing the taxonomy, and are based on the DL $\mathcal{SHIQ}$ [49] that is an extension of

$\mathcal{ALC}$ with a variety of expressive means that turned out to be quite useful [70]; $\mathcal{SHIQ}$ is discussed in detail in Section 5. Despite these additional expressive means, $\mathcal{SHIQ}$ is of the the same worst-case complexity as $\mathcal{ALC}$, namely Ex-PTIME-complete [83]. This high complexity implies that, in the worst-case, the computation might take far too much time. However, the algorithms in these DL-based systems proved to be amenable to a wide range of optimisations, as a consequence of which these systems behave surprisingly well in many realistic applications [40, 41, 37, 48].

The suitability of DLs as ontology languages has been highlighted by their role as the foundation for several web ontology languages, including OIL [28], DAML+OIL [43], and OWL, a newly emerging ontology language standard being developed by the W3C Web-Ontology Working Group.[1] All of these languages have a syntax based on RDF Schema, but the basis for their design is a combination of the DLs $\mathcal{SHIQ}$ (mentioned above) and $\mathcal{SHOQ}(D)$ [46]. Both are DLs that were designed with the goal to find a good compromise between expressiveness and the complexity of reasoning.

3   Standard Expressive Means in DLs

To give the reader an impression of what DLs are, we present a variety of expressive means that are commonly used in DLs and discuss, if appropriate, their modal logic equivalent and their influence on the computational complexity. For a detailed description of the relationship between modal and description logics, see [73, 23]: $\mathcal{ALC}$ (without TBoxes) is a notational variant of the multi modal logic $\mathbf{K}_n$ [38]. To see the connection between $\mathbf{K}_n$ and $\mathcal{ALC}$, it suffices to view *elements* of a DL interpretation domain as *worlds* in a Kripke structure, roles as modal parameters, universal value restrictions as box formulae, and existential restrictions as diamond formulae. Then, for example, it can be easily seen that $A \sqcap \exists r.(C \sqcup \forall s.D)$ is equivalent to $A \wedge \langle r \rangle (C \vee [s]D)$.

TBoxes were introduced in Section 1.1, and it was mentioned in Section 2 that they are divided into a *background knowledge* part and a *definitorial* part. Some DLs only allow for the definitorial part and possibly require this part to be free of "definitorial cycles". Now reasoning w.r.t. acyclic concept definitions can be reduced to pure concept reasoning: one can either use a (sub-optimal) technique, called *unfolding*, which reduces reasoning w.r.t. acyclic concept definition to pure concept reasoning [59], or use more direct techniques [54]. As a result of the latter, it turned out that, for a variety of logics, reasoning w.r.t. acyclic concept definitions is as complex as pure concept reasoning [54].

In modal logics, the closest relative to a TBox is the *universal role*, a role that is interpreted as $\Delta^{\mathcal{I}} \times \Delta^{\mathcal{I}}$; for more details about this relationship, see [55].

Number Restrictions are an expressive means rather popular in DLs: they are present in almost all implemented DL systems. They are concepts of the form $(\geqslant nr.C)$ (atleast restriction) or $(\leqslant nr.C)$ (atmost restriction), for n a non-negative

---

[1] http://www.w3.org/2001/sw/WebOnt/

integer, r a role, and C a (possibly complex) concept, and are interpreted as follows:

$$(\geqslant nr.C)^{\mathcal{I}} = \{d \in \Delta^{\mathcal{I}} \mid \#\{e \in C^{\mathcal{I}} \mid (d, e) \in r^{\mathcal{I}}\} \geq n\},$$
$$(\leqslant nr.C)^{\mathcal{I}} = \{d \in \Delta^{\mathcal{I}} \mid \#\{e \in C^{\mathcal{I}} \mid (d, e) \in r^{\mathcal{I}}\} \leq n\},$$

where $\#M$ denotes the cardinality of a set M. They can be used, e.g., to described pipes as those connections having exactly one input and one output (we use $(= nr.C)$ as an abbreviation for $(\geqslant nr.C) \sqcap (\leqslant nr.C)$):

Connection $\sqcap$ $(= 1\text{hasComp.Input})$ $\sqcap$ $(= 1\text{hasComp.Output})$

In their simpler form, number restrictions only allow for the concept $\top$ in the place of C above. A further restrictions only allows for 2 in atleast restrictions and 1 in atmost restrictions. Finally, *features* are role names that are to be interpreted as partial functions—they can be viewed as a "globalised" version of a simple form of number restrictions. Number restrictions rarely seem to have effects on the complexity of DLs: for a variety of logics, extending them with number restrictions does not change their complexity, even if such an extension yields the loss of the *finite model property* (see Section 5.4 for a more detailed discussion). For example, when extended with number restrictions, $\mathcal{ALC}$ remains in PSPACE [84] and $\mathcal{ALC}$ with TBoxes remains in EXPTIME, even if further extended with other expressive means such as inverse roles (see below) [83].

Number restrictions are known in modal logics as *graded modalities* [29], whereas features play an important role in dynamic logic: they are syntactic variants of *deterministic programs* [13].

Nominals are, in their simplest form, special concept names that are to be interpreted as singleton sets. For example, the concept $\exists\text{partOf.BrentSpar}$ describes those objects that are part of the oil platform Brent Spar provided that BrentSpar is a nominal. In DLs, a weak form of nominals, ABoxes ("A" for assertional), are widely known and used: using individual names $a, b, \ldots$, we can assert that an an individual is an instance of a concept C by $a : C$ and that two individuals are related via a role r by $\langle a, b \rangle : r$. Interpretations associate, additionally, an element of the interpretation domain with each individual name. Please note that individual names are only to be used in assertions, in contrast to nominals that can be used in the place of concepts in concepts. Whereas ABox consistency is often as complex as satisfiability of concepts [72], extending a description logic with nominals often increases its complexity. For example, $\mathcal{ALC}$ with inverse roles (see below) is PSPACE-complete, but becomes EXPTIME-complete when extended with inverse roles [1]. If, additionally, number restrictions are present, the complexity leaps from EXPTIME-completeness to NEXPTIME-completeness [83]. A reason for this increase in complexity might be that nominals destroy the *tree model property* [87]: a logic enjoys the tree model property if every satisfiable concept/formula has a model whose relational structure forms a tree. For example, for nominals N_1 and N_2, the concept $N_1 \sqcap \exists r.(N_2 \sqcap \exists r.N_1)$ only has models with a cycle of length two.

Nominals originate in *hybrid* logic [63, 1, 2], and are known in DLs as an elegant and powerful generalisation of ABoxes.

Inverse Roles In various applications, one wants to use both "directions" of a role, e.g., one wants to use both hasPart and isPartOf. To model these roles adequately, i.e., to ensure that $\langle x, y \rangle \in$ hasPart$^{\mathcal{I}}$ iff $\langle y, x \rangle \in$ isPartof$^{\mathcal{I}}$, some description logics provide *inverse roles*: for r a role name, r^- is an inverse role, which is interpreted as $(r^-)^{\mathcal{I}} = \{ \langle y, x \rangle \mid \langle x, y \rangle \in r^- \}$.

A variety of DLs can be extended with inverse roles without affecting their computational complexity: examples are $\mathcal{ALC}$ with or without TBoxes and possibly with number restrictions [21, 83]. However, there are counter-examples such as $\mathcal{ALC}$ with concrete domains, which becomes NExpTime-complete when extended with inverse roles [56] or $\mathcal{ALC}$ with nominals and without TBoxes, which becomes ExpTime-complete when extended with inverse roles [1].

Inverse roles are closely related to the tense logic "past" modality [67, 78, 88] and are syntactic variants of converse programs in dynamic logics [81, 86].

Transitive Roles are special role names $r \in \mathbf{R}_+ \subseteq \mathbf{R}$ that are to be interpreted as *transitive relations* [68]. Transitive roles can be used to model transitive relations such as isAncestorOf or isPartOf. Another way to extend DLs with transitivity is to allow for the *transitive closure operator* on roles, i.e., to allow for roles r^* in the place of roles [3, 24], where $(r^*)^{\mathcal{I}}$ is to be interpreted as the transitive closure of $r^{\mathcal{I}}$. We will discuss the expressiveness of transitive roles in more detail in Section 5.1.

Adding transitive roles to $\mathcal{ALC}$ without TBoxes yields a DL whose reasoning problems are still PSpace-complete [68], whereas adding the transitive closure operator on roles yields an ExpTime-complete logics [30]. Transitive roles are notational variants of transitive accessibility relations in modal logics [38], whereas a transitive closure operator is also present in the dynamic logic PDL [30], which is a notational variant of $\mathcal{ALC}$ with regular role expressions [73].

Boolean Operator on Roles So far, we considered DLs with full Boolean operators on concepts, but no Boolean operators on roles. In DLs, Boolean operators on roles are mostly restricted to intersection [26], or to union and difference [24, 19]. They are interpreted in the obvious way, i.e., $(r \sqcap s)^{\mathcal{I}} = r^{\mathcal{I}} \cap s^{\mathcal{I}}$, etc., and are an interesting expressive means. For example, role negation allows to express the so-called *window* operator from modal logic [32]. The window operator can be viewed as the dual of universal value restrictions: an instance of $\forall$connectedTo.Pipe is connected *only* to pipes, whereas an instance of the concept $\forall$Pipe.connectedTo using the window operator is connected to *all* pipes. It can be easily seen that the concept $\forall\neg$connectedTo.$\neg$Pipe using role negation is equivalent to $\forall$Pipe.connectedTo. For a complete description of the (mostly dramatic) effects of adding Boolean operators on roles to the computational complexity of $\mathcal{ALC}$, see [57].

In dynamic logic, union of programs is present in all logics allowing for regular programs [30], and Boolean operators on modalities are discussed, e.g., in [32].

Role Inclusion Axioms Another expressive means on roles are role inclusion axioms, which are of the form $r \sqsubseteq s$ for r, s roles, and force interpretations to map r to a sub-relation of s. Such axioms can be used, for example, to introduce a sub-role hasComponent of hasPart. In the presence of inverse roles, role hierarchies can be used to enforce symmetric roles using $r^- \sqsubseteq r$ and $r \sqsubseteq r^-$.

It should be noted that role hierarchies are a weak form of role intersection: replacing each role expression $r_1 \sqcap r_2$ with a new role name $r_{1,2}$ and adding $r_{1,2} \sqsubseteq r_1$ and $r_{1,2} \sqsubseteq r_2$ to the role hierarchy yields a "weakened" form of intersection since $r_{1,2}^{\mathcal{I}} \subseteq r_1^{\mathcal{I}} \cap r_2^{\mathcal{I}}$. Moreover, for s a transitive role, the role inclusion axioms $r \sqsubseteq s$ yields a weakened form of the transitive closure: s is interpreted as *some* transitive role containing r, whereas r^* is interpreted as *the smallest* transitive role containing r. The latter observation implies that pure concept satisfiability of $\mathcal{ALC}$, when extended with both transitive roles and role inclusion axioms, becomes EXPTIME-hard [68].

General Role Inclusion Axioms (g-RIAs) are a generalisation of the above role inclusion axioms to the form $r_1 \ldots r_m \sqsubseteq s_1 \ldots s_n$ for r_i, s_j role names [47]. A model of such an axiom satisfies $r_1^{\mathcal{I}} \circ \ldots \circ r_m^{\mathcal{I}} \subseteq s_1^{\mathcal{I}} \circ \ldots \circ s_n^{\mathcal{I}}$, where $\circ$ denotes the composition of binary relations. *Role value maps*, i.e., concepts of the form $r_1 \ldots r_m \Rightarrow s_1 \ldots s_n$ with the semantics

$$(r_1 \ldots r_m \Rightarrow s_1 \ldots s_n)^{\mathcal{I}} = \{x \mid \forall y. \langle x, y \rangle \in r_1^{\mathcal{I}} \circ \ldots \circ r_m^{\mathcal{I}} \Rightarrow \langle x, y \rangle \in s_1^{\mathcal{I}} \circ \ldots \circ s_n^{\mathcal{I}}\},$$

can be viewed as a "local" form of g-RIAs. Both constructors have dramatic effects on the decidability of a description logic: it was shown in [74] that subsumption of a very weak DL becomes undecidable when extended with role value maps. DLs with g-RIAs are closely related to *grammar logics* [25, 10, 11], i.e., the multi modal logic $\mathbf{K}_n$ with accessibility relations being constrained by a grammar: a production rule of the form $s_1 \ldots s_n \rightarrow r_1 \ldots r_m$ can be viewed as a notational variant of the g-RIA $r_1 \ldots r_m \sqsubseteq s_1 \ldots s_n$ enforcing models to interpret $r_1 \ldots r_m$ as s sub-relation of $s_1 \ldots s_n$. Since each context-free grammar can be transformed into an equivalent one in Chomsky normal form, and $\mathcal{ALC}$ becomes undecidable with context-free grammars [10, 11], the satisfiability of $\mathcal{ALC}$-concepts w.r.t. g-RIAs of the form $r_1 r_2 \sqsubseteq s$ is undecidable.

Fixpoint Operators are the first expressive means mentioned here that are not first order definable, and they are known in DLs in at least three forms: a restricted form includes the transitive closure operator on roles [3, 24] (see above) and an operator that allows to enforce that a role is interpreted as a well-founded relation [19]. Secondly, general least and greatest fixpoints operators in DLs [20] are notational variants of the fixpoint operators in the μ-calculus [51]. Thirdly, *cyclic* concept definitions such as

$$\texttt{Device} \doteq \texttt{TechThing} \sqcap \neg\texttt{Connection} \sqcap \forall\texttt{connectedTo.Connection}$$
$$\texttt{Connection} \doteq \texttt{TechThing} \sqcap \neg\texttt{Device} \sqcap \forall\texttt{connectedTo.Device}$$

can be read with least or greatest *fixpoint semantics* [59, 4]: in contrast to the *descriptive* semantics, which takes into account *all* fixpoints of such GCIs, one might chose to take into account only the least or the greatest fixpoints.

A variety of EXPTIME-complete description and modal logics exist that have some form of fixpoints, e.g. the dynamic logic PDL [30, 62] or $\mathcal{DLR}_\mu$, a generalisation of the μ-calculus with n-ary relations [20].

4   Introduction to Tableau Algorithms for DLs

For several expressive DLs, there exist efficient tableau-based implementations that decide satisfiability of concepts w.r.t. a TBox [41, 37]. In the following, we will give an intuitive description of description logic tableau algorithms; for an extensive survey of tableau algorithms for description logics, see, e.g., [8]. In general, they work on trees whose nodes stand for individuals of an interpretation. Nodes are labelled with sets of concepts, namely those they are assumed to be an instance of. Edges between nodes are labelled with role names or sets of role names, namely those that hold between the corresponding individuals.

Intuitively, to decide the satisfiability of a concept C, a tableau algorithm starts with an instance x_0 of C, i.e., a tree consisting of a root node x_0 with C as its node label (written $\mathcal{L}(x_0) = \{C\}$). Then the algorithm breaks down concepts in node labels syntactically, thus inferring new constraints on the model of C to be built, and possibly generating new individuals, i.e., new nodes. For example, if $(D \sqcap E) \in \mathcal{L}(y)$ has already been inferred, it adds D and E to $\mathcal{L}(y)$. For $\exists r.F \in \mathcal{L}(y)$, it generates a new r-successor node of y, say z, and sets $\mathcal{L}(z) = \{F\}$. If a node y has some r-successor z and it finds $\forall r.G \in \mathcal{L}(y)$, then G is added to $\mathcal{L}(z)$. Finally, in the presence of a TBox $\mathcal{T}$, it adds, for each GCI $C_i \sqsubseteq D_i \in \mathcal{T}$, and for each node y, the concept $(\neg C_i \sqcup D_i)$ to $\mathcal{L}(y)$. Now, for logics with disjunctions, various tableau algorithms non-deterministically choose whether to add D or E to $\mathcal{L}(y)$ for $(D \sqcup E) \in \mathcal{L}(y)$. The answer behaviour is as follows: if this "completion" can be carried out exhaustively without encountering a node with both a concept and its negation in its label—a so-called *clash*,[2] then the algorithm answers that the input concept was satisfiable, and unsatisfiable, otherwise.

Thus, disjunctions are often treated non-deterministically. Avoiding this non-determinism in a way that is more efficient than naive back-tracking proves to be a complicated task for many logics—to the best of our knowledge, the algorithm in [27] is the only known worst-case optimal tableau algorithm for an EXPTIME-complete description logic. In contrast, the tableau algorithm implemented in state-of-the-art DL systems such as FaCT and Racer is 2NEXPTIME even though the underlying logic is EXPTIME-complete [49, 83]. Despite this sub-optimality, these tableau algorithms allow for a set of well-known efficient optimisations, so that they perform much better in practice than their worst-case complexity suggests; see [41, 49, 36, 37, 48] for descriptions of these optimisations. An interesting open question is whether an implementation of a worst-case optimal algorithm would behave better in practice—so far, only implementations of worst-case sub-optimal algorithms exist.

Since we are talking about decision procedures, *termination* is an important issue. Even though tableau algorithms for some inexpressive DLs terminate "automatically", this is not the case for more expressive ones. For example, consider the algorithm sketched above on the input concept A and TBox $\{A \sqsubseteq \exists r.A\}$:

---

[2] We assume that the description logic in question is propositionally closed, and we can thus work on concepts in negation normal form.

it would create an infinite r-chain of nodes with labels $\{A, \exists r.A\}$. To guarantee termination, the tableau algorithm needs to be stopped explicitly. Intuitively, the processing of an element z is stopped if all "relevant" concepts in the label of z are also present in the label of an "older" element z'. In this case, z' is said to *block* z. The definition of "relevant" has to be chosen carefully since it is crucial for the correctness of the algorithm [3, 49] and for the efficiency of the implementation [48, 39].

Correctness of DL tableau algorithms are often proved as follows: first, termination is proved by, roughly spoken, showing that the algorithm builds a (tree) structure of bounded size in a monotonic manner. Soundness is proved by constructing a model (or an abstraction of a model) of the input concept (and TBox) in case that the algorithm stops without having generated a clash. Completeness can be proved by using a model of the input concept (and TBox) to steer the application of the non-deterministic rules and proving that no clash occurs using this control.

Summing up, tableau algorithms are successfully used in state-of-the-art implementations, and many well-understood optimisations are available. However, they involve special techniques to ensure termination and avoid non-determinism, and are thus rarely optimal for logics complete for deterministic complexity classes.

4.1   Other Reasoning Techniques

For several expressive description and modal logics, there exist optimal automata-based algorithms that decide satisfiability (and thus subsumption) of concepts w.r.t. a TBox [89, 82, 88, 20, 57, 71]: for a concept C and a TBox $\mathcal{T}$, we define an automaton $\mathcal{A}_{C,\mathcal{T}}$ which accepts exactly the (abstractions of) models of C and $\mathcal{T}$. Thus, the satisfiability problem is reduced to the emptiness problem of automata. In summary, automata-based approaches often allow for a very elegant and natural translation of a logic and provide EXPTIME upper complexity bounds and are thus optimal for EXPTIME-hard logics. Equally important, they handle infinite structures and non-determinism implicitly.

For certain DLs that are *not* propositionally closed such as the one used in the system CLASSIC [60], one can use a reasoning technique called *structural subsumption*: roughly speaking, to decide the subsumption between two concepts C and D, both concepts are transformed into a certain normal form C' and D', and then subsumption can be decided by a syntactic comparison of C' and D', see Section 2.3.1 of [6]. This technique yields a polynomial time decision procedure for a sub-Boolean fragment of $\mathcal{ALC}$ with number restrictions, but seems to be applicable only to DLs without disjunction and existential restrictions.

Finally, the successful resolution-based theorem prover SPASS was modified into a decision procedure for expressive modal and description logics, then called MSPASS [50]. Interestingly, it is well-suited for DLs extending $\mathcal{ALC}$ with Boolean operators on roles and can be extended to n-ary description logics [33].

5   DLs with Expressive Operators on Roles

In various ontology applications such as engineering or medicine, *aggregated objects* play a central role, that is, objects that are composed of various parts, which again can be composite, etc. It is natural to describe an aggregated object by means of its parts and, vice versa, to describe parts by means of the aggregate they belong to. For example, the following statements describe a control rod and a reactor core by means of their parts and wholes:

ControlRod $\dot{\sqsubseteq}$ Device ⊓ ∃partOf.ReactorCore
ReactorCore $\dot{\sqsubseteq}$ Device ⊓ ∃hasPart.ControlRod ⊓ ∃partOf.NuclReactor

In contrast to, for example, the relation likes, the part-whole relation has a variety of properties; for a complete collection of these properties, we refer to [77]. Most importantly, the general part-whole relation is a strict partial order, i.e., it is *transitive* and *asymmetric* (and hence irreflexive). Moreover, an aggregated object has at least two parts where none is a part of the other. Next, we might consider to assume that two objects consisting of the same parts are identical. As a last example, we might assume the existence of atoms, i.e., indivisible objects of which all other objects are composed. This is equivalent to assuming that hasPart is well-founded.

Besides the properties mentioned above, it might be useful to distinguish various sub-relations of the part-whole relation such as, for example, the relation between a *component* and its *composite* (e.g. between a motor and the car the motor is in), the relation between *stuff* and an *object* containing this stuff (e.g. between metal and a car), or the relation between a *member* and a *collection* it belongs to (e.g. between a tree and the forest this tree belongs to) [91, 34].

In this section, we describe expressive means relevant for the representation of aggregated objects and the development of the DLs $\mathcal{SHIQ}$ and $\mathcal{RIQ}$.

5.1   Adding Transitivity

Coming back to representing aggregated objects in ontologies using DLs, we observe that the DL $\mathcal{ALC}$ provides no means to express that a relation is *transitive*. For example, in $\mathcal{ALC}$, the concept

Device ⊓ ∃hasPart.(ReactorCore ⊓ ∃hasPart.ControlRod)

is *not* subsumed by Device ⊓ ∃hasPart.ControlRod, although the first concept is a specialisation of the second one under the assumption that hasPart is interpreted as a transitive relation.

Thus the adequate modelling of aggregated objects asks for the extension of $\mathcal{ALC}$ with some form of transitivity. As mentioned in Section 3, there are at least two such possible extensions. After investigating their expressive power and complexity, we have chosen the "cheaper" possibility: by $\mathcal{S}$, we refer to the description logic $\mathcal{ALC}$ extended with transitive roles.[3] Obviously, $\mathcal{S}$ provides the

---

[3] The logic $\mathcal{S}$ has previously been called $\mathcal{ALC}_{R+}$, but this becomes too cumbersome when adding letters to represent additional features.

means to represent the general part-whole relation as a transitive relation by asserting that partOf is a transitive role. Additionally, since S has a tree model property, all satisfiable concepts and TBoxes have a model in which partOf is interpreted as a strict partial order.

Tableau algorithm for S A naive extension of the tableau algorithm for $\mathcal{ALC}$ sketched in Section 4 to transitive roles does not necessarily terminate: assume the algorithm is started with the concept $C_0 := C \sqcap \exists r.C \sqcap \forall r.(\exists r.C)$ for r a transitive role. After some rule applications, the algorithm has generated three nodes, x, y, and z where y is an r-successor of x, z is an r-successor of y, and $C_0, \forall r.(\exists r.C) \in \mathcal{L}(x)$, $\exists r.C \in \mathcal{L}(y)$, and $C \in \mathcal{L}(z)$. Since r is a transitive role, we could make z an r-successor of x, but this would destroy the tree structure that turned out to be quite useful. Instead, we do something which has the same effect: we add $\forall r.(\exists r.C)$ to $\mathcal{L}(y)$. More precisely, if $\forall r.C \in \mathcal{L}(x)$ and x has an r-successor y, we add both C and $\forall r.C$ to y's label. In this case, this yields $\exists r.C \in \mathcal{L}(z)$. It can easily be seen that the repeated application of this modification builds an infinite r-chain, and thus leads to non-termination. To re-gain termination without corrupting soundness or completeness of the algorithm, we use the blocking technique mentioned in Section 4: we stop generating new successors of a node z in case there is another node z' with $\mathcal{L}(z) \subseteq \mathcal{L}(z')$. In this case, we say that z' *blocks* z, and we can build a model by "merging" z and z' (and all other nodes which z' blocks), thus building a finite, possibly cyclic model.

5.2 Further Adding Inverse Roles

When modelling aggregated objects using S and using both partOf and hasPart, we might end up with an inadequate representation in the following sense: for example, extending the TBox in the beginning of Section 5 with

$$\text{NuclReactor} \sqcap \exists \text{hasPart.Faulty} \sqsubseteq \text{Dangerous},$$

we would assume that ControlRod⊓Faulty is subsumed by ∃partOf.Dangerous w.r.t. to this TBox—which is only the case if partOf were *the inverse* of hasPart, i.e., if $\langle x, y \rangle \in \text{hasPart}^{\mathcal{I}}$ iff $\langle y, x \rangle \in \text{partOf}^{\mathcal{I}}$. Thus we extend S with inverse roles, which yields the DL called $S\mathcal{I}$ and allows to describe both objects by means of the wholes they belong to and by means of the parts they have. Substituting hasPart by partOf$^-$ in the last example yields a TBox with respect to which ControlRod ⊓ Faulty is indeed subsumed by ∃partOf.Dangerous.

Tableau algorithm for $S\mathcal{I}$ Intuitively, we can extend the tableau algorithm for S as follows to yield a decision procedure for satisfiability of $S\mathcal{I}$-concepts: if $\forall r.C \in \mathcal{L}(w)$, instead of adding C only to r-successors, we also add C to $\mathsf{Inv}(r)$-predecessors.[4] For example, for the concept $\exists r^-.(C \sqcap \forall r.B) \in \mathcal{L}(x)$, we would first create an r^--successor y of x with $C \sqcap \forall r.B \in \mathcal{L}(y)$. For $\forall r.B \in \mathcal{L}(y)$, we would

---

[4] To avoid considering roles such as r^{--}, we use $\mathsf{Inv}(r)$ to denote r^- if r is a role name and s if $r = s^-$ for a role name s.

then add B to $\mathcal{L}(x)$ since x is an r-predecessor of y. Moreover, the blocking condition has to be more strict: for z' to block z, they must have identical labels, i.e., $\mathcal{L}(z) = \mathcal{L}(z')$. Finally, blocking becomes necessarily "dynamic": in the presence of inverse roles, node labels influence each other up and down the completion tree. Thus the label of a node x blocking some node y further down the tree can change due to some of its other predecessors, the node labels of x and y become different, and we must "unblock" them.

This tableau algorithm decides satisfiability (and thus subsumption) of $\mathcal{SI}$-concepts w.r.t. TBoxes. Moreover, we were able to prove that, in the absence of a TBox and employing a certain strategy and a more intricate blocking condition, it uses polynomial space only. This is one example for the fact that the definition of the blocking condition is not only crucial for the correctness of the algorithm, but also for its complexity. As a consequence, $\mathcal{ALC}$ without TBoxes and with transitive and inverse roles is of the same complexity as pure $\mathcal{ALC}$, namely PSPACE-complete [49].

5.3   Further Adding Role Inclusion Axioms

To represent, beside the general part-whole relation, certain sub-part-whole relations such as "is a component of" or "is an ingredient of", we can use role inclusion axioms [42].

A *role inclusion axiom* is an expression of the form $r \sqsubseteq s$, where r and s are (possibly inverse) roles. A *role hierarchy* is a finite set of role inclusion axioms. An interpretation $\mathcal{I}$ *satisfies* a role hierarchy $\mathcal{R}$ iff $r^{\mathcal{I}} \subseteq s^{\mathcal{I}}$ for each $r \sqsubseteq s$ in $\mathcal{R}$. Such an interpretation is called a *model* of $\mathcal{R}$. Satisfiability and subsumption w.r.t. role hierarchies are defined in the obvious way. $\mathcal{SHI}$ is the extension of $\mathcal{SI}$ with role hierarchies.

Adding role hierarchies to $\mathcal{SI}$ has mainly two consequences: firstly, we can introduce (possibly transitive—depending on the additional relation) role names such as hasComp or hasIngredient and add role inclusion axioms hasComp $\sqsubseteq$ hasPart and hasIngredient $\sqsubseteq$ hasPart. This turns out to be quite useful in various applications since it allows for a concise and natural description not only of aggregated objects.

Secondly, $\mathcal{SHI}$ (as well as $\mathcal{SH}$ and $\mathcal{SHIQ}$) has the expressive power for the *internalisation* of TBoxes [3, 45]. This technique polynomially reduces reasoning w.r.t. a general TBox to pure concept reasoning as follows. We introduce a new transitive role name $u \in \mathbf{R}_{+}$ and specify that u is a super-role of all roles and their respective inverses. This implies that, in connected models, u behaves like a universal role, i.e., u relates all elements of the interpretation domain; cf. Section 3. Since each satisfiable $\mathcal{SHI}$ concept is satisfiable in a connected model, it can be shown that a concept C is satisfiable w.r.t. $\{C_i \sqsubseteq D_i \mid 1 \leq i \leq n\}$ iff $\exists u.C \sqcap \forall u. \bigsqcap_{1 \leq i \leq n} (C_i \Rightarrow D_i)$ is satisfiable.

Tableau algorithm for $\mathcal{SHI}$ Basically, the extension of the $\mathcal{SI}$ tableau algorithm to $\mathcal{SHI}$ involves an adaption of the notion of an "r-successor" to take into account role hierarchies [45]: if y is an r-successor of x and $r \sqsubseteq s$ is in the role hierarchy, then y is also an s-successor of x. An analogous adaption for predecessors is also required in the presence of inverse roles, and transitive roles require a further, rather complex adaption of the propagation of universal value restrictions $\forall r.C$. Moreover, the correctness proof of the tableau algorithm becomes more complex since the tree structure the algorithm works on does no longer correspond to the relational structure that is to be built in case that the algorithm answers "satisfiable": this is already the case in the presence of transitive roles, but becomes more notable if, additionally, role hierarchies are taken into account.

5.4 Further Adding Number Restrictions

In general, when describing the relevant concepts of an application domain, it seems to be natural to describe an object by restricting the number of objects it is related to via a certain relation. For example, the following are concept definitions for pipes and forks:

$$\text{Pipe} \doteq \text{Connection} \sqcap (= 1\,\text{partOf}^-\,\text{Input}) \sqcap (= 1\,\text{partOf}^-\,\text{Output})$$
$$\text{Fork} \doteq \text{Connection} \sqcap (= 1\,\text{partOf}^-\,\text{Input}) \sqcap (\geq 2\,\text{partOf}^-\,\text{Output})$$

Before adding number restriction to $\mathcal{SHI}$, we have to define *simple* roles since only simple roles are allowed in number restrictions—without that restriction, satisfiability of $\mathcal{SHI}$ extended with number restrictions is undecidable [49].

A (possibly inverse) role is called *simple* if it is neither transitive nor has a transitive sub-role. $\mathcal{SHIQ}$ is obtained from $\mathcal{SHI}$ by allowing, additionally, for concepts of the form $(\geq ns.C)$ and $(\leq ns.C)$ for n a non-negative integer, s a simple role, and C a $\mathcal{SHIQ}$-concept. The semantics of number restrictions is given in Section 3.

In contrast to $\mathcal{SHI}$, $\mathcal{SHIQ}$ lacks the finite model property. That is, there are concepts that are satisfiable, but only in *infinite* models. For example, for r a transitive role and $s \sqsubseteq r$, each model of the following concept contains an infinite, acyclic s-chain: $\neg A \sqcap \exists s.A \sqcap \forall r.((\exists s.A) \sqcap (\geq 1s^-.\top))$.

As mentioned in Section 2, state-of-the-art DL reasoners such as FaCT and Racer implement tableau algorithms for $\mathcal{SHIQ}$ [41, 37]. Thus, $\mathcal{SHIQ}$ forms the logical basis of ontology editors Rice and Oiled [22, 12], and of the intelligent conceptual modelling tool Icom [31].

Tableau algorithm for $\mathcal{SHIQ}$ It is not difficult to see that, in the presence of number restrictions, we have to add two new rules to our tableau algorithm:

1. if $(\geq nr.C) \in \mathcal{L}(x)$ and x has less than n r-neighbours with C in their label, then generate these missing r-neighbours and set their labels to $\{C\}$.
2. if $(\leq nr.C) \in \mathcal{L}(x)$ and x has more than n r-neighbours with C in their label, then merge some of them, so that only n remain.

However, this is not sufficient. Firstly, such a naive extension might easily yield a "yo-yo" effect: for example, if applied to a node x with $(\geq 3r.C \sqcap D), (\leq 2r.C) \in \mathcal{L}(x)$, the above tableau algorithm would generate three r-successors y_i with

$C \sqcap D \in \mathcal{L}(y_i)$, break down the conjunctions in $C \sqcap D \in \mathcal{L}(y_i)$, and then notice that there are too many r-successors y_i of x with $C \in \mathcal{L}(y_i)$ for $(\leqslant 2r.C) \in \mathcal{L}(x)$. Thus two of them would be merged into a single one. Now there are not enough r-successors for $(\geqslant 3r.C \sqcap D)$, so one would be generated, and so on, thus leading to non-termination. To re-gain termination, we can use, for example, an explicit inequality relation $\neq$ that prevents nodes that were introduced for one $(\geqslant nr.C)$ from being merged again later. Moreover, we extend the notion of a "clash" to cases where $(\leqslant nr.C) \in \mathcal{L}(x)$ and x has more than $n \neq$-distinct r-successors with C in their label.

Secondly, consider the concept $C := (\geqslant 3r.B) \sqcap (\leqslant 1r.A) \sqcap (\leqslant 1r.\neg A)$. So far, for $C \in \mathcal{L}(x)$, the tableau algorithm would generate three r-successors y_i of x with $\{B\} = \mathcal{L}(y_i)$, and stop with the answer "C is satisfiable". However, the concept C is obviously unsatisfiable: the algorithm's unsoundness is due to its ignorance of which of the y_i are instances of A and which are instances of $\neg A$. To overcome this problem, we add a third rule

3. if $(\leqslant nr.C) \in \mathcal{L}(x)$ and y is an r-neighbour of x, then non-deterministically add C or $\neg C$ to $\mathcal{L}(y)$.

Thirdly, one also needs to modify the blocking condition—otherwise, the algorithm would still be unsound. Roughly spoken, the $\mathcal{SHIQ}$ blocking condition involves two pairs of subsequent nodes whose labels must coincide pairwise. Together, these three modifications indeed yield a decision procedure for the satisfiability of $\mathcal{SHIQ}$ [49].

Interestingly, the first proposal of the $\mathcal{SHIQ}$ blocking condition was so strict that it delayed blocking severely, thus enlarging the search space for a model dramatically and degrading the performance of FaCT. Investigating the soundness and completeness proof of the $\mathcal{SHIQ}$ tableau algorithm more closely, we were able to devise an new blocking condition which still ensures soundness, completeness, and termination, but was less strict [48]. Intuitively, node labels have only to be equal for "relevant concepts" in the respective nodes, a fact that made the formulation of the new blocking condition rather intricate. However, an empirical evaluation of the new tableau algorithm in FaCT showed that this more intricate but less strict blocking condition payes off: it improves performance up to two orders of magnitude.

Concerning worst-case complexity, both the original and the optimised $\mathcal{SHIQ}$ tableau algorithm are far from being optimal: in the worst case, they run in 2NExptime, whereas satisfiability of $\mathcal{SHIQ}$-concepts is known to be in Exp-Time, even with numbers in number restrictions coded in binary [83]. Despite this worst-case sub-optimality, its implementation in the FaCT and Racer systems behave surprisingly well in practice [48, 37]. However, the worst-case complexity implies that there exist rather small example inputs for which these systems need so much time that they are practically not terminating [14].

5.5   Further Adding More Expressive Role Inclusion Axioms

Although $\mathcal{SHIQ}$ is rather expressive, there is a common phenomenon that $\mathcal{SHIQ}$ is not able to express, and that would be useful in many applications, especially for those involving aggregated objects. This phenomenon is sometimes coined *propagation of properties*: for example, one wants to express that a fracture located in the shaft of the femur (which is a division of the femur) is a fracture located in the femur. Or one might want to express that the owner of a thing also owns the parts of this thing. The importance of this expressive means is illustrated by the fact that the Grail DL [44, 66], which was designed for medical terminologies, is able to express these propagations (although it is quite weak in other respects). In two other medical terminology applications, rather complex work arounds to represent propagations can be found: SEP-triplets[5] in [76] and right-identities in [79]. Finally, the CycL language provides the transfersThro statement for similar propagations [53]. So far and to the best of our knowledge, none of these systems were proven to handle these propagations in a sound and complete way.

It is rather straightforward to extend $\mathcal{SHIQ}$ to allow for the propagation of properties: obviously, it suffices to extend role hierarchies to the general role inclusion axioms, see Section 3. For the first example, one would introduce an axiom hasLocation∘divisionOf $\dot{\sqsubseteq}$ hasLocation and, indeed, w.r.t. this axiom,

Fracture ⊓ ∃hasLocation.(Shaft ⊓ ∃divisionOf.Femur)

is subsumed by Fracture ⊓ ∃hasLocation.Femur. For the second example, one would introduce an axiom owns ∘ hasPart $\sqsubseteq$ owns and, w.r.t. this axiom,

∃owns.(Bicycle ⊓ ∃hasPart.SuspensionFork)

is subsumed by ∃owns.SuspensionFork.

As mentioned in Section 3, results in grammar and description logics imply that extending $\mathcal{ALC}$ with role inclusion axioms of the form $r \circ s \dot{\sqsubseteq} t$ yields a logic for which satisfiability and subsumption are undecidable [10, 11, 90]. However, for expressing propagation of properties, we only need axioms of the form $r \circ s \dot{\sqsubseteq} s$ or $s \circ r \dot{\sqsubseteq} s$ [44, 65]. Unfortunately, extending $\mathcal{SHIQ}$ with this restricted form of axioms still yields an undecidable logic [47].

One way to re-gain decidability would be to restrict the underlying logic $\mathcal{SHIQ}$. Since we have argued that, especially for the representation of aggregated objects, the concept- and role-forming operators of $\mathcal{SHIQ}$ are crucial, we have chosen a different approach, namely to further restrict the role inclusion axioms: further restricting role hierarchies to not contain "affecting cycles" of length greater than one finally yields a decidable logic. Roughly speaking, "affecting" is the transitive closure of the relation "directly affecting", and r directly affects s if $r \circ s \dot{\sqsubseteq} s$, $s \circ r \dot{\sqsubseteq} s$, or $r \dot{\sqsubseteq} s$ is contained in the role hierarchy. A role hierarchy

---

[5] *SEP-triplets* are used both to compensate for the absence of transitive roles in $\mathcal{ALC}$, and to express the propagation of properties across a distinguished "part-of" role.

containing no "affecting" cycles of length greater than one is called *acyclic*, and the extension of $\mathcal{SHIQ}$ with acyclic role hierarchies is called $\mathcal{RIQ}$.

In $\mathcal{RIQ}$, we can model the propagation of properties as mentioned above, and the restriction to acyclicity does not seem to be too severe since non-trivial cycles seem to indicate modelling flaws [65].

Tableau algorithm for $\mathcal{RIQ}$ The tableau algorithm for $\mathcal{RIQ}$ [47] involves two pre-processing steps that transform the role hierarchy into a more explicit and manageable structure. Firstly, acyclic role hierarchies are unfolded in a similar way as acyclic TBoxes can be unfolded [59], thus making all implicit implications explicit. As a result of this unfolding, we obtain, for each role name r, a regular expressions τ_r on role names. Secondly, we construct, for each τ_r, a non-deterministic finite automata $\mathcal{A}_r$ which accepts $L(\tau_r)$.

Then, in the tableau rules, we add three rules

1. if $\forall r.C \in \mathcal{L}(x)$, then we add $\forall \mathcal{A}_r.C$.
2. if $\forall \mathcal{A}.C \in \mathcal{L}(x)$ and x has an s-successor y, then we add $\forall \mathcal{A}'.C$ to $\mathcal{L}(y)$ for each automaton $\mathcal{A}'$ that is the result of $\mathcal{A}$ reading s, i.e., $\mathcal{A}'$ is obtained from $\mathcal{A}$ by simply changing the initial state to a state that is reachable from $\mathcal{A}$'s initial state by an s transition.
3. if $\forall \mathcal{A}.C \in \mathcal{L}(x)$ and $\varepsilon \in L(\mathcal{A})$, then add C to $\mathcal{L}(x)$.

The pre-processing together with these three rules can be shown to yield a decision procedure for $\mathcal{RIQ}$.

This tableau algorithm for $\mathcal{RIQ}$ is implemented successfully in FaCT. The additional overhead introduced by using automata in tableau rules seems to pay off since it does not degrade the performance of FaCT, yields a more readable algorithm, and can draw additional inferences like the medical one above [47].

References

1. C. Areces, P. Blackburn, and M. Marx. A road-map on complexity for hybrid logics. In *Proc. of CSL'99*, vol. 1683 of *LNCS*. Springer-Verlag, 1999.
2. C. Areces, P. Blackburn, and M. Marx. The computational complexity of hybrid temporal logics. *Logic Journal of the IGPL*, 8(5), 2000.
3. F. Baader. Augmenting concept languages by transitive closure of roles: An alternative to terminological cycles. In *Proc. of IJCAI-91*, Sydney, 1991.
4. F. Baader. Using automata theory for characterizing the semantics of terminological cycles. *Annals of Mathematics and Artificial Intelligence*, 18(2–4):175–219, 1996.
5. F. Baader, S. Brandt, and R. Küsters. Matching under side conditions in description logics. In B. Nebel, ed., *Proc. of IJCAI-01*. Morgan Kaufmann, 2001.
6. F. Baader, D. Calvanese, D. McGuinness, D. Nardi, and P. F. Patel-Schneider, eds. *The Description Logic Handbook: Theory, Implementation, and Applications*. Cambridge University Press, 2003.
7. F. Baader, E. Franconi, B. Hollunder, B. Nebel, and H.J. Profitlich. An empirical analysis of optimization techniques for terminological representation systems, or: Making KRIS get a move on. *Applied Artificial Intelligence*, 4:109–132, 1994.

8. F. Baader and U. Sattler. An overview of tableau algorithms for description logics. *Studia Logica*, 69:5–40, 2001. An abridged version appeared in *Tableaux 2000*, vol. 1847 of LNAI, 2000. Springer-Verlag.

9. F. Baader and A.-Y. Turhan. On the problem of computing small representations of least common subsumers. In *Proc. of KI 2002*, vol. 2479 of *LNAI*. Springer-Verlag, 2002.

10. M. Baldoni. *Normal Multimodal Logics: Automatic Deduction and Logic Programming Extension*. PhD thesis, Dipartimento di Informatica, Università degli Studi di Torino, Italy, 1998.

11. M. Baldoni, L. Giordano, and A. Martelli. A tableau calculus for multimodal logics and some (un)decidability results. In *Proc. of TABLEAUX-98*, vol. 1397 of *LNAI*. Springer-Verlag, 1998.

12. S. Bechhofer, I. Horrocks, C. Goble, and R. Stevens. OilEd: a reason-able ontology editor for the semantic web. In *Proc. of DL 2001*. CEUR (http://ceur-ws.org/), 2001.

13. M. Ben-Ari, J.Y. Halpern, and A. Pnueli. Deterministic propositional dynamic logic: finite models, complexity and completeness. *J. of Computer and System Science*, 25:402–417, 1982.

14. D. Berardi, D. Calvanese, and G. De Giacomo. Reasoning on UML Class Diagrams using Description Logic Based Systems. In *Proc. of the KI'2001 Workshop on Applications of Description Logics*. CEUR (http://ceur-ws.org/), 2001.

15. T. Berners-Lee, J. Hendler, and O. Lassila. The semantic Web. *Scientific American*, 284(5):34–43, 2001.

16. P. Borst, H. Akkermans, and J. Top. Engineering ontologies. *International Journal of Human-Computer Studies*, 46:365–406, 1997.

17. S. Brandt, R. Küsters, and A.-Y. Turhan. Approximation and difference in description logics. In *Proc. of KR-02*. Morgan Kaufmann, 2002.

18. D. Calvanese. *Unrestricted and Finite Model Reasoning in Class-Based Representation Formalisms*. PhD thesis, Dip. di Inf. e Sist., Univ. di Roma "La Sapienza", 1996.

19. D. Calvanese, G. De Giacomo, and M. Lenzerini. Structured objects: Modeling and reasoning. In *Proc. of DOOD-95*, vol. 1013 of *LNCS*, pages 229–246, 1995.

20. D. Calvanese, G. De Giacomo, and M. Lenzerini. Reasoning in expressive description logics with fixpoints based on automata on infinite trees. In *Proc. of IJCAI-99*. Morgan Kaufmann, 1999.

21. D. Calvanese, G. De Giacomo, M. Lenzerini, and D. Nardi. Reasoning in expressive description logics. In A. Robinson and A. Voronkov, eds., *Handbook of Automated Reasoning*. Elsevier Science Publishers, 1999.

22. R. Cornet. Rice ontology editor. Homepage at http://www.b1g-systems.com/ronald/rice/.

23. G. De Giacomo and M. Lenzerini. Boosting the correspondence between description logics and propositional dynamic logics (extended abstract). In *Proc. of AAAI-94*. AAAI Press, 1994.

24. G. De Giacomo and M. Lenzerini. Tbox and Abox reasoning in expressive description logics. In *Proc. of KR-96*. Morgan Kaufmann, 1996.

25. S. Demri. The complexity of regularity in grammar logics and related modal logics. *J. of Logic and Computation*, 11(6), 2001.

26. F. Donini, M. Lenzerini, D. Nardi, and W. Nutt. The complexity of concept languages. In *Proc. of KR-91*, Boston, MA, USA, 1991.

27. F. M. Donini and F. Massacci. Exptime tableaux for *ALC*. *Artificial Intelligence*, 124(1):87–138, 2000.

28. D. Fensel, F. van Harmelen, I. Horrocks, D. McGuinness, and P. F. Patel-Schneider. OIL: An ontology infrastructure for the semantic web. *IEEE Intelligent Systems*, 16(2):38–45, 2001.

29. K. Fine. In so many possible worlds. *Notre Dame Journal of Formal Logics*, 13:516–520, 1972.

30. M. J. Fischer and R. E. Ladner. Propositional dynamic logic of regular programs. *J. of Computer and System Science*, 18:194–211, 1979.

31. E. Franconi and G. Ng. The i.com tool for intelligent conceptual modelling. In *Working Notes of the ECAI2000 Workshop KRDB2000*. CEUR (http://ceur-ws.org/), 2000.

32. G. Gargov, S. Passy, and T. Tinchev. Modal environment for Boolean speculations. In D. Skordev, ed., *Mathematical Logic and Applications*, pages 253–263. Plenum Publ. Co., New York, 1987.

33. L. Georgieva, U. Hustadt, and R. A. Schmidt. Hyperresolution for guarded formulae. *J. of Symbolic Logic*, 2003. To appear.

34. P. Gerstl and S. Pribbenow. Midwinters, end games and bodyparts. *International Journal of Human-Computer Studies*, 43:847–864, 1995.

35. T. R. Gruber. Towards Principles for the Design of Ontologies Used for Knowledge Sharing. In N. Guarino and R. Poli, eds., *Formal Ontology in Conceptual Analysis and Knowledge Representation*, 1993. Kluwer Academic Publishers.

36. V. Haarslev and R. Möller. Consistency testing: The RACE experience. In *Proc. TABLEAUX 2000*, vol. 1847 of *LNAI*. Springer-Verlag, 2000.

37. V. Haarslev and R. Möller. RACER system description. In *IJCAR-01*, vol. 2083 of *LNAI*. Springer-Verlag, 2001.

38. J. Y. Halpern and Y. Moses. A guide to completeness and complexity for modal logic of knowledge and belief. *Artificial Intelligence*, 54:319–379, 1992.

39. J. Hladik. Implementation and optimisation of a tableau algorithm for the guarded frament. In *Proc. TABLEAUX 2002*, vol. 2381 of *LNAI*. Springer-Verlag, 2002.

40. I. Horrocks. *Optimising Tableaux Decision Procedures for Description Logics*. PhD thesis, University of Manchester, 1997.

41. I. Horrocks. Using an Expressive Description Logic: FaCT or Fiction? In *Proc. of KR-98*. Morgan Kaufmann, 1998.

42. I. Horrocks and G. Gough. Description logics with transitive roles. In M.-C. Rousset, R. Brachmann, F. Donini, E. Franconi, I. Horrocks, and A. Levy, eds., *Proc. of DL'97*, pages 25–28, 1997.

43. I. Horrocks, P. F. Patel-Schneider, and F. van Harmelen. Reviewing the design of DAML+OIL: An ontology language for the semantic web. In *Proc. of AAAI-02*, 2002.

44. I. Horrocks, A. Rector, and C. Goble. A description logic based schema for the classification of medical data. In *Working Notes of the ECAI-96 Workshop KRDB-96*. CEUR (http://ceur-ws.org/), 1996.

45. I. Horrocks and U. Sattler. A description logic with transitive and inverse roles and role hierarchies. *J. of Logic and Computation*, 1999.

46. I. Horrocks and U. Sattler. Ontology reasoning in the $\mathcal{SHOQ}(D)$ description logic. In B. Nebel, ed., *Proc. of IJCAI-01*. Morgan Kaufmann, 2001.

47. I. Horrocks and U. Sattler. Decidability of $\mathcal{SHIQ}$ with complex role inclusion axioms. In *Proc. of CADE-19*. LNAI, Springer-Verlag, 2003 (to appear).

48. I. Horrocks and U. Sattler. Optimised reasoning for $\mathcal{SHIQ}$. In *Proc. of ECAI 2002*, 2002.

49. I. Horrocks, U. Sattler, and S. Tobies. Practical reasoning for expressive description logics. In *Proc. of LPAR'99*, vol. 1705 of *LNAI*. Springer-Verlag, 1999.

50. U. Hustadt and R. A. Schmidt. MSPASS: Modal reasoning by translation and first-order resolution. In *Proc. TABLEAUX 2000*, vol. 1847 of *LNAI*, pages 67–71. Springer-Verlag, 2000.

51. D. Kozen. Results on the propositional μ-calculus. In *Automata, Languages and Programming, 9th Colloquium*, vol. 140 of *LNCS*. Springer-Verlag, 1982.

52. R. Küsters. *Non-Standard Inferences in Description Logics*, vol. 2100 of *LNAI*. Springer-Verlag, 2001.

53. D. B. Lenat and R. V. Guha. *Building Large Knowledge-Based Systems*. Addison Wesley Publ. Co., Reading, Massachussetts, 1989.

54. C. Lutz. Complexity of terminological reasoning revisited. In *Proc. of LPAR'99*, LNAI, pages 181–200. Springer-Verlag, 1999.

55. C. Lutz. *The Complexity of Description Logics with Concrete Domains*. PhD thesis, RWTH Aachen, 2002.

56. C. Lutz. Description logics with concrete domains—a survey. In *Advances in Modal Logics Vol. 4*. World Scientific Publishing Co. Pte. Ltd., 2003.

57. C. Lutz and U. Sattler. The complexity of reasoning with boolean modal logics. In F. Wolter, H. Wansing, M. de Rijke, and M. Zakharyaschev, eds., *Advances in Modal Logics 3*. CSLI Publications, Stanford, 2001.

58. R. Molitor. *Unterstützung der Modellierung verfahrenstechnischer Prozesse durch Nicht-Standardinferenzen in Beschreibungslogiken*. PhD thesis, RWTH Aachen, 2000.

59. B. Nebel. *Reasoning and Revision in Hybrid Representation Systems*, vol. 422 of *LNAI*. Springer-Verlag, 1990.

60. P. Patel-Schneider, D. McGuinness, R. Brachman, L. Resnick, and A. Borgida. The CLASSIC knowledge representation system: Guiding principles and implementation rationale. *SIGART Bulletin*, 2(3):108–113, 1991.

61. P. F. Patel-Schneider. Undecidability of subsumption in NIKL. *Artificial Intelligence*, 39:263–272, 1989.

62. V. R. Pratt. Models of program logics. In *Proceedings of the 20th Annual Symposium on Foundations of Computer Science*, San Juan, Puerto Rico, 1979.

63. A. Prior. *Past, Present and Future*. Oxford University Press, 1967.

64. Protégé. Homepage at http://protege.stanford.edu/, 2003.

65. A. Rector. Analysis of propagation along transitive roles: Formalisation of the galen experience with medical ontologies. In *Proc. of DL 2002*. CEUR (http://ceur-ws.org/), 2002.

66. A. Rector, S. Bechhofer, C. A. Goble, I. Horrocks, W. A. Nowlan, and W. D. Solomon. The GRAIL concept modelling language for medical terminology. *AI in Medicine*, 9:139–171, 1997.

67. N. Rescher and A. Urquhart. *Temporal Logic*. Springer-Verlag, 1971.

68. U. Sattler. A concept language extended with different kinds of transitive roles. In *Proc. of KI'96*, vol. 1137 of *LNAI*. Springer-Verlag, 1996.

69. U. Sattler. *Terminological knowledge representation systems in a process engineering application*. PhD thesis, RWTH Aachen, 1998.

70. U. Sattler. Description logics for the representation of aggregated objects. In W. Horn, ed., *Proc. of ECAI-2000*. IOS Press, Amsterdam, 2000.

71. U. Sattler and M. Y. Vardi. The hybrid μ-calculus. In *IJCAR-01*, vol. 2083 of *LNAI*. Springer-Verlag, 2001.

72. A. Schaerf. Reasoning with individuals in concept languages. *Data and Knowledge Engineering*, 13(2):141–176, 1994.

73. K. Schild. A correspondence theory for terminological logics: Preliminary report. In *Proc. of IJCAI-91*. Morgan Kaufmann, 1991.

74. M. Schmidt-Schauss. Subsumption in KL-ONE is undecidable. In *Proc. of KR-89*, Boston (USA), 1989.

75. M. Schmidt-Schauß and G. Smolka. Attributive concept descriptions with complements. *Artificial Intelligence*, 48(1):1–26, 1991.

76. S. Schulz and U. Hahn. Parts, locations, and holes - formal reasoning about anatomical structures. In *Proc. of AIME 2001*, vol. 2101 of *LNAI*. Springer-Verlag, 2001.

77. P. M. Simons. *Parts. A study in Ontology*. Oxford: Clarendon, 1987.

78. E. Spaan. The complexity of propositional tense logics. In M. de Rijke, ed., *Diamonds and Defaults*. Kluwer Academic Publishers, 1993.

79. K.A. Spackman. Managing clinical terminology hierarchies using algorithmic calculation of subsumption: Experience with SNOMED-RT. *J. of the Amer. Med. Informatics Ass.*, 2000. Fall Symposium Special Issue.

80. R. Stevens, I. Horrocks, C. Goble, and S. Bechhofer. Building a bioinformatics ontology using OIL. *IEEE Inf. Techn. in Biomedicine.*, 6(2):135–141, 2002.

81. R. S. Streett. Propositional dynamic logic of looping and converse is elementarily decidable. *Information and Computation*, 54:121–141, 1982.

82. Robert S. Streett and E. Allen Emerson. An automata theoretic decision procedure for the propositional μ-calculus. *Information and Computation*, 81:249–264, 1989.

83. S. Tobies. *Complexity Results and Practical Algorithms for Logics in Knowledge Representation*. PhD thesis, RWTH Aachen, 2001.

84. S. Tobies. PSPACE reasoning for graded modal logics. *J. of Logic and Computation*, 11(1):85–106, 2001.

85. M. Uschold, M. King, S. Moralee, and Y. Zorgios. The enterprise ontology. *The Knowledge Engineering Review*, 13, 1998.

86. M. Y. Vardi. The taming of converse: Reasoning about two-way computations. In *Proc. of the 4th Workshop on Logics of Programs*, vol. 193 of *LNCS*. Springer-Verlag, 1985.

87. M. Y. Vardi. Why is modal logic so robustly decidable? In N. Immerman and P. G. Kolaitis, eds., *Descriptive Complexity and Finite Models*, volume 31 of *DIMACS*. American Mathematical Society, 1997.

88. M. Y. Vardi. Reasoning about the past with two-way automata. In *Proc. of ICALP'98*, vol. 1443 of *LNCS*. Springer-Verlag, 1998.

89. M. Y. Vardi and P. Wolper. Automata-theoretic techniques for modal logics of programs. *J. of Computer and System Science*, 32:183–221, 1986.

90. M. Wessel. Obstacles on the way to qualitative spatial reasoning with description logics: Some undecidability results. In *Proc. of DL 2001*. CEUR (http://ceur-ws.org/), 2001.

91. M.E. Winston, R. Chaffin, and D. Herrmann. A taxonomy of part whole relations. *Cognitive Science*, 11:417–444, 1987.

Computing the Least Common Subsumer in the Description Logic $\mathcal{EL}$ w.r.t. Terminological Cycles with Descriptive Semantics

Franz Baader

Theoretical Computer Science, TU Dresden, D-01062 Dresden, Germany
baader@inf.tu-dresden.de

Abstract. Computing the least common subsumer (lcs) is one of the most prominent non-standard inference in description logics. Baader, Küsters, and Molitor have shown that the lcs of concept descriptions in the description logic $\mathcal{EL}$ always exists and can be computed in polynomial time. In the present paper, we try to extend this result from concept descriptions to concepts defined in a (possibly cyclic) $\mathcal{EL}$-terminology interpreted with descriptive semantics, which is the usual first-order semantics for description logics. In this setting, the lcs need not exist. However, we are able to define possible candidates P_k $(k \geq 0)$ for the lcs, and can show that the lcs exists iff one of these candidates is the lcs. Since each of these candidates is a common subsumer, they can also be used to approximate the lcs even if it does not exist. In addition, we give a sufficient condition for the lcs to exist, and show that, under this condition, it can be computed in polynomial time.

1 Introduction

Computing the least common subsumer of concepts can be used in the bottom-up construction of description logic (DL) knowledge bases. Instead of defining the relevant concepts of an application domain from scratch, this methodology allows the user to describe the concept to be defined by examples, which are themselves given as concepts.[1] These examples are then generalized into a new concept by computing their least common subsumer (i.e., the least concept description in the available description language that subsumes all these concepts). The knowledge engineer can then use the computed concept as a starting point for the concept definition. Another application of the least common subsumer computation is structuring of DL knowledge bases. In fact, in many cases these knowledge bases are rather "flat" in the sense that their subsumption hierarchy is not deep and that a given concept may have a huge number of direct descendants in this hierarchy. To support browsing such hierarchies, one would like to introduce meaningful intermediate concepts, and this can again be facilitated by computing

---

[1] If the examples are not given as concepts, but as individuals in a DL ABox, then one must first generalize the individuals into concepts by computing their most specific concept [4, 8].

A. de Moor, W. Lex, and B. Ganter (Eds.): ICCS 2003, LNAI 2746, pp. 117–130, 2003.

the lcs of subsets of the direct descendants of concepts with many descendants. These applications (and how formal concept analysis can be employed in this context) are described in more detail in [3].

The least common subsumer (lcs) in DLs with existential restrictions was investigated in [5]. In particular, it was shown there that the lcs in the small DL $\mathcal{EL}$ (which allows conjunctions, existential restrictions, and the top-concept) always exists, and that the binary lcs can be computed in polynomial time. In the present paper, we try to extend this result from concept descriptions to concepts defined in a (possibly cyclic) $\mathcal{EL}$-terminology interpreted with descriptive semantics, which is the usual first-order semantics for description logics.

The report [2] considers cyclic terminologies in $\mathcal{EL}$ w.r.t. the three types of semantics (greatest fixpoint (gfp), least fixpoint (lfp), and descriptive semantics) introduced by Nebel [9], and shows that the subsumption problem can be decided in polynomial time in all three cases. This is in strong contrast to the case of DLs with value restrictions. Even for the small DL $\mathcal{FL}_0$ (which allows conjunctions and value restrictions only), adding cyclic terminologies increases the complexity of the subsumption problem from polynomial (for concept descriptions) to PSPACE. The main tool in the investigation of cyclic definitions in $\mathcal{EL}$ is a characterization of subsumption through the existence of so-called simulation relations, which can be computed in polynomial time [7].

The characterization of subsumption in $\mathcal{EL}$ w.r.t. gfp-semantics through the existence of certain simulation relations on the graph associated with the terminology can be used to characterize the lcs via the product of this graph with itself [1]. This shows that, w.r.t. gfp-semantics, the lcs always exists, and the binary lcs can be computed in polynomial time. (The n-ary lcs may grow exponentially even in $\mathcal{EL}$ without cyclic terminologies [5].)

In the present paper, we concentrate on the lcs w.r.t. cyclic terminologies in $\mathcal{EL}$ with descriptive semantics. Here things are not as rosy as for gfp-semantics. We will show that, in general, the lcs need not exist (Section 4.1). We then introduce possible candidates P_k ($k \geq 0$) for the lcs, and show that the lcs exists iff one of these candidates is the lcs (Section 4.2). Finally, we give a sufficient condition for the lcs to exist, and show that, under this condition, it can be computed in polynomial time (Section 4.3).

Before we can start presenting the new results, we must first introduce $\mathcal{EL}$ with cyclic terminologies as well as the lcs (Section 2), and recall the important definitions and results from [2] (Section 3). Full proofs for the results presented in this paper can be found in [1].

2   Cyclic TBoxes and Least Common Subsumers in $\mathcal{EL}$

Concept descriptions are inductively defined with the help of a set of *constructors*, starting with a set N_C of *concept names* and a set N_R of *role names*. The constructors determine the expressive power of the DL. In this report, we restrict the attention to the DL $\mathcal{EL}$, whose concept descriptions are formed using the constructors top-concept ($\top$), conjunction (C & D), and existential restriction

Table 1. Syntax and semantics of $\mathcal{EL}$-concept descriptions and TBox definitions.

name of constructor	Syntax	Semantics
concept name $A \in N_C$	A	$A^{\mathcal{I}} \subseteq \Delta^{\mathcal{I}}$
role name $r \in N_R$	r	$r^{\mathcal{I}} \subseteq \Delta^{\mathcal{I}} \times \Delta^{\mathcal{I}}$
top-concept	$\top$	$\Delta^{\mathcal{I}}$
conjunction	$C \ \& \ D$	$C^{\mathcal{I}} \cap D^{\mathcal{I}}$
existential restriction	$\exists r.C$	$\{x \in \Delta^{\mathcal{I}} \mid \exists y : (x,y) \in r^{\mathcal{I}} \wedge y \in C^{\mathcal{I}}\}$
concept definition	$A \equiv D$	$A^{\mathcal{I}} = D^{\mathcal{I}}$

$(\exists r.C)$. The semantics of $\mathcal{EL}$-concept descriptions is defined in terms of an *interpretation* $\mathcal{I} = (\Delta^{\mathcal{I}}, \cdot^{\mathcal{I}})$. The domain $\Delta^{\mathcal{I}}$ of $\mathcal{I}$ is a non-empty set of individuals and the interpretation function $\cdot^{\mathcal{I}}$ maps each concept name $A \in N_C$ to a subset $A^{\mathcal{I}}$ of $\Delta^{\mathcal{I}}$ and each role $r \in N_R$ to a binary relation $r^{\mathcal{I}}$ on $\Delta^{\mathcal{I}}$. The extension of $\cdot^{\mathcal{I}}$ to arbitrary concept descriptions is inductively defined, as shown in the third column of Table 1.

A *terminology* (or *TBox* for short) is a finite set of concept definitions of the form $A \equiv D$, where A is a concept name and D a concept description. In addition, we require that TBoxes do not contain *multiple definitions*, i.e., there cannot be two distinct concept descriptions D_1 and D_2 such that both $A \equiv D_1$ and $A \equiv D_2$ belongs to the TBox. Concept names occurring on the left-hand side of a definition are called *defined concepts*. All other concept names occurring in the TBox are called *primitive concepts*. Note that we allow for cyclic dependencies between the defined concepts, i.e., the definition of A may refer (directly or indirectly) to A itself. An interpretation $\mathcal{I}$ is a model of the TBox $\mathcal{T}$ iff it satisfies all its concept definitions, i.e., $A^{\mathcal{I}} = D^{\mathcal{I}}$ for all definitions $A \equiv D$ in $\mathcal{T}$.

The semantics of (possibly cyclic) $\mathcal{EL}$-TBoxes we have just defined is called *descriptive semantic* by Nebel [9]. For some applications, it is more appropriate to interpret cyclic concept definitions with the help of an appropriate fixpoint semantics. However, in this paper we will restrict our attention to descriptive semantic (see [1] for a treatment of subsumption in $\mathcal{EL}$ w.r.t. greatest fixpoint, least fixpoint, and descriptive semantics, and [2] for a treatment of the lcs w.r.t. greatest fixpoint semantics).

Definition 1. *Let $\mathcal{T}$ be an $\mathcal{EL}$-TBox and let A, B be defined concepts occurring in $\mathcal{T}$. Then, A is subsumed by B w.r.t. descriptive semantics ($A \sqsubseteq_{\mathcal{T}} B$) iff $A^{\mathcal{I}} \subseteq B^{\mathcal{I}}$ holds for all models $\mathcal{I}$ of $\mathcal{T}$.*

On the level of concept descriptions, the least common subsumer of two concept descriptions C, D is the least concept description E that subsumes both C and D. An extensions of this definition to the level of (possibly cyclic) TBoxes is not completely trivial. In fact, assume that A_1, A_2 are concepts defined in the TBox $\mathcal{T}$. It should be obvious that taking as the lcs of A_1, A_2 the least defined concept B in $\mathcal{T}$ such that $A_1 \sqsubseteq_{\mathcal{T}} B$ and $A_2 \sqsubseteq_{\mathcal{T}} B$ is too weak since the lcs would then strongly depend on other defined concepts that are already present in $\mathcal{T}$.

However, a second approach (which might look like the obvious generalization of the definition of the lcs in the case of concept descriptions) is also not quite satisfactory. We could say that the lcs of A_1, A_2 is the least concept description C (possibly using defined concepts of $\mathcal{T}$) such that $A_1 \sqsubseteq_\mathcal{T} C$ and $A_2 \sqsubseteq_\mathcal{T} C$. The drawback of this definition is that it does not allow us to use the expressive power of cyclic definitions when constructing the lcs.

To avoid this problem, we allow the original TBox to be extended by new definitions when constructing the lcs. We say that the TBox $\mathcal{T}_2$ is a *conservative extension* of the TBox $\mathcal{T}_1$ iff $\mathcal{T}_1 \subseteq \mathcal{T}_2$ and $\mathcal{T}_1$ and $\mathcal{T}_2$ have the same primitive concepts and roles. Thus, $\mathcal{T}_2$ may contain new definitions $A \equiv D$, but then D does not introduce new primitive concepts and roles (i.e., all of them already occur in $\mathcal{T}_1$), and A is a new concept name (i.e., A does not occur in $\mathcal{T}_1$). The name "conservative extension" is justified by the fact that the new definitions in $\mathcal{T}_2$ do not influence the subsumption relationships between defined concepts in $\mathcal{T}_1$.

Lemma 1. *Let* $\mathcal{T}_1, \mathcal{T}_2$ *be* $\mathcal{EL}$-*TBoxes such that* $\mathcal{T}_2$ *is a conservative extension of* $\mathcal{T}_1$, *and let* A, B *be defined concepts in* $\mathcal{T}_1$ *(and thus also in* $\mathcal{T}_2$*). Then* $A \sqsubseteq_{\mathcal{T}_1} B$ *iff* $A \sqsubseteq_{\mathcal{T}_2} B$.

Definition 2. *Let* $\mathcal{T}_1$ *be an* $\mathcal{EL}$-*TBox containing the defined concepts* A, B, *and let* $\mathcal{T}_2$ *be a conservative extension of* $\mathcal{T}_1$ *containing the new defined concept* E. *Then* E *in* $\mathcal{T}_2$ *is a* least common subsumer *of* A, B *in* $\mathcal{T}_1$ *w.r.t. descriptive semantics (lcs) iff the following two conditions are satisfied:*

1. $A \sqsubseteq_{\mathcal{T}_2} E$ *and* $B \sqsubseteq_{\mathcal{T}_2} E$.
2. *If* $\mathcal{T}_3$ *is a conservative extension of* $\mathcal{T}_2$ *and* F *a defined concept in* $\mathcal{T}_3$ *such that* $A \sqsubseteq_{\mathcal{T}_3} F$ *and* $B \sqsubseteq_{\mathcal{T}_3} F$, *then* $E \sqsubseteq_{\mathcal{T}_3} F$.

In the case of concept descriptions, the lcs is unique up to equivalence, i.e., if E_1 and E_2 are both least common subsumers of the descriptions C, D, then $E_1 \equiv E_2$ (i.e., $E_1 \sqsubseteq E_2$ and $E_2 \sqsubseteq E_1$). In the presence of (possibly cyclic) TBoxes, this uniqueness property also holds (though its formulation is more complicated).

Proposition 1. *Let* $\mathcal{T}_1$ *be an* $\mathcal{EL}$-*TBox containing the defined concepts* A, B. *Assume that* $\mathcal{T}_2$ *and* $\mathcal{T}_2'$ *are conservative extensions of* $\mathcal{T}_1$ *such that*

- *the defined concept* E *in* $\mathcal{T}_2$ *is an lcs of* A, B *in* $\mathcal{T}_1$;
- *the defined concept* E' *in* $\mathcal{T}_2'$ *is an lcs of* A, B *in* $\mathcal{T}_1$;
- *the sets of newly defined concepts in respectively* $\mathcal{T}_2$ *and* $\mathcal{T}_2'$ *are disjoint.*

For $\mathcal{T}_3 := \mathcal{T}_2 \cup \mathcal{T}_2'$, *we have* $E \equiv_{\mathcal{T}_3} E'$ *(i.e.,* $E \sqsubseteq_{\mathcal{T}_3} E'$ *and* $E' \sqsubseteq_{\mathcal{T}_3} E$*).*

3   Characterizing Subsumption in $\mathcal{EL}$ with Cyclic Definitions

In this section, we recall the characterizations of subsumption w.r.t. descriptive semantics developed in [2]. To this purpose, we must represent TBoxes by description graphs, and introduce the notion of a simulation on description graphs.

3.1 Description Graphs and Simulations

Before we can translate $\mathcal{EL}$-TBoxes into description graphs, we must normalize the TBoxes. In the following, let $\mathcal{T}$ be an $\mathcal{EL}$-TBox, N_{def} the defined concepts of $\mathcal{T}$, N_{prim} the primitive concepts of $\mathcal{T}$, and N_{role} the roles of $\mathcal{T}$. We say that the $\mathcal{EL}$-TBox $\mathcal{T}$ is *normalized* iff $A \equiv D \in \mathcal{T}$ implies that D is of the form

$$P_1 \ \& \ \ldots \ \& \ P_m \sqcap \exists r_1.B_1 \ \& \ \ldots \ \& \ \exists r_\ell.B_\ell,$$

for $m, \ell \geq 0$, $P_1, \ldots, P_m \in N_{prim}$, $r_1, \ldots, r_\ell \in N_{role}$, and $B_1, \ldots, B_\ell \in N_{def}$. If $m = \ell = 0$, then $D = \top$.

As shown in [2], one can (without loss of generality) restrict the attention to normalized TBox. In the following, we thus assume that all TBoxes are normalized. Normalized $\mathcal{EL}$-TBoxes can be viewed as graphs whose nodes are the defined concepts, which are labeled by sets of primitive concepts, and whose edges are given by the existential restrictions. For the rest of this section, we fix a normalized $\mathcal{EL}$-TBox $\mathcal{T}$ with primitive concepts N_{prim}, defined concepts N_{def}, and roles N_{role}.

Definition 3. *An $\mathcal{EL}$-description graph is a graph $\mathcal{G} = (V, E, L)$ where*

- *V is a set of nodes;*
- *$E \subseteq V \times N_{role} \times V$ is a set of edges labeled by role names;*
- *$L: V \to 2^{N_{prim}}$ is a function that labels nodes with sets of primitive concepts.*

The TBox $\mathcal{T}$ can be translated into the following $\mathcal{EL}$-description graph $\mathcal{G}_\mathcal{T} = (N_{def}, E_\mathcal{T}, L_\mathcal{T})$:

- *the nodes of $\mathcal{G}_\mathcal{T}$ are the defined concepts of $\mathcal{T}$;*
- *if A is a defined concept and $A \equiv P_1 \ \& \ \ldots \ \& \ P_m \sqcap \exists r_1.B_1 \ \& \ \ldots \ \& \ \exists r_\ell.B_\ell$ its definition in $\mathcal{T}$, then*
  - *$L_\mathcal{T}(A) = \{P_1, \ldots, P_m\}$, and*
  - *A is the source of the edges $(A, r_1, B_1), \ldots, (A, r_\ell, B_\ell) \in E_\mathcal{T}$.*

Simulations are binary relations between nodes of two $\mathcal{EL}$-description graphs that respect labels and edges in the sense defined below.

Definition 4. *Let $\mathcal{G}_i = (V_i, E_i, L_i)$ $(i = 1, 2)$ be two $\mathcal{EL}$-description graphs. The binary relation $Z \subseteq V_1 \times V_2$ is a simulation from $\mathcal{G}_1$ to $\mathcal{G}_2$ iff*

(S1) $(v_1, v_2) \in Z$ *implies* $L_1(v_1) \subseteq L_2(v_2)$; *and*
(S2) *if* $(v_1, v_2) \in Z$ *and* $(v_1, r, v_1') \in E_1$, *then there exists a node* $v_2' \in V_2$ *such that* $(v_1', v_2') \in Z$ *and* $(v_2, r, v_2') \in E_2$.

We write $Z: \mathcal{G}_1 \overset{\sim}{\to} \mathcal{G}_2$ to express that Z is a simulation from $\mathcal{G}_1$ to $\mathcal{G}_2$.

It is easy to see that the set of all simulations from $\mathcal{G}_1$ to $\mathcal{G}_2$ is closed under arbitrary unions. Consequently, there always exists a greatest simulation from $\mathcal{G}_1$ to $\mathcal{G}_2$. If $\mathcal{G}_1, \mathcal{G}_2$ are finite, then this greatest simulation can be computed in polynomial time [7]. As an easy consequence of this fact, the following proposition is proved in [2].

Proposition 2. *Let $\mathcal{G}_1, \mathcal{G}_2$ be two finite $\mathcal{EL}$-description graphs, v_1 a node of $\mathcal{G}_1$ and v_2 a node of $\mathcal{G}_2$. Then we can decide in polynomial time whether there is a simulation $Z: \mathcal{G}_1 \overset{\sim}{\to} \mathcal{G}_2$ such that $(v_1, v_2) \in Z$.*

$$B = B_0 \overset{r_1}{\to} B_1 \overset{r_2}{\to} B_2 \overset{r_3}{\to} B_3 \overset{r_4}{\to} \cdots$$
$$Z\downarrow \quad Z\downarrow \quad Z\downarrow \quad Z\downarrow$$
$$A = A_0 \overset{r_1}{\to} A_1 \overset{r_2}{\to} A_2 \overset{r_3}{\to} A_3 \overset{r_4}{\to} \cdots$$

Fig. 1. A (B, A)-simulation chain.

$$B = B_0 \overset{r_1}{\to} B_1 \overset{r_2}{\to} \cdots \overset{r_{n-1}}{\to} B_{n-1} \overset{r_n}{\to} B_n$$
$$Z\downarrow \quad Z\downarrow \quad \qquad Z\downarrow$$
$$A = A_0 \overset{r_1}{\to} A_1 \overset{r_2}{\to} \cdots \overset{r_{n-1}}{\to} A_{n-1}$$

Fig. 2. A partial (B, A)-simulation chain.

3.2   Subsumption w.r.t. Descriptive Semantics

W.r.t. gfp-semantics, A is subsumed by B iff there is a simulation $Z: \mathcal{G}_T \overset{\rightharpoonup}{\sim} \mathcal{G}_T$ such that $(B, A) \in Z$ (see [2]). W.r.t. descriptive semantics, the simulation Z must satisfy some additional properties for this equivalence to hold. To define these properties, we must introduce some notation.

Definition 5. *The path p_1: $B = B_0 \overset{r_1}{\to} B_1 \overset{r_2}{\to} B_2 \overset{r_3}{\to} B_3 \overset{r_4}{\to} \cdots$ in $\mathcal{G}_T$ is Z-simulated by the path p_2: $A = A_0 \overset{r_1}{\to} A_1 \overset{r_2}{\to} A_2 \overset{r_3}{\to} A_3 \overset{r_4}{\to} \cdots$ in $\mathcal{G}_T$ iff $(B_i, A_i) \in Z$ for all $i \geq 0$. In this case we say that the pair (p_1, p_2) is a (B, A)-simulation chain w.r.t. Z (see Figure 1).*

If $(B, A) \in Z$, then (S2) of Definition 4 implies that, for every infinite path p_1 starting with $B_0 := B$, there is an infinite path p_2 starting with $A_0 := A$ such that p_1 is Z-simulated by p_2. In the following we construct such a simulating path step by step. The main point is, however, that the decision which concept A_n to take in step n should depend only on the partial (B, A)-simulation chain already constructed, and *not* on the parts of the path p_1 not yet considered.

Definition 6. *A partial (B, A)-simulation chain is of the form depicted in Figure 2. A selection function S for A, B and Z assigns to each partial (B, A)-simulation chain of this form a defined concept A_n such that (A_{n-1}, r_n, A_n) is an edge in $\mathcal{G}_T$ and $(B_n, A_n) \in Z$.*

Given a path $B = B_0 \overset{r_1}{\to} B_1 \overset{r_2}{\to} B_2 \overset{r_3}{\to} B_3 \overset{r_4}{\to} \cdots$ and a defined concept A such that $(B, A) \in Z$, one can use a selection function S for A, B and Z to construct a Z-simulating path. In this case we say that the resulting (B, A)-simulation chain is *S-selected.*

Definition 7. *Let A, B be defined concepts in T, and $Z: \mathcal{G}_T \overset{\rightharpoonup}{\sim} \mathcal{G}_T$ a simulation with $(B, A) \in Z$. Then Z is called (B, A)-synchronized iff there exists a selection function S for A, B and Z such that the following holds: for every infinite S-selected (B, A)-simulation chain of the form depicted in Figure 1 there exists an $i \geq 0$ such that $A_i = B_i$.*

We are now ready to state the characterization of subsumption w.r.t. descriptive semantics proved in [2].

Theorem 1. *Let $\mathcal{T}$ be an $\mathcal{EL}$-TBox, and A, B defined concepts in $\mathcal{T}$. Then the following are equivalent:*

1. $A \sqsubseteq_{\mathcal{T}} B$.
2. *There is a (B, A)-synchronized simulation $Z: \mathcal{G}_{\mathcal{T}} \overset{\rightarrow}{\sim} \mathcal{G}_{\mathcal{T}}$ such that $(B, A) \in Z$.*

In [2] it is also shown that, for a given $\mathcal{EL}$-TBox $\mathcal{T}$ and defined concepts A, B in $\mathcal{T}$, the existence of a (B, A)-synchronized simulation $Z: \mathcal{G}_{\mathcal{T}} \overset{\rightarrow}{\sim} \mathcal{G}_{\mathcal{T}}$ with $(B, A) \in Z$ can be decided in polynomial time.

Corollary 1. *Subsumption w.r.t. descriptive semantics in $\mathcal{EL}$ can be decided in polynomial time.*

4   The lcs w.r.t. Descriptive Semantics

Deriving a characterization of the lcs (w.r.t. descriptive semantics) from Theorem 1 is not straightforward. First, we will show that, w.r.t. descriptive semantics, the lcs of two concepts defined in an $\mathcal{EL}$-TBox need not exist. Subsequently, we will introduce possible "candidates" P_k ($k \geq 0$) for the lcs, and show that the lcs exists iff one of these candidates is the lcs. Finally, we will give a sufficient condition for the existence of the lcs.

4.1   The lcs Need Not Exist

Theorem 2. *Let $\mathcal{T}_1 := \{A \equiv \exists r.A, \ B \equiv \exists r.B\}$. Then, A, B in $\mathcal{T}_1$ do not have an lcs.*

Proof. Assume to the contrary that $\mathcal{T}_2$ is a conservative extension of $\mathcal{T}_1$ and that the defined concept E in $\mathcal{T}_2$ is an lcs of A, B in $\mathcal{T}_1$. Let $\mathcal{G}_2 = (V_2, E_2, L_2)$ be the description graph induced by $\mathcal{T}_2$.

First, we show that there cannot be an infinite path in $\mathcal{G}_2$ starting with E. In fact, assume that

$$E = E_0 \overset{r_1}{\rightarrow} E_1 \overset{r_2}{\rightarrow} E_2 \overset{r_3}{\rightarrow} \cdots$$

is such an infinite path. Since $A \sqsubseteq_{\mathcal{T}_1} E$, there is an (E, A)-synchronized simulation $Z_1: \mathcal{G}_2 \overset{\rightarrow}{\sim} \mathcal{G}_2$ such that $(E, A) \in Z_1$. Consequently, the corresponding selection function S_1 can be used to turn the above infinite chain issuing from E into an (E, A)-simulation chain. Since the only edge with source A is the edge (A, r, A), this simulation chain is actually of the form

$$
\begin{array}{ccccccccc}
E = & E_0 & \overset{r}{\rightarrow} & E_1 & \overset{r}{\rightarrow} & E_2 & \overset{r}{\rightarrow} & E_3 & \overset{r}{\rightarrow} \cdots \\
& Z_1\downarrow & & Z_1\downarrow & & Z_1\downarrow & & Z_1\downarrow & \\
& A & \overset{r}{\rightarrow} & A & \overset{r}{\rightarrow} & A & \overset{r}{\rightarrow} & A & \overset{r}{\rightarrow} \cdots
\end{array}
$$

Since Z_1 is (E, A)-synchronized with selection function S_1, this implies that there is an index j_1 such that $E_{j_1} = A$, and thus $E_i = A$ for all $i \geq j_1$.

Analogously, we can show that there is an index j_2 such that $E_{j_2} = B$, and thus $E_i = B$ for all $i \geq j_2$. Since $A \neq B$, this is a contradiction. Thus, we know that there is a positive integer n_0 such that every path in $\mathcal{G}_2$ starting with E has length $\leq n_0$.

Second, we define conservative extensions $\mathcal{T}_n'$ $(n \geq 1)$ of $\mathcal{T}_2$ such that the defined concept F_n in $\mathcal{T}_n'$ is a common subsumer of A, B:

$$\mathcal{T}_n' := \mathcal{T}_2 \cup \{F_n \equiv \exists r.F_{n-1}, ..., F_1 \equiv \exists r.F_0, F_0 \equiv \top\}.$$

It is easy to see that $A \sqsubseteq_{\mathcal{T}_n'} F_n$ and $B \sqsubseteq_{\mathcal{T}_n'} F_n$.

Third, we claim that, for $n > n_0$, $E \not\sqsubseteq_{\mathcal{T}_n'} F_n$. In fact, the path

$$F_n \xrightarrow{r} F_{n-1} \xrightarrow{r} F_{n-2} \xrightarrow{r} \cdots \xrightarrow{r} F_0$$

has length n, and thus it cannot be simulated by any path starting with E. This shows that $E \not\sqsubseteq_{\mathcal{T}_n'} F_n$, and thus contradicts our assumption that E in $\mathcal{T}_2$ is the lcs of A, B in $\mathcal{T}_1$. $\square$

4.2   Characterizing When the lcs Exists

Given an $\mathcal{EL}$-TBox $\mathcal{T}_1$ and defined concepts A, B in $\mathcal{T}_1$, we will define for each $k \geq 0$ a conservative extension $\mathcal{T}_2^{(k)}$ of $\mathcal{T}_1$ containing a defined concept P_k, and show that A, B have an lcs iff there is a k such that P_k is the lcs of A, B. To prove this result, we will need a slight modification of Theorem 1. However, this modified theorem follows easily from the the proof of Theorem 1 given in [2].

Definition 8. *(i) We call a selection function S nice iff it satisfies the following two conditions:*

1. *It is memoryless, i.e., its result A_n depends only on $B_{n-1}, A_{n-1}, r_n, B_n$, and not on the other parts of the partial (B, A)-simulation chain.*
2. *If $B_{n-1} = A_{n-1}$, then its result A_n is just B_n.*

(ii) The simulation relation Z is called strongly (B, A)-synchronized iff there exists a nice selection function S for A, B and Z such that the following holds: for every infinite S-selected (B, A)-simulation chain of the form depicted in Figure 1 there exists an $i \geq 0$ such that $A_i = B_i$.

Corollary 2. *Let $\mathcal{T}$ be an $\mathcal{EL}$-TBox, and A, B be defined concepts in $\mathcal{T}$. Then the following are equivalent:*

1. *$A \sqsubseteq_{\mathcal{T}} B$.*
2. *There is a strongly (B, A)-synchronized simulation $Z: \mathcal{G}_{\mathcal{T}} \overset{\rightarrow}{\sim} \mathcal{G}_{\mathcal{T}}$ such that $(B, A) \in Z$.*

Strongly (B, A)-synchronized simulations satisfy the following property:

Lemma 2. *Let $\mathcal{T}$ be an $\mathcal{EL}$-TBox containing at most n defined concepts, A, B be defined concepts in $\mathcal{T}$, and $Z \colon \mathcal{G}_T \overset{\rightarrow}{\sim} \mathcal{G}_T$ be a strongly (B, A)-synchronized simulation relation. Consider an infinite S-selected (B, A)-simulation chain of the form depicted in Figure 1. Then there exists an $m < n^2$ such that $B_m = A_m$.*

Obviously, the lemma also holds for finite S-selected (B, A)-simulation chains, provided that they are long enough, i.e., of length at least n^2.

Now, let $\mathcal{T}_1$ be an $\mathcal{EL}$-TBox, let $\mathcal{G}_{\mathcal{T}_1} = (N_{def}, E_{\mathcal{T}_1}, L_{\mathcal{T}_1})$ be the corresponding description graph, and let A, B be defined concepts in $\mathcal{T}_1$ (i.e., elements of N_{def}). W.r.t. gfp-semantics, the node (A, B) in the product $\mathcal{G} := \mathcal{G}_{\mathcal{T}_1} \times \mathcal{G}_{\mathcal{T}_1}$ of $\mathcal{G}_{\mathcal{T}_1}$ with itself yields the lcs of A, B [1]. The nodes of $\mathcal{G}$ are pairs (u, v) of nodes of $\mathcal{G}_{\mathcal{T}_1}$; there is an edge $((u, v), r, (u', v'))$ in $\mathcal{G}$ iff $(u, r, u') \in E_{\mathcal{T}_1}$ and $(v, r, v') \in E_{\mathcal{T}_1}$; and the label of (u, v) in $\mathcal{G}$ is $L_{\mathcal{T}_1}(u) \cap L_{\mathcal{T}_1}(v)$.

W.r.t. descriptive semantics, the product graph $\mathcal{G}$ as a whole cannot be part of the lcs of A, B since it may contain cycles reachable from (A, B), which would prevent the subsumption relationship between A and (A, B) to hold. Nevertheless, the lcs must "contain" paths in $\mathcal{G}$ starting with (A, B) up to a certain length k. In order to obtain these paths without also getting the cycles in $\mathcal{G}$, we make copies of the nodes in $\mathcal{G}$ on levels between 1 and k. Actually, we will not need nodes of the form (u, u) since they are represented by the nodes u in $\mathcal{G}_{\mathcal{T}_1}$.

To be more precise, we define

$$\mathcal{P}_k := \{(A, B)^0\} \cup \{(u, v)^n \mid u \neq v, (u, v) \in N_{def} \times N_{def} \text{ and } 1 \leq n \leq k\}.$$

For $p = (u, v)^n \in \mathcal{P}_k$ we call (u, v) the node of p and n the level of p.

The edges of $\mathcal{G}$ induce edges between elements of $\mathcal{P}_k$. To be more precise, we define the set of edges $E_{\mathcal{P}_k}$ as follows: $(p, r, q) \in E_{\mathcal{P}_k}$ iff the following conditions are satisfied:

- $p, q \in \mathcal{P}_k$;
- $p = (u, v)^n$ for some $n, 0 \leq n \leq k$;
- $q = (u', v')^{n+1}$;
- $(u, r, u') \in E_{\mathcal{T}_1}$ and $(v, r, v') \in E_{\mathcal{T}_1}$;

Note that the graph $(\mathcal{P}_k, E_{\mathcal{P}_k})$ is a directed acyclic graph. The only element on level 0 is $(A, B)^0$.

The label of an element of $\mathcal{P}_k$ is the label of its node in the product graph $\mathcal{G}$, i.e., if $p = (u, v)^n \in \mathcal{P}_k$, then

$$L_{\mathcal{P}_k}(p) = L_{\mathcal{T}_1}(u) \cap L_{\mathcal{T}_1}(v).$$

We are now ready to define an $\mathcal{EL}$-description graph $\mathcal{G}_2^{(k)}$ whose corresponding TBox $\mathcal{T}_2^{(k)}$ is a conservative extension of $\mathcal{T}_1$, and which contains a defined concept P_k that is a common subsumer of A, B.

Definition 9. *For all $k \geq 0$, we define $\mathcal{G}_2^{(k)} := (V_2^{(k)}, E_2^{(k)}, L_2^{(k)})$ where*

- $V_2^{(k)} := N_{def} \cup \mathcal{P}_k$;

- $L_2^{(k)} = L_{\mathcal{T}_1} \cup L_{\mathcal{P}_k}$, i.e.

$$L_2^{(k)}(v) := \begin{cases} L_{\mathcal{T}_1}(v) & \text{if } v \in N_{def} \\ L_{\mathcal{P}_k}(v) & \text{if } v \in \mathcal{P}_k \end{cases}$$

- $E_2^{(k)}$ consists of the edges in $E_{\mathcal{T}_1}$ and $E_{\mathcal{P}_k}$, extended by some additional edges from $\mathcal{P}_k$ to N_{def}:

$$E_2^{(k)} := E_{\mathcal{T}_1} \cup E_{\mathcal{P}} \cup \{(p, r, w) \mid p = (u, v)^n \in \mathcal{P}_k, \ w \in N_{def}, \text{ and} \\ (u, r, w) \in E_{\mathcal{T}_1} \text{ and } (v, r, w) \in E_{\mathcal{T}_1}\}.$$

Let $\mathcal{T}_2^{(k)}$ be the $\mathcal{EL}$-TBox such that $\mathcal{G}_2^{(k)} = \mathcal{G}_{\mathcal{T}_2^{(k)}}$. It is easy to see that $\mathcal{T}_2^{(k)}$ is a conservative extension of $\mathcal{T}_1$.

Lemma 3. $A \sqsubseteq_{\mathcal{T}_2^{(k)}} (A, B)^0$ and $B \sqsubseteq_{\mathcal{T}_2^{(k)}} (A, B)^0$.

Proof. To prove $A \sqsubseteq_{\mathcal{T}_2^{(k)}} (A, B)^0$, it is enough to show that there exists an $((A, B)^0, A)$-synchronized simulation $Z: \mathcal{G}_{\mathcal{T}_2^{(k)}} \overset{\rightarrow}{\sim} \mathcal{G}_{\mathcal{T}_2^{(k)}}$ such that $((A, B)^0, A) \in Z$. We define the relation Z as follows:

$$Z := \{(p, u) \mid p \in \mathcal{P}_k, u \in N_{def}, \text{ and the node of } p \text{ is of the form } (u, v)\} \cup \\ \{(u, u) \mid u \in N_{def}\}.$$

In [1] it is shown that Z is indeed an $((A, B)^0, A)$-synchronized simulation such that $((A, B)^0, A) \in Z$. $\square$

What we want to show next is that every common subsumer of A, B also subsumes $(A, B)^0$ in $\mathcal{T}_2^{(k)}$ for an appropriate k. To make this more precise, assume that $\mathcal{T}_2$ is a conservative extension of $\mathcal{T}_1$, and that F is a defined concept in $\mathcal{T}_2$ such that $A \sqsubseteq_{\mathcal{T}_2} F$ and $B \sqsubseteq_{\mathcal{T}_2} F$. For $\mathcal{G}_{\mathcal{T}_2} = (V_2, E_2, L_2)$, this implies that there is

- an (F, A)-synchronized simulation relation $Y_1: \mathcal{G}_{\mathcal{T}_2} \overset{\rightarrow}{\sim} \mathcal{G}_{\mathcal{T}_2}$ with selection function S_1 such that $(F, A) \in Y_1$, and
- an (F, B)-synchronized simulation relation $Y_2: \mathcal{G}_{\mathcal{T}_2} \overset{\rightarrow}{\sim} \mathcal{G}_{\mathcal{T}_2}$ with selection function S_2 such that $(F, B) \in Y_2$.

By Corollary 2 we may assume without loss of generality that the selection functions S_1, S_2 are nice. Consequently, if $k = |V_2|^2$, then Lemma 2 shows that the selection functions S_1, S_2 ensure synchronization after less than k steps.

In the following, let $k := |V_2|^2$. In order to have a subsumption relationship between $(A, B)^0$ in $\mathcal{T}_2^{(k)}$ and F, both must "live" in the same TBox. For this, we simply take the union $\mathcal{T}_3$ of $\mathcal{T}_2^{(k)}$ and $\mathcal{T}_2$. Note that we may assume without loss of generality that the only defined concepts that $\mathcal{T}_2^{(k)}$ and $\mathcal{T}_2$ have in common are the ones from $\mathcal{T}_1$. In fact, none of the new defined concepts in $\mathcal{T}_2^{(k)}$ (i.e., the elements of $\mathcal{P}_k$) lies on a cycle, and thus we can rename them without changing the meaning of these concepts. (Note that the characterization of subsumption given in Theorem 1 implies that only for defined concepts occurring on cycles their actual names are relevant.) Thus, $\mathcal{T}_3$ is a conservative extension of both $\mathcal{T}_2^{(k)}$ and $\mathcal{T}_2$.

Lemma 4. $(A, B)^0 \sqsubseteq_{\mathcal{T}_3} F$

Proof. We must show that there is an $(F, (A, B)^0)$-synchronized simulation relation $Y \colon \mathcal{G}_{\mathcal{T}_3} \overset{\sim}{\to} \mathcal{G}_{\mathcal{T}_3}$ such that $(F, (A, B)^0) \in Y$. The definition of this simulation is based on the "product" of Y_1 and Y_2:

$$Y := \{(u, p) \mid (u, v_1) \in Y_1 \text{ and } (u, v_2) \in Y_2$$
$$\text{where } (v_1, v_2) \text{ is the node of } p \in \mathcal{P}_k \,\} \quad \cup$$
$$\{(u, v) \mid v \in N_{def} \text{ and } (u, v) \in Y_1\}.$$

In [1] it is shown that Y is indeed an $(F, (A, B)^0)$-synchronized simulation such that $(F, (A, B)^0) \in Y$. $\qquad \square$

In the following, we assume without loss of generality that the TBoxes $\mathcal{T}_2^{(k)}$ ($k \geq 0$) are renamed such that they share only the defined concepts of $\mathcal{T}_1$. For example, in addition to the upper index describing the level of a node in $\mathcal{P}_k$ we could add a lower index k. Thus, $(u, v)_k^n$ denotes a node on level n in $\mathcal{P}_k$. For $k \geq 0$, we denote $(A, B)_k^0$ by P_k.

Theorem 3. *Let $\mathcal{T}_1$ be an $\mathcal{EL}$-TBox and A, B defined concepts in $\mathcal{T}_1$. Then A, B in $\mathcal{T}_1$ have an lcs iff there is a k such that P_k in $\mathcal{T}_2^{(k)}$ is the lcs of A, B in $\mathcal{T}_1$.*

Proof. The direction from right to left is trivial. Thus, assume that $\mathcal{T}_2$ is a conservative extension of $\mathcal{T}_1$ and that P in $\mathcal{T}_2$ is the lcs of A, B. We define $k := n^2$ where n is the number of defined concepts in $\mathcal{T}_2$. Let $\mathcal{T}_3$ be the union of $\mathcal{T}_2$ and $\mathcal{T}_2^{(k)}$, where we assume without loss of generality that the only defined concepts shared by $\mathcal{T}_2$ and $\mathcal{T}_2^{(k)}$ are the ones in $\mathcal{T}_1$. Then Lemma 4 shows that $P_k \sqsubseteq_{\mathcal{T}_3} P$.

Since P_k is a common subsumer of A, B by Lemma 3, the fact that P is the least common subsumer of A, B implies that subsumption in the other direction holds as well: $P \sqsubseteq_{\mathcal{T}_3} P_k$. Thus, P and P_k are equivalent, and this implies that P_k is also an lcs of A, B. $\qquad \square$

In [1] it is also shown that the concepts P_k form a decreasing chain w.r.t. subsumption, and that P_k is the lcs of A, B iff it is equivalent to P_{k+i} for all $i \geq 1$.

Example 1. Let us reconsider the TBox $\mathcal{T}_1$ defined in Theorem 2. In this case, the TBoxes $\mathcal{T}_2^{(k)}$ are basically of the form[2]

$$\mathcal{T}_1 \cup \{P_k \equiv \exists r.(A, B)_k^1, \ (A, B)_k^1 \equiv \exists r.(A, B)_k^2, \ \ldots, \ (A, B)_k^{k-1} \equiv \exists r.(A, B)_k^k\},$$

and it is easy to see that there always is a strict subsumption relationship between P_k and P_{k+1} (since P_{k+1} requires an r-chain of length $k + 1$ whereas P_k only requires one of length k).

---

[2] We have restricted the attention to elements of $\mathcal{P}_k$ that are reachable from P_k.

The following is an example where the lcs exists.

Example 2. Let us consider the following TBox

$$\mathcal{T}_1 := \{A \equiv \exists r.A \sqcap \exists r.C, \; B \equiv \exists r.B \sqcap \exists r.C, \; C \equiv \exists r.C\}.$$

In this case, $k = 0$ does the job, and thus the lcs of A, B is P_0:

$$\mathcal{T}_2^{(0)} := \mathcal{T}_1 \cup \{P_0 \equiv \exists r.C\}.$$

In fact, it is easy to see that the path $P_0 \xrightarrow{r} C \xrightarrow{r} C \xrightarrow{r} \cdots$ can simulate any path starting with some P_ℓ for $\ell \geq 1$. Since the infinite paths starting with P_ℓ must eventually also lead to C (after at most ℓ steps), this really yields a synchronized simulation relation.

4.3   A Sufficient Condition for the Existence of the lcs

If we want to use the results from the previous subsection to compute the lcs, we must be able to decide whether there is an index k such that P_k is the lcs of A, B, and if yes we must also be able to compute such a k. Though we strongly conjecture that this is possible, we have not yet found such a procedure. For this reason, we must restrict ourself to give a *sufficient condition* for the lcs of two concepts defined in an $\mathcal{EL}$-TBox to exist.

As before, let $\mathcal{T}_1$ be an $\mathcal{EL}$-TBox, let $\mathcal{G}_{\mathcal{T}_1} = (N_{def}, E_{\mathcal{T}_1}, L_{\mathcal{T}_1})$ be the corresponding description graph, and let A, B be defined concepts in $\mathcal{T}_1$ (i.e., elements of N_{def}). We consider the product $\mathcal{G} := \mathcal{G}_{\mathcal{T}_1} \times \mathcal{G}_{\mathcal{T}_1}$ of $\mathcal{G}_{\mathcal{T}_1}$ with itself. Let $\mathcal{G} = (V, E, L)$.

Definition 10. *We say that (A, B) is synchronized in $\mathcal{T}_1$ iff, for every infinite path $(A, B) = (u_0, v_0) \xrightarrow{r_1} (u_1, v_1) \xrightarrow{r_2} (u_2, v_2) \xrightarrow{r_3} \cdots$ in $\mathcal{G}$, there exists an index $i \geq 0$, such that $u_i = v_i$.*

For example, in the TBox $\mathcal{T}_1$ introduced in Theorem 2, (A, B) is *not* synchronized. The same is true for the TBox defined in Example 2. As another example, consider the TBox $\mathcal{T}_1' := \{A' \equiv \exists r_1.A' \sqcap \exists r.C, \; B' \equiv \exists r_2.B' \sqcap \exists r.C, \; C \equiv \exists r.C\}$. In this TBox, (A', B') is synchronized.

Lemma 5. *Assume that (A, B) is synchronized in $\mathcal{T}_1$, and let $k := |N_{def}|^2$. Then, for every path $(A, B) = (u_0, v_0) \xrightarrow{r_1} (u_1, v_1) \xrightarrow{r_2} (u_2, v_2) \xrightarrow{r_3} \cdots \xrightarrow{r_k} (u_k, v_k)$ in $\mathcal{G}$ of length k, there exists an index $i, 0 \leq i \leq k$ such that $u_i = v_i$.*

As an easy consequence of this lemma we obtain that $k = |N_{def}|^2$ is such that P_k is the lcs of A, B (see [1] for the proof). Thus, the lcs of A, B in $\mathcal{T}_1$ always exists, provided that (A, B) is synchronized in $\mathcal{T}_1$. Our construction of the TBox $\mathcal{T}_2^{(k)}$ is obviously polynomial in k and the size of $\mathcal{T}_1$. Since k is also polynomial in the size of $\mathcal{T}_1$, the size of $\mathcal{T}_2^{(k)}$ is polynomial in the size of $\mathcal{T}_1$.

Theorem 4. *Let T_1 be an $\mathcal{EL}$-TBox, and let A, B be defined concepts in T_1 such that (A, B) is synchronized in T_1. Then the lcs of A, B in T_1 always exists, and it can be computed in polynomial time.*

Example 2 shows that the lcs may exist even if (A, B) is not synchronized in T_1. Thus, this is a sufficient, but not necessary condition for the existence of the lcs. We close this section by showing that this sufficient condition can be decided in polynomial time.

Proposition 3. *Let T_1 be an $\mathcal{EL}$-TBox, and let A, B be defined concepts in T_1. Then it can be decided in polynomial time whether (A, B) is synchronized in T_1.*

Proof. As before, consider the product $\mathcal{G} := \mathcal{G}_{T_1} \times \mathcal{G}_{T_1}$ of $\mathcal{G}_{T_1}$ with itself. Let $\mathcal{G} = (V, E, L)$. We define

$$W_0 := \{(u, u) \mid (u, u) \in V\},$$
$$W_{i+1} := W_i \cup \{(u, v) \mid (u, v) \in V \text{ and all edges with source } (u, v) \text{ in } \mathcal{G}$$
$$\text{lead to elements of } W_i\}, \text{ and}$$
$$W_\infty := \bigcup_{i \geq 0} W_i.$$

Obviously, W_∞ can be computed in time polynomial in the size of $\mathcal{G}$. In [1] it is shown that (A, B) is synchronized in T_1 iff $(A, B) \in W_\infty$. From this, the proposition immediately follows. $\square$

5 Related and Future Work

Cyclic definitions in $\mathcal{EL}$ w.r.t. the three types of semantics introduced by Nebel [9] were investigated in [2]. (A short version of this paper is submitted for publication at another conference.) It was shown that the subsumption problem remains polynomial in all three cases. The main tool in the investigation of cyclic definitions in $\mathcal{EL}$ is a characterization of subsumption through the existence of so-called simulation relations on the graph associated with an $\mathcal{EL}$-terminology.

The characterization of subsumption in $\mathcal{EL}$ w.r.t. gfp-semantics was used in [1] to characterize the lcs w.r.t. gfp-semantics via the product of this graph with itself. This shows that, w.r.t. gfp semantics, the lcs always exists, and that the binary lcs can be computed in polynomial time. The characterization of subsumption w.r.t. gfp-semantics can also be extended to the instance problem in $\mathcal{EL}$. This was used in [1] to show that the most specific concept in $\mathcal{EL}$ with cyclic terminologies interpreted with gfp-semantics always exists, and can be computed in polynomial time. (These results on the lcs and msc in $\mathcal{EL}$ w.r.t. gfp-semantics are submitted for publication at another conference.)

Subsumption is also polynomial w.r.t. descriptive semantics [2]. For the lcs, descriptive semantics is not that well-behaved: the lcs need not exist in general. In addition, we could only give a sufficient condition for the existence of the lcs. If this condition applies, then the lcs can be computed in polynomial time. Thus, one of the main technical problems left open by the present paper is the question how to characterize the cases in which the lcs exists w.r.t. descriptive semantics, and to determine whether in these cases it can always be computed in polynomial time. Another problem that was not addressed by the present paper is the question of how to characterize and compute the most specific concept w.r.t. descriptive semantics.

It should be noted that there are indeed applications where the expressive power of the small DL $\mathcal{EL}$ appears to be sufficient. In fact, SNOMED, the Systematized Nomenclature of Medicine [6] uses $\mathcal{EL}$ [10, 11].

References

1. F. Baader. Least common subsumers, most specific concepts, and role-value-maps in a description logic with existential restrictions and terminological cycles. LTCS-Report 02-07, TU Dresden, Germany, 2002. See http://lat.inf.tu-dresden.de/research/reports.html. A short version will appear in *Proc. IJCAI 2003*.
2. F. Baader. Terminological cycles in a description logic with existential restrictions. LTCS-Report 02-02, Dresden University of Technology, Germany, 2002. See http://lat.inf.tu-dresden.de/research/reports.html. Some of the results in this report will also be published in *Proc. IJCAI 2003*.
3. F. Baader and R. Molitor. Building and structuring description logic knowledge bases using least common subsumers and concept analysis. In *Proc. ICCS 2000*, Springer LNAI 1867, 2000.
4. Franz Baader and Ralf Küsters. Computing the least common subsumer and the most specific concept in the presence of cyclic $\mathcal{ALN}$-concept descriptions. In *Proc. KI'98*, Springer LNAI 1504, 1998.
5. Franz Baader, Ralf Küsters, and Ralf Molitor. Computing least common subsumers in description logics with existential restrictions. In *Proc. IJCAI'99*, 1999.
6. R.A. Cote, D.J. Rothwell, J.L. Palotay, R.S. Beckett, and L. Brochu. The systematized nomenclature of human and veterinary medicine. Technical report, SNOMED International, Northfield, IL: College of American Pathologists, 1993.
7. Monika R. Henzinger, Thomas A. Henzinger, and Peter W. Kopke. Computing simulations on finite and infinite graphs. In *36th Annual Symposium on Foundations of Computer Science*, 1995.
8. Ralf Küsters and Ralf Molitor. Approximating most specific concepts in description logics with existential restrictions. In *Proc. KI 2001*, Springer LNAI 2174, 2001.
9. Bernhard Nebel. Terminological cycles: Semantics and computational properties. In John F. Sowa, editor, *Principles of Semantic Networks*. Morgan Kaufmann, 1991.
10. K.A. Spackman. Managing clinical terminology hierarchies using algorithmic calculation of subsumption: Experience with SNOMED-RT. *J. of the American Medical Informatics Association*, 2000. Fall Symposium Special Issue.
11. K.A. Spackman. Normal forms for description logic expressions of clinical concepts in SNOMED RT. *J. of the American Medical Informatics Association*, 2001. Symposium Supplement.

Creation and Merging of Ontology Top-Levels

Bernhard Ganter[1] and Gerd Stumme[2]

[1] Institute for Algebra, TU Dresden, D–01062 Dresden, Germany
ganter@math.tu-dresden.de
[2] Institute for Applied Informatics and Formal Description Methods (AIFB),
University of Karlsruhe, D–76128 Karlsruhe, Germany
www.aifb.uni-karlsruhe.de/WBS/gst

Abstract. We provide a new method for systematically structuring the top-down level of ontologies. It is based on an interactive, top–down knowledge acquisition process, which assures that the knowledge engineer considers all possible cases while avoiding redundant acquisition. The method is suited especially for creating/merging the top part(s) of the ontologies, where high accuracy is required, and for supporting the merging of two (or more) ontologies on that level.

1   Introduction

Ontologies have been established as a means for conceptually structuring domains of interest and are increasingly used for knowledge sharing. Efficient support in the creation, maintenance and interoperability of ontologies is an essential aspect for their success. Manual ontology engineering using conventional editing tools without support is difficult, labor intensive and error prone. Therefore, several systems and frameworks for supporting the knowledge engineer in the ontology engineering task have been proposed (see Section 2). These approaches often rely on syntactic and semantic matching heuristics which are derived from the behavior of ontology engineers when confronted with the task of creating and manipulating ontologies, i. e. human behaviour is simulated. Although some of them locally use different kinds of logics for comparisons, they do not provide a formal guarantee that the knowledge engineer has considered all relevant aspects in the acquisition phase.

We propose a new method, called ONTEX (Ontology Exploration), for supporting the two tasks of creating and merging ontologies. It relies on the knowledge acquisition technique of Attribute Exploration [5] as developed in the mathematical framework of Formal Concept Analysis [21, 7]. ONTEX guarantees that the knowledge engineer considers all relevant combinations of concepts for the creation/merging process but avoids redundant acquisition.

The first task we address is the *creation of a new ontology*. When ontologies have to be created from scratch, the user needs some guidance how to start. Especially the very first decisions have strong impact on the result, as they determine the overall structure of the ontology. Creation of ontologies from scratch is usually performed top–down. First the most general concepts of the ontology

A. de Moor, W. Lex, and B. Ganter (Eds.): ICCS 2003, LNAI 2746, pp. 131–145, 2003.

are selected. More specific concepts are then added by classifying them in the already present structure. ONTEX supports one part of this creation process, namely the structuring of the conceptual hierarchy on the top–level concepts, and the creation of new concepts on the next level by providing suggestions based on an interactive exploration of the existing structure.

The second task we address is *ontology merging*. With the growing usage of ontologies, the problem of overlapping ontologies occurs more often and becomes critical. It is impossible in practice to provide a single ontology satisfying all users with regard to coverage, precision, actuality, and individualization. Hence the balance between the two conflicting objectives of providing a common knowledge core on the one hand and a representation which reflects closely the respective view of every of the users on the other hand has to be maintained. A solution to this problem is to provide multiple ontologies. But when the need for communication arises, the different ontologies have to be made compatible. Compatibility can be obtained by *merging* the ontologies into a single one. Merging two ontologies means creating a new ontology in a semi-automatic manner by merging concepts of the source ontologies. ONTEX provides an interactive knowledge acquisition technique for merging of the top–level concepts, where design decisions have the most impact on the overall structure of the target ontology.

The use of ONTEX guarantees that the knowledge engineer considers all relevant possibilities both for the creation and for the merging task. However, this guarantee is paid with a certain workload for the knowledge engineer, making it applicable only to relatively small parts of the ontologies at hand. Therefore we propose a two–step, top–down approach. The first step aims at reliably creating/merging the top–part(s) of the ontologies with high accuracy, using ONTEX. In the second step, any heuristics–based approach can be used for creating the remainder of the target ontology with less user interaction. This two–step approach allows for high accuracy for the design of the top–level ontology, which has large impact on the global structure of the resulting ontology, as it is more difficult to modify in a later phase than local decisions on a lower level of detail. On the other hand, it restricts the comparatively high workload on the user to the first, critical phase.

In this paper, we restrict ourselves to the construction of the concept hierarchy. The extension of this approach to relations (based on [23]) is planned for the immediate future. Our approach allows to make use of any background knowledge encoded in propositional logic; especially of axioms which come along with source ontologies that are to be merged.

This paper is organized as follows. In Section 3, we briefly introduce some basic definitions concentrating on a formal definition of what an ontology is and recall the basics of Formal Concept Analysis. In Section 4, the approach is sketched. Technical details are given in Section 5. It is illustrated in Section 6 by an example. Section 7 discusses how the approach is adopted to the task of ontology merging, and illustrates it by an example. In Section 2, related work is discussed. Section 8 summarizes the paper and concludes with an outlook on future work.

2   Related Work

A first approach for supporting the merging of ontologies is described in [8]. There, several heuristics are described for identifying corresponding concepts in different ontologies, e. g. comparing the names and the natural language definitions of two concepts, and checking the closeness of two concepts in the concept hierarchy.

The OntoMorph system [3] offers two kinds of mechanisms for translating and merging ontologies: syntactic rewriting supports the translation between two different knowledge representation languages, semantic rewriting offers means for inference-based transformations. It explicitly allows to violate the preservation of semantics in trade-off for a more flexible transformation mechanism.

In [12] the Chimaera system is described. It provides support for merging of ontological terms from different sources, for checking the coverage and correctness of ontologies and for maintaining ontologies over time. Chimaera offers a broad collection of functions, but the underlying assumptions about structural properties of the ontologies at hand are not made explicit.

Protégé-2000 [13] is a knowledge acquisition tool and ontology editor which can be used for creating ontologies. Prompt [14] is an algorithm for ontology merging and alignment embedded in Protégé 2000. It starts with the identification of matching class names. Based on this initial step an iterative approach is carried out for performing automatic updates, finding resulting conflicts, and making suggestions to remove these conflicts.

OilEd [2] is an editor for ontologies based on the description language FaCT. OntoEdit [17] is an ontology editor based on Frame Logic. These tools make use of inferences for checking the consistency and for deriving new facts from the knowledge base.

About merging, there is also much related work in the database community, especially in the area of federated database systems. The work closest to our approach is described in [15]. There, Formal Concept Analysis is applied to a related problem, namely database schema integration.

FCA–MERGE [19] is a technique for merging ontologies based on Formal Concept Analysis. It offers a structural description of the overall merging process with an underlying mathematical framework. It relies on the existence of instances annotated in both ontologies, and provides alternatively a way to use documents as such instances. Concepts are suggested to be merged iff they have the same extent.

Efficient support for creating ontologies from scratch is topic of current research. In [9–11], for instance, text mining techniques have been discussed for supporting the user in creating ontologies.

Most of the tools described above offer extensive functionalities, often based on syntactic and semantic matching heuristics, which are derived from the behaviour of ontology engineers when confronted with the task of creating/merging ontologies. OntoMorph, Chimarea, OilEd, and OntoEdit use a (description) logics based approach that influences the creation and merging process locally, e. g. checking subsumption relationships between terms and checking for inconsistencies. However, none of these approaches offers a guarantee that all relevant relationships have been considered for modeling during the acquisition phase.

3   Basic Notions

In this section, we briefly introduce some basic definitions. We thereby concentrate on a formal definition of ontologies and recall the basics of Formal Concept Analysis.

3.1   Ontologies

There is no common formal definition of what an ontology is. However, most approaches share a few core items: concepts, a hierarchical IS-A-relation, and further relations. For sake of generality, we do not discuss more specific features like constraints, functions, or axioms here. We follow the definition provided in [19]:

Definition: A *(core) ontology* is a tuple $\mathcal{O} := (C, \mathtt{is\_a}, R, \sigma)$, where C is a set whose elements are called *concepts*, $\mathtt{is\_a}$ is a partial order on C (i.e., a binary relation $\mathtt{is\_a} \subseteq C \times C$ which is reflexive, transitive, and antisymmetric), R is a set whose elements are called *relation names* (or *relations* for short), and $\sigma\colon R \to C^+$ is a function which assigns to each relation name its arity.

As said above, the definition considers the core elements of most languages for ontology representation only. It is possible to map the definition to most types of ontology representation languages.

In this paper, we allow additionally a set $\mathcal{A}$ of propositional logic axioms describing dependencies between the concepts.

3.2   Formal Concept Analysis

We recall the basics of Formal Concept Analysis (FCA) as far as they are needed for this paper. A more extensive overview is given in [7]. To allow a mathematical description of concepts as being composed of extensions and intensions, FCA starts with a *formal context*:

Definition: A *formal context* is a triple $\mathbb{K} := (G, M, I)$, where G is a set of *objects*, M is a set of *attributes*, and I is a binary relation between G and M (i.e. $I \subseteq G \times M$). $(g, m) \in I$ is read "*object g has attribute m*".

From a formal context, (formal) concepts can be derived:

Definition: For $A \subseteq G$, we define $A^I := \{m \in M \mid \forall g \in A\colon (g, m) \in I\}$ and, for $B \subseteq M$, we define $B^I := \{g \in G \mid \forall m \in B\colon (g, m) \in I\}$.

A *formal concept* of a formal context (G, M, I) is defined as a pair (A, B) with $A \subseteq G$, $B \subseteq M$, $A^I = B$ and $B^I = A$. The sets A and B are called the *extent* and the *intent* of the formal concept (A, B). The *subconcept–superconcept relation* is formalized by

$$(A_1, B_1) \leq (A_2, B_2) :\Longleftrightarrow A_1 \subseteq A_2 \quad (\Longleftrightarrow B_1 \supseteq B_2) .$$

The set of all formal concepts of a context $\mathbb{K}$ together with the partial order $\leq$ is always a complete lattice,[1] called the *concept lattice* of $\mathbb{K}$ and denoted by $\mathfrak{B}(\mathbb{K})$.

In what follows, we will make use of the fact that a set $B \subseteq M$ of attributes is a concept intent iff $B = B^{II}$.

A possible confusion might arise from the double use of the word 'concept' in FCA and in ontologies. This comes from the fact that FCA and ontologies are two models for the concept of 'concept' which arose independently. In order to distinguish both notions, *we will always refer to the FCA concepts as 'formal concepts'. The concepts in ontologies are referred to just as 'concepts' or as 'ontology concepts'.* There is no direct counter-part of formal concepts in ontologies. Ontology concepts are best compared to FCA attributes, as both can be considered as unary predicates on the set of objects.

4   Creating Ontologies with OntEx

Our approach is based on the knowledge acquisition technique called *Attribute Exploration* [5]. For a given set of ontology concepts, it determines the lattice of all conjunctions of these concepts. In an interactive process, it asks the user questions of the kind "Is the conjunction of the concepts c_1, c_2, ..., and c_n a subconcept of all of the concepts c'_1, c'_2, ..., and c'_m?" with $n, m \geq 1$. The user can either accept or reject the subsumption. If she rejects, she has to provide a counter-example, i.e., a new object [or a new concept] which belongs to the extent of [is a subconcept of, resp.] all concepts c_1, c_2, ..., and c_n but not to at least one of the concepts c'_1, c'_2, ..., and c'_m. In this way, the list of subsumptions as well as the list of counter-examples grows iteratively, until all pairs of concepts are either in a subconcept–superconcept–relation or there is a counter-example prohibiting this.

The set of counter–examples can be empty at the beginning, or it can already contain some elements, if they are known from the beginning. The same holds for the subsumptions. If subsumptions are known from the beginning, they may be entered before starting the exploration.[2] The algorithm also allows to add further background knowledge expressible in propositional logic. A piece of information is for instance that two (or more) concepts are mutually exclusive.

---

[1] I.e., for each set of formal concepts, there exists always a greatest common subconcept and a least common superconcept.

[2] This is especially important for the merging task, where each ontology already comes along with its own subsumption hierarchy.

The ONTEX approach can be split into three steps:

1. initialization of the exploration contexts,
2. the exploration process,
3. further processing.

In the initialization step, the user has to provide an initial set of concepts she considers to be relevant. How to obtain this initial set is beyond the scope of this paper. It may be determined by other knowledge acquisition techniques as for instance described in [18]. Then one formalizes both the background knowledge (especially the known subsumptions) and the (possibly empty) set of counter–examples. The exploration process comprises the exploration dialogue with the user, consisting of questions as described above. At the end of this process, the lattice of all conjunctions of the input concepts is determined. It contains all input concepts and some new concepts constructed during the process. In the third phase, the user can modify the resulting hierarchy using any ontology editor, eventually supported by some heuristic approach as described in Section 2.

5   The OntEx Method

In this section, we discuss in detail the three steps for creating the concept hierarchy of the (top–level of the) ontology in detail. An example is provided in the next section.

In order to simplify notations, we will not distinguish between objects (instances) and sub-concepts as counter-examples in what follows. In practice, of course this distinction will be tracked, and will be used as additional information for further processing of the exploration results.

5.1   Initialization of the Exploration Contexts

In the initialization step, we formalize both the background knowledge (especially the known subsumptions) and the (possibly empty) set of counter-examples in form of formal contexts. We assume that the user has already fixed some initial set C of ontology concepts (without necessarily any hierarchy information on it) which she considers important. Further concepts may be added at later iterations of the process.

In order to simplify notations, we will not encode the background knowledge in form of axioms, but rather as a list of potential examples.[3] This list contains thus all combinations of attributes which are not excluded by the background

---

[3] Here and in what follows we will not address performance issues – which are of course important – but rather try to keep the explanation as simple as possible. The performance of attribute exploration with background knowledge is analyzed in detail in [6]. There it is shown that the fact that the exploration is restricted to implicational logic (whereas the background knowledge may be given by arbitrary propositional formulae) makes the approach computationally feasible.

knowledge. It is stored as a context $\mathbb{P} := (G, C, I)$, called *frame context*, where C is the initial set of concepts. The set G consists at the first stage of dummy concepts. They will be replaced by concrete concept names (or will be deleted) during the exploration. There is one $g \in G$ for each attribute combination $B \subseteq C$ with $\{g\}^I = B$ unless the combination is known to be impossible according to the eventually given background knowledge.[4]

If we already know of some instances of ontology concepts at the very beginning, then we code this information in two more contexts, $\mathbb{O}^+ := (O, C, I^+)$ and $\mathbb{O}^? := (O, C, I^?)$ with $O \subseteq G$. $\mathbb{O}^+$ encodes which of the objects are known to belong to which concepts, i.e., $(o, c) \in I^+$ iff we know that object o belongs to concept c. $\mathbb{O}^?$ encodes which of the objects are not known not to belong to certain concepts, i.e., $(o, c) \in I^?$ iff we cannot exclude that object o belongs to concept c. In this encoding, we have $I^+ \subseteq I \cap (O \times C) \subseteq I^?$.

5.2   The Exploration Process

In the exploration process, the set G of possible objects will decrease, and the set O of counter–examples will increase, based on the answers given by the user, until the implicational theory of the contexts is equal. Then the exploration process ends; and the resulting implicational theory is the one of the target ontology. In particular, it describes the subsumption hierarchy of the target ontology.

At each step in the interactive process, ONTEX checks if there is any set $X \subseteq C$ of concepts with $X = X^{II} \neq X^{??}$.[5] If so, then such a set is chosen and the user is asked if the implication $\bigwedge X \to \bigwedge(X^{??})$ is universally valid for the ontology (see details below). If no such X exists then the implicational theory of the ontology is completely determined, and there are no more "open questions". The concept lattice of $\mathbb{P}$ will contain the full concept order. But even more: we then know all cases where a conjunction of concepts subsumes another concept. The conjunctions can be understood as *implicit* encodings of new concepts, and will be used later as seeds for new concepts for the target ontology.

When there still exists a set $X \subseteq C$ of concepts with $X = X^{II} \neq X^{??}$, then 'the next such set' is chosen[6] and the user is asked if the conjunction of all concepts in X is a subconcept of all concepts in $X^{??}$ (i.e., if $\bigwedge X \to \bigwedge(X^{??})$ is universally valid). Then the user has to provide a counter-example or has to accept the implication.[7]

Each single input to the exploration process can be given in form of a clause $\bigwedge A \to \bigvee N$ (with $A, N \subseteq C$), which is marked either as *valid* or as *non-valid*.

---

[4] The fact that in the worst case $\mathrm{card}(G) = 2^{\mathrm{card}(C)}$ holds, indicates the importance of formalizing as much background knowledge as possible in order to reduce the size of $\mathbb{P}$.

[5] We write $X^?$ and X^+ instead of $X^{I^?}$ and X^{I^+}, resp. (see Section 3.2).

[6] In [4] (see also [7]), an efficient strategy is described.

[7] Depending on the underlying ontology language, the user can be supported in this task by some additional inference mechanism, for instance a description logics inference engine (see [1]).

The input of a "valid" clause (i.e., the partial acceptance of a subsumption suggested by the system) will cause modifications of $\mathbb{P}$, $\mathbb{O}^+$, and of $\mathbb{O}^?$. From $\mathbb{P}$ all objects not consistent with $\bigwedge A \to \bigvee N$ are removed, and the partial description of the objects $o \in O$, encoded by $\mathbb{O}^+$ and $\mathbb{O}^?$, is updated according to $\bigwedge A \to \bigvee N$.

A clause $\bigwedge A \to \bigvee N$ which is marked as "non–valid" is interpreted as a partial description of a counter–example, i.e. of a new object $o \in O$ with $\{o\}^+ := A$ and $\{o\}^? := C \setminus N$. The consistency of the description of the given counter–example with the background knowledge and the results already obtained has to be checked. This can be done using standard methods of propositional logics and will not be described in this paper. If the user says that the counter–example is a concept, then it may additionally be added to the set C. In this case, it will be considered during the subsequent exploration phase. This optional extension of the set C of concepts provides higher accuracy, as also all combinations with this new concept are considered, but it extends the duration of the knowledge acquisition.

When the user has given his answer, and the exploration contexts are modified as described, the 'next' set $X \subseteq C$ with $X = X^{II} \neq X^{??}$ (according to the modified contexts) is determined[8] and the user is asked the next question.

In the worst case (i.e., when all subsumptions are rejected without exception) and with the worst answering strategy, the number of questions is exponential in the cardinality of the initial set of concepts. In practice, however, this worst–case complexity is far from being reached, since there are numerous dependencies between the concepts. On the other hand, the underlying theory guarantees that the number of accepted subsumptions is minimal, and can thus not be outperformed by any other technique.

5.3   Further Processing

Having finished the exploration process, we have now completely determined the subsumption order on the initial set of concepts, and we know all constraints between these concepts which can be described in implicational logic. This includes in particular the information which combinations of concepts any instances can have, and which combinations are excluded. This information is added to the set of axioms of the ontology.

At this point the highly accurate part of the creation of the target ontology is finished. Now heuristics–based approaches may take over. They may be selected out of the list of tools as described in Section 2.

An additional aspect resulting from our approach is that the possible combinations (conjunctions) of initial concepts are seeds for new concepts. A name for a newly generated concept can be derived from its minimal generators (i.e., the minimal subsets of the set of initial concepts whose conjunctions are equal to the new concept) as described in detail in [19]. The knowledge engineer may accept

---

[8] The technique described in [4,7] guarantees that the order on the sets $X \subseteq C$ is compatible with the modification of the contexts.

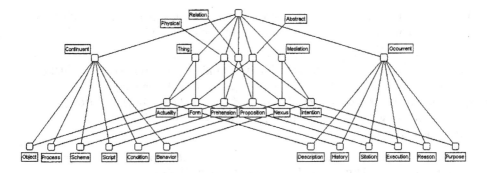

Fig. 1. Sowa's top–level ontology. The bottom concept is not shown for sake of readability.

some or all of the new concepts. Accepting all of them has the advantage that the resulting ontology is a lattice which allows for computation with the concepts (i. e., for any given set of concepts, the computation of its unique least common superconcept and its unique greatest common subconcept; and the computation of implications (functional dependencies) between the concepts). On the other hand, the resulting ontology may become too large. It depends on the individual application how this trade–off will be resolved.

6   An Example Application

In [16], J. F. Sowa elaborated a top–level ontology, based on a careful study of the ontology work of the philosophers Heraclitus, Aristotle, Kant, Peirce, Husserl, Whitehead, and Heidegger. Using the three distinctions 'Physical — Abstract' of Heraclitus, 'Thing — Relation — Mediation' of Peirce, and 'Continuant — Occurrent' of Whitehead, Sowa established the top–level ontology shown in Figure 1. In this section, we show how ONTEX could have helped Sowa in this approach.[9]

We initialize the exploration by setting $C := \{\text{PHYSICAL, ABSTRACT, THING,}$ RELATION, MEDIATION, CONTINUANT, OCCURRENT$\}$, $G := \mathfrak{P}(C)$ (i. e., we do not exclude any combinations of concepts at the beginning), and $I := \{(g, c) \in G \times C \mid c \in g\}$. Finally, we let $O := \emptyset$ and $I^+ := I^? := \emptyset$. The contexts are now determined, and the exploration dialogue can begin:

The empty set is the first set satisfying $X = X^{II} \neq X^{??}$. For $X = \emptyset$, we have $X^{II} = \emptyset$ and $X^{??} = C$.

Q: Is the conjunction of no concepts at all (i. e., the top concept) a subconcept of all of the concepts PHYSICAL, ABSTRACT, THING, RELATION, MEDIATION, CONTINUANT, and OCCURRENT?

---

[9] In [22], the same example has been used to explain 'Simply Implicational Theories', a formalism for supporting empirical theory building within the framework of Formal Concept Analysis.

A: No. A counter–example is the new concept OBJECT, which is subconcept of the top concept, but beside that only of the concepts CONTINUANT, THING, and PHYSICAL.

The set O is extended by OBJECT,[10] and I^+ and $I^?$ are extended to {(OBJECT, CONTINUANT),(OBJECT,THING),(OBJECT,PHYSICAL)}. One might also add the new concept OBJECT to the set C, but we do not make use of this option here.[11]

We choose again $X := \emptyset$, since now $X^{II} = \emptyset$ and $X^{??} =$ {CONTINUANT, THING, PHYSICAL}.

Q: Is the conjunction of no concepts (i. e., the top concept) a subconcept of all of the concepts PHYSICAL, THING, and CONTINUANT?

A: No. A counter–example is the new concept PURPOSE, which is subconcept of the top concept, but beside that only of the concepts OCCURRENT, MEDIATION, and ABSTRACT (and is hence not subconcept of any of the concepts mentioned).

The set O is extended by PURPOSE, and I^+ and $I^?$ are extended by the three tuples (PURPOSE,OCCURRENT), (PURPOSE,MEDIATION), and (PURPOSE,ABSTRACT).

Now we have $X^{II} = \emptyset = X^{??}$ for $X = \emptyset$, hence the next set has to be chosen. With the strategy described in [5], we obtain $X =$ {OCCURRENT} (with $X^{II} =$ {OCCURRENT} and $X^{??} =$ {OCCURRENT, MEDIATION, ABSTRACT}).

Q: Is OCCURRENT a subconcept of all of the concepts MEDIATION and ABSTRACT?

A: No. A counter–example is the new concept DESCRIPTION, which is subconcept of OCCURRENT, but beside that only of the concepts THING and PHYSICAL (and is hence not subconcept of any of the concepts mentioned).

Again O, I^+, and $I^?$ are extended according to the rules. The exploration continues this manner. The first question we encounter which will be accepted is the following:

Q: Is the conjunction of RELATION and MEDIATION a subconcept of all other concepts?

A: Yes (because these two concepts are considered to be disjoint and their conjunction is thus equal to the "absurd" bottom element).

Now the set G is decreased by subtracting all its elements g with {RELATION, MEDIATION$\subseteq g$; and from the relation I are subtracted all tuples having these sets as first component.

This way, the exploration takes on, until no sets X remain fulfilling the condition $X = X^{II} \neq X^{??}$. After 23 questions in total, we obtain the concept hierarchy shown in Figure 2.

---

[10] In the set G, OBJECT is identified with {CONTINUANT, THING, PHYSICAL}$\in G$.

[11] Choosing this option would lead to some more questions, and would in the end result in the additional observation that OBJECT is in fact the conjunction (i. e., the greatest common subconcept) of the three concepts CONTINUANT, THING, and PHYSICAL, and not just an arbitrary subconcept of them.

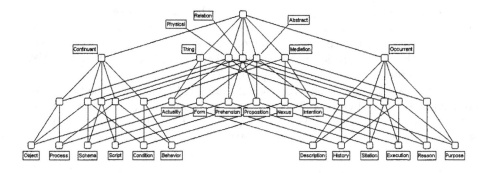

Fig. 2. The result of ONTEX based on the categories of Heraclitus, Peirce, and White-head. The bottom concept is not shown for sake of readability.

In the first level, we see the concepts we started with. In the second and third level, we see the concepts which were given as counter-examples. We can for instance see that BEHAVIOR is a subconcept of the three concepts CONTIN-UANT, ABSTRACT, and MEDIATION. (If we had decided to add BEHAVIOR to the set C during the exploration, we would additionally know that it is indeed equal to their greatest common subsumer.) The ten concepts without label are formal concepts in the sense of FCA. If we discard them, then we obtain exactly Sowa's top–level ontology in Figure 1. However, they can also be used as further concepts of the top–level ontology. It is then up to the knowledge engineer to provide names for them. If all these concepts are accepted, we have the additional advantage that one can compute with the concepts, since the resulting concept hierarchy is a lattice: each pair of concepts has a meet (i. e., a unique greatest common subconcept) and a join (i. e., a unique least common superconcept).

In the diagram, we can see that none of the concepts we started with is subsumed by any other one. By studying the diagram closer, we discover that all possible combinations of concepts out of each of the sets of categories of Heraclitus, Peirce, and Whitehead are realized as a new concept. Hence the sets of categories of the three philosophers are truly orthogonal to each other. According to Sowa (see Figure 1), though, the combination of one category of Whitehead with one category of either Heraclitus or Peirce (e. g., OCCURRENT and MEDIATION) is not a concept of its own. This is of course a philosophical question, and cannot be solved by our algorithm. However, ONTEX makes these potential concepts explicit and accessible to discussion.

7   Ontology Merging with OntEx

In this section, we discuss the use of ONTEX for ontology merging. The main dif-ference to the task of creating an ontology from scratch is the initialization phase. The input of the merging process are two[12] ontologies $\mathcal{O}_1 := (C_1, \texttt{is\_a}_1, R_1, \sigma_1)$

---

[12] The approach is applicable to more than two ontologies simultanuously.

and $\mathcal{O}_2 := (C_2, \text{is\_a}_2, R_2, \sigma_2)$, eventually together with two sets $\mathcal{A}_1$ and $\mathcal{A}_2$ of propositional logics axioms. The output is then a target ontology $\mathcal{O}_T := (C_T, \text{is\_a}_T, R_T, \sigma_T)$ together with a set of axioms $\mathcal{A}_T$. As said above, we will not deal with the exploration of the relations here. Their treatment by ONTEX is subject to further research. We simply let $R_T := R_1 \cup R_2$.

We illustrate the approach with a small example. It consists of the two mini-ontologies shown in Figure 3. The two ontologies come along with two sets of axioms, namely $\mathcal{A}_1 := \{\emptyset \to \text{ROOT1}, \neg(\text{AREA1} \wedge \text{HOTEL1})\}$ and $\mathcal{A}_2 := \{\emptyset \to \text{ROOT2}, \neg(\text{REGION2} \wedge \text{ACCOMMODATION2})\}$.

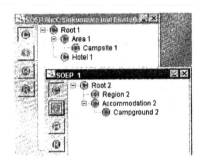

Fig. 3. The two examples of source ontologies.

7.1 Initialization of the Exploration Contexts

We have to initialize the frame context $\mathbb{P} := (G, C, I)$ and the two contexts $\mathbb{O}^+ := (O, C, I^+)$ and $\mathbb{O}^? := (O, C, I^?)$ according to the given knowledge. Let first of all $C := C_1 \dot\cup C_2$, $O := \emptyset$, and $I^+ := I^? := \emptyset$. The set G first contains $2^{\text{card}(C)}$ objects such that, for each $B \subseteq C$, there exists exactly one $g \in G$ with $\{g\}' = B$.

Then the sets $\mathcal{A}_1$ and $\mathcal{A}_2$ of axioms are turned into a set of clauses marked as valid resp. invalid (using standard logic techniques). For each pair $(c, c') \in (C_1 \times C_1) \cup (C_2 \times C_2)$ of concepts in the two ontologies where c is an immediate subconcept of c' (i.e., a lower cover in the is_a hierarchy), we add $c \to c'$ as valid and $c' \to c$ as invalid clause. The contexts $\mathbb{P}$, $\mathbb{O}^+$ and $\mathbb{O}^?$ are then modified as described in the fourth paragraph of Section 5.2. If the two ontologies (or one of them) come along with instances, then they are added to the set O, and I^+ and $I^?$ are updated as described for new counter–examples in Section 5.2. With the three contexts initialized this way, the exploration starts.

7.2 The Exploration Process

The exploration dialogue follows the same rules as described in Section 5.2. For our example, this means that the first question is as follows:

Q: Is HOTEL1 a subconcept of all of REGION2, CAMPGROUND2 and ACCOM-
MODATION2?
A: No. A counter–example is 'Raffles Hotel'.

The sets O and I^+ are extended accordingly: The set O is extended by the object
'Raffles Hotel', and the relations I^+ and $I^?$ by $\{$'Raffles Hotel'$\}\times\{$ROOT1, HO-
TEL1, ROOT2, ACCOMMODATION2, $\}$.

Q: Is HOTEL1 a subconcept of ACCOMMODATION2?
A: Yes.

Hence the set G of the frame context $\mathbb{P}$ is pruned by deleting all objects $g \in G$
where $\{g\}'$ contradicts HOTEL1 $\to$ ACCOMMODATION2 .

The exploration continues until all open questions are solved. Finally, the set
$\mathcal{A}_T$ of axioms of the target ontology is obtained (without any further user interac-
tion) by $\mathcal{A}_T := \mathcal{A}_1 \cup \mathcal{A}_2 \cup \{c_1 \leftrightarrow c_2 \mid c_1 \in C_1$ and $c_2 \in C_2$ are merged in the target
ontology$\}$. In our example, we have

$$\mathcal{A}_T := \mathcal{A}_1 \cup \mathcal{A}_2 \cup \{\text{ROOT1} \leftrightarrow \text{ROOT2}, \text{CAMPSITE1} \leftrightarrow \text{CAMPGROUND2}\} \ .$$

The is_a_T hierarchy is shown in Figure 4, together with the first counter–
example acquired during the exploration process. In total, there are two more
counter–examples: 'Yosemite Camping', being a counter–example to both CAMP-
SITE1 $\to$ REGION2 and CAMPGROUND2 $\to$ HOTEL1; and 'Tuscany', being a
counter–example to REGION2 $\to$ HOTEL1.

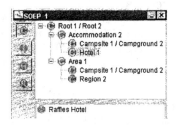

Fig. 4. The target ontology obtained by ONTEX.

7.3 Further Processing

For this third step, all remarks made in Section 5.3 are valid. Additionally,
ONTEX may have determined pairs of concepts of the source ontologies which
are merged in the target ontology (as ROOT1/ROOT2 and CAMPSITE1/CAMP-
GROUND2 in our example). In the preprocessing step, these concepts can serve as
starting point for further merging of concepts and relations by addressing next
concepts which are close to them in the source ontologies. This approach is for
instance realized in Prompt [14] (but with a simpler, more straightforward way
of obtaining the initially merged concepts).

8   Conclusion and Future Work

In this paper, we presented OntEx (Ontology Exploration), a technique for creating/merging ontologies with high accuracy. The technique is suited especially for small parts of the ontologies — especially its top parts — where high quality results are required. Its results can be used as starting point for heuristics–based approaches for dealing with the remainder of the ontologies.

In the paper, we described the three steps of the technique: the initialization of the exploration contexts, the exploration process, and the further processing. We also discussed how the approach can be used for merging (the top levels of) ontologies. The paper described the underlying assumptions and discussed the overall methodology.

Future work includes the closer integration of the method in the KAON ontology engineering environment.[13] It is also planned to extend this approach to the creation/merging of the ontology relations by combining the Attribute Exploration technique with Rule exploration [23], which allows for first order literals instead of binary attributes. Here, decidability will be an important issue, hinting at the use of some description logic.

References

1. F. Baader: Computing a minimal representation of the subsumption lattice of all conjunctions of concept defined in a terminology. In: G. Ellis, R. A. Levinson, A. Fall, V. Dahl (eds.): *Proc. Intl. KRUSE Symposium*, August 11–13, 1995, UCSC, Santa Cruz 1995, 168–178
2. S. Bechhofer, I. Horrocks, C. Goble, R. Stevens: OilEd: A Reason-able Ontology Editor for the Semantic Web, In: *KI–2001: Advances in Artificial Intelligence*, LNAI **2174**, Springer, Heidelberg 2001, 396–408
3. H. Chalupsky: OntoMorph: A translation system for symbolic knowledge. *Proc. KR '00*, Breckenridge, CO, USA, 471–482.
4. B. Ganter: Algorithmen zur Formalen Begriffsanalyse. In: B. Ganter, R. Wille, K. E. Wolff (eds.): *Beiträge zur Formalen Begriffsanalyse*, B. I.–Wissenschaftsverlag, Mannheim 1987, 241–254
5. B. Ganter: Attribute Exploration with Background Knowledge. *TCS* **217**(2), 1999, 215–233
6. B. Ganter, R. Krauße: *Pseudo models and propositional Horn inference*. Preprint MATH-AL-15-1999, TU Dresden 1999
7. B. Ganter, R. Wille: *Formal Concept Analysis: mathematical foundations*. Springer, Berlin–Heidelberg 1999
8. E. Hovy: Combining and standardizing large-scale, practical ontologies for machine translation and other uses. *Proc. 1st Intl. Conf. on Language Resources and Evaluation*, Granada, Spain, May 1998.
9. A. Mädche: *Ontology Learning for the Semantic Web*. PhD thesis, Universität Karlsruhe. Kluwer, Dordrecht 2002
10. A. Mädche, S. Staab: Semi-automatic engineering of ontologies from Text. *Proc. 12th Intl. Conf. on Software Engineering and Knowledge Engineering (SEKE'2000)*

---

[13] kaon.semanticweb.org

11. A. Mädche, S. Staab: Discovering conceptual relations from text. *Proc. 14th European Conference on Artificial Intelligence (ECAI 2000)*, IOS Press, Amsterdam, 2000
12. D. L. McGuinness, R. Fikes, J. Rice, and S. Wilder: An environment for merging and testing large Ontologies. *Proc. KR '00*, 483–493.
13. N. F. Noy, R. Fergerson, M. Musen: The Knowledge Model of Protégé-2000: Combining Interoperability and Flexibility, *Proc. EKAW 2000*, LNCS **1937**, Springer, Heidelberg 2000, 17–32
14. N. F. Noy, M. A. Musen: PROMPT: algorithm and tool for automated ontology merging and alignment. *Proc. AAAI '00*, 450–455
15. I. Schmitt, G. Saake: Merging inheritance hierarchies for database integration. *Proc. CoopIS'98*, IEEE Computer Science Press, 322–331.
16. J. F. Sowa: *Knowledge Representation: Logical, Philosophical, and Computational Foundations*. Brooks Cole Publ., Pacific Grove, CA, 2000
17. S. Staab, A. Mädche: Ontology engineering beyond the modeling of concepts and relations. *Proc. ECAI'2000 workshop on application of ontologies and problem-solving methods*, IOS Press, Amsterdam 2000
18. S. Staab, H.-P. Schnurr, R. Studer, Y. Sure: Knowledge Processes and Ontologies. *IEEE Intelligent Systems* **16**(1), 2001
19. G. Stumme, A. Mädche: FCA–Merge: Bottom-Up Merging of Ontologies. *Proc. 17th Intl. Conf. on Artificial Intelligence (IJCAI '01)*. Seattle, WA, USA, 2001, 225–230
20. Y. Sure, C. Boyens: OntoKick – Ignition for Ontologies. Poster Session at: *WI-IF 2001 — 5th International Conference WirtschaftsInformatik*, Sep. 19–21, 2001, Augsburg, Germany
21. R. Wille: Restructuring lattice theory: an approach based on hierarchies of concepts. In: I. Rival (ed.): *Ordered sets*. Reidel, Dordrecht 1982, 445–470
22. S. Strahringer, R. Wille, U. Wille: Mathematical Support for Empirical Theory Building. In: H. Delugach, G. Stumme (Eds.): *Conceptual Structures: Broadening the Base*. Proc. ICCS '01. LNAI **2120**, Springer, Heidelberg 2001, 169–186
23. M. Zickwolff: *Rule Exploration: First Order Logic in Formal Concept Analysis*. PhD thesis, TH Darmstadt 1991

Dynamic Type Creation
in Metaphor Interpretation and Analogical Reasoning:
A Case-Study with WordNet

Tony Veale

Department of Computer Science,
University College Dublin, Belfield, Dublin 6, Ireland.
Tony.veale@UCD.ie
http://www.cs.ucd.ie/staff/tveale/home/

Abstract. Metaphor and Analogy are perhaps the most challenging phenomena for a conceptual representation to facilitate, since by their very nature they seek to stretch the boundaries of domain description and dynamically establish new ways of determining inter-domain similarity. This research considers the problem of how a conceptual system structured around a central taxonomy can dynamically create new categories or types to understand novel metaphors and analogies. This theoretical perspective yields a practical method for dynamic type creation within WordNet.

1 Introduction

This paper considers the dynamic nature of the representations required to facilitate two of the most knowledge-hungry processes in a conceptual system, metaphor interpretation and analogical reasoning. These processes are interesting because they often exploit the latent similarities between domains that have not been explicitly represented in the underlying conceptual structure [8], thus revealing the inadequacies of such structures. In particular, because metaphors and analogies are used to create new ways of thinking about familiar things, they reveal the essential fluidity of the categories we use to structure the world [9]. This fluidity contrasts sharply with the rigidity of the taxonomies that have been traditionally employed to organize our category systems [4].

Taxonomies have, since antiquity [1], provided a systematic means of hierarchical decomposition of knowledge, whereby a domain is successively dissected via differentiation into smaller pockets of related concepts. Rich differentiation leads to effective clustering, so that similar concepts become localized to the same region of the taxonomy. This locality not only makes the categorial similarity of different ideas easier to assess computationally, it also means that the elements of a domain tend to be clustered around the same parent types, which can thus act as indices into the domain for effective analogical mapping. Indeed, the first account of metaphor as a conceptual process, as offered by Aristotle in his *Poetics* [1], was wholly taxonomic.

A. de Moor, W. Lex, and B. Ganter (Eds.): ICCS 2003, LNAI 2746, pp. 146–159, 2003.

In the Aristotelian scheme, two concepts can be metaphorically or analogically connected if a common taxonomic parent can be found to unify them both. The crucial role of a central taxonomic backbone in organizing knowledge survives today in such large-scale ontologies as Cyc [2], a common-sense ontology for general reasoning, and WordNet [3], a psycholinguistically motivated lexical database of English. The Aristotelian view of taxonomic metaphor also continues to exert considerable influence in computer theories, as demonstrated by [4] and [5].

Yet, if a taxonomy is to be a driving force in the understanding of metaphor and analogy, it must anticipate every possible point of comparison between every pair of domains. However, even to suggest that such an exhaustive taxonomy is possible – and the idea certainly raises grave concerns about tractability – would be to diminish the role of metaphor as a tool for affecting change in our category systems. To resolve this contradiction, authors such as Eileen Cornell Way [4] have argued for the importance of a dynamic type hierarchy (DTH) as a taxonomic backbone for conceptual structure. Such a taxonomy would dynamically reveal new types in response to appropriate metaphors. For example, Way [4] gives as an example "Nixon is the submarine of world politics", and suggests that this metaphor is resolved by the dynamic type *ThingsWhichBehaveInASecretOrHiddenManner*. However, as useful as a dynamic hierarchy would be for metaphor, Way does not suggest an empirical means of constructing such a DTH, which effectively leaves the issue of exhaustiveness, and all it entails for computational tractability, unresolved.

This paper describes an empirical means of constructing a DTH that dynamically generates new taxonomic types in response to challenging analogies and metaphors. The underlying static type hierarchy is provided by WordNet 1.6, while dynamic types are extracted when needed from the textual glosses provided by the designers of WordNet[1]. In addition, we identify an important class of taxonomic type we dub an "analogical pivot", and show how types in existing taxonomies like WordNet and Cyc, which contain relatively few such pivots naturally, can be automatically upgraded into pivots by the addition of dynamic types, further facilitating the processes of analogical retrieval and mapping.

2 Analogical Pivots

Taxonomic systematicity implies that related or analogous domains should be differentiated in the same ways, so that similarity judgments in each domain will be comparable. But in very large taxonomies, this systematicity is often lacking. For example, in WordNet 1.6, the concept {*alphabet*}[2] is differentiated culturally into {*Greek_alphabet*} and {*Hebrew_alphabet*}, but the concept {*letter, alphabetic_character*} is not

---

[1] Note that WordNet does not enforce a type/instance distinction, and so uses the *isa* relation to denote both subset relations and membership relations.

[2] Each concept in WordNet is unambiguously denoted by a *synset* of synonymous words that can all be used to denote the same underlying concept. We do not list all the members of a synset when it clear what entity/type is being denoted.

similarly differentiated into {*Greek_letter*} and {*Hebrew_letter*}. Rather, every letter of each alphabet, such as {*alpha*} and {*aleph*}, is located under exactly the same hypernym, {*letter, alphabetic_character*}. This means that on structural grounds alone, each letter is equally similar to every other letter, no matter what alphabet they belong to (e.g., *alpha* is as similar to *aleph* as it is to *beta*). But more than similarity judgment is impaired: crucially, a lack of systematicity and symmetry undermines another core rationale of taxonomic structure, the ability to recognize analogies and metaphors. For instance, the structure of WordNet 1.6 is not sufficiently differentiated for analogies like "what is the Jewish gamma?" *(gimel)* or "who is the Viking Ares" *(Tyr)*.

Consider the analogical compound "Hindu Zeus" and how one might interpret it using WordNet. The goal is to find a counterpart for the source concept Zeus (the supreme deity of the Greek pantheon) in the target domain of Hinduism. In WordNet 1.6, {*Zeus*} is a daughter of {*Greek_deity*}, which is turn is a daughter of {*deity, god*}. Now, because WordNet also defines an entry for {*Hindu_deity*}, it requires just a simple composition of ideas to determine that the "Hindu Zeus" will be daughter of the type {*Hindu_deity*}. More generally, we simply find the lowest parent of the head term ("Zeus") that, when concatenated with the modifier term ("Hindu") or some synonym thereof, yields an existing WordNet concept. We dub this type, here {*deity, god*}, the pivot of the analogy, since the mapping process can use this pivot to construct a target counterpart of the source concept that significantly narrows the space of possible correspondences. So the Hindu counterpart of Zeus is a daughter of {*Hindu_deity*}, and the precise one can be chosen on the basis of other types (such as {*supreme_deity*}) that encode finer differentiating criteria. Fig. 1 presents a schematic view of this process.

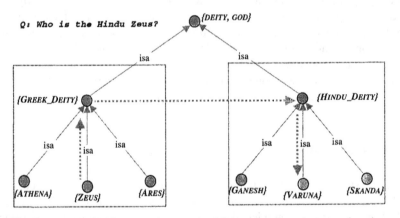

Fig. 1. An example of taxonomic structure driving an analogy between domains

Compare this approach with the conventional one of taxonomic reconciliation, due to Aristotle[1], in which two entities or types can be considered analogous if they share a common superordinate. This approach still finds considerable traction in computational models today (e.g., see [4,5]), but it is easily trivialized: in a well designed taxonomy, any two entities or types will always share at least one superordinate (even if it is the root type), and so any two concepts will always be potential analogues in such a system. The current approach uses a much stricter notion of taxonomic analogy: two types are potentially analogous if they each possess superordinates that are themselves analogous differentiations of the same direct parent (the pivot of the analogy). The approach is also constructive: it indicates how the target counterpart of the pivot, {*Hindu_deity*} is to be constructed from the source analogue (Zeus). Thus, Zeus and Varuna are analogous because {*Greek_deity*} and {*Hindu_deity*} are analogous, by virtue of being different domain sub-types of the same pivot. This constraint is the taxonomic equivalent of the squaring rule described in [6] to ensure that there is structural support for every analogical mapping.

2.1 Conceptual Association

More formally, an analogical pivot is an interior type of a taxonomy whose daughter types (hyponyms) are explicitly differentiated to extend into different domains of knowledge. Pivots thus sit at the junctures of different domains and act as the most effective signposts into those domains when performing an analogical mapping. So the key to the mapping process is to first locate the pivot of the analogy and then follow the labeled signposts into the target domain. For robust understanding, every synonym of the target modifier should be considered when following these signposts, so that "Hindoo Zeus" or even "Hindustani Zeus" will point the same target taxonomy. However, for the mapping to effectively navigate from a pivot in this way, we require a more powerful notion than simple synonymy.

We thus broaden the notion of synonymy to that of symmetric associativity. By definition, synonyms are symmetric associates of each other since one can be substituted for the other without loss of meaning (e.g., "Moslem" and "Muslim"). We broaden this notion to include terms that are so closely correlated in meaning that one can be used as a metonymic proxy for the other (e.g., "Muslim", "Islam", and "Koran"). This associativity can be determined statistically from a corpus, but a simpler and more principled method involves using the WordNet entries themselves.

Definition: The *symmetric associates* of a word X comprises the set of syno-   (1)
nyms of X, as well as the set of each word Y that appears in a definition/gloss
of a sense of X such that X also appears in the definition of an individual
sense of Y

Thus "Islam" is a symmetric associate of "Muslim" since the former occurs in a definition of the latter and vice versa. Similarly by this reckoning, the symmetric associates of "Hindu" are {*Hindu, Hindoo, Hinduism, Hindustan Hindustani, Trimurti*}, where "Trimurti" denotes a triad of divinities in Hindu mythology. By using the symmetric associations of the target to differentiate the pivot, the mapping process can understand analogies as allusive as "Trimurti Zeus".

The above approach is still very fragile as natural pivots like {*deity, god*} are extremely rare in WordNet, since it has not been explicitly constructed for analogical purposes. As noted earlier, the WordNet concept {*letter, alphabetic_character*} is not culturally differentiated, so a mapping cannot be constructed for "Jewish delta" → {*Hebrew_letter*}. Furthermore, even when pivots do exist to facilitate the mapping, what is produced is a target hypernym rather than a specific domain counterpart. One still needs to go from {*Hindu_deity*} to {*Varuna*} (like Zeus a supreme cosmic deity, but of Hinduism), or from {*Hebrew_letter*} to {*daleth*} (like "delta" the fourth letter, but of the Hebrew alphabet).

3   Creating Dynamic Types

Both problems can be solved by adding additional differentiating structure to the taxonomy. These new types will dynamically dissect the taxonomy in novel ways as new metaphors and analogies attempt to establish points of comparison between domains. The dynamic creation of these types will, in the process, convert the parents of these types into the analogical pivots that are needed to direct the interpretation of analogies and metaphors from the source to the target domain.

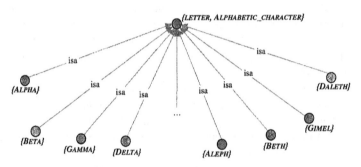

Fig. 2. The impoverished *{letter, alphabetic_character}* taxonomy in WordNet 1.6

For example, the creation of new types like {*Greek_letter*} and {*Hebrew_letter*} will transform {*letter, alphabetic_character*} into a pivot that extends into the Greek and Hebrew domains. Types such as these act as signposts from the pivot into specialized areas of the taxonomy and thus allow the first cut of the analogical mapping to occur. In contrast, other dynamic types may be less ambitious: a new type like {*1ˢᵗ_letter*} will unite just two concepts, {*alpha*} and {*aleph*}, and a new type like {*thunder_deity*} will also unite just two concepts, {*thor*} and {*donar*}. However, these lower-level types allow for finer-grained mapping within the target domain once the appropriate area of the taxonomy has been identified using the pivot.

Fig. 2 illustrates the situation as found in WordNet 1.6 as it concerns the representation of the {*letter, alphabetic_character*} domain. Note how no differentiating structure exists within the taxonomy to allow an analogizer even to discriminate the letters of one alphabet from the other, which makes mapping of domain counterparts impossible. Compare this structure with that of Fig. 3, which illustrates the most desirable

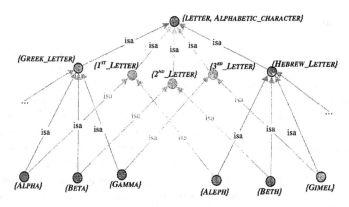

Fig. 3. The taxonomic structure of *{letter, alphabetic_character}* becomes a richly structured lattice when enriched with a variety of new types like *{Greek_letter}* and *{1ˢᵗ_letter}*

state of the taxonomy from an analogy and metaphor perspective. In this reworking, the {*letter, alphabetic_character*} concept is not only differentiated by domain into {*Greek_letter*} and {*Hebrew_letter*}, it is simultaneously differentiated by relative letter position. This structure is sufficient to allow a near-isomorphic mapping to be generated from one alphabet to another (with the exception of extra letters that have no true analogue in the other domain), by first mapping from one alphabet system to another, and then mapping from one relative position to another.

We do not refer to new types like {*1ˢᵗ_letter*}, or even {*Greek_letter*}, as pivots; rather, it is their existing parent, {*letter, alphabetic_character*}, that becomes a pivot after these types have been added. Once a type has been sufficiently differentiated by a number of sub-types, it can act as a sign-posted crossroads between multiple domains, and thus facilitate precise mappings of entities from those domains. When enough pivots are in place, a taxonomy becomes a decision lattice for metaphor and analogy. In Fig. 3 above, only one entity lies at the intersection of {*Hebrew_letter*} and {*1ˢᵗ_letter*}, only one at the intersection of {*Hebrew_letter*} and {*2ⁿᵈ_letter*}, and so on, so with this lattice, a 1-to-1 mapping of entities from one alphabet to another can be generated. It is also possible to convert such a lattice into a decision tree by labeling the arcs of the taxonomy appropriately. One such decision tree for a fragment of the {*deity, god*} sub-taxonomy in WordNet 1.6 is illustrated in Fig. 4.

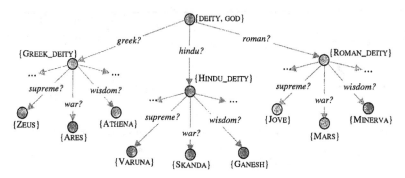

Fig. 4. A decision-tree perspective on the sub-taxonomy for *{deity, god}* in WordNet

3.1 Type Creation as Feature Reification

Enhancing the differentiating power of WordNet is essentially a task of feature reification. WordNet (like other taxonomies, such as Cyc [2]) expresses some of its structure explicitly, via isa-links, and some of it implicitly, in textual glosses intended for human rather than machine consumption. Fortunately, these glosses are consistent enough to permit automatic extraction of structural features (e.g., see [7], who extract lateral connections between concepts from these glosses). What is needed is a means to recognize the word features in these glosses with the most analogical potential, so that they may be lifted to create new taxonomic types. The noun sense glosses of WordNet 1.6 collectively contain over 40,000 unique content words, but clearly only a small fraction of these words can be profitably reified. We thus employ two broad criteria to identify the words worth reifying, *differentiation potential* and *alignment potential*:

Definition: A lemmatized word-form has *differentiation potential* if it occurs   (2)
in more than one gloss, but not in too many glosses (e.g., more than 1000).
Additionally, there must be a precedent for using the word as an explicit differentiator in at least one existing taxonomic entry

Definition: A word-form has *alignment potential* if it can be found in multi-   (3)
ple places in the taxonomy at the same relative depth from a pivot

Consider the word "wisdom", which occurs in 20 different WordNet glosses, enough to demonstrate cross-domain potential but not too many to suggest over-generalization. Additionally, there is a WordNet precedent, {*wisdom_tooth*}, for its explicit use as a differentiator. And of the concepts that "wisdom" is used to gloss, at least three – {*Athena*}, {*Ganesh*} and {*Minerva*} – are grand-daughters of the concept {*deity, god*}. As shown in Fig. 5., the word "wisdom" thus has alignment potential relative to the concept {*deity, god*}, suggesting that "wisdom" can be reified to create a new taxonomic concept {*wisdom_deity*}.

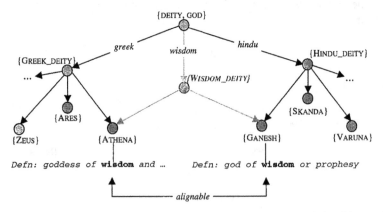

Fig. 5. Analysis of the gloss for *{Athene}* suggests that the word-form "wisdom" has analogical potential, since it is alignable with another use in *{Ganesh}*

How does one identify the potential pivot against which alignability is measured? In general, any interior type of the taxonomy can be a potential pivot, but from a practical perspective, it makes sense to only consider the atomic types that have not already been differentiated. Thus, {*deity, god*} is a potential pivot but {*Greek_deity*} is not, since the latter is already domain specific. We thus assume the following:

Definition: A hypernym X is a potential pivot relative to a hyponym Y if X is   (4)
the lowest, undifferentiated (atomic) hypernym of Y

Thus, when we consider the word forms in the gloss of {*Athena*}, alignability will be determined relative to the concept {*deity, god*} rather than {*greek_deity*}, so that any reification that is performed will create a new differentiation of the former. With an appropriate reverse-index of gloss words to the concepts that are defined by them, this makes the identification of alignable features very efficient. The system simply needs to examine each concept reachable via the index entry for the word and consider only those at the same relative depth from the potential pivot.

4   Roles and Relations

Analogies that are interpreted in taxonomic terms have the advantage of being conceptually justified via the process of type inclusion, since the source and target concepts are understood relative to the same common super-type. There is thus a sound conceptual basis for considering the target to be an analogue of the source. Furthermore, the extent of this basis can be quantified numerically by considering how far one must ascend in the taxonomy to find this common super-type.

Yet this advantage is bought at the price of symmetry: because taxonomic interpretations are created on the basis of commonalities, they tend to be highly symmetric, while the most creative interpretations tend to be highly asymmetric [8]. Insightful metaphors and analogies help to inform and enrich the target concept by imposing the highly-developed relational structure of the source onto the less-developed target [9]. Without this imposition of relational structure, metaphor is reduced to the status of a fanciful but redundant way of referring to existing concepts, which, in WordNet at least, already have linguistic labels. Systematic projection of structure from one domain to another is the basis of the structure-mapping approach to analogy and metaphor [6], which determines correspondences between entities based on their relative positions in a larger relational structures.

It thus becomes necessary to determine the relational structure of each concept if a full metaphoric/analogical interpretation is to be generated. Of course, one can build these representations by hand, but tedium aside, the resulting structures would have little theoretical force, since any conceptual theory can be made to appear tractable if one has complete freedom to hand-craft its representations. For this reason, we prefer instead to extract such representations automatically from independent linguistic data. There are two broad sources of data for this task: one can look to external corpora and attempt to mine relational patterns from large quantities of raw text, or one can look to the lexicographer glosses that accompany most concepts in WordNet. In either case, some type of parsing must be done on the linguistic data to extract relational patterns, and the reliability of extraction will depend crucially on the robustness of the parsing method used. We choose to mine the glosses, since their direct association with specific concepts resolves some of the problems of lexical ambiguity that can arise.

We describe here an extraction technique that balances coverage with quality: by attempting to extract a relatively narrow slice of the relational structure inherent in WordNet glosses, we can be confident of quite high levels of competence. We thus concentrate on relations that pertain to the agency/telicity of a concept, both because they can be extracted reliably and because they tend to capture the most intrinsic, behavioral aspects of a concept that are most likely to be projected by a metaphor.

4.1   Extraction of Relational Structure

Our extraction technique combines a knowledge of the derivational morphology of English with the taxonomic structure of WordNet, so that extracted relations are both linguistically and conceptually sound. The key here is that the agent-telic aspects of a concept are often expressed using nominalized verbs that implicitly encode relational structure, such as "observer" (from "observe") and "specializer" (from "specialize"), and these nominalization patterns can be captured in a small number of highly productive morphology rules. For example, concepts such as {*geologist*} and {*linguist*} are defined with glosses that explicitly invoke the term "specializer", while {*witness*} is defined relative to the term {*observer*}. Now, WordNet 1.6 provides two senses for "specializer", of types {*doctor*} and {*expert*}, both of which are sub-types of {*person*}. The concepts {*geologist*} and {*linguist*} are also sub-types of {*person*}, strongly suggesting that "specializer" is an appropriate telic relation for each. Note how the WordNet taxonomy plays a key role in recognizing this relation. If "specializer" did not have a sense that was compatible with {*person*}, it would be rejected as a relation, so ultimately, extraction depends crucially on the taxonomic metaphor.

It is also straightforward to morphologically derive the patient form of "specializer" as "specialism", allowing a system to conclude that both {*geologist*} and {*linguist*} have the relational structure *specializer_of:specialism* (we leave the particular sense of "specializer" under-specified for maximal metaphoric reuse). Now, while strongly telic nouns like "specializer" are often used in WordNet glosses, the underlying verbs themselves are even more frequent. For example, the concepts {*surgeon*} and {*pastry_cook*} are both provided with glosses that use the word "specializes", but using the same morphology rules in reverse, the corresponding nominalization "specializer" can be found. In this way both concepts are receive the relational structure *specializer_of:specialism*.

Using morphological rules in conjunction with taxonomic type checking, a large quantity of agent-telic relations can be robustly extracted from glosses with the simplest of shallow parses. Broad clues to the syntactic form of the gloss (such as active versus passive voice) can be derived from a combination of keyword analysis and inflectional morphology. The passive voice causes a relational arc to be inverted, as in the case of {*dupe*}, whose gloss is *"a person who is swindled or tricked"*. The resulting relational structure is thus: *of_swindler:swindler* ∧ *of_trickster:trickster*.

The glosses of many WordNet concepts suggest a metonymic relational structure. Consider the gloss assigned to the concept {*diary, journal*}: *"a daily record of (usually private) experiences and observations"*. The morphology of the word "diary" itself yields the agentive relation *of_diarist:diarist*, while nominalization rules suggest the additional relations *of_experience:experience*, *recorder_of:recording* and *observer_of:observation*. However, doubt is cast upon the latter two by subsequent taxonomic analysis, which reveals that {*diary, journal*}, a sub-type of {*communication*}, is not compatible with either {*recorder*} or {*observer*}, both sub-types of {*person*}. Nonetheless, these relations are not rejected, since the suggested patients, {*recording*} and {*observation*}, like {*diary, journal*}, are sub-types of {*communication*}, suggesting that a diary can be seen as a kind of metonym for the observer/recorder (as evoked by the familiar address "dear diary"). The concept {*diary, journal*} therefore yields the relational structure *meta_observer_of:observation* ∧ *meta_recorder_of:recording* ∧ *of_experience:experience*.

4.2 Projection of Relational Structure

The projection of relational structure can be performed either literally or figuratively. In a literal interpretation, the relational structure of the source is simply instantiated with the target concept, so for example, a literal "travel diary" is a diary that contains travel recordings and travel observations. In contrast, figurative interpretations first attempt to find a target domain correspondence for the source concept, and then project the relational structure of the source onto this counterpart [6]. For instance, WordNet suggests {*passport*} as a figurative reference for "travel diary" since both are kinds of travel document. Projecting the relational structure of {*diary, journal*} onto {*passport*} causes the latter to be seen as a journal of travel observations and experiences, and indeed, many travelers retain old passports for this very purpose. In structure-mapping terms, the more productive the metaphor, the greater the amount of relational structure that is projected. Furthermore, the more systematic and apt the metaphor, the greater the isomorphism between the structure that is projected from the source onto the target [6, 9].

4.3 Aptness and Projection

Metaphors are most apt when projection highlights a latent relational structure that already exists in the target concept [8]. For example, the compound "pastry surgeon" can be understood taxonomically as referring to {*pastry_cook*}, since like {*surgeon*} it is a sub-type of {*person*}. While this seems quite appropriate, the taxonomic approach arrives at the precisely the same interpretation even when the compound is "pastry astrologer" or "pastry hostage". To see why {*surgeon*} is more apt than {*astrologer*} as a source concept, one need look no further than their relational structures. WordNet 1.6 defines a surgeon as a *"physician who specializes in surgery"*, while a pastry cook is glossed as *"a chef who specializes in pastry"*. Both {*surgeon*} and {*pastry_cook*} contain the relation *specializer_of:specialism*, and it is precisely this relation that is highlighted by the metaphor. In contrast, the concept {*astrologer*} causes the relations *astrologer_of:astrology*, *forecaster_of:forecast* and *visionary_of:vision* to be projected, and neither of these relations exist in the {*pastry_cook*} structure.

Metaphors also appear more apt when they systematically evoke, or connect into, established modes of metaphoric thought [9]. Consider the compound "political mechanic": many different concepts can be reached from "political" that prove to be taxonomically compatible with the concept {*mechanic*}, among them {*political_leader*}, {*political_scientist*} and {*machine_politician*}. However, the extracted structure of {*mechanic*} contains the relation *machinist_of:machine*, whose surface similarity with {*machine_politician*} is highly suggestive. More interestingly, however, the instantiated structure for "political mechanic" thus becomes *machinist_of:political_machine*, where {*political_machine*} is a conventional metaphor already established within WordNet. This marks "political mechanic" as a systematic outgrowth of an established metaphor, making it seem all the more appropriate. Comparable systematicity is exhibited by the compounds "political chemist", which relationally connects to {*political_science*}, "political missionary", which in WordNet 1.6 connects to {*political_program*}, and "political torchbearer", which connects to {*political_campaign*}.

5   Empirical Analysis

Dynamic types are created in the context of specific metaphor interpretation or analogical reasoning tasks. For example, types like {*Hebrew_letter*} and {*Greek_letter*} are created in response to specific analogies, such as "What is the Jewish delta?". However, to test the applicability of the type creation process, we have pre-applied this type creation process to 69,780 unique noun senses in WordNet 1.6, whose glosses collectively contain 35,397 unique unlemmatized content words.

Now, because of the strict reification criteria for feature-lifting from glosses (see definitions 2 and 3), only 2806 of these content words are reified, to add 9822 new types, like {*cheese_dish*}, to WordNet. These types serve to differentiate 2737 existing concepts in WordNet, such as {*dish*}, transforming these concepts into analogically-useful pivots. In total, 18922 noun concepts (27% of the sample) are connected to the new types, via the addition of 28,998 new isa-links to WordNet. Each dynamic type thus serves to unite an average of 3 daughters apiece. But in a subsequent pass over the new types, 1258 additions (or 12.8%) were culled because they did not sufficiently differentiate their parents to be worthwhile. For example, the type {*Greek_gorgon*} is worthless since all known gorgons are Greek.

A review of the other 87.2% of differentiators reveals that WordNet is being dissected in new and useful ways, both from the perspective of simple similarity judgments (e.g., the new types achieve a fine-grained clustering of similar ideas) and from the perspective of analogical potential. Overall, the most differentiating feature is "Mexico", which serves to differentiate 34 different pivots (such as *{dish}*, to group together *{taco}*, *{burrito}* and *{refried_beans}*), while the most differentiated pivot is *{herb, herbaceous_plant}*, which is differentiated into 134 sub-categories (like *{prickly_herb}*). To consider just a few other domains: sports are differentiated into team sports, ball sports, court sports, racket sports and net sports; constellations are divided according to northern and southern hemispheric locales; food dishes are differentiated according to their ingredients, into cheese dishes, meat dishes, chicken dishes, rice dishes, etc.; letters are differentiated both by culture, giving Greek letters and Hebrew letters, and by relative position, so that "alpha" is both a *{1ˢᵗ_letter}* and a *{Greek_letter}*, while "Aleph" becomes both a *{1ˢᵗ_letter}* and a *{Hebrew_letter}*; and deities are further differentiated to yield *{war_deity}*, *{love_deity}*, *{wine_deity}*, *{sea_deity}*, *{thunder_deity}*, *{fertility_deity}*, and so on.

Table 1 presents a cross-section of the various sub-domains of *{deity, god}* in Word-Net as they are organized by dynamic types such as *{supreme_deity}*. Where a mapping is unavailable for cultural reasons, N/A is used to fill the corresponding cell. In two cases, marked by (*), an adequate mapping could not be generated when one was culturally available; in the case of *{Odin}*, this is due to the gloss provided by Word-Net 1.6, which defines Odin as a "ruler of the Aesir" rather than the supreme deity of his pantheon; in the case of *{Apollo}*, a Greco-Roman deity, the failure is due to this entity being solely defined as a Greek deity in WordNet 1.6.

Dynamic types primarily increase the precision, rather than the recall rate, of analogical mapping. Consider again the alphabet mapping task, in which the 24 letters of the Greek alphabet are mapped onto the 23 letters of the Hebrew alphabet (as represented in WordNet), and vice versa. The recall rate for the Hebrew to Greek letter task, for both dynamic and static WordNet hierarchies, is 100%, while for the reverse task, Greek to Hebrew, it is 96% (since Greek contains an extra letter). However, the precision of the static hierarchy is only 4%, since every letter of the target alphabet appears equally similar as a candidate mapping (Fig. 2), while for the dynamic hierarchy it is 96% (Greek to Hebrew) and 100% (vice versa).

Table 1. Mappings between sub-domains of the type *{deity, god}* in WordNet 1.6

Common Basis	Greek	Roman	Hindu	Norse	Celtic
supreme	Zeus	Jove	Varuna	Odin *	N/A
wisdom	Athena	Minerva	Ganesh	N/A	Brigit
beauty, love	Aphrodite	Venus	Kama	Freyja	Arianrhod
Sea	Poseidon	Neptune	N/A	N/A	Ler
fertility	Dionysus	Ops	N/A	Freyr	Brigit
queen	Hera	Juno	Aditi	Hela	Ana
war	Ares	Mars	Skanda	Tyr	Morrigan
hearth	Hestia	Vesta	Agni	N/A	Brigit
moon	Artemis	Diana	Aditi	N/A	N/A
sun	Apollo	Apollo *	Rahu	N/A	Lug

The data of Table 1 allows for 20 different mapping tasks in the deities domain (Greek to Roman, Roman to Hindu, etc.). The average recall rate of the dynamic hierarchy is 61%, since some pantheons are less fleshed out than others (e.g., the Norse to Hindu mapping has a precision of just 30% for this reason). For the static hierarchy, average recall is significantly lower at 34%, since many concepts (such as Varuna and Aphrodite) are not indexed on the appropriate terms due to poorly defined glosses (e.g., Varuna is defined as "supreme cosmic deity" in WordNet 1.6, with no explicit reference of Hinduism). Average precision for the dynamic hierarchy is 93.5%, with the loss of 6.5% precision due to the items marked (*) in Table 1. In contrast, average precision for the static hierarchy is just 11.5%, and would be lower still if incomplete glosses did not limit the number of incorrect answers that the static hierarchy approach can retrieve.

6 Conclusions

Manually constructed representations on the ambitious scale of WordNet and Cyc are naturally prone to problems of incompleteness and imbalance. The one-size-fits-all nature of the task results in a taxonomy that is often too undifferentiated for precise similarity judgments and too lopsided to support metaphor and analogical mapping. A symptom of this incompleteness is the fact that English glosses or commentaries provide the ultimate level of differentiation, so that one cannot truly differentiate two concepts without first understanding what the glosses mean. The goal of this work is to lift the implicit discriminators out of the flat text of the glosses and insert them into the taxonomy proper, as dynamic types that will facilitate finer similarity judgments and richer analogical mappings.

It is interesting to note that WordNet 1.6 already contains lexical entries for some dynamic types, such as {war_god} and {sea_god}, but fails to relate them to the appropriate sub-types such as {Ares} and {Poseidon}. This may be due to simple oversight, but is more likely a symptom of WordNet's tendency toward single-inheritance (even though multi-inheritance is supported, and actually used in a relatively small number of noun concepts). With single inheritance, the taxonomist must decide between differentiating a concept on the basis of culture (e.g., Greek) or on that of function (e.g., war), rather than simply doing both. Whatever the true cause, dynamic types do more than facilitate metaphor and analogy, but actually repair deficiencies in an existing static taxonomy.

References

1. Hutton, J.: Aristotle's Poetics. Norton, New York (1982)
2. Lenat, D., Guha, R.V.: Building Large Knowledge-Based Systems. Addison Wesley (1990)
3. Miller, G. A.: WordNet: A Lexical Database for English. Communications of the ACM, Vol. 38 No. 11 (1995)

4. Way, E. C.: Knowledge Representation and Metaphor. Studies in Cognitive systems, Kluwer Academic Publishers (1991)
5. Fass, D: An Account of Coherence, Semantic Relations, Metonymy, and Lexical Ambiguity Resolution. In: Small, S. I, Cottrell, G. W., Tanenhaus, M.K. (eds.): Lexical Ambiguity Resolution: Perspectives from Psycholinguistics, Neuropsychology and Artificial Intelligence. Morgan Kaufman, San Mateo California (1988)
6. Veale, T., Keane, M. T.: The Competence of Sub-Optimal Structure Mapping on Hard Analogies. The proceedings of IJCAI'97, the Int. Joint Conference on Artificial Intelligence, Nagoya, Japan. Morgan Kaufman, San Mateo California (1997)
7. Harabagiu, S. M., Miller, G. A., Moldovan, D. I.: WordNet 2 - A Morphologically and Semantically Enhanced Resource. The Proceedings of the ACL SIGLEX Workshop: Standardizing Lexical Resources. Maryland, USA (1999)
8. Ortony, A.: The role of similarity in similes and metaphors. In: Ortony, A. (ed.): Metaphor and Thought. Cambridge University Press : Cambridge, U.K (1979)
9. Lakoff, G., Johnson, M.: Metaphors we live by. University of Chicago Press: Chicago (1980)
10. Pustejovsky, J.: The generative lexicon. Computational Linguistics, Vol. 17 No. 4 (1991)

Correction and Extension of WordNet 1.7

Philippe Martin

Distributed System Technology Centre
Griffith University, PMB 50 Gold Coast MC, QLD 9726 Australia
philippe.martin@gu.edu.au

Abstract. This article presents the transformation of the noun-related part of WordNet into a genuine "lexical ontology" to support knowledge representation, sharing and retrieval within a knowledge base or on the Web, i.e. to support "knowledge creation and communication". The corrections and extensions are documented at http://www.webkb.org/doc/wn/ and the ontology is downloadable in various formats. Web users can also search and extend the ontology via the WebKB-2 knowledge server.

1 Introduction

WordNet [1] is a lexical database that connects English words to "synonym sets" (each "synset" represents one of the meanings of the words in the set) and organizes the synsets by semantic links, e.g. specialization and partOf links. WordNet is increasingly interpreted and exploited as a lexical ontology (i.e. a set of categories connected by links having a formal semantics) despite its shortcomings for this purpose.

A natural language ontology derived from WordNet and other sources could support or enhance various kinds of applications, e.g. query expansion and answering [5], machine translation [8], and knowledge representation, sharing or brokering [3] [4]. In [11], I argued that the Semantic Web (as I understand it) cannot be achieved without at least one natural language ontology that can be extended by people and permit to give categories from different ontologies some shared meaning. [11] also details how the knowledge server WebKB-2 exploits WordNet 1.7 and its extensions for guiding and checking knowledge representation, and for permitting Web users to share or retrieve knowledge, and further extend or correct the shared ontology if necessary. (Protocols and naming conventions prevent lexical and semantic conflicts).

This article introduces extensions and corrections of the noun-related part of WordNet 1.7 to transform it into a lexical ontology usable for knowledge-based applications, and especially the *manual* representation of natural language sentences. (Much more would be needed to support natural language parsing)[1]. No claim is made that this ontology is sufficient to support the inter-operation

---

[1] Only the noun-related part of WordNet 1.7 is used because the use of categories representing the meanings of verbs, adverbs or adjectives has several drawbacks: (i) using such categories with quantifiers has no real meaning (e.g. "any transformation" or "3 transformation" has a meaning but "any transform" and "3

A. de Moor, W. Lex, and B. Ganter (Eds.): ICCS 2003, LNAI 2746, pp. 160–173, 2003.

of fully automatic software agents, e.g. for e-commerce or database integration purposes. [6] shows that such inter-operations have strong requirements and, in the general case, are not likely to be fully supported by ontologies anytime soon.

This article first explains why short and intuitive identifiers were generated for each WordNet category, and illustrates the lexical corrections. Second, it explains how types (1st-order categories) were distinguished from individuals (0th-order categories), and hence how WordNet specialization links were differentiated into subtype and instance links. Third, it introduces the top-level ontology of concept and relation types into which the top-level categories of WordNet were inserted to support the construction of normalized (i.e. better *retrievable*) knowledge statements and certain *semantic checks* on the ontology and the statements. Fourth, it illustrates the kinds of problems that led to the removal or modification of links in WordNet. Fifth, it details the kinds of additions (links, schemas, annotations) made to some WordNet categories.

2   Category Identifier Generation and Lexical Corrections

A category may have many names (the elements of the "synset" in WordNet) that may be shared by other categories, but should have at least one "identifier" to refer to it uniquely. In WebKB-2, a category identifier is allowed to be a URL or an e-mail address, but for readability reasons, is most often composed of a short identifier for the user (or source) that created the category, and a *key name* distinguishing the category from other ones created by the same user. For example, wn#car refers to a WordNet category for the noun "car", while pm#car may represent a different notion for the user pm. WebKB-2 allows the prefix "wn" to be dropped, and the category creator may specify other names by appending them to the identifier; thus, #car__auto__automobile refers to the same category as wn#car. (Such categories may also be referred and accessed from outside WebKB-2 via a URL. The reader is encouraged to access http://www.webkb.org and browse from the categories referred to in this article).

WordNet has at least two internal identifiers for each category, e.g. the category for "Friday" has for identifiers 12558316 and friday%1:28:00::. While some applications re-use them, others (such as [2]) generated their own identifiers by concatenating names or using suffixes, e.g. Inessential$Nonessential and Cell_1. However, for knowledge representation, exchange and sharing purposes, category identifiers should be concise and clear to permit readable text-based knowledge statements (graphical interfaces should not be required and are not necessarily the best device to enter, display, debug or maintain a large amount of knowledge [11]; category identifiers should also be usable within *controlled languages*). Hence, for these purposes, each category should have at least one identifier composed of a common and unambiguous word or expression for its

---

transform"has not), (ii) organizing these categories by generalization links is difficult or impossible, (iii) these categories are (non-defined) shortcuts for more explicit constructions using categories for nouns. More details and rationales can be found in [11].

meaning, *and as little else as possible*. This means that one of the category names should be used as key name, if possible with no suffix. This was possible for 92% of WordNet categories related to nouns: *only* 5944 WordNet categories (out of 74,488) have been given a key name with a suffix. The list of these categories and the used identifier generation algorithm is accessible from [15].

So is the list of my 353 lexical corrections: 28 modifications of category annotations, 248 category names added, and 77 manual re-orderings of category names. Here is an example showing how the corrections were documented:

```
#wn07834480|German_citizen__German (^ $("German_citizen" has been added
    as key name; the original annotation was: ''a native or inhabitant of
    Germany")$ a person of German nationality^)
```

This format is used by WebKB-2 for saving the KB in a backup file. 07834480 is the WordNet identifier, `German_citizen` the added key name (since "German" also refers to a language), `German` the original name, (^...^) the category annotation, and $(...)$ a sub-annotation which WebKB-2 does not show to end-users.

To conclude, this work provides Web users a shared formal vocabulary to mark up their documents or the meanings of words in their documents, or to use in their knowledge statements. If a word meaning is missing, a user may easily add it to the ontology via WebKB-2, thus permitting other people to retrieve and re-use his/her categories or statements.

3   Explicitation of Individuals

Distinguishing 1st-order types from their instances ("individuals"), is important for knowledge representation, inferencing and checking. Individuals cannot have specializations, i.e. subtypes or instances. Certain individuals, often called continuants or endurants [2], can change in time without being viewed as different individuals (i.e. without loosing their identity), e.g. individuals for persons or cities. Specializing such individuals according to time might be tempting, e.g. `pm#ParisIn1995`, but better avoided: statements (facts or definitions) about individuals should represent dates and durations in an explicit way using contexts.

Distinguishing types from individuals is not always obvious. For example, [2] asserts that the WordNet category `#karate` should be an individual, but there are various kinds of karate, and furthermore, since `#karate` is a subtype of `#activity` [2], each individual practice of karate may be considered as an instance. Anything that may be specialized, or has various occurrences, or comes in different variants or versions should be represented as a type rather than an individual; otherwise, knowledge representation possibilities and accuracy are reduced. For example, any doctrine, book, language, alphabetic character, code, diploma, sport or recurring situation should rather be represented as a type. The first character of the alphabet has many variants (e.g. its uppercase and lowercase variants) and billions of instances (occurrences) in books. An alternative view would be to consider that in certain cases a variant is not a subtype and an occurrence is not an instance, and then use different links or relations to represent this information. However, in this alternative model, information would be more complex to describe, and inferencing more complex to implement.

I chose the simplest model. However, since people often wish to use certain types without quantifiers, as if they were individuals (e.g. in English, the nouns "Monday" and "Polish" are rarely used with an article, i.e. a quantifier), WebKB-2 allows it in Frame-CG (FCG) and Formalized-English [10] (both extend and simplify the Conceptual Graph Linear Form (CGLF)) on the condition that the category has no subtype, no instance and is not a subtype of pm#physical_entity nor #time_period.

[15] lists the 6211 individuals that I manually isolated: typically, time periods, persons, organizations, places and battles. To do so, I first translated all WordNet specialization links as subtype links. Then, since WordNet categories are grouped by theme within the WordNet database files, I operated a careful but relatively quick "search and replace" of subtype links into instance links in the zones where individuals could appear. I double checked this work on categories having a name with a capitalized first letter. Here is an example in the FO notation (which is also derived from CGLF; ^' represents the instanceOf link and 'P' the WordNet partOf link): #Neolithic_Age ^ #time_period, P #Stone_Age;

To sum up, formal information was added to WordNet categories (consistent with the original meanings of these categories) while adopting an approach that maximise re-use possibilities. For knowledge sharing and inferencing purposes, I also argue against the use of instance links between types (i.e. against the introduction of second-order types and second-order statements) *when* subtype links can be used instead. Indeed, subtype links are easier to use for structuring categories, and then to exploit. The logical interpretation of statements using types of different orders may also be difficult and they are not commonly exploited by inference engines. Over-uses of the instance link are frequent. For example, the TAP KB [14] categorizes certain types of magazines or books as instances of a second-order type tap#product_type which has no other supertype than rdfs#class. Even if it had, the use of a first-order type such as #product permits much more comparison with (or connection or inheritance of constraints from) other types, hence more retrieval and checking possibilities.

4   Top-Level Ontology

WordNet has not been built for knowledge representation purposes, nor apparently according to basic taxonomy building principles and with consistency checking tools. As noted in [2], types and individuals are not distinguished; the annotation of a category is not to be relied on as it may be contradicted by specializations of this category; direct specializations often have heterogeneous levels of generality; role types (e.g. #student) are not distinguished from natural types (e.g. #person) and may generalize them. I also found that (i) specialization links are sometimes used where "location" or "similar" links should be used, (ii) the "part" and "member" links between types are not used in a consistent way (most seem to mean that all instances of the source type have for part/member at least one instance of the destination type, but this is not *always* the case), (iii) some of these transitive links are redundant, and (iv) exclusion links are sometimes broken, i.e. some exclusive categories have common specializations. Table 1 shows the uppermost WordNet categories for nouns and some of their direct subtypes. The lack of structure is clear.

Table 1. WordNet 1.7 top categories for nouns *(brackets enclose exclusive subtypes)*

#human_action__act__human_activity > #action #nonaccomplishment #leaning #assumption #rejection #forfeit {#activity #inactivity} #wearing #judgment ...
#state > #skillfulness #cognitive_state #cleavage.state #medium.state #condition #condition.state #conditionality #state_of_affairs #relationship #relationship.state ...
#event > #might-have-been #nonevent #happening #social_event #miracle.event #Fall;
#phenomenon > #natural_phenomenon #levitation #metempsychosis #outcome #process ...
#entity > #self-contained_entity #whole_thing #living_thing #cell #causal_agent #holy_of_holies #physical_object #location #depicted_object #unnamed_thing #sky ...
#group__grouping > #arrangement #straggle #kingdom.group #biological_group #biotic_community #human_race #people #social_group #aggregation #edition.group ...
#possession > #belongings #territorial_dominion #white_elephant.possession #transferred_property #circumstances #assets #treasure.possession #liabilities;
#psychological_feature > #cognition #motivation #feeling;
#abstraction > #time #space #attribute #relation #measure #set;

This work seems the first to have isolated individuals, generated intuitive category identifiers, corrected and documented a large number of problems, and permitted Web users to further extend and correct this ontology. No attempt to bring more structure to the whole of WordNet was made, as this would probably take many years. However, like others, this work inserts the top-level categories of WordNet into a better structured top-level ontology ("top-level" simply means "general" without arbitrary notions of "primitiveness" or "depth"; time is the main limiting factor since, outside particular applications, the more specialized the categories, the less their (re)structuration is likely to be useful). In 1994, Sensus [8] was created by manually merging the WordNet top-level into Ontos and the Generalized Upper Model, and then semi-automatically merging WordNet with the Longmann Dictionary of Contemporary English. Sensus was created for machine translation purposes. At the same period, for knowledge acquisition and representation purposes, I extended Sowa's first top-level ontology [12] and used it for structuring WordNet 1.5 top-level [9]. In 2001, for the Semantic Web and other knowledge sharing purposes, the OntoClean ontology and methodology was used to re-structure WordNet 1.6 top-level [2]. In October 2002, I integrated the last version of the OntoClean ontology, DOLCE (D17) [16] into WebKB-2 ontology but found most of the 40 DOLCE top categories *too specific* to specialize them with WordNet categories. The next section presents two examples.

4.1   Minimizing Re-categorization

Example 1. OntoClean/DOLCE distinguishes "qualities" (like size, color, redness, smell and duration) from "quales" (quality regions/spaces, i.e. categories of values for qualities, e.g. #red, #past_times and #Greenwich_Mean_Time [2]). They specialize the exclusive categories dolce#quality and dolce#region__quale. However, in WordNet, such categories (about 8900) are inter-related by specialization links, e.g. #red__redness specializes #chromatic_color and #color, while #past_times specializes #time. Hence, specializing the types dolce#quality and dolce#region by WordNet categories, as suggested in [2], is problematic: (i) this classification has to be done for most of the 8900 categories, *not just for their most general categories*; (ii) a great number of WordNet specialization links have to be *broken*, hence this structure is *lost* and the meaning of a great number of WordNet categories is *modified*; (iii) it is often difficult to decide whether a WordNet category should be *interpreted* as a quality or as a quale; as opposed to [2], I consider #Greenwich_Mean_Time, #work_time and #red as quality types (the authors of [2] argue for the representation of red and other adjectives for colors as quales, but #red (i.e. #red__redness) represents the meaning of the nouns "red" and "redness"). In my integration of WordNet, I added or refined but *not removed or modified* links – except for 306 (out of 74,488) in order to fix inconsistencies. From an Ontoclean perspective, this is possible by interpreting most of the above cited 8900 categories as qualities. However, I have not explicitly categorized their upper types as specializations of dolce#quality in order to permit WebKB-2 users to classify certain WordNet categories as subtypes of dolce#region when this does not introduce inconsistencies. I have generalized these upper types, plus dolce#quality and dolce#region, by the type pm#attribute_or_measure (this name is due to the fact that things I call "measures" may specialize the things that are often called "attributes").

Here is a statement in FCG showing how knowledge representation can be done in an intuitive and normalizing way with the interpretation of WordNet attributes or measures as qualities: [a #car, #color: a #red, #weight: 900 #kg]. In Formalized-English [10]: — there exists a #car that has for #color some #red and for #weight 900 #kg. Both #red and #kg are quantified (KIF definitions for the FCG numerical quantifiers are given in [10]). As in Ontoseek [3] (a WordNet-based knowledge retrieval system built by the team that designed OntoClean), the types #color and #weight are used as if they were relation types. WebKB-2 checks that these types specialize pm#thing_that_can_be_seen_as_a_relation and respectively generalize #red and #kg. No quale is explicitly referred to in this statement. If #red and #kg were categorized as quales, more complex statements would have to be written, e.g.: [a #car, #color:(a #color, pm#measure: a #red), #weight: (a #weight, pm#measure: 900 #kg)]. Checking this graph would also be more complex and would require additional information on categories acceptable as measures for colors and weight.

Example 2. In [2], #substance is subtype of dolce#amount_of_matter which is exclusive with dolce#physical_object However, #substance has many subtypes which are also subtypes of dolce#physical_object. An example is #olive.relish which specializes #fruit(#physical_object) and #relish(#condiment, #substance). Another example is #glass_wool, subtype of #artifact (#physical_object) and #insulator (#substance). Since these WordNet links do not appear as clear mistakes, it seems that in [2], #substance has been over-interpreted (or adapted) to fit the meaning of dolce#amount_of_matter. Instead, I categorized #substance (along with other types such as #physical_part and #building_block) as subtype of pm#physical_part_or_substance which, like dolce#physical_object and dolce#amount_of_matter, is a direct subtype of pm#physical_entity. Since this last type covers both substances and physical objects, it may be seen as an adequate candidate for classifying a "statue of clay". It may also be used for signatures of relations, e.g. relations representing physical attributes such as color or mass (although as hinted in Example 1, this is discouraged in WebKB).

4.2   Summary of the Approach and Its Results

The distinctions made by DOLCE and other top-level ontologies are important and their integration may be used by knowledge servers to guide the users to represent knowledge in more precise and re-usable ways. The precision of DOLCE categories and their associated constraints are also intended to ease the automatic matching of categories from (Semantic Web) ontologies independently developed but re-using the DOLCE ontology. Although this precision makes the current set of DOLCE categories difficult to use for structuring WordNet top-level (and other distinctions are also required), it is valuable (including the distinction between qualities and quales, although in my approach I am more interested in the more general distinction between concept types that can be used for relations and those that can be used as destinations of those relations; Section 6 will show how I have prefered to make the distinction).

Table 2 presents a summary of top-level types[2]. Many types from Word-Net, DOLCE and Sowa are shown. The catch-all WordNet categories #entity and #abstraction do not appear but their direct subtypes have been categorized in various places. Most of the upper types, e.g. pm#spatial_entity or pm#description, are common and relatively intuitive categories that are required for the signatures of the relation types (Table 3). These concept types have been given constraints (mainly exclusion links) and prototypes (e.g. typical relations) that are inherited by their numerous WordNet specializations.

The relation types proposed by WebKB-2 are mainly for primitive binary relations and intended to support an explicit and normalized way of representing natural language sentences (in [11], I give rationales against the use of non-binary relations and complex relations, e.g. relations representing processes). I also integrated argumentation relations and the relations of DAML, RDF,

---

[2] To be compatible with most other top-level ontologies, the uppermost type has for subtypes all other categories, including relation types and second-order types. Hence, Table 2 does not just present concept types. Although this is not certain, it does not seem that the top types of Sowa's ontology and DOLCE are concept types only.

Table 2. Some of the 160 top-level types in WebKB-2

/: complementOf link;   {(...)}: close subtype partition;   {...}: open subtype partition

pm#thing__something__universal__top_type__T (ˆany object is instance of this typeˆ)
> {(pm#situation pm#entity)}   {(pm#thing_playing_some_role sowa#independent_thing)}
  {(sowa#physical_thing sowa#abstract_thing)}   {(sowa#continuant sowa#occurrent)},
  {(suo#physical suo#abstract)}   {pm#individual pm#1st_order_type pm#2nd_order_type},
/ daml#nothing,   = daml#thing suo#entity sowa#entity dolce#entity__ALL;

pm#situation (ˆsomething that "occurs" in a real/imaginary region of time and spaceˆ)
> {(pm#state pm#process)}   {(dolce#stative dolce#event)}
  pm#phenomenon sowa#process sowa#situation #event pm#situation_playing_some_role,
= dolce#perdurant__occurence__PD suo#process;

pm#state > #state #feeling pm#state_playing_some_role;

pm#process (ˆsituation that makes a change during some period of timeˆ)
> pm#event pm#problem_solving_process #unconscious_process
  #cognitive_process #human_action pm#process_playing_a_role;

pm#entity (ˆsomething that can be "involved" in a situationˆ)
> {(pm#spatial_object pm#nonspatial_object)} {(pm#undivisible_entity pm#divisible_entity)}
  dolce#endurant pm#entity_playing_some_role;

pm#spatial_object > pm#space dolce#physical_endurant sowa#object, = suo#object;

pm#space > dolce#feature #space #location #natural_enclosure #expanse #sky #shape;

dolce#physical_endurant > {(pm#physical_entity dolce#feature)};

pm#physical_entity
> {dolce#physical_object dolce#amount_of_matter} pm#physical_part_or_substance;

dolce#physical_object
> {(dolce#agentive_physical_object dolce#non_agentive_physical_object)};

dolce#agentive_physical_object > pm#living_entity #living_thing #cell;

dolce#non_agentive_physical_object > pm#dead_entity #physical_object;

pm#physical_part_or_substance
> #part.physical_object #physical_part #building_block #substance;

pm#nonspatial_object (ˆe.g. knowledge, motivation, language, measureˆ)
> pm#psychological_entity pm#collection dolce#abstract
  {pm#description_content/medium/container pm#attribute_or_measure};

pm#psychological_entity > dolce#mental_object #psychological_feature;

pm#collection > #group #set dolce#set dolce#arbitrary_sum pm#structured_ADT
              sowa#structure pm#type;

pm#description_content/medium/container > {pm#description pm#description_container};

pm#description > pm#description_content pm#description_medium sowa#form;

pm#description_content__information (ˆe.g. a narration, an hypothesisˆ)
> sowa#proposition sowa#intention dolce#fact kads#role rdf#description
  #code.laws #subject_matter #written_material #public_knowledge
  #cognitive_factor #perception.cognition #cognitive_content #history.cognition;

pm#description_medium (ˆe.g. a syntax, a language, a script, a structureˆ)
> #structure #communication #language_unit #symbolic_representation;

pm#description_container > pm#document_element #representation_container;

Table 3. Some of the 150 primitive relation types in WebKB-2

/: complementOf link;   ˆ: instanceOf link;   (...): signature;   ?: any type;   *: 0 or more types

```
pm#relation__related_with (*) (ˆtype for any relation (unary, binary, ..., *-ary)ˆ)
 > {pm#relation_from_situation pm#relation_from_spatial_object pm#relation_from_type
    pm#relation_from_description_content/medium/container} {dc#Type dc#Description} kif#subst
    pm#relation_from_collection {pm#relation_to_collection pm#relation_to_time_measure}
    pm#attributive_relation {pm#different pm#ordering_relation} pm#relation_for_an_application,
 / pm#thing,  ˆ rdf#property,  = suo#relation;

   pm#relation_from_situation (pm#situation,*) > pm#relation_from_situation_to_time_measure
      > pm#relation_from_situation_to_situation pm#case_relation pm#within_group;

      pm#relation_from_situation_to_time_measure (pm#situation,pm#time_measure)
       > pm#time pm#duration pm#from_time pm#until_time pm#before_time;

      pm#relation_from_situation_to_situation (pm#situation,pm#situation) > pm#later_situation;

         pm#later_situation (pm#situation,pm#situation) > pm#next_situation pm#consequence;

   pm#case_relation__thematic_relation (pm#situation,*)
      > pm#doer/object/result/place pm#experiencer pm#recipient pm#relation_from_process_only;

      pm#doer/object/result/place (pm#situation,*)
       > pm#doer/object/result pm#place pm#from/to_place;

         pm#doer/object/result (pm#situation,*) > pm#agent pm#initiator pm#object/result;

            pm#agent__doer (pm#situation,pm#entity) > pm#organizer pm#participant;

               pm#organizer (pm#situation,pm#causal_entity);

            pm#initiator (pm#situation,pm#causal_entity);

            pm#object/result (pm#situation,?) > pm#object pm#instrument pm#result;

               pm#object__patient__theme (pm#situation,?) > pm#input pm#input_output;

               pm#instrument (pm#situation,pm#entity);

         pm#from/to_place (pm#process,pm#spatial_object)
          > pm#from_place pm#to_place pm#via_place pm#path;

      pm#experiencer (pm#situation,pm#causal_entity);

      pm#recipient (pm#situation,pm#entity) > pm#beneficiary;

      pm#relation_from_process_only (pm#process,?) > pm#purpose pm#triggering_event
         pm#ending_event pm#precondition pm#postcondition pm#input pm#input_output
         pm#sub_process pm#method pm#from/to_place pm#process_attribute;

   pm#relation_from_spatial_object__relation_from_a_spatial_object (pm#spatial_object,*) > pm#location;

      pm#location (pm#spatial_object,pm#spatial_object) > pm#address pm#on pm#above
      pm#in pm#near pm#interior pm#exterior pm#before_location;

   pm#relation_from_description_content/medium/container (pm#description_content/medium/container,*)
    > pm#relation_from_description pm#version dc#Coverage dc#Contributor dc#Source dc#Publisher
      dc#Rights pm#authoring_time pm#author dc#Language dc#Format pm#description_instrument
      pm#description_object pm#physical_support pm#rhetorical_relation pm#argumentation_relation;

      pm#relation_from_description (pm#description,*)
       > pm#descr_container pm#logical_relation pm#contextualizing_relation;

   pm#different__different_from (?,?) > daml#different_individual_from pm#exclusive_class, / pm#equal;
```

RDFS, Dublin Core and the core of KIF. Table 3 shows the overall organization, although it also deepens in the case relations. The grouping by source category proved to be the cleanest and most intuitive structure, and WebKB-2 exploits it when generating menus to guide knowledge representation.

The "Suggested Upper Merged Ontology" (SUMO) [17] has similarities with WebKB-2 ontology in the sense that it has mappings with categories of Word-Net 1.6, and includes some spatial and case relations, and various top concept types from various top-level ontologies, e.g. Sowa's last top-level ontology. Its integration into WebKB-2 ontology has begun. Some elements of the CYC top-level [18] may also be added in the future (only some elements since on the one hand, there is already some overlap, and on the other hand different approaches have been adopted in CYC, e.g. it includes many non-binary relations and relations representing processes).

To conclude, WebKB-2 and its ontology may help people avoid the difficult task of finding, integrating and extending adequate ontologies, especially top-level ontologies (a task that some Semantic Web researchers, seem to think the knowledge providers to the future Semantic Web are able to do, have the time to do and will *have to* do! [7]). Instead, the WebKB-2 user is simply supposed to find adequate categories by typing words and browsing from the proposed categories for these words, and then fill cascading menus adapted to the categories s/he selected or entered. Knowledge precision and normalization is encouraged by the various proposed distinctions, the adopted approach (e.g. the proposed basic binary relations) and the proposed notations (e.g. their extended quantifiers).

5   Semantic Corrections

Up to March 2003, 117 links have been removed, and the types or destinations of 198 links have been modified. Of these 315 links, 41 were redundant and about 240 inconsistent with other links. Most of the inconsistencies were automatically detected thanks to exclusion links in WebKB-2 top-level ontology. For example, some categories in WordNet were classified as *both human action and causal agent, instrument or result of action* (e.g. #relaxant and #interpretation) or of communication medium/content (e.g. #epilog and #thanksgiving), or *as both communication medium/content and* physical entity (e.g. #book_jacket) or attribute (e.g. #academic_degree). Some WordNet specialization links were also used instead of "member" links, (e.g. between types for species and genus of species). Similarly, WordNet does not have "location", "similarTo" and "identity" links, and uses subtype links instead of location links (e.g. many city/regions where battles have occured were classified both as city/regions and battles), similarTo links (e.g. for a Greek god and its Roman counterpart) and identity links (WordNet introduces a few categories to represent obsolete names).

Redundancy was detected by exploiting the transitivity of specialization, part and member links. Only the combination of exclusion and specialization links was exploited to detect inconsistencies or redundancies. More could be done. For example, the fact that "if t2 specializes t1, and t1 is member of t0, then t2 is member of t0" should be exploited to detect more redundant links, e.g. in WordNet both #dog and its subtype #hound_dog are member of #pack.animal_group. Negative constraints such as "if t2 specializes t1, then t2

Table 4. Examples of corrections

<: subtypeOf;   1: location; $(...)$: sub-annotation
#wn12347769|Payne's_gray (^$('<' #blue removed since exclusive with #pigment, subtype of #substance)$ any pigment that produces a grayish to dark grayish blue^)   < #pigment;
#wn07130190|Anglia (^$('<' #England replaced by '=')$ the Latin name for England^) = #England;
#wn07799755|Mancunian (^$('<' Manchester replaced by '1')$ a resident of Manchester^) < #English_person, 1 #Manchester;
#wn05168522|transmission (^$('<' #communicating replaced by '<' #communication since the subtypes of this category indicate that it represents a transmission medium, not a process)$ communication by means of transmitted signals^)   < #communication;

cannot be linked by any other kind of link to t1" have not been exploited either but it does not seem that WordNet 1.7 has many problems of this kind.

The corrections I made are documented [15]. Table 4 shows some examples in the format used for saving the KB of WebKB-2 in a backup text file.

6   Additions

Up to March 1993 (and apart from the connections of WordNet upper categories to my top-level concept types), I have added 161 links, 17 during the integration of WordNet to WebKB-2 and 144 later when using the ontology for representing knowledge (thus, this excludes the 3000 specializations of WordNet categories that I created for specific applications/domains, e.g. information technology). About 65 of these links connect WordNet categories, while 90 connect a WordNet category to a new specialization. Table 5 shows some examples.

I also added sub-annotations to guide or check knowledge entering. For example, since I do not want to distinguish between qualities and quales *using subtype links*, the subtypes of pm#attribute_or_measure that represent values need to be distinguished in another way to prevent them being used within relations (or proposed in WebKB-2 generated menus for relations). Hence, I checked all

Table 5. Examples of link additions

>: subtype;   ~: similar;   1: location;   p: part;   //: comment
#yellow > pm#blond_color;   #agency > pm#real_estate_agency;
#name > pm#previous_surname pm#middle_name;
//"(pm)" explicits the creator of the link; this is needed between WordNet types
#region > #dry_land (pm);   #mass > #mass_unit (pm);
#city > #capital_city (pm);   #male > #male_person (pm);
#Tasmania 1 #Tasmanian_Island (pm);   #Great_Britain p #England #Wales (pm);
#length > #distance (pm) #distance.size (pm); #Venus.Roman_deity ~ #Aphrodite (pm);

the subtypes of pm#thing_that_can_be_seen_as_a_relation) and added the string
$(value)$ in the annotations of about 1300 of them. (It should be noted that
individuals are representing values and hence such sub-annotations are not required for them). The string $(artificial)$ was also added in the annotations of
WordNet categories that were found unfit for knowledge representation purposes,
generally because they had a lexical rather than semantic character. Table 6 gives
some examples.

Table 6. Examples of value/artificial categories

#dark_red (^$(value)$ a red that reflects little light^)	#then (^$(artificial)$ that time; "we will arrive before then"^)
#gram__gramme__gm__g (^$(value)$ metric unit of weight equal to one thousandth of a kg^)	#thing.action (^$(artificial)$ action "how do you do such a thing?"^)
#west_by_south__WbS (^$(value)$ the compass point that is one point south of due west^)	#thing.happening (^$(artificial)$ an event; "something happened ..."^)
#Monday__Mon (^$(value)$ the second day of the week; the first working day^)	#tonight (^$(artificial)$ the present or immediately coming night^)
#andante (^$(value)$ moderately slow tempo^)	
#mealtime (^$(value)$ time for eating a meal^)	

Finally, I entered statements representing the most common relations that
are or may be associated to certain categories. I call them schemas. Table 7
shows an example in FCG. WebKB-2 exploits such schemas to generate menus
that help users to search or represent knowledge. Figure 1 shows an example
based on the schema in Table 7 and where only this schema is exploited because
of its sub-annotation $(no inheritance)$. The sub-annotation $(explore)$ in a
relation annotation directs WebKB-2 to present the subtypes of the type used
for the relation, in a select menu (except for the subtypes marked as "value" or
"artificial"). The '+' symbols in the menus permit the user to access a sub-menu
to detail relations from/to any destination object s/he has entered; in other
words, menus can be cascaded to guide the entering of a query or statement.

Table 7. Example of schema

```
[any #flight (^$(no inheritance)$^),
    pm#from_place: a pm#spatial_object,        pm#to_place: a pm#spatial_object,
    #day_of_the_week: a #day_of_the_week,      pm#via_place: a pm#spatial_object,
    pm#departure_time: a pm#time_measure,      pm#arrival_time: a pm#time_measure,
    may have for pm#relation_from_situation (^$(explore)$^): a pm#thing,
    pm#agent: an #airplane_pilot,     may have for pm#experiencer: several #passenger
](pm);
```

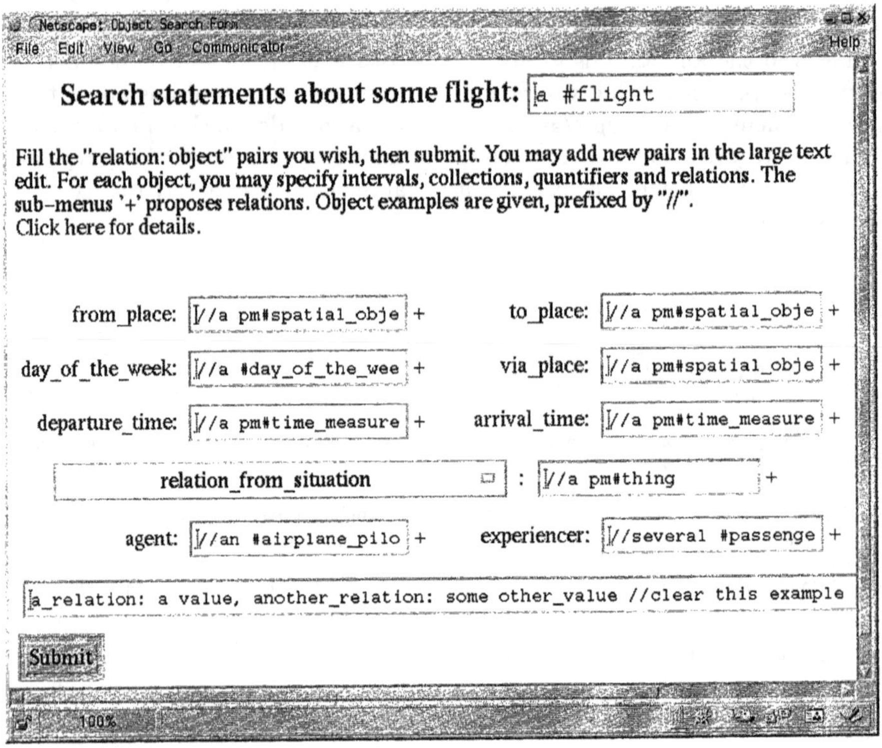

Fig. 1. A generated menu to help searching flights in the knowledge base.

7   Conclusion

The noun-related part of WordNet has been transformed into a genuine "lexical ontology" usable as a component in various knowledge-based applications: metadata registries, Yellow-pages like catalogs, query expansion, Semantic Web, etc. The focus was to guide and ease the representation, retrieval and sharing of general knowledge. This involved the generation of readable and unambiguous identifiers, the extraction of individuals, the merge with various top-level ontologies, and the correction of lexical and semantic problems. The result ontology is downloadable, browsable and extendible by anyone at http://www.webkb.org/.

Although I structured the top-level of WordNet and added a few links in other parts, the direct specializations of nearly all WordNet categories remain quite heterogeneous, with few exclusion links, and without distinction between role types and natural types. This lack of structure may be a problem for certain applications but fixing it may be as difficult as creating a better WordNet from scratch. Another problem is that distinctions in WordNet seem to have often been made not simply on semantic grounds but also on lexical grounds, thus leading to a multiplicity of "artificial" categories or categories that should be connected but are not. A few categories have been marked as "artificial" but many more would need to be similarly marked, or connected by specialization links, to improve knowledge normalization and retrieval.

The next step is to integrate other ontologies from the IEEE Standard Upper Ontology library [19], in particular the Suggested Upper Merged Ontology (SUMO), and the DAML Ontology Library [20], in particular the CIA World Factbook. In the mapping that has been done between the SUMO and WordNet, one SUMO categories is often linked to several WordNet categories. That will give cues to find and mark many WordNet categories as "artificial".

References

1. G. Miller. Wordnet: a Lexical Database for English. *Communications of the ACM*, 11:39–41, 1995. *http://www.cogsci.princeton.edu/~wn/*
2. A. Gangemi, N. Guarino and A. Oltramari. Restructuring Wordnet's Top-Level. *AI Magazine*, 40(5):235–244, fall 2002.
3. N. Guarino and G. Vetere. Ontoseek: Content-based Access to the Web. *IEEE Intelligent Systems*, 14(3):70–80, October 1999.
4. A. Puder and K. Romer. Generic Trading Service in Telecommunication Platforms. In *Proceedings of ICCS'97*. LNAI 1257:551–565, August 1997.
5. C. Kwok, O. Etzioni and D. Weld. Scaling Question Answering to the Web. *ACM Transactions on Information Systems*, 19(3), July 2001.
6. R. Colomb. Impact of Semantic Heterogeneity on Federating Databases. *The Computer Journal*, 40(5):235–244, 1997.
7. J. Hendler. Agents and the Semantic Web. *IEEE Intelligent Systems*, 16(2), 2001.
8. K. Knight and S. Luk. Building a Large-Scale Knowledge Base for Machine Translation. In *Proceedings of AAAI'94*, 773–778, Seattle, USA, July 1994.
9. P. Martin. Using the Wordnet Concept Catalog and a Relation Hierarchy for Knowledge Acquisition. In *Proceedings of Peirce'95*, Santa Cruz, CA, August 1995. *http://www.webkb.org/doc/papers/peirce95/*
10. P. Martin. Knowledge Representation in CGLF, CGIF, KIF, Frame-CG and Formalized-English. In *Proceedings of ICCS'02*, LNAI 2393:77–91. *http://www.webkb.org/doc/papers/iccs02/*
11. P. Martin. *Knowledge Representation, Sharing and Retrieval on the Web*. Web Intelligence (Eds.: N. Zhong, J. Liu, Y. Yao). Springer-Verlag, January 2003. *http://www.webkb.org/doc/papers/wi02/*
12. J. Sowa. *Conceptual Structures: Information Processing in Mind and Machine*. Addison-Wesley, Reading, MA, 1984.
13. J. Sowa. *Knowledge Representation: Logical, Philosophical, and Computational Foundations*. Brooks Cole Publishing Co., Pacific Grove, CA, 2000. See also *http://users.bestweb.net/ sowa/ontology/*
14. R. Guha and R. McCool. TAP: a system for integrating web services into a global knowledge base. 2002. *http://tap.stanford.edu/*
15. The WordNet 1.7 integration documentation. *http://www.webkb.org/doc/wn/*
16. The DOLCE Ontology. *http://wonderweb.semanticweb.org/*
17. The Suggested Upper Merged Ontology. *http://ontology.teknowledge.com/*
18. The CYC Top-level Ontology. *http://www.cyc.com/cyc-2-1/cover.html*
19. The SUO Library. *http://suo.ieee.org/refs.html*
20. The DAML Library. *http://www.daml.org/ontologies/*

Representing Temporal Ontology
in Conceptual Graphs

Pavel Kocura

Department of Computer Science
Loughborough University
Loughborough, Leicestershire LE11 3TU
England
P.Kocura@lboro.ac.uk

Abstract. The paper addresses the analysis and rational reconstruction of the semantic structure and logic of the temporal aspects of the knowledge representation formalism of Conceptual Graphs. The approach identifies the associated temporal concepts and relations, and analyses the behaviour of the atomic components, such as individual identifiers, type and relational labels. The definitions and illustrative examples use the notation of Peirce logic.

1 Introduction

Modern information processing places emphasis on the time dimension. Temporal information is essential in departmental data marts and enterprise data warehouses, in financial trading or general marketing databases, in the databases of pharmaceutical, bio-chemical or physical laboratories, or in criminal intelligence analysis systems. Such data can be used, for example, to predict future movements of share, commodity or currency prices (time series analysis). More advanced tasks require the representation of and reasoning with temporal qualitative knowledge. This includes formulating and enforcing temporal semantic constraints. Conceptual Graphs (CGs) provide a powerful and flexible system for building ontologies with rich temporal semantics.

In this paper, we address aspects of temporal representation in CGs different to those tackled by other authors, e.g. Esch [1] or Moulin [2]. In our view, the temporal aspects of CGs range over several levels of complexity. We will show that temporal information is not limited to the values of 1st-order objects, but applies also to predicates, and that the determination of the semantic correctness of a temporally qualified statement depends on the interplay between the 1st and 2nd order components.

2 The Temporal Dimension of Conceptual Graphs

2.1 The Structure of a Conceptual Graphs Model

A CGs-based model consists of the following structured components [3], each of which may have, depending on the domain, a temporal dimension:

A. de Moor, W. Lex, and B. Ganter (Eds.): ICCS 2003, LNAI 2746, pp. 174–187, 2003.
© Springer-Verlag Berlin Heidelberg 2003

- Set of unique individual identifiers: surrogates for individual objects of the domain;
- Type inheritance hierarchy: monadic predicates representing properties of domain objects;
- Relation inheritance hierarchy: generally n-adic predicates representing relations between and among domain objects;
- Definitions of non-primitive types and relations;
- Schemata and/or prototype definitions of types and relations (optional);
- Positive canonical models (semantic 'upper bound'), which determine the 'compulsory' conditions on the links of each type or relation to other types and relations;
- Negative canonical models (semantic 'lower bound'), which specify exceptions to the positive canonical models for each type or relation;
- Domain-specific rules that extend beyond pattern-oriented canonical models.

We can enter conceptual graphs into the knowledge base interactively, download them from an external source, or generate them automatically, as hypotheses, by multiple joins or merges. Each graph entered as new input or query has to be checked for compliance with the system's ontology which, for a dynamic domain, has to embody a semantic model of time. The knowledge base has to be tested, after each input, against this model. This semantic checking mechanism combines CGs projection with Peirce-logic-based reasoning [5]. The semantic checks apply the ontology models and rules to the growing or changing CGs knowledge base. This requires a principled representation of temporal information.

As an illustration of time-related aspects of a CGs model, we might use Frege's proverbial sentence *The present King of France is bald*. Giving him a name, John, and defining baldness as fewer than 10 hairs per cm^2 of scalp, we can represent this, first *sans* temporal information, as follows:

Fig. 1. The French King's scalp hair density is less than 10 hairs per cm^2

Looking at the concept [French_King:John], we can surmise that it has, implicitly, four temporal parameters describing:

- When France had a king;
- when John was alive and, if true
- when John was the French king;
- that John's purported reign as the French king is current.

We feel that these four items do not have the same importance. Whether it is currently true that John is the French King is just one bit of 'disembodied' information (assuming he exists and France is a monarchy), when John reigned gives us a longer-term view, *if* and *when* John lived is relevant for any statement about him (potentially many facts). Whether France has currently a King is part of the ontology of French Constitution. This 'weighting' will determine what temporal information we represent and at which level or in which part of the knowledge base, and how we will access it. At object level, a conceptual relation may have at most one piece of temporal information associated with it (validity), and one more at higher-order level (existence). Concepts are more complex, consisting of a type label, referent and conformity relation. Each of these has its own temporal aspect. The temporal dimension of concepts or relations may be irrelevant within the time scale of the domain. This usually concerns concepts representing attribute values, e.g. the relation AVERAGE_SCALP_HAIR_DENSITY. We may safely assume that the *general property* (2nd order) of having hair is 'timeless' – it has no temporal dimension. However, the *actual* hair density for a specific individual (1st order) is time-dependent. It changes, and we may want to show the progress of John's alopecia over time:

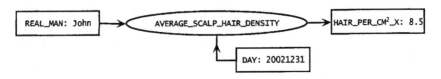

Fig. 2. On the last day of 2002, John boasted 8.5 hairs per cm^2 of his scalp

The second concept in the graph in Fig.2, [HAIR_PER_CM^2_X:8.5], has no relevant temporal aspects. There are no possible instants in the time scope of our domain (or any domain) when number 8.5 (or any other number) did not exist. The type label HAIR_PER_CM^2_X describes a property that has existed as long as any sort of hairs were in existence (this also includes hairs on animals, insects, plants, bacteria, etc.). This time interval easily subsumes our domain. The relationship between the type label HAIR_PER_CM^2_X and the referent, 8.5, is not logical in principle, i.e. the type label is not a predicate. The concept does not imply conformity between an identifier and a type, but rather a 'data type declaration', which has no temporal meaning associated with it. Thus, the Fig.2 graph contains one concept where each component has a valid temporal dimension, one relation with only one temporal aspect of the possible two, and the components of the last concept are temporally dimensionless. To better illustrate our treatment of temporal information in CGs, let us paraphrase Frege's example by the following graph:

Fig. 3. Mary, the common-law wife of John, the current King of France, is an astronaut

It still does not include explicit temporal components, but unlike the 2nd concept in Fig.2, the 2nd concept of the Fig.3 graph can and must be, eventually, temporally specified. Provided John insists on using his royal title, how could we verify his claim? We could ask him to bring his 'royal diploma,' witnesses of his coronation, or get any other independent corroboration. We would have to go through such a rigmarole each time anybody would claim to be the current French monarch. Similarly, we would have to search for evidence of John's marriage to Mary, and for Mary being an astronaut. Fortunately, we can avoid some or all of the expensive search or inference. We know that our contemporary John cannot be the French King because France is not currently a monarchy – the type label FRENCH_KING is not a valid predicate – it does not exist in our valid ontology.

Translating our intuitive temporal semantic check into CGs-speak, the same way we perform automatic semantic checks for the violation of positive and negative canonical models [3], we can verify the temporal consistency of any statement as long as we have the relevant temporal information for each part of each component of our CGs model. For an elementary conceptual graph consisting of a single concept, or for a relation linking two or more concepts, the temporal scope is given by the

1. existence in the domain of the object represented by the concept referent;
2. existence in the domain of the property represented by the type label of the concept;
3. existence in the domain of the property represented by the linking relation;
4. the time interval when the property was applicable to the referent representing the object, i.e. the temporal scope of the conformity of the referent to the type label;
5. the time interval during which the given objects were in the given relationship, i.e. when the relation linking the objects (concepts) was actually valid.

Only if there is a non-empty intersection of the temporal scopes in (1) and (2) does it make sense to try to determine the scope in (3), (4) or (5) and thus verify the temporal validity of the whole model and, possibly, the validity of it for a specific time point.

2.2   Temporal Dimension of Referent-Type Conformity

Let us assume, for the moment, that the interval during which an object existed in the domain and the interval during which a property existed in the domain form a non-empty intersection that subsumes the interval throughout which the object had the property. We can assume a primitive 2nd-order relation CONFORM4 that links an object and a property to the start and end times of the time interval during which the object has the property, as shown Fig.4.

The representation in Fig.4 does not make temporal information explicit in a 1st-order graph, as it is done in Fig.1. Should we need direct access to the time data, which is often the case with role types, we can use the relation in Fig.5.

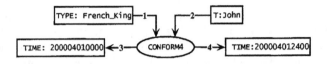

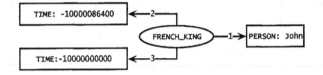

Fig. 4. John was the French King for one day, 1st April, 2000

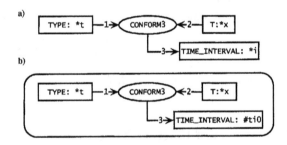

Fig. 5. John was the King of France from midnight to midnight on 1st January, 2003

The referents of the [TIME] concepts in Fig.5 show the 'computer time'. This can be converted into the required format, e.g. 01/01/2003.

The role type FRENCH_KING is represented in Fig.5 as a 'hanging' monadic relation with temporal validity - effectively a ternary relation. In contrast, PERSON is a natural type to which John conforms throughout his living existence. If the system differentiates between natural and role types, temporal aspects of conformity to natural types need not be explicitly recorded.

In CONFORM4, the type and the identifier are the 'inputs', and the time concepts the 'outputs'. We can simplify this by using the triadic relation CONFORM3, which replaces the time points with a time interval. This is a 'virtual' relation, used here only to avoid multitudes of time points in our examples.

Fig. 6. a) Object x conformed to type t during time interval i; b) It is not true that any object conforms to any type during the empty interval

The interval concepts are also 'virtual'. The type TIME_INTERVAL implies an uncountably infinite set of individual identifiers. This would violate the principles of representational parsimony, cause major computational problems, and we would still have to use explicit [TIME] concepts to compare time-interval magnitudes, ordering, intersections, etc.

The graph in Fig.4 splits a concept into its type label and its referent. Using a role type that does not clearly imply a potential empty interval, could we simplify this by using the relation THRU, as shown next?

*[NETWORK_MANAGER: John]→(THRU)→[TIME_INTERVAL:*i].

? John was a network manager throughout the time interval i

Not really. The main cause of the graph's semantic unsoundness is that the relation (THRU) needs simultaneous access to both parts of the concept [NETWORK_MANAGER:John] - the type label and the referent. In this example, (THRU) links only to the referent, as originally defined by Sowa for the Φ relation [8]. The type label of [NETWORK_MANAGER:John] is entirely outside the domain of (THRU), which is essentially triadic, synonymous to the relation (CONFORM3).

Sowa shows [7, p269] that, for type labels with dynamic conformity, coreferent links may introduce implicit semantic, temporal inconsistency, as in

*[BABY:Tom]- - - -[ADULT:Tom].

Sowa solves this by putting the concept (or graph) in question into the referent of an aparently 2nd order concept [Situation], which is then linked to the relevant temporal information, as in

? [Situation: [Baby: Tom]]→(PTim) →[Year:1976].

We introduce our treatment of this problem in the next section. Note that the type label SITUATION is not a predicate in the strict sense of the term, i.e. we cannot or do not care to assign a truth-value to an instantiated concept [SITUATION]. It is rather a 'data type' declaration or, ultimately, an encapsulation of a temporal context.

2.3 Temporal Context

Sowa's type label SITUATION implies multiple meanings and lacks well-defined semantics. We provisionally use the notion of a *temporal context*, [TC], which 'corrals' the temporal values associated with a given graph. A graph is correct with respect to temporal semantics if the intervals associated with its temporal context form a non-empty intersection. The equivalence in Fig.7 illustrates the relationship between the relations (THRU), (CONFORM3) and the [TC] concept.

The notation in the graph in Fig.7, and further, on is a shorthand for logical equivalence, i.e. ¬(A ∧ ¬B) and ¬(B ∧ ¬A). The dashed links represent lines of identity. You may notice that the coreferent links point to specific parts of the concept box – the type label or the referent. In some cases, this provides a connection between 1st-order and higher-order representation.

Similar to Sowa's SITUATION, the type label TC does not correspond to a 2nd-order monadic predicate – it is used here to point out the components whose temporal information is being reasoned about.

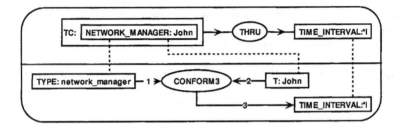

Fig. 7. John conformed to the type NETWORK_MANAGER during interval i IFF the temporal context of John being a manager lasted through interval i

2.4   Temporal Validity of Individual Concept Referents

An individual identifier is associated with the instants of the 'creation' and 'destruction' of its domain object. Obviously, for an object that exists throughout the life of the domain, such information would be superfluous. However, for dynamically created objects, a child being born or a freshly baked cake, say, their identifiers and creation times must be added to the model. The destruction of an object (the cake being eaten) is reflected in the model not by destroying the identifier, but rather by making it invalid for any relation that occurs after the object's demise. Subsequently, only historic queries can use the 'retired' referent. We can represent the existence of an object, using (CONFORM3), [TC] and (THRU), as conformity to the universal type T:

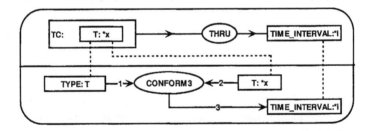

Fig. 8. Object existence as conformity to the universal type

2.5   Temporal Dimension of Type Labels

In a real-world domain, properties may emerege or disappear, and in its model, corresponding predicates may gain or lose validity. In the very long run, any predicate will have a temporal dimension. Even natural types will evolve and become extinct. For example, the type MAMMAL became valid relatively late in geological time. The type TZAR_OF_RUSSIA has not been legally valid since 1917. Certain factions in Russia are promoting Monarchy, so the predicate might become valid again. Dynamic properties characterize role types. Individuals may acquire or discard role predicates of their own accord, or the change may be imposed on them. For example, imagine that John's career progressed from FITTER to LINE_MANAGER, NETWORK_MANAGER and eventually to CEO. Assume that these relation labels exist throughout the time scope of the knowledge base, and John's birth and demise fall outside this interval. We may model the temporal validity of domain properties (types) in the same way as when modelling the existence of a domain object:

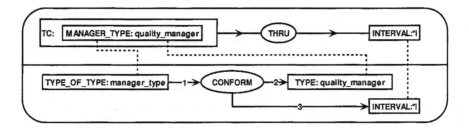

Fig. 9. The type QUALITY_MANAGER existed throughout interval i (the company did not have one before a certain date, and abolished the position eventually)

2.6 Temporally Exclusive Type Labels

A domain ontology must include temporal constraints on its role predicates. These constraints specify globally which type labels can or cannot apply to the same individual at the same time, e.g. the same serviceman cannot be simultaneously a corporal and a five-star general. Let us call such types *temporally exclusive*. We may specify sets of temporally exclusive type labels, e.g. [MILITARY_RANKS_SET] for exclusive military rank types, TYPE $>...>$ MILITARY_RANK, such as [MILITARY_RANK:captain], etc. For any single individual, the intersection of the intervals associated with temporally exclusive type labels will be the empty interval (Fig.10, 11).

Now we can specify rules of temporal ordering or similar constraints on role types, e.g. the sequence for a person is BABY, TODDLER, SCHOOLCHILD,...,TEENAGER, ADULT, etc. We can also formulate rules prohibiting a 'reverse' transition from a later to an earlier stage, e.g. from ADULT to TODDLER.

2.7 Conjunction of Temporal Contexts

When asserting multiple concepts, we reconcile their temporal contexts by constructing a new temporal context whose interval is the intersection of the intervals of the contexts of the inputs, as shown in Fig.13.

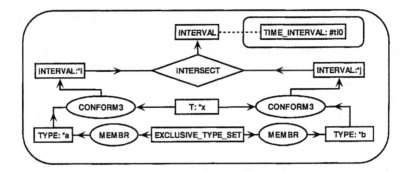

Fig. 10. If type a and type b belong to an exclusive type set then the intersection of their conformity intervals for the same individual is the empty interval

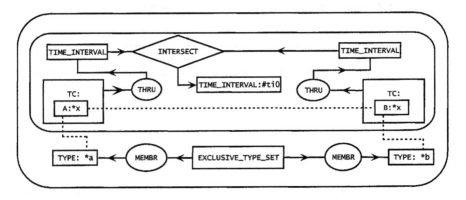

Fig. 11. If type a and type b are elements of an exclusive type set then the intersection of the temporal contexts of any individual conforming respectively to these types is the empty interval

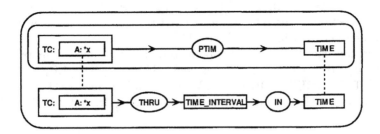

Fig. 12. If a temporal context is valid through an interval and time t is in that interval, then the context is valid at time t

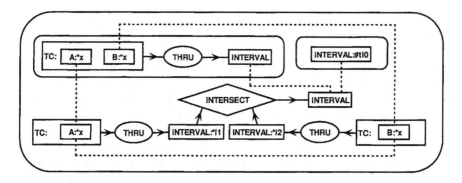

Fig. 13. IF A(x) is valid during interval i1, B(x) during interval i2, and the intersection of i1 and i2 is a non-empty interval i3, THEN {A(x) & B(x)} is valid during i3

2.8   Temporal Dimension of Conceptual Relations

Relations with a temporal dimension can be viewed as states. For example, unless people's affections last for ever, in [MAN:John]→(LOVE)→[WOMAN:Mary], (LOVE) might be expanded into a state, i.e.

[MAN:John]←(EXPR)←[LOVE]→(THEME)→[WOMAN:Mary].

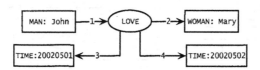

Fig. 14. John loved Mary from 1st till the 2nd May, 2002

This distinction blurs if we augment (LOVE) with explicit temporal information (Fig.14). Both the concept and relation versions of LOVE can represent different times when John loved Mary, e.g. they were childhood sweethearts, then parted, met again, got married, and so on. However, only the concept (state) version can add additional information, e.g. how or why John loved Mary, etc. Relations requiring only temporal specification should not be converted into states: creating and then 'deactivating' individual identifiers for multiple transient states goes against the spirit of representational parsimony.

An example by Sowa, taken from the CGs mailing list, suggests augmenting the dyadic relation (On) with "arbitrarily many additional arguments" as in:

(On)-
-1->[Cat]
-2->[Mat]
-3->[Time: <2001, Jan, 11, 16:12:05 UCT>]
-4->[Location: <73W 53' 35", 41N 12' 27">]
-5->[Reporter: 'John F. Sowa'].

The concept [REPORTER] in the graph above is obviously not a regular 1st-order argument of the (ON) relation. It is not part of the model representing cats, mats, 'on-ness', etc., and could not access the arguments it allegedly reports on. The type label REPORTER does not do the job it is meant to do, i.e. tell us who supplied the information about the cat. That function would have been done by the 5th arc of the (ON) relation, if it were 1st order, that is. However, the 5th arc stands for a meta-level relation, such as, say,

$$[MAN:J.F.Sowa]\leftarrow(ORIGINATOR)\leftarrow[GRAPH].$$

The time concept attached to the 3rd arc contains a complex referent, which combines 7 distinct components: *year, month, day of the month, hours, minutes, seconds and the time standard, UTC.* As syntactic sugar, it is fine. However, for theoretical and implementation reasons, we eschew concepts with complex referents. The Φ operator, as it was defined, cannot map complex referents into FOPL. We can hide the complexity of the referent of the time concept by using 'computer time', i.e. a single integer representing time elapsed from a fixed time point in the past, or that will elapse before reaching a future point. The reasoning mechanism then recalculates, on the fly, each such referent into the required format. Alternatively, we may use a 6-adic relation whose arcs specify the appropriate data types of the relevant temporal values:

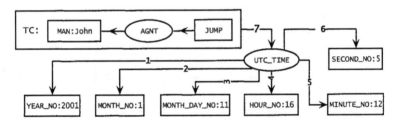

Fig. 15. On 11/01/2001, at 16:12:05, John jumped

The type labels of the temporal data type concepts above read *year-number, month-number*, etc. The standard time prefix, in this case UTC, for the time relation makes it possible to compare data originating from different time zones and convert them.

Clumping so much information into a single relation may not be such a good idea, as the same relation label could potentially appear in multiple versions, each with a different arity. On the other hand, 'multi-adic' relations provide a good vehicle for established, regularly repeating patterns of knowledge that would be elsewhere represented by frames. Assuming there is no other information that could be attached to the spatial relation (ON), we augment it only with the temporal information (Fig.16). Notice that any such relation will behave similarly to our primitive triadic relation (CONFORM3).

In Fig.17, the time intervals would only be non-empty if the cat's sitting on the mat was associated with a sub-interval of both Linda's being a cat and the

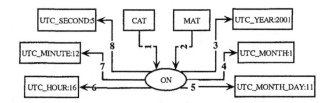

Fig. 16. "Binary" relation (ON) with expanded temporal information

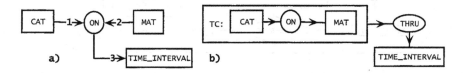

Fig. 17. a) ON ternary, b) ON binary: A cat is on a mat during an interval

mat's existence. In graph a) of Fig.17, we do not consider the temporal contexts for [CAT:Linda] and [MAT]. We may not need them if the type labels CAT and MAT exist, and Linda and the mat conform to their respective natural type labels throughout the life of the model.

3   Temporal Scenarios

In a model of any specific domain, the objects, types, relations and their conformities may be temporally either *static* or *dynamic*. A temporally static object, type, relation, conformity or validity always exists in the model. A dynamic object, predicate or validity may arise or terminate. We need to analyse the dynamic properties of domain elements. Given all the possible combinations - three items, each with two options - there are eight hypothetical static-dynamic scenarios some of which, however, are logically impossible or of little practical interest. Let us illustrate the last scenario, *Dynamic Referents - Dynamic Predicates- Dynamic Conformity*, by resolving the French King's problem.

To establish the truth-value of the graph in Fig.18, we have to determine first whether its temporal semantics is valid, namely, there is:

1. An intersection of John's life with the existence of French Monarchy;
2. an intersection of Mary's life with the reign of astronauts;
3. an intersection of John's and Mary's relationship with the time when the institution of Common Law Marriage was legal;
4. an intersection of all these intervals above.

Only then does it make sense to ask about the truth-value of the statement. This means that, without taking into consideration its temporal semantics, Frege's old example does not make sense.

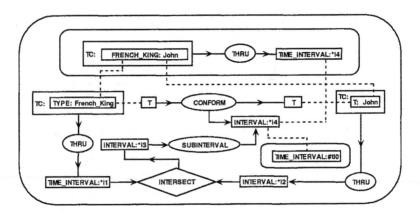

Fig. 18. If the type FRENCH_KING is valid through interval i1, the individual John exists through interval i2, both of which intersect to form i3, which has a non-empty subinterval i4 during which FRENCH_KING::John, then the temporal context for [FRENCH_KING:John] is valid through i4

4 Conclusions

We have attempted to analyse the temporal structure of a CGs-based model, and went some way towards a system that enables us to determine its temporal consistency. In particular, we have analysed the representation and semantic consistency of graphs whose referents, types and n-adic relations evince dynamic behaviour. The issue of dynamic properties has been also studied by Guarino and Welty, e.g. [6], whose notion of rigid properties roughly corresponds to natural types, that of non-rigid properties broadly corresponds to role types, and their dependency constraint may be seen as part of our canonical models [3]. However, their approach addresses other issues than temporal aspects of such predicates.

This project is by no means finished: in particular, the temporal structure and logic of the meta-level components of a CGs model, e.g. of the inheritance hierarchies, definitions and the canonical models will require further intensive study. It appears that these components, together with the analysis of temporal scenarios provide natural starting points in constructing the temporal dimension of a domain ontology. The different version of the representation and semantic structure have been tested using a Peirce logic theorem prover [5], experimenting progressively with more complex models. This work has highlighted a number of interesting problems concerning the underlying theory and its implementation, e.g. how to use the domain ontology to minimize the computation necessary to determine the semantic or temporal validity of a model. The approach described in the paper has also been used to reconstruct, in terms of CGs, the Situation and Event Calculi. Such work, however, is beyond the scope of this paper and will be presented at another opportunity.

References

1. John W. Esch and Timothy E. Nagle: Representing Temporal Intervals Using Conceptual Graphs. In: Laurie Gerholz and Peter Eklund (eds): Proc: Fifth Annual Workshop on Conceptual Structures, pp 43-52, Boston, USA and Stockholm, Sweden, 1990.
2. Moulin, Bernard & Dumas, Stéphanie: The Temporal Structure of a Discourse and Verb Tense Determination. In: Proc. Int.Conf. on Conceptual Structures, ICCS'94, College Park, Maryland, USA, August 16-20, 1994, LNCS 835 Springer 1994, ISBN 3-540-58328-9, pp 45-68.
3. Kocura, P.: Conceptual Graphs Canonicity and Semantic Constraints, In: Proc. 4th Int.Conf. on Conceptual Structures (b), ICCS'96, Sydney, Australia, ISBN 0 7334 1387 0, pp 133-145.
4. Kocura, P.: Semantics of Attribute Relations in Conceptual Graphs, Conceptual Structures: Logical, Linguistic and Computational Issues, In: Ganter, B., Mineau, G.W. (eds.): Proc. 8th Int.Conf. on Conceptual Structures, ICCS 2000, Darmstadt, August 2000, LNAI 1867, Springer, ISBN 3-540-67859-X, pp 235-248.
5. Heaton, J.E. and Kocura, P.: Open World Theorem Prover for Conceptual Graphs, In: Stumme, G. (ed.): Working with Conceptual Structures, Proc. 8th International Conference on Conceptual Structures, Shaker-Verlag, Aachen, ICCS 2000, Darmstadt, Germany, August 2000, pp 215-228, ISBN 3-8265-7669-1.
6. Guarino, Nicola and Chris Welty: A Formal Ontology of Properties. In: Dieng, R., and Corby, O., (eds): Proc. EKAW-2000: The 12th International Conference on Knowledge Engineering and Knowledge Management. Springer-Verlag, LNCS Vol. 1937:97-112. October 2000.
7. Sowa, John F.: Knowledge Representation: Logical, Philosophical and Computational Foundations, Brooks/Cole 2000, ISBN 0-534-94965-7.
8. Sowa, John F.: Conceptual Structures: Information Processing in Mind and Machine, Addison-Wesley, 1984, ISBN 0-201-14472-7.

Time Dimension, Objects, and Life Tracks.
A Conceptual Analysis

Karl Erich Wolff[1] and Wendsomde Yameogo[2]

[1] Department of Mathematics and Science, Darmstadt University of Applied Sciences
Schoefferstr. 3, D-64295 Darmstadt, Germany
wolff@mathematik.tu-darmstadt.de,
[2] Department of Computer Science, Darmstadt University of Applied Sciences
Schoefferstr. 3, D-64295 Darmstadt, Germany
wendsomde@yameogo.com

Abstract. The purpose of this paper is to clarify in the framework of Temporal Concept Analysis the distinction between the notion of the dimension of time and the one-dimensionality of life tracks of simple objects. The reader is led to the central problems by many examples.

1   Introduction: Problems in the Representation of Time

At first we mention some basic problems arising in the representation of temporal phenomena. These problems led the authors to the introduction of conceptual structures for an appropriate description of processes. The mathematical background is Temporal Concept Analysis [23–27] which is based on Formal Concept Analysis [8, 9, 20–22].

1.1   The Zenonian Paradox of the Flying Arrow

One of the most famous problems in the representation of temporal phenomena is the Zenonian paradox of the flying arrow [2, Physics, 239b30ff] which is in each moment of time at rest - but is flying. The classical solution of this paradox is based on the representation of space and time by the continuum of the real numbers; the arrow may be described by a single point x which moves along a differentiable curve in euclidean 3-space, and the derivation at a time point t yields the non-zero velocity of the arrow. That is a nice and very successful mathematical model and many people believe that the reality is like that - but how can we know that? Let us assume that we construct a virtual reality, simply by making a movie of a flying arrow. Then there are only finitely many pictures, each of them shown for a moment, and in each moment the arrow is at rest.

Therefore, it is impossible to decide by finitely many measurements whether the flying arrow is at rest or flying in each moment. A formal representation of moments is one of the leading ideas in the construction of conceptual time systems (see section 2 and 3).

A. de Moor, W. Lex, and B. Ganter (Eds.): ICCS 2003, LNAI 2746, pp. 188–200, 2003.

1.2 Aristotle and Kant: Change and Movement of Objects

Aristotle [2, Categories, 2a] has carefully discussed space and time as two of his ten categories, he studied [2, Physics, 200bff] change and movement, continuum and infinity, place and even empty place, bodies and objects, time order and the time point "now". In his discussion of the *object-space-time-process* [2, Physics, 231-241] he studies the conceptual foundations of descriptions of objects in space and time.

Immanuel Kant [15, Transzenentale Ästetik, 2-4] called space and time "necessary imaginations a priori" (*"notwendige Vorstellungen a priori"*), either infinite, space of dimension three and time of dimension one. In his transcendental discussion of the concept of time Kant states [15, Transzendentale Ästhetik, §5]:

Hier füge ich noch hinzu, daß der Begriff der Veränderung und, mit ihm, der Begriff der Bewegung (als Veränderung des Orts) nur durch und in der Zeitvorstellung möglich ist: daß, wenn diese Vorstellung nicht Anschauung (innere) a priori wäre, kein Begriff, welcher es auch sei, die Möglichkeit einer Veränderung, d.i. einer Verbindung kontradiktorisch entgegengesetzter Prädikate (z.B. das Sein an einem Orte und das Nichtsein eben desselben Dinges an demselben Orte) in einem und demselben Objekte begreiflich machen könnte.

Our translation:

The concept of change and, together with it, the concept of motion (as a change of the place) is possible only by and within an idea of time: that, if that idea would not be an (internal) perception a priori, no concept whatsoever, could make comprehensible the possibility of a change, i.e. a connection of contradictory opposite predicates (e.g. being at a place and not being of just the same thing at the same place) in one and the same object.

We shall give a formal description of a "conceptual time system with actual objects and a time relation" (CTSOT) which takes into account conceptually what Kant describes. That is done by defining objects using an identification and a "changement (or time) relation" among observed "actual objects" or "phenomena".

1.3 Poincaré and Einstein: Moments, Places, and Inaccuracy

Contradicting Kant's 'a priori' Poincaré [10] pointed out that mathematical theories should be used conventionally. That led to the possibility of using not only points (t, x, y, z) of the euclidean 4-space for the representation of a moment t and a place (x, y, z), but also to represent multi-dimensional places $(x_1, \ldots, x_n)$ ("states") or (preparing our own notions) "situations" $(t, x_1, \ldots, x_n)$ as points in a multi-dimensional euclidean space. Then the difference vector between two states (or situations) seems to be a good description of a "movement". But then the objects are not represented formally, only their states (or situations), but even not that, since a spatial object usually needs a little bit more than just a single point as its volume. Therefore, physicist like to talk about "infinitely small particles". The difficulties which arise from that simple representation of moments, places and objects can be seen from the following text by Albert Einstein [5, p. 415]:

Zur örtlichen Wertung eines in einem Raumelement stattfindenden Vorganges von unendlich kurzer Dauer (Punktereignis) bedürfen wir eines Cartesischen Koordinatensystems, d.h. dreier aufeinander senkrecht stehender, starr miteinander verbundener, starrer Stäbe, sowie eines starren Einheitsmaßstabes. ... Die Geometrie gestattet, die Lage eines Punktes bezw. den Ort eines Punktereignisses durch drei Maßzahlen (Koordinaten x, y, z) zu bestimmen.... Für die zeitliche Wertung eines Punktereignisses bedienen wir uns einer Uhr, die relativ zum Koordinatensystem ruht und in deren unmittelbarer Nähe das Punktereignis stattfindet. Die Zeit des Punktereignisses ist definiert durch die gleichzeitige Angabe der Uhr.

Our translation:

For the spatial evaluation of an event of infinitesimal duration (point event) in a spatial element we need a cartesian coordinate system, i.e. three pairwise orthogonal, rigidly connected, rigid rods, and a rigid unit measure.... The geometry allows for the determination of the position of a point respectively the place of a point event by three numbers (coordinates x, y, z).... For the temporal measurement of a point event we use a clock which is at rest relative to the coordinate system such that the point event happens very near to it. The time of the point event is defined by the simultaneous time state of the clock.

Albert Einstein was aware of these difficulties even two years earlier. In his famous paper [4] (1905) "Zur Elektrodynamik bewegter Körper" where he introduced the special theory of relativity he wrote in the footnote on page 893:

Die Ungenauigkeit, welche in dem Begriffe der Gleichzeitigkeit zweier Ereignisse an (annähernd) demselben Orte steckt und gleichfalls durch eine Abstraktion überbrückt werden muß, soll hier nicht erörtert werden.

Our translation:

The inaccuracy which lies in the concept of simultaneity of two events at (about) the same place and which has to be bridged also by an abstraction, shall not be discussed here.

The acceptance of a scientific treatment of inaccuracy by formal methods grew very slowly during the twentieth century because of the huge success of classical continuous mathematics in the natural sciences as well as the progress of discrete methods in computer sciences. But it remained a conceptual gap between continuous and discrete theories. Now, this gap can be bridged by the methods of Formal Concept Analysis.

1.4 Discrete Representations of Time and States

The usual representation of the movements of figures on a chess board is a typical example of a discrete representation of a temporal process in some finite space. In many of these discrete problems we need only a discrete time representation, usually chosen as the set of integers with an appropriate structure, sometimes just with the usual ordering, sometimes with an ordered semigroup (or similar) structure.

Automata theory [1, 3] is based on the notions of states and transitions without an explicit time description. Mathematical System Theory [14, 16, 17] has tried to combine the notion of a state with an explicit time description, but without success as Zadeh [30, p. 40] states:

To define the notion of state in a way which would make it applicable to all systems is a difficult, perhaps impossible, task. In this chapter, our modest objective is to sketch an approach that seems to be more natural as well as more general than those employed heretofore, but still falls short of complete generality.

The first author [23] has introduced conceptual time systems with an explicit general time description such that the notion of state can be defined as generalizing the usual notions of states in physics [13] as well as in automata theory [1, 3]. Before explaining the main ideas we first discuss the problems around granularity in the next subsection.

1.5 Reasoning through Granularities

In the following we use the term "granularity" (inspired by Lotfi Zadeh) in connection with descriptions using sets of subsets of a given set, for example partitions. A typical example is the granularity in time described by days, weeks, months and years. In colloquial speech we say that the statement "In July Walter took a flight from Little Rock to Frankfurt" implies the statement "In the summer Walter travelled from the US to Germany". To describe such a reasoning through granularities we need a formal representation of granularity. There are many approaches in the literature, the two most famous ones, namely Fuzzy Theory [31, 32] and Rough Set Theory [18] are in some sense just special cases of Conceptual Scaling Theory [8] as shown by the first author [28, 24].

The importance of granularity lies also in the fact that the notion of "state" depends on the chosen granularity. That is obvious from the observation that the number of states of a system grows (usually rapidly) with a refinement of the granularity.

Based on a clear conceptual system description and the notion of states of such a system we can introduce life tracks of objects (e.g. in the state space) which leads us to distinguish the time dimension of a system from the usually one-dimensional life tracks of objects. That will be discussed in the next sections.

2   Temporal Concept Analysis

Temporal Concept Analysis was introduced by the first author [23–27] as the theory of temporal phenomena described with tools of Formal Concept Analysis [8, 9, 20]. In the following we assume that the reader knows the basic facts in Formal Concept Analysis and Conceptual Scaling Theory. For an introduction we refer to [21, 22].

The development of Temporal Concept Analysis started with the introduction of conceptual time systems [23] such that *states* could be defined as object concepts of time points. Then *transitions* in conceptual time systems with a time relation were introduced in [26]. That led to the definition of the life track of such a system. In a third step the idea that each object should be described by a conceptual time system with a time relation led to the definition of a CTSOT, a conceptual time system with *actual objects* and a time relation [27]. That led to the "map reconstruction theorem" which roughly says that each automaton can be represented by a CTSOT, proving that the notions of states and transitions in a CTSOT "cover" the notions of states and transitions in automata theory. Therefore, CTSOTs provide a common generalization of (possibly continuous) physical systems as well as discrete temporal systems.

To continue our discussion about the relation between objects, space and time we study the Venus example from [25], demonstrating a typical object construction and the representation of life tracks of objects.

2.1   Example: Morning Star - Evening Star

In the following example we demonstrate the scientific process of theory extension based on some background theory. An important step in theory extension is the construction of new objects. The following Table 1 reports some observations of "luminous phenomena" at the sky.

Table 1. Observations of stars

Observation	Day	Daytime	Space	Brightness
1	Monday	morning	east	luminous
2	Monday	evening	west	luminous
3	Tuesday	morning	east	luminous
4	Tuesday	evening	west	luminous

For example, the observation labeled "1" states that at Monday morning in the east a luminous phenomenon was observed.

We represent this table formally as a many-valued context and describe the background theory of the meaning of the values by conceptual scales. Using nominal scales as in [25] the derived context [25] has the concept lattice represented by the line diagram in Figure 1.

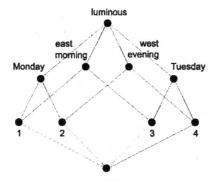

Fig. 1. A line diagram of the concept lattice of the nominally scaled many-valued context in Table 1

Now, we give a short description of the main idea for the introduction of objects and their life tracks. If we assume that the four observations of Table 1 are "caused" by a single object, called "Venus", then it seems to be clear to draw the "life track" of Venus as in Figure 2. But until now we did not represent any

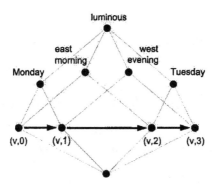

Fig. 2. The life track of Venus

ordering in our time description. Therefore, together with the introduction of an object we should represent formally that this object may be observed differently in different observations. Hence, each observation tells us something about the "actual object" observed during that observation. Therefore we introduce four actual objects for Venus, called (v,0), (v,1), (v,2), (v,3) and introduce a binary relation among the actual objects, usually a sequence of actual objects, in our example: $(v,0) \rightarrow (v,1) \rightarrow (v,2) \rightarrow (v,3)$.

We interpret the labels 0,1,2,3 in the actual objects as the "eigentime" (also called "proper time" in physics) of the object Venus. The common letter "v" characterizes the object Venus, an arrow of the time relation, for example $(v,0) \rightarrow (v,1)$, is interpreted as "the change of (v,0) to (v,1)". Therefore, this time relation seems to be a suitable formal representation of the often used term "arrow of time" in physics [11, 12]. It also represents nicely the ideas of Kant [15] mentioned in section 1.2.

If we interpret the four observations of Table 1 as being caused by two stars, say the "Morning Star" and the "Evening Star" then we would draw the "life tracks" of them as in Figure 3. The corresponding conceptual time system with actual objects and a time relation (CTSOT) has the following data table:

Its time relation is given by : $(ms, 0) \rightarrow (ms, 1), (es, 0) \rightarrow (es, 1)$.

In the following section we introduce the time dimension of a CTSOT.

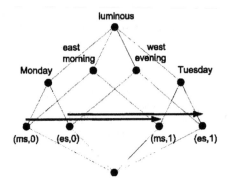

Fig. 3. The life tracks of Morning Star and Evening Star

3   Time Dimension and Life Tracks

After having demonstrated some examples we now need the formal definition of a CTSOT to introduce the time dimension of a CTSOT.

3.1   Basic Definitions

We just recall the basic definitions and relate them to the given examples.

Definition 1. *'conceptual time system'*
Let G be an arbitrary set (of time objects) and $T := ((G, M, W, I_T), (S_m | m \in M))$ and $C := ((G, E, V, I), (S_e | e \in E))$ scaled many-valued contexts (on the same object set G). Then the pair (T, C) is called a conceptual time system on G. T is called the time part and C the event part of (T, C). The scales S_m and S_e of the time and event part describe the chosen granularity strucure for the values in these many-valued contexts.

Example 1. Table 1 is the data table of a conceptual time system where $G :=$ $\{1, 2, 3, 4\}$, the two columns labeled "day" and "day time" represent the many-valued context of the time part, the two columns labeled "space" and "brightness" represent the many-valued context of the event part. The used nominal scales are described in [25].

Definition 2. *'conceptual time systems with actual objects and a time relation'*
Let P be a set (of 'persons', or 'objects', or 'particles') and G a set (of 'points of time') and $\Pi \subseteq P \times G$ a set (of 'actual objects'). Let (T, C) be a conceptual time system on Π and $R \subseteq \Pi \times \Pi$. Then the tuple (P, G, Π, T, C, R) is called a conceptual time system (on $\Pi \subseteq P \times G$) with actual objects and a time relation R, shortly a CTSOT. For each object $p \in P$ the set $p^{\Pi} := \{g \in G | (p, g) \in \Pi\}$ is called the (eigen)time of p in P (which is the intent of p in the formal context (P, G, Π)). Then the set $R_p := \{(g, h) | ((p, g), (p, h)) \in R\}$ is called the set of R–transitions of p and the relational structure (p^{Π}, R_p) is called the (eigen)time structure of p.

Table 2. The data table of a conceptual time system with actual objects

Actual Object	Day	Daytime	Space	Brightness
(ms,0)	Monday	morning	east	luminous
(es,0)	Monday	evening	west	luminous
(ms,1)	Tuesday	morning	east	luminous
(es,1)	Tuesday	evening	west	luminous

Example 2. Table 2 is the data table of a CTSOT where $P := \{ms, es\}$ is the set of Morning Star and Evening Star, $G := \{0, 1\}$ the set of points of time. In this example $\Pi = P \times G$ and $R := \{((ms, 0), (ms, 1)), ((es, 0), (es, 1))\}$. The relational structure $(\{0, 1\}, \{(0, 1)\})$ is the common eigentime structure of the Morning Star and the Evening Star. The conceptual time system of this CTSOT is described in the previous example.

Definition 3. *'life track of an object'*
Let (P, G, Π, T, C, R) be a CTSOT, and $p \in P$. Then for any mapping $f : \{p\} \times p^{\Pi} \to X$ (into some set X) the set $f = \{((p, g), f(p, g)) | g \in p^{\Pi}\}$ is called the f-life track (or f-trajectory or f-life line) of p. The two most useful examples for such mappings are the object mappings γ and γ_C of the derived contexts $K_T | K_C$ and K_C of the conceptual time system (T, C, R) on Π, each of them restricted to the set $p \times p^{\Pi}$ of actual objects. They are called the life track of p in the situation space and the life track of p in the state space respectively.

Example 3. The life tracks of "Morning Star" and "Evening Star" in the situation space (where the time and the event part is represented) are shown in Figure 3.

3.2 Dimensions

Definition 4. *'order dimension of an ordered set'*
An ordered set $(P, \leq)$ has order dimension $dim(P, \leq) = n$ if and only if it can be order embedded in a direct product of n chains and n is the smallest number for

which this is possible ([9, p. 236]). As usual, a chain is an ordered set $(M, \leq)$ *such that for any two elements* $a, b \in M$ $a \leq b$ *or* $b \leq a$.

Clearly, the order dimension of the ordered set $(\Re, \leq)$, the set of real numbers and their usual ordering, equals 1 and the n-fold direct product of this ordered set, the usual real n-dimensional space with the product order $(\Re^n, \leq)$ has order dimension n.

Definition 5. *'order dimension of a scaled many-valued context'*
The order dimension of a scaled many-valued context $((G, M, W, I), (S_m | m \in M))$ *is the order dimension of the concept lattice of its derived context.*

Definition 6. *'scale dimension of a scaled many-valued context'* *The scale dimension of a scaled many-valued context* $((G, M, W, I), (S_m | m \in M))$ *is the sum of the order dimensions of the concept lattices of the scales* $S_m (m \in M)$.

Since the concept lattice of the derived context of a scaled many-valued context can be order embedded in the direct product of the concept lattices of the scales (see [9, p. 77]) we obtain:

Lemma 1. *The order dimension of a scaled many-valued context is less or equal to its scale dimension.*

Definition 7. *'time dimension of a conceptual time system'*
Let (T, C) *be a conceptual time system. The time dimension of* (T, C) *is the order dimension of the time part* T. *Analogously, the (event or) space dimension of* (T, C) *is the order dimension of the event part* C.

Example 4. To demonstrate that these dimensions of a conceptual time system are a useful and for practical purposes relevant notion we study a formal representation of the following story:

A participant p of an international conference, let's say the ICCS 2003, travels on Sunday morning from Darmstadt to Dresden, arrives at Dresden in the afternoon, visits the conference until Wednesday afternoon, then he has to travel back to Darmstadt arriving there in the evening.

In Figure 4 we have drawn (in a simplified way) the life tracks of the participant p in the state space, in the 2-dimensional time scale, and in the direct product of these two lattices, namely the situation space. To sketch the precise construction we represent the data table of the corresponding CTSOT with a single person p in Table 3: The time relation R on the set of actual persons is given by $(p, 0) \to (p, 1) \to (p, 2) \to (p, 3)$. The two attributes "day" and "day time" of the time part T and the attribute "town" of the event part C are scaled by ordinal scales of order dimension one. Therefore, the scale dimension of the time part T is two. Since the derived context of the time part of this CTSOT has a chain as its concept lattice the time dimension equals one. Clearly, if we would extend the CTSOT by a new actual object, say $(p, 3.0)$ which was at Wednesday morning in Dresden and change the time relation to $(p, 0) \to (p, 1) \to (p, 2) \to (p, 3.0) \to (p, 3)$ then the time dimension would be two. That demonstrates why we are often interested only in the scale dimension.

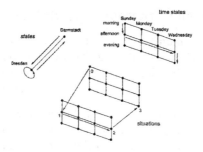

Fig. 4. Life tracks of a journey to Dresden: time scale dimension = 2

Table 3. The data table of the Dresden journey

actual person	day	daytime	town
(p,0)	Sunday	morning	Darmstadt
(p,1)	Monday	afternoon	Dresden
(p,2)	Tuesday	afternoon	Dresden
(p,3)	Wednesday	evening	Darmstadt

3.3 The Unique-State Theorem

A very popular statement, namely that each system is at each point of time in exactly one state can be made precise for CTSOTs and can be proven easily.

Definition 8. *Let (P, G, Π, T, C, R) be a CTSOT and $(p, g) \in \Pi$. We say that p is at time point g in the situation s (or that the actual object (p, g) is in the situation s) iff $s = \gamma((p, g))$; (p,g) is in the time state s iff $s = \gamma_T((p, g))$, and (p,g) is in the state s iff $s = \gamma_C((p, g))$.*

Theorem 1. *'unique-state theorem'*
Let (P, G, Π, T, C, R) be a CTSOT. Then each object $p \in P$ is at each time point g of its eigentime in exactly one situation, in exactly one time state and in exactly one state.

Proof. Let $(p, g) \in \Pi$, then clearly $\gamma((p, g))$ is the only situation s such that p is at the time point g in the situation s. In the same way the rest of the theorem can be proved.

Remark: We do not know any formal theory where this popular theorem was stated or even proved. It seems to be that some people use that statement as a kind of description of a state.

It seems to be obvious that as a consequence of the unique-state theorem the eigentime structure of an object looks like a (graph theoretic) chain, for example $0 \rightarrow 1 \rightarrow 2$. That is not true as shown in the next section.

3.4   Branching Time and Branching Objects?

The following example demonstrates a CTSOT with a single object and a "branching" time relation. It describes an "abstract letter", that is a class of "identical" copies of an original letter. As usual, we say that the abstract letter is sent from A to B iff a copy of the original letter is sent from A to B.

Example 5. An abstract letter is sent on day 0 from place A to place B arriving there on day 1, and on day 0 from A to the place C arriving there on day 2. Then the letter is sent on day 1 from B to the place D and on day 2 from C to D either arriving on day 3.

Some of the information is represented by the CTSOT of Table 4 with an ordinal scale for the attribute "day" and a nominal scale for the "place". The time relation is given by $(letter, 0) \rightarrow (letter, 1) \rightarrow (letter, 3)$ and $(letter, 0) \rightarrow (letter, 2) \rightarrow (letter, 3)$ Clearly, this "branching" abstract letter can be described by its copies as in Table 5 where the time relation on the actual objects is given by $(copy1, 0) \rightarrow (copy1, 1) \rightarrow (copy1, 3)$ and $(copy2, 0) \rightarrow (copy2, 2) \rightarrow (copy2, 3)$.

Table 4. A "branching" abstract letter

actual objects	day	place
(letter,0)	0	A
(letter,1)	1	B
(letter,2)	2	C
(letter,3)	3	D

Table 5. Two copies as "simple" objects

actual objects	day	place
(copy1,0)	0	A
(copy1,1)	1	B
(copy1,3)	3	D
(copy2,0)	0	A
(copy2,2)	2	C
(copy2,3)	3	D

Definition 9. *'simple objects'*

An object p of a CTSOT is called simple iff the restriction of the time relation R on the actual objects of p, i.e. $R \cap (\{p\} \times p^{\Pi})^2$, has a chain $c_p := (\{p\} \times p^{\Pi}, \leq_p)$ as its reflexive and transitive closure.

Let p be a simple object of a CTSOT and $f : \{p\} \times p^{\Pi} \rightarrow X$ a mapping into some set X. Then the f-life track $\{((p,g), f(p,g)) | g \in p^{\Pi}\}$ of p can be ordered in a trivial way isomorphically to c_p by the following definition of an order relation $\leq_f$ on $f : ((p,g), f(p,g)) \leq_f ((p,h), f(p,h)) :\Leftrightarrow (p,g) \leq_p (p,h)$. Hence the ordered life-track $(f, \leq_f)$ of a simple object p is one-dimensional in the sense that it is a chain (which is isomorphic to c_p). Note that f need not be injective.

4 Conclusion and Future Research

It is shown that Conceptual Time Systems with Actual Objects and a Time Relation (CTSOT) allow for distinguishing clearly between the time dimension of a conceptual time system and the order dimension of the chain c_p of a simple object of a CTSOT. The shown existence of non-simple objects and their applicability to ideas like an "abstract letter" open up interesting problems in the conceptual representation of central notions in other branches of science like for example in computer science where "information" and "object oriented programming" and its time-conceptual foundations (as developed by the second author in [29]) now can be investigated from a formal conceptual base. Clearly, also many fundamental problems in physics [11–13] and Temporal Logic [7, 19] will be studied using these methods.

References

1. Arbib, M.A.: Theory of Abstract Automata. Prentice Hall, Englewood Cliffs, N.J., 1970.
2. Aristoteles: Philosophische Schriften in sechs Bänden. Felix Meiner Verlag Hamburg 1995.
3. Eilenberg, S.: Automata, Languages, and Machines. Vol. A. Academic Press 1974.
4. Einstein, A.: Zur Elektrodynamik bewegter Körper. Annalen der Physik 17 (1905): 891-921.
5. Einstein, A.: Über das Relativitätsprinzip und die aus demselben gezogenen Folgerungen. In: Jahrbuch der Radioaktivität und Elektronik 4 (1907) : 411-462.
6. Einstein, A.: The collected papers of Albert Einstein. Vol. 2: The Swiss Years: Writings, 1900-1909. Princeton University Press 1989.
7. Gabbay, D.M., I. Hodkinson, M. Reynolds: Temporal Logic - Mathematical Foundations and Computational Aspects. Vol.1, Clarendon Press Oxford 1994.
8. Ganter, B., R. Wille: Conceptual Scaling. In: F.Roberts (ed.) Applications of combinatorics and graph theory to the biological and social sciences,139-167. Springer Verlag, New York, 1989.
9. Ganter, B., R. Wille: Formal Concept Analysis: mathematical foundations. (translated from the German by Cornelia Franzke) Springer-Verlag, Berlin-Heidelberg 1999.
10. Gottwald, S. (ed.): Lexikon bedeutender Mathematiker. Bibliographisches Institut 1990.
11. Hawking, S.: A Brief History of Time: From the Big Bang to Black Holes. Bantam Books, New York 1988.
12. Hawking, S., Penrose, R.: The Nature of Space and Time. Princeton University Press, 1996.
13. Isham, C.J.: Time and Modern Physics. In: Ridderbos, K. (ed.): Time. Cambridge University Press 2002, 6-26.

14. Kalman, R.E., Falb, P.L., Arbib, M.A.: Topics in Mathematical System Theory. McGraw-Hill Book Company, New York, 1969.

15. Kant, I.: Kritik der reinen Vernunft. In: Weischedel, W. (ed.): Immanuel Kant - Werke in sechs Bänden. Band II, Insel Verlag, Wiesbaden 1956 (first edition 1781).

16. Lin, Y.: General Systems Theory: A Mathematical Approach. Kluwer Academic/ Plenum Publishers, New York, 1999.

17. Mesarovic, M.D., Y. Takahara: General Systems Theory: Mathematical Foundations. Academic Press, London, 1975.

18. Pawlak, Z.: Rough Sets: Theoretical Aspects of Reasoning About Data. Kluwer Academic Publishers, 1991.

19. van Benthem, J.: Temporal Logic. In: Gabbay, D.M., C.J. Hogger, J.A. Robinson: Handbook of Logic in Artificial Intelligence and Logic Programming. Vol. 4, Epistemic and Temporal Reasoning. Clarendon Press, Oxford, 1995, 241-350.

20. Wille, R.: Restructuring lattice theory: an approach based on hierarchies of concepts. In: Rival, I.(ed.):Ordered Sets. Reidel, Dordrecht-Boston 1982, 445-470.

21. Wille, R.: Introduction to Formal Concept Analysis. In: G. Negrini (ed.): Modelli e modellizzazione. Models and modelling. Consiglio Nazionale delle Ricerche, Instituto di Studi sulli Ricerca e Documentatione Scientifica, Roma 1997, 39-51.

22. Wolff, K.E.: A first course in Formal Concept Analysis - How to understand line diagrams. In: Faulbaum, F. (ed.): SoftStat '93, Advances in Statistical Software 4, Gustav Fischer Verlag, Stuttgart 1994, 429-438.

23. Wolff, K.E.: Concepts, States, and Systems. In: Dubois, D.M. (ed.): Computing Anticipatory Systems. CASYS'99 - Third International Conference, Liège, Belgium, 1999, American Institute of Physics, Conference Proceedings 517, 2000, pp. 83-97.

24. Wolff, K.E.: A Conceptual View of Knowledge Bases in Rough Set Theory. In: Ziarko, W., Yao, Y. (eds.): Rough Sets and Current Trends in Computing. Second International Conference, RSCTC 2000, Banff, Canada, October 16-19, 2000, Revised Papers, 220-228.

25. Wolff, K.E.: Temporal Concept Analysis. In: E. Mephu Nguifo & al. (eds.): ICCS-2001 International Workshop on Concept Lattices-Based Theory, Methods and Tools for Knowledge Discovery in Databases, Stanford University, Palo Alto (CA), 91-107.

26. Wolff, K.E.: Transitions in Conceptual Time Systems. In: D.M.Dubois (ed.): International Journal of Computing Anticipatory Systems vol. 11, CHAOS 2002, p.398-412.

27. Wolff, K.E.: Interpretation of Automata in Temporal Concept Analysis. In: U. Priss, D. Corbett, G. Angelova (eds.): Integration and Interfaces. Tenth International Conference on Conceptual Structures. Lecture Notes in Artificial Intelligence, 2393, Springer-Verlag(2002), 341-353.

28. Wolff, K.E.: Concepts in Fuzzy Scaling Theory: Order and Granularity. 7th European Congress on Intelligent Techniques and Soft Computing, Aachen 1999. Fuzzy Sets and Systems 132 (2002) 63-75.

29. Yameogo, W.: Time Conceptual Foundations of Programming. Master Thesis. Department of Computer Science at Darmstadt University of Applied Sciences, 2003.

30. Zadeh, L.A.: The Concept of State in System Theory. In: M.D. Mesarovic: Views on General Systems Theory. John Wiley & Sons, New York, 1964, 39-50.

31. Zadeh, L.A.; Fuzzy sets. Information and Control 8, 1965, 338 - 353.

32. Zadeh, L.A.; The concept of a linguistic variable and its application to approximate reasoning. Part I: Inf. Science 8,199-249; Part II: Inf. Science 8, 301-357; Part III: Inf. Science 9, 43-80, 1975.

Representing Time and Modality in Narratives with Conceptual Graphs

Henrik Schärfe and Peter Øhrstrøm

Department of Communication, Aalborg University, Denmark
{scharfe, poe}@hum.auc.dk

Abstract. In this paper we investigate principles of the representation of time and modality in terms of conceptual graphs. Based on the works of A. N. Prior, and C. S. Peirce, we suggest a formalism that takes into account the notion of 'Now', as a starting point for developing a formalism that can account for comprehension of texts involving temporal descriptions. We analyze the temporal structure of two short narratives and represent the structure using branching time models and Conceptual Graphs. We argue that the suggested CG representation is in fact more in accordance with Prior's approach to time than Prior's own idea of branching time.

1 Introduction

Since John Ellis McTaggart (1866-1925), philosophers and logicians have discussed the tensions between the uses of two important kinds of distinctions in the description of temporal reality. According to McTaggart positions in time can be distinguished in two ways. In the first place "each position is Earlier than some and Later than some of the others", and "in the second place, each position is either Past, Present or Future. The distinctions of the former class are permanent; those of the latter are not. If M is earlier than N, it is always earlier. But an event, which is now present, was future, and will be past." [13 ch. xxxiii §306]. The distinctions referring to tenses are called A-logical, whereas the distinctions involving 'earlier' and 'later' are called B-logical. The two kinds of distinctions can obviously also be seen as two description languages and as two kinds of temporal representation. According to the B-logical point of view time is represented as a (partially) ordered set of instants or durations. According to the A-logical point of view time is represented as a logical system of tense-operators corresponding to past and future. Since A.N. Prior's first development of modern temporal logic in the 1950s and 1960s, logicians have been discussing the relations between A- and B-logic. Prior and his followers have argued that A-logic (often called: tense-logic) is fundamental and that the notions of B-logic should be derived from the notions of A-logic. Other logicians have argued that the notions of B-logic should be regarded as fundamental. According to these B-theorists, tenses should be considered as mere meta-linguistic abbreviations. Some of the B-theorists have tried to reduce A-logic (tense logic) to a by-product of the introduction of some practical definitions into a fundamental B-logic. In other words, they see tense operators as nothing but a handy way of summarising the properties of the before-after relations. In their opinion, the B-concepts determine the proper understanding of time and

A. de Moor, W. Lex, and B. Ganter (Eds.): ICCS 2003, LNAI 2746, pp. 201–214, 2003.

reality, and they deem tenses to have no independent epistemological or experiential status. Many B-theorists would like to refer to Reichenbach's famous description of the tenses from 1947 [22] (For a detailed analysis of the discussion between A- and B-theorists, we refer to [29].)

In this paper we intend to discuss how time and modality in narratives can be represented in terms of conceptual graphs (CGs). In section 2 we shall critically investigate the kinds of temporal representation suggested by John Sowa and other writers from the CG community. In section 3 it is argued that Prior is right in insisting on the A-logical kind of temporal representation. In section 4 we intend to discuss the use of branching time models in the analysis of narratives and we shall show that important elements of Prior's temporal logic call for rather new constructions in the development of temporal representation. In section 5 it is argued that the study of conceptual graphs can give rise to another kind of graphical representation which from an A-logical point of view is in fact much more natural than branching time models.

2   CG Representations of Time and Modality

Following the suggestions of Charles Sanders Peirce, John Sowa [25: 294 ff] presents *possibility* as a primitive relation, which is in principle a dyadic relation between the actual and modal contexts. When possibility is presented as a monadic relation this is simply due to leaving the actual context implicit. It can be argued that Peirce wanted to treat the tenses very much like the notions of possibility and necessity (see [27]), although it should also be admitted that Peirce probably never developed this idea in detail. Treating tenses like Peircean modality would amount to choosing an A-logical approach. John Sowa, however, has chosen the B-logical approach to time and tense. Discussing the proper representation of the sentence 'Brutus stabbed Caesar' he states that the past relation involved in 'stabbed' is not a primitive. According to Sowa [25: 207] it can be defined in the following way:

```
Past = [Time: #now]←(Succ)←[Time]←(PTim)←[Situation: λ]
```

The idea is that tenses have to be defined in terms of B-logical notions such as temporal order, represented by the relation (Succ), and instants or points in time, represented by the relation (PTim). The definition also involves a reference to the current time, #now, which according to Sowa is neither a constant nor a variable, but an indexical that has to be resolved using information from the context.

Sowa has also considered another way of representing time [25: 114 ff.]. According to this approach time should be primarily represented without instants. The fundamental notion in temporal representation should rather be the interval or the duration. This approach can be traced back to John Buridan (ca. 1295-1358) who discussed the problems related to 'the duration of the present' (see [29: 43 ff.]), and in the 20th century it has been reformulated using symbolic logic by A.G. Walker [26], C.L. Hamblin [9], and James Allen [2], [3]. Using this approach it becomes quite obvious how temporal relations such as (STARTS), (DURING), and (FINISHES) can be represented. These relations will simply correspond to the three qualitatively different ways in which one duration can be included in another duration. See also

[25: 508]. However, the traditional approach to temporal representation using instants and the non-traditional approach using intervals or durations are both B-logical. In both cases time is understood as a set of elements, which are at least partially ordered by some before/after-relation. In both cases the tense-logical relations are conceived as derived relations.

Obviously, Prior's tense-logical operators can be criticized in various ways, and it may be questioned whether tenses can fully capture all aspects of the tenses in natural language. For instance, Daniel Côté and Bernard Moulin [6] may be right in claiming that there is no obvious and direct way of expressing all verb tenses in roman languages such as French, Spanish or Italian in terms of the past and the future operators. We may very well have to develop a more sophisticated apparatus if we want to deal with the semantics of these natural language structures. But there can be no doubt that past and future operators are needed in order to reflect the human experience of the passing of time. If temporality is only described in terms of 'before', 'after' and other B-logical notions, something very crucial is missing.

We suggest that Prior's approach to time and tense is used in the CG representation of the temporal aspects of reality. This means that tense operators are seen as primitives and that instants are understood as derived from tense logic. Intuitively, Prior's way of doing this means that an instant should be understood as the conjunction of all propositions true at the instant in question [20].

3 The Priority of A-specifications in Narratives

According to the B-theorists, truth should be understood as relative to temporal instants. In their opinion, a statement is true 'at an instant', and consequently the idea of 'nowness' is irrelevant within logic. A.N. Prior rejected the idea of temporal instants as something primitive and objective. In fact, he claimed the reality of tenses:

> So far, then, as I have anything that you could call a philosophical creed, its first article is this: I believe in the reality of the distinction between past, present, and future. I believe that what we see as a progress of events is a progress of events, a coming to pass of one thing after another, and not just a timeless tapestry with everything stuck there for good and all [5: 47].

Prior argued that "what we know when we know that the 1960 final examinations are over can't be just a timeless relation between dates because this isn't the thing we're pleased about when we're pleased that the examinations are over." [20: 42] In this way he pointed out that many A-statements cannot be reduced to B-statements without demonstrable loss of meaning. Consider, for example, this statement, uttered at 1 p.m. on a given date: 'Fortunately, my consultation with the dentist is over'. According to the B-theory the statement should be reduced into this: 'Fortunately, the consultation with the dentist is before 1 p.m. (on the given date)'. This B-statement, however, exhibits no semantic equivalence at all with the A-statement, which is uttered at 1 p.m. While it is possible for a person by means of the A-statement to express the reason for his or her joy and relief, the B-statement merely states that the temporal order between two instants is fortunate: 'Fortunately, t_1 is before t_2!'

In the B-reduction of the A-statement valuable semantic content pertinent to the reason for one's joy and relief - the fact that *it is over* - has been lost. Similarly, the B-theoretical reduction of an A-statement about the future would fail to capture such semantic content, which could give one reason to experience fear, expectation or excitement. On this background it must be reasonable to reject the B-theoretical procedure of reduction [19].

It was Prior's conviction that tense logic was not merely a formal language together with rules for purely syntactic manipulations. It also embodied a crucial ontological and epistemological point of view according to which "the tenses (it will be, it was the case) are primitive; only present objects exist." [21: 116] Within Prior's A-logic the times (instants), which are fundamental in the B-logic, are understood as rather special propositions, which are only true once. In fact, all B-notions can be presented in terms of A-notions, but not vice versa, since there is no B-logical equivalent to the A-logical "now". Some philosophers have rejected Prior's analysis and conclusions. One of them is D.H. Mellor [14], who has strongly rejected the idea of tensed facts. But even Mellor has to acknowledge that tense is indispensable for conscious communication in general [14: 88]. So although he does not think that tenses are important for a satisfactory account for the temporal reality, he has to admit that no user of language can do without tenses.

If we regard as texts all artifacts intended to communicate meaning (including books, movies and games; but excluding transmitters, design objects, and urban structures) it is fairly obvious that the act of comprehension presupposes a *now*, which is the time of perceiving. The act of comprehending meaningful communication is possible due to our ability to perceive a *now* in relation to a *past* (retrospection), and in relation to a *future* (anticipation). Thus, we comprehend a communicative structure by understanding new information as contingent upon previously perceived information, and by anticipating this or that future outcome of the communication. This is true of scientific as well as of narrative discourse, even though the means and pragmatic rationale may differ in terms of precision and granularity of intention. It is simply hard to describe the act of reading / seeing / listening / playing without the notion of a *now* from which past occurrences and future developments are understood. In other words: retrospection and anticipation works by means of A-logical notions as long as the communication is proceeding. But it is equally obvious that when we are dealing with retrospective text comprehension, the perspective of the receiver may change from an inside view to an outside view, as we probe event structures to grasp the signification and ramification of particular events. From this outside perspective, B-logical notions of *before* and *after* can be used just as natural as A-notions are used in describing comprehension based on empathy.

The lesson that can be learned form Prior's analysis it certainly not that B-notions are useless in the resentation of the temporal structure of narratives. Prior's point is just that a careful analysis shows that the B-notions can be obtained (derived) from the A-notions, which from a philosophical and conceptual point of view are fundamental for the very understanding of time. However, the derivation of the B-notions is very useful as many temporal aspects of narratives can be stated much easier with B-notions that without such notions.

The idea of seeing texts from the 'inside' has among others been advocated by Marie-Laure Ryan. She labels the phenomenon *'fictional recentering'* [23: ch. 1]. The intuition is that (literary) texts invite us to pretend that we are partaking in the communication that goes in the textual world. This amounts to perceiving the text from the inside, one bit at the time, which in terms require a narratological now. Discourse analysis as such, however, depends on the ability to see a text from the 'outside' as well. Thus, discourse analysis frequently makes use of plot-graphs and the like, often depicted as branching time models, by which B-logical notions are prevalent. Ryan's contribution is no exception in this regard [23: ch. 7, 8, and 10].

This could indicate a hermeneutical alternation between *part*, using A-notions and *whole*, using B-notions. In light of this we can describe text comprehension as a process in which internal representations, employing A- and B-notions, become figure and ground for each other respectively, but admittedly, starting with and from an A-theoretical vocabulary. Because of this, and because B-notions can be derived from A-notions, the analytical process in dealing with text comprehension is likely to benefit from a theoretical rendering of the A-theoretical position.

The study of time-relations is of special interest in narratology because, obviously, narratives intrinsically are used to convey sequences of events ordered by temporal relations. In the field known as *possible worlds narratology* [7] [15], modal operators are employed to describe wishes, intentions, and obligations pertaining to characters involved in the unfolding of the plot [23]. If we maintain that knowledge has a primary function if the description of narrative development [24], the question of temporal relations in narratives can be reformulated as the epistemic question of 'who knows what when'. A necessary step towards answering such problems consists of dealing with the time-relations between *event structure* and *discourse structure*; a problem that been studied extensively during the last decades. By event structure we refer to the order in which the events are thought to occur; this is sometimes referred to as 'told time'; and by discourse structure, we refer to the order in which the events are presented in the text. This is sometimes referred to as 'telling time'.

In his seminal work 'Narrative Discourse' [8] Genette makes three kinds of distinctions regarding this subject: those of *order* (e.g. retrospection and anticipation), *duration* (e.g. summary and real-time), and *frequency* (e.g. repetition). In this paper we shall confine ourselves to investigate matters of order. Following Genette, it is commonly accepted to distinguish between *normal sequence* – where story and discourse have the same order, and *anachrony*, in which the two most important instances are *analepsis* – going back in time, and *prolepsis* – going forward in time. These two terms we shall formalize by means of a past-operator and a future-operator.

Furthermore, we will adhere to the distinctions made by Margolin / Herman regarding the ordering of events. According to them, a set of events can be ordered in at least 4 ways: *Full* or *unequivocal* ordering, where it is decidable how any two events are related, *random* ordering, where every possible arrangement is probable, *Alternative* or *multiple* ordering, where one order is more likely than another, and *partial* ordering, where "Some elements of the set can be uniquely sequenced relative to all others, some only relative to some others, and some to none" [10: 213]. This rendition obviously favors a B-logical approach, but as a perspective on time-relations as such, the distinction can also be expressed in terms of A-logical notions. These temporal notions and their cognates have been the subject of further refinements by several theorists, see for instance: [4: 63], [12], [17]. Here we shall focus on partial orderings of events, and leave certain other elements out of considerations. Our examples have been selected accordingly.

In the following sections we shall analyze the temporal structure of two fables by Aesop [1], using branching time models and Conceptual Graphs. For the linear CG representation we use the PROLOG+CG syntax [11].

4 Branching Time Models and Event Structures in Narratives

Prior wanted to develop an indeterministic tense-logic. Prior related his belief in real freedom to the concept of branching time. He would agree that the determinist sees time as a line, and the indeterminist sees it as a system of 'forking paths'. Inspired by some ideas from Saul Kripke, Prior worked out the formal details of several different systems, which constitute different and even competing interpretations of the idea of branching time. Eventually, he incorporated the idea of branching into the concept of time itself. Like all Prior's branching time models, this model is forward branching and backwards linear. This means that for every event there may be several future possibilities, but there is just one past. Consider for the first example this fable:

> Fable No 177: The Bull and the Calf.
> A BULL was striving with all his might to squeeze himself through a narrow passage, which led to his stall. A young Calf came up, and offered to go before and show him the way by which he could manage to pass. "Save yourself the trouble," said the Bull; "I knew that way long before you were born."

This quite simple narrative holds a rather complex temporal exposition consisting of two possible futures (prolepsis) as well as an analepsis. If we for the moment disregard the setting, and a couple of minute details, the discourse structure can be said to consists of three events: 1) the calf offers a future act, 2) the bull refuses and introduces an alternative future, and 3) the bull motivates his answer with a reference to the past. A branching time model of the event structure appears as follows. Please note that for didactic reasons, the events in figs. 1 and 2 are numbered left to right, favoring the actualized strand of events:

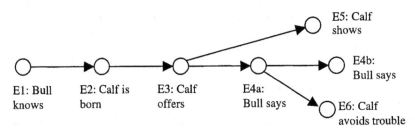

Fig. 1. Branching Time model of Fable No 177

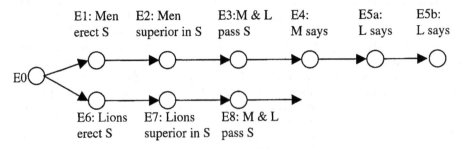

Fig. 2. Branching Time Model for Fable No 29

In his development of the formal idea of branching time A.N. Prior used an idea of 'a state', which differs from the idea of a 'snapshot' of things, as they are right at a certain instant. In his opinion a complete description of the present instant should also contain information about all past states of things. Using this idea of the present instant (and of instants as such) it becomes clear that there can be no branching into the past. This was very important to Prior. He had a strong commitment to idea of asymmetry between past and future, and in his opinion, one of the most important differences between the past and the future is that:

> ...once something has become past, it is, as it were, out of our reach - once a thing has happened, nothing we can do can make it not to have happened. But the future is to some extent, even though it is only a very small extent, something we can make for ourselves... [5: 48].

It has to be admitted, however, that the representation of the idea of branching time in terms of the diagram in figure 1 appears to be rather B-like. This means that the branching time system is viewed from the outside. In this figure there is no representation of the present. A shortcoming that becomes crucial when we are considering different perspectives on a matter, as in the text above, or, perhaps even more clear when we consider counterfactual utterances as in the following example:

Fable No. 29: The Man and the Lion
A MAN and a Lion travelled together through the forest. They soon began to boast of their respective superiority to each other in strength and prowess. As they were disputing, they passed a statue carved in stone, which represented "a Lion strangled by a Man." The traveller pointed to it and said: "See there! How strong we are, and how we prevail over even the king of beasts." The Lion replied: "This statue was made by one of you men. If we Lions knew how to erect statues, you would see the Man placed under the paw of the Lion." - One story is good, till another is told.

Similar to the first fable, the text presents an instance of analepsis, but in addition, we also have an example of counterfactual reasoning: The idea that a lion might have erected the statue, branches the event structure into two:
The truth of this counterfactual statement is obviously crucial for the understanding of the story. The elements of surprise and also of counterfactual reasoning cannot be satisfactorily represented without primitive notions of past and future. Again, we note the incapability of the model to illustrate how the total structure is rendered.

5   CG Representation as Alternative to Branching Time Models

Branching Time Models are powerful tools when they are used to survey a complex temporal structure. However, they do not indicate at all that the story is told (or read), and consequently the model becomes a static picture of the text as a whole. The A-logical perspective is absolutely required if we are to represent the process of telling and of comprehending. In the story about the man and lion the essence of the first crucial claim can be stated with CGs in the following way:

```
fab29(Event3,
[act: pass-
  -agnt->[animate: {man, lion}],
  -obj->[artifact: statue]-chrc->[description: carved]-
       -in->[natural_resource: stone],
              -thme->[phrase = [state: strangled]-
                            -ptnt->[animal:lion],
                            -agnt->[animate: man]]]).
```

From this basis it is deduced (as does the man) that at least one statue shows that man prevails over lions. Therefore the statue in the world seems to support the claim that man prevails over lions. However, the wise lion in the fable brings in the fact: that the statue was made by men, whereby an event further in the past is evoked i.e.:

```
fab29(Event5a,
[act: reply]-
  -agnt->[animal: lion],
  -thme->[proposition = [Event1 =
    [artifact:statue]<-obj-[act:make]-agnt->[animate:man]]]).
```

And the lion adds that there could also have been statues made by lions i.e. such statues, he claims, would have been quite different. They would in fact support the claim that lions prevail over men i.e.:

```
fab29(Event5b,
[act: reply]-
  -agnt->[animal: lion],
  -thme->[proposition = [if = [Event6=
    [act: know]-
      -agnt->[lion],
      -thme->[act: erect]-obj->[artifact: statue]]]<-cndt-
        [than = [Event8 =
          [act:see]-
            -agnt->[human: you],
            -thme->[proposition =
                      [state:placed]-
                        -ptnt->[man],
                        -loc->[spatial_relation:_under]<-loc-
                        [bodypart:paw]<-poss-[animal:lion]]]]]).
```

From this it can be deduced that if lions had erected the statue, then it would have supported the claim that lions prevail over men.

Logical Form of Fable No 177

Because text comprehension takes as its starting place a series of events characterized by 'now-ness', the discourse structure can be formally represented by successively added graphs, each of which consists of a 'now' and of some sub-graphs embedded in the context of that now. The sub-graphs are given the contexts of 'Past' and 'Future' respectively, according to the logical occurrence of the events that they refer to. Recall the first fable about the bull and the calf. The discourse structure of this fable can be represented by three statements, corresponding, in form, to the following three graphs:

```
[Event3 =    [Future = [Event5]]].
[Event4a =   [Future = [Event6]]].
[Event4b =   [Past = [Event2]-and->[Past = [Event1]]]].
```

From this, the discourse-time sequence can easily be rendered as: E3, E5, E4a, E6, E4b, E2, and E1. The event structure, however, is somewhat more complicated to provide. Because, as mentioned above, each event carries within it the history of preceding events (and of possible futures), the conception of the event structure must carefully be accumulated by means of the unfolding of discourse structure. As the discourse progresses, more and more information is conveyed until, at the end, the structure is completed.

In [27] and in [28] a different representation in terms of Charles Sanders Peirce's existential graphs has been suggested. Using this alternative representation we find that the temporal relations can be illustrated in the following manner, where a solid enclosure stands for something in a possible future, and a dashed enclosure stands for the past:

The first of these three figures shows that Event5 is a future, when Event3 is the present. The second figure shows that Event4a has a future and a past future. The third figure shows the entire event structure of fable No 177 assuming that 4b is present. The figure as a whole illustrates the way in which the temporal information is accrued as the text progresses. In the linear notation below, we have represented this information in the context of numbered 'histories': [Hist1]...[Hist3], signifying the fact that the process of comprehending a text entails a movement through a set of 'histories', until the entire temporal structure is rendered.

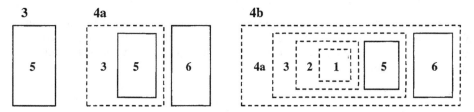

Fig. 3. CG Models of an unfolding temporal structure

The first figures can be logically derived from the last one in the unfolding temporal structure shown in Fig. 3. We will, however, leave a detailed exposition of this possibility to further investigations.

This analysis reveals another problem, namely the need for temporal reference in order to represent the fact that the possible futures suggested by the calf and by the bull respectively, occur at same pseudo-time. In order to represent this fact, we introduce a relation called 'time', which links an event to a time interval. In the linear notation below, it is clearly stated that the (possible) futures Event5 and Event6 are 'scheduled' to occur at the same time.

```
[Hist1 =
[Event3]-and->[Future = [Event5]]-time->[int:x]]].

[Hist2 =
[Event4a]-
  -and->[Future = [Event6]]-time->[int:x],
  -and->[Past = [Event3]-and->[Future =
                              [Event5]]-time->[int:x]]].

[Hist3 =
[Event4b]-and->
[Past = [Event4a]-
  -and->[Future = [Event6]]-time->[int:x],
  -and->[Past = [Event3]-
          -and->[Future = [Event5]]-time->[int:x],
          -and->[Past = [Event2]-and->[Past = [Event1]]]]]]].
```

Logical Form of Fable No 29

Similar, the fable of the lion and the traveler can be said to consist of four statements: E3, E4, E5a, and E5b from fig. 2. Hence, a representation of the discourse structure can be rendered as:

```
[Event3].
[Event4 =  [Past = [Event2]]].
[Event5a = [Past = [Event1]]].
[Event5b = [Past =
    [Event0]-and->[Future =
        [Event6]-and->[Future =
            [Event7]-and->[Future =
                [Event8]]]]]].
```

And the unfolding of the event structure can be rendered as a collection of four histories, beginning with the simple Hist1, and ending with the rather complex Hist4, which comprises the entire temporal structure including the counterfactual condition mentioned in Event5b:

```
[Hist1 = [Event3]].
```

```
[Hist2 =
[Event4]-and->[Past =
    [Event3]-and->[Past =
        [Event2]]-time->[int:x]]].

[Hist3 =
[Event5a]-and->[Past =
    [Event4]-and->[Past =
        [Event3]-and->[Past =
            [Event2]-and->[Past =
                [Event1]]-time->[int:y]]
            -time->[int:x]]
        -time->[int:z]]].
```

```
[Hist4 =
[Event5b]-and->[Past =
   [Event5a]-and->[Past =
     [Event4]-and->[Past =
       [Event3]-and->[Past =
         [Event2]-and->[Past =
           [Event1]-and->[Past =
             [Event0]-and->[Future =
               [Event6]-and->[Future =
                 [Event7]-and->[Future =
                   [Event8]]-time->[int:z]]
                 -time->[int:x]]
               -time->[int:y]]
           -time->[int:u]]
         -time->[int:y]]
       -time->[int:x]]
   -time->[int:z]]].
```

We suggest that the representation of branching time in terms of conceptual graphs is much more appropriate from a tense-logical point of view that the traditional branching time model (as in figure 1 and 2). Clearly, the representation in terms of conceptual graphs presupposes that the present state of affairs is taken into account as such i.e. the graphical representation makes it clear which event is the present one (5b in figure 4). In this way the representation becomes clearly A-logical.

By collecting the graphs displaying the histories of the unfolding narrative, the temporal structure of the story telling may be represented using the context 'told' in the following way:

```
Fable29=
           [told:Hist4-
           -and->[told:Hist3-
             -and->[told:Hist2-
               -and->[told:Hist1]]]].
```

where the stories (Hist1,Hist2,Hist3, and Hist4) can be defined in terms of the events introduced above.

In the above definition of Fable29 it is assumed that the full story is told. One may wonder if the story could just as well have been defined by using a context corresponding to what will be told in the future. But such a procedure turns out to be very problematic. This can be documented by taking Prior's findings concerning the relation between time and existence into account. In *Time and Modality* he proposed a system called 'Q' which was specifically meant to be a 'logic for contingent beings' [18: 41 ff.]. System Q can deal with a certain kind of propositions, which are not always 'statable'. The point is that the communication potential is growing in the sense that the story gradually obtains the ability to formulate more kinds of statements. Some statements which were unstatable in the beginning of the story, become statable simply because the vocabulary is growing with new names and an extended ontology during the process of story telling. For instance: in the lion fable, the concept type [artifact] would ordinarily be defined as a man-made object, but in the course of the events, this definition is challenged by the lion's speculation that in some possible worlds, some artifacts could be brought about by lions.

5b

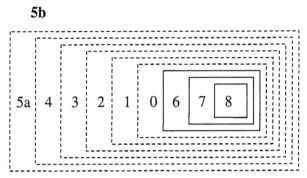

Fig. 4. The Lion's perspective

It should also be noted that it might be possible to develop the suggested CG representation of time and modality further. In fact it seems that C.S. Peirce himself had an idea, which is very relevant in this context. In Lecture IV of his "Lowell Lectures of 1903", Peirce stated as follows:

> As we have seen, although the universe of existential fact can only be conceived as mapped upon a surface by each point of the surface representing a vast expanse of fact, yet we can conceive the facts [as] sufficiently separated upon the map for all our purposes; and in the same sense the entire universe of logical possibilities might be conceived to be mapped upon a surface. Nevertheless, in order to represent to our minds the relation between the universe of possibilities and the universe of actual existent facts, if we are going to think of the latter as a surface, we must think of the former as three-dimensional space in which any surface would represent all the facts that might exist in one existential universe. [16: CP. 4.514]

Following this Peircean idea a diagram like the above fig. 4 would have to be interpreted as a three-dimensional space seen from above. The present (Now) corresponding to 5b would represent 'the ground level'. Each broken line would mean going one step deeper, whereas each full line would mean going one step higher. If this three-dimensional structure is observed from the 'side' and not from 'above' as in fig. 4, some kind of branching structure will appear. For this reason it can be argued that the CG structure, which we have suggested here, and Prior's branching time model are really just different aspects of the same thing.

6 Conclusion

We have argued that Prior's A-logical approach to temporal representation has some important advantages as compared with the B-logical approach suggested by John Sowa and many others. We have demonstrated that an A-logical approach can be formulated in terms of conceptual graphs (CGs). By using CGs it is in fact possible to represent the temporal structure of a narrative in a way, which is much more in accordance with Prior's A-logical approach than Prior's own branching time models.

In this way it is also possible to account for another important feature of the kind of temporal representation needed for narratives, namely the fact that the ontology may be changing during the process of story telling. Obviously, there is still a lot to be done in order to develop a full-fledged A-logical theory for temporal representation with CGs. In particular, the proper incorporation of Prior's ideas of statability (formulated in his system Q) has to be studied in further details.

References:

1. Aesop, *Aesop's Fables - Translated by George Fyler Townsend*, Available at http://classics.mit.edu/Aesop/fab.mb.txt.
2. Allen, J.F., *Maintaining Knowledge about Temporal Intervals*. Communications of the ACM, 1983: p. vol. 26 823-843.
3. Allen, J.F., *Towards a General Theory of Action and Time*. Artificial Intelligence 23, 1984.
4. Chatman, S., *Story and Discourse*. 1980: Cornell University Press.
5. Copeland, J., ed. *Logic and Reality: Essays in Pure and Applied Logic in Memory of Arthur Prior*. 1995, Oxford University Press.
6. Côté, D. and B. Moulin, *Refining Sowa's conceptual graph theory for text generation*, in *Proceedings of the Third International Conference on Industrial and Engineering Applications of Artificial Intelligence and Experts Systems*. 1990, ACM Press: New York. p. 528-537.
7. Doležel, L., *Narrative Modalities*. Journal of Literary Semantics, 1976(5): p. 5-15.
8. Genette, G., *Narrative Discourse*. 1980: Cornell University Press.
9. Hamblin, C.L., *Instants and Intervals*, in *The Study of Time*, J.T. Fraser, F.C. Haber, and G.H. Müller, Editors. 1972, Springer-Verlag: Berlin. p. 324-331.
10. Herman, D., *Story Logic*. 2002: University of Nebraska Press.
11. Kabbaj, A. and M. Janta-Polczynski, *From PROLOG++ to PROLOG+CG*, in *Conceptual Structures: Logical, Linguistic, and Computational Issues. ICCS 2000*, B. Ganter and G.W. Mineau, Editors. 2000, Springer.
12. Margolin, U., *Of What is Past, Is Passing, or to Come: Temporality, Aspectuality, Modality, and the Nature of Literary Narrative*, in *Narratologies*, D. Herman, Editor. 1999, Ohio State University Press. p. 142-166.
13. McTaggart, J.E., *The Nature of Existence*. 1927, Cambridge.
14. Mellor, D.H., *Real Time*. 1981: Cambridge University Press.
15. Pavel, T., *Fictional Worlds*. 1986: Harvard University Press.
16. Peirce, C.S., *Past Masters. The Collected Papers of Charles Sanders Peirce, I-VIII 1931-1966 Cambridge: Harvard University Press*, C. Hartshorne, P. Weiss, and A.W. Burks, Editors. 1992, InteLex Corporation.
17. Prince, G., *A Dictionary of Narratology*. 1987: University of Nebraska Press.
18. Prior, A.N., *Time and Modality*. 1957, Oxford.
19. Prior, A.N., *Thank Goodness That's Over*. Philosophy, 1959. **34**: p. 12-17.
20. Prior, A.N., *Papers on Time and Tense*, ed. P. Hasle, et al. 2002: Oxford University Press.
21. Prior, A.N. and K. Fine, *Worlds, Times and Selves*. 1977: Duckworth.
22. Reichenbach, H., *Elements of Symbolic Logic*. 1947, New York.
23. Ryan, M.-L., *Possible Worlds, Artificial Intelligence, and Narrative Theory*. 1991: Indiana University Press.
24. Schärfe, H. *Searching for Narrative Structures*. in *AAAI 2002 Spring Symposium*. 2002. Stanford, California: AAAI Press, Menlo Park, California.
25. Sowa, J., *Knowledge Representation*. 2000: Brooks Cole Publishing Co.
26. Walker, A.G., *Durées et instants*. La Revue Scientifique, 1947(No. 3266): p. 131 ff.

27. Øhrstrøm, P., *Existential Graphs and Tense Logic*, in *Conceptual Structures: Knowledge Representation as Interlingua. Lecture Notes in Artificial Intelligence 115*, P. Eklund and G. Ellis, Editors. 1996, Springer-Verlag. p. 203-217.
28. Øhrstrøm, P., *Temporal Logic as a Tool for the Description of the Narrativity of Interactive Multimedia Systems*, in *Virtual Interaction: Interaction in Virtual Inhabited 3D Worlds*, L. Qvortrup, Editor. 2000, Springer.
29. Øhrstrøm, P. and P. Hasle, *Temporal Logic - From Ancient Ideas to Artificial Intelligence.* 1995: Kluwer Academic Publishers.

Modeling and Representing Structures for Analyzing Spatial Organization in Agronomy

Jean-Luc Metzger, Florence Le Ber, and Amedeo Napoli

UMR 7503 LORIA, B.P. 239,
54506 Vandœuvre-lès-Nancy
{metzger, leber, napoli}@loria.fr
http://www.loria.fr/equipes/orpailleur/

Abstract. In this paper, we present the knowledge-based system ROSA working on spatial and functional organizations of farms in agronomy. The goal of this system is twofold: validation of the domain model and analysis of the functional organization with respect to the spatial organization of farms. First, we propose and detail a model of the domain based on the so-called "spatial organization graph" (SOG). Then, we show how description logics are used to represent and to manipulate these spatial organization graphs. The reasoning in the ROSA system is mainly based on case-based reasoning, and involves also concept classification and spatial reasoning. Finally, we describe the manipulation of spatial organization graphs using case-based reasoning, and we discuss the overall approach before concluding.

1 Introduction

In this paper, we present a knowledge-based system called ROSA for *Reasoning about Organization of Space in Agriculture*. This system is currently under development in collaboration with agronomists. The goal of the system is twofold: validation of the current agronomic model of space occupation and organization, and analysis of the functional organization of the farmland. The reasoning of the system is based on the assumption that there exists a strong relation between the spatial and functional organization of agricultural work. Moreover, following the principles of case-based reasoning, and given a number of reference analysis cases, the system has to help the agronomists to analyze new problems regarding occupation and organization of the lands. Several original research results have emerged from this study, that are presented in the following.

In a first step, a model of the domain knowledge has been proposed, in accordance with agronomists. This model is based on *spatial organization graphs* or SOG with labeled vertices and edges. Relying on these spatial organization graphs, cases of spatio-functional organization of farmland have been designed: they mainly consist of a description of the land use, and an associated explanation linking spatial organization and functioning.

In a second step, the SOGs and the cases have been represented with a knowledge representation formalism, namely the description logic (DL) system

A. de Moor, W. Lex, and B. Ganter (Eds.): ICCS 2003, LNAI 2746, pp. 215–228, 2003.

RACER [7]. Thus, the reasoning in the ROSA system relies on an original combination of classification (in the description logic sense), case-based reasoning [15, 16] and qualitative spatial reasoning [13].

The paper is organized as follows. We first present the context of this study and the modeling problem. Then, we discuss the knowledge representation step and we propose several examples. Then, we detail the case-based reasoning process with the help of a complete example (case retrieval and adaptation). Finally, we discuss the present work, compare with related works, and conclude.

2   Modeling the Spatial and Functional Organization of a Farm

2.1   The Agronomic Context

Agronomists of INRA SAD[1] analyze and model land use and farm practices to answer problems of environment and space management. They perform farm inquiries in various regions, and study both their spatial and functional organizations [11].

We work with agronomists who especially study two types of environment problems. In the Causses region (south of France), they focus on scrub invasion. In Lorraine (north-east of France) the problem is water quality. Those two types of problems are closely linked to land use and more precisely to the spatial and functional organizations of farm territories. For example, in Lorraine, water pollution is linked to corn, that is used for cattle feeding and that is cropped mostly on draining limestone plateaus. For acquiring knowledge on the relationships between spatial and functional organizations of farm territories, agronomists have done farm inquiries in Lorraine and Causses region. They also have used several documents and information like geographical maps, economical data, etc. and their own expert knowledge. Finally, they have produced synthetic maps of farm territories, called *choremes*, that express both the spatial organization and the functioning of the farms considered.

A choreme describes a spatial organization of fields, buildings, roads, but also contains an interpretation part, that agronomists are able to reveal. For example, the figure 1 shows a choreme modeling a farm in Lorraine. The buildings of the farm are outside of the village (which is in up left side on the drawing); grasslands[2] are around buildings; and cereals fields are distributed on all side of the village territory. Analyzing this figure, an agronomist is able to single out spatial structures and to interpret them with respect to the farm functioning [12].

2.2   Modeling Domain Knowledge

According to our objectives, we have to define spatio-functional cases, that are spatial structures associated with functional explanations. Therefore, we

---

[1] INRA is the French research institute for agriculture and agronomy. SAD is a research department dealing with farm systems and rural development.

[2] Grasslands can be dairy cows paddock, heifer paddock and permanent meadow

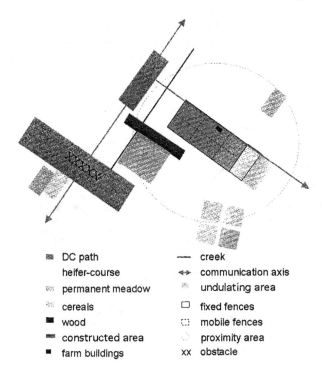

Fig. 1. A choreme modeling a farm in Lorraine (East region of France).

▩	DC path	—	creek
	heifer-course	↔	communication axis
▩	permanent meadow	░	undulating area
░	cereals	▢	fixed fences
■	wood	⸬	mobile fences
▬	constructed area	⸫	proximity area
■	farm buildings	xx	obstacle

have used the synthetic information of the choremes. We have worked with the agronomists to transform the choremes into the so-called *spatial organization graphs* (SOG). A SOG is a bipartite graph composed of vertices and edges. A vertex is labeled with a term referring to a spatial entity or a spatial relation. An edge relates a spatial entity and a spatial relation and is labeled with a term referring to the role played by the spatial entity within the relation. Two examples of SOGs describing a part of the organization of the farm described on figure 1 are given in figure 3.

This interdisciplinary work, based on SOG, has lead to three main results. Firstly, the agronomists have improved the graphical representation of choremes. Secondly, we have modeled the concepts used by the agronomists to describe farm spatial and functional organizations. Finally, we have defined a set of spatio-functional cases.

The agronomists use several terms for describing spatial and functional aspects of farm organizations. They speak about *land use* (cereals, dairy cow meadows, temporary pastures), *morpho-geological types* (plateau, stony, draining), *cattle* (dairy cows, heifers, ewes), *spatial relations* between fields or buildings (is near, separate, isolate, is an obstacle), etc. The same term can relate to various concepts, depending on the farm system or the region considered. We have thus defined two hierarchical domain models, that can be considered as domain ontologies, one for Lorraine and one for Causses region. Figure 2 shows a part of the Lorraine domain model. A third hierarchical model has been designed for spatial relations. This hierarchical model relies on spatial theories described in [2, 19].

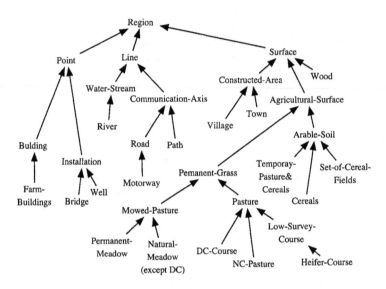

Fig. 2. The domain model $\mathcal{H}_\mathcal{L}$ describing the Lorraine region (partial view): DC means dairy cows (providing milk), NC nurse cows (young cows).

2.3   Case Modeling

A *spatio-functional organization case* is a referent piece of knowledge used for analyzing new farms. A case relates to a particular spatial structure and its functional interpretation. A case is modeled with an *explained graph*, denoted by E-SOG, associating a SOG and an *explanation*. An explanation is a textual statement about the functional interpretation of the current spatial structure. Actually, a case is restricted to a part of a particular farm: in this way, a farm can be described by several cases, at various scales, and with the same general context, i.e. the region, the farming system, etc.

Figure 3 shows two cases examples from the farm depicted in figure 1. The E-SOG denoted by A is a model of a particular spatial structure with the following associated explanation: "The farmer has inserted a meadow between the wood and the cereal field to protect the cereals from dampness, shade, and animals coming from the wood". The E-SOG A is composed of four vertices: wood, cereals and meadow refer to spatial entities while isolate refers to a spatial relation with functional aspects. These aspects are given by the labels associated with the edges between the vertices, namely isolated and isolating.

The E-SOG denoted by B describes another spatial structure with another explanation: "The farmer can use grasslands far from the farm to park animals that do not need too much survey".

E-sog A

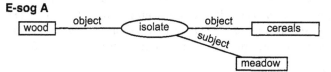

Explanation: the farmer left a strip of grass (the meadow) to protect cereals from moisture, shade and wild animals induced by wood.

E-sog B

Explanation: as heifers do not need too much survey, they can be parked far from the farm buildings.

Fig. 3. Two case representing a part of the farm showed in the choreme of figure 1.

These examples show that the spatial organization of a farm is modeled with several E-SOGs. The different E-SOGs can be combined into a general SOG that describes the whole structure of a farm. The explanations can be combined accordingly. As shown in the following, every E-SOG is considered as a reference case and is used for analyzing new farms. Hence, the set of E-SOGs constitutes the farm case base.

3 The Representation of Spatial Knowledge

3.1 The Underlying Description Logic System

The present research work is a continuation of a previous research work on the representation of spatial structures and topological relations [14, 13]. The knowledge representation formerly used was an object-based representation system, with representation primitives similar to those of description logics. For the present work, the use of a description logic system, namely RACER [7], is a natural choice for following and exploiting the previous research results. Moreover, the knowledge representation formalism has to enable the representation and the manipulation of complex spatial structures modeled by SOGs and E-SOGs.

Below, we briefly introduce description logics and the RACER system. A DL system allows the representation of knowledge using descriptions that can be concepts, roles or individuals[1]. A concept represents a set of individuals and is composed of roles representing properties of the concept and relations with other concepts. Descriptions of concepts and roles are built according to a set of constructors (e.g. and, or, not,...). Concepts can be either primitive or defined, and are organized within a hierarchy using the subsumption relation. The concept hierarchy (also called TBox) corresponds to the ontological knowledge representation. Beside the concept hierarchy, a set of assertions in which individuals are involved forms the ABox of the DL system. An assertion can be a concept or a role instantiation. The two main reasoning procedures (at least in our case) are classification and instantiation. The former is used to classify a concept in the concept hierarchy according to the search of its most specific subsumers and most general subsumees. The latter is used for finding the concepts of which a given individual may be an instance (this operation is also called *concept recognition*).

The RACER DL system is used for implementing the ROSA system. It provides a rich set of constructors for descriptions, and the means for exploiting a concept hierarchy and a set of assertions. The RACER system presents several noticeable advantages:

- In the present work, there are no more procedural needs such as the ones existing in the previous research works (image computing [14]). Thus, and according to the representation primitives (in object-based knowledge representation) discussed in [14], the used of a full-fledged DL system such as RACER is natural. Moreover, RACER is implemented with a JAVA API and this is a valuable extension regarding the LISP-based system used in [14].
- The RACER system and assistance are actually available. The set of constructors and the possibilities associated to the system are very complete and can fulfill the most of our needs. Recall that RACER implements the $\mathcal{SHIQ}$ DL [6], one of the most expressive DL at the moment.
- Finally, the subsumption test (on which is based classification) and the instantiation test in RACER have a satisfying behavior with respect to our needs, i.e. complex structures can be tested in a reasonable time (although the subsumption test is a NP-hard problem in worst case). Furthermore, it has to be noticed that concrete domain (number, string) are also available, together with an interface with XML.

3.2 The Representation of Spatial Knowledge in the ROSA System

The TBox of the ROSA system includes a hierarchy of concepts denoted by $\mathcal{H}c$ and a hierarchy of role denoted by $\mathcal{H}r$. The hierarchy $\mathcal{H}c$ includes the Sog concept representing the SOGs in a general way, as a list of vertices whose type is the Vertex concept. In the following, concepts are written with the first letter in capitals and instances in normal size letters. A vertex may be a spatial entity whose type is the Region concept, or a spatial relation between entities whose type is the Relation concept. It can be noticed that only a list of vertices is recorded for a Sog; edges are retrieved if necessary through the vertices of type Relation to which they are attached.

The E-Sog concept, is a specialization of the Sog concept and encloses an explanation (that at present is a string of type Explanation). A particular SOG is represented as an instance of the concept Sog or E-Sog. For example, the E-SOG denoted by A on figure 3 is represented as an instance of E-Sog with:

- a list of vertices constituted by instances of Wood, Cereal, Meadow linked to each other by an instance of the relation Isolate,
- an instance of the Explanation concept enclosing the text from the explanation given in the figure 3.

The specific SOGs and E-SOGs are introduced in the ROSA system through an interface providing dialog boxes. The system is then in charge of building the corresponding instances in the ABox of RACER. Moreover, the complete domain knowledge is recorded within an XML format for persistent representation thanks to a particular DTD (Document Type Definition) [5] that has been extended for the manipulation of instances.

At present, the knowledge base of the ROSA system is made of the TBox including $\mathcal{H}c$ and $\mathcal{H}r$ (the domain model) and the ABox, including the case base and the complete descriptions of nine farms (four from Causses and five from Lorraine). The spatial organization of a farm is described by an instance of Sog including an average of forty vertices (both for Region and Relation). The functioning of a farm relies on an average of ten cases that are all instances of E-Sog, with an average of five vertices for each of them. The domain models of the Lorraine and Causses regions include about forty concepts. The hierarchical model of spatial relations contains forty concepts too. These eighty concepts form the TBox of the system that is used to classify SOG and E-SOG individuals. Thus, the elements in the knowledge base (TBox and ABox) provide the knowledge on spatial organization and functioning that is used for solving new analysis problems, as explained in the next section.

4   The Reasoning in the ROSA System

The ROSA system is mainly used for solving an analysis problem regarding spatial organization and functioning. The reasoning in the ROSA system is based on classification and on case-based reasoning (CBR). In the following, we detail the CBR process that involve classification of spatial structures and relations as well (and thus spatial qualitative reasoning).

4.1   The Basic CBR Process in the ROSA System

In the ROSA system, the analysis of a new farm is based on the CBR process. Recall that CBR is based on three main operations : retrieval, adaptation and memorization. Given a new problem to be solved, denoted by target, the system searches in the case base for a pair (srce,Sol(srce)) where the srce problem is similar to the target problem. The solution of the target problem is built by adapting the solution Sol(srce) of the source problem. Finally, the new pair (target,Sol(target)) can be memorized in the case base if the new case is of general interest.

In our framework, the target problem corresponds to the analysis of a new farm and relies on the adaptation of a source explanation. A source case is a

pair (srce,explanation) which is nothing else than an E-SOG. Let target
be a new analysis problem, i.e. the SOG description of a new farm to be analyzed. The retrieval operation consists in searching for one or more source cases
(srce,explanation) that provide an explanation for the target problem. In
practice, the retrieval operation consists in an exploration of the case base to
find the source cases matching a part of the target problem. The case retrieval
is based on the strong and smooth classification principles introduced in [15, 16].

Below, we first present a detailed example based on strong classification, and
then we explain the principle of smooth classification and give an example. Let
us consider the target problem denoted by target-C on figure 4. The retrieval
process tries to build a *similarity path* between target-C and a source problem, here srce-L. The similarity path is made of a generalization path between
srce-L and G-sup, and a specialization path between G-sup and target-C,
in order to fulfill the equation srce-L ⊑ G-sup ⊒ target-C. Actually, srce-L
and target-C are instances, and the subsumption operation holds on concepts.
Thus, the first operation is an instantiation (concept recognition) where the concepts of both srce-L and target-C are recognized. In the example on figure 4,
the instances farm-building and heifer-paddock are generalized into the
concepts Farm-Building and Heifer-Paddock. Then the second operation is
the search for the most specific subsumer of both concepts.

The details are as follows. The instances srce-L and target-C are generalized into concepts Sog-L and Sog-C then the matching process is applied
on these two graphs. In the present example (see figure 4), the matching puts
in correspondence the two vertices Far in Sog-L and Sog-C. Two adjacent
vertices are identical, i.e. Farm-Building, but the Heifer-Paddock vertex in
Sog-L does not match the Ewe-Paddock vertex in Sog-C. Nevertheless, there
exists a concept generalizing both Heifer-Paddock and Ewe-Paddock, namely
Low-Survey-Paddock (actually, heifer and ewe are the same type of cattle with

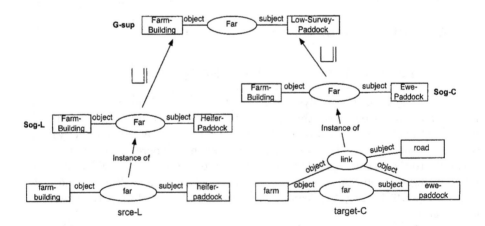

Fig. 4. A similarity path between the SOG srce-L and the SOG target-C.

low survey needs). Thus, the most specific subsumer G-sup can be built and as well the similarity path between srce-L end target-C.

The adaptation of the solution Sol(srce-L) into the target solution Sol(target-C) is done according to the similarity path srce-L $\sqsubseteq$ Sog-L $\sqsubseteq$ G-sup $\sqsupseteq$ Sog-C $\sqsupseteq$ target-C. Given that Heifer-Paddock in Sog-L has been generalized into Low-Survey-Paddock, and that Low-Survey-Paddock is a subsumer of EwePaddock, the suggested adaptation is to replace "heifers" in the explanation attached to the Sog-L with "ewes" (actually, the explanation is attached to the E-SOG corresponding to srce-L and has been introduced in figure 3). The new explanation can be read as follows: "as ewes have low survey needs, they can be parked far away from the farm buildings".

When the agronomist is not satisfied with the explanation produced by the system, he can ask for a matching involving a specific relation, e.g. the link relation in our example. If a similarity path can be designed for the relation, a complementary explanation can be added to the actual explanation set. If no similarity path and thus no explanation can be provided by the system, then the agronomist may complete by himself the explanation set and even propose a new pair (srce,explanation) as a source case.

4.2 The Combination of Reasoning Formalisms in the ROSA System

In order to enhance the efficiency of the ROSA system, e.g. for taking into account more problems, or for proposing more solutions to the agronomist, the CBR process may rely on the so-called *smooth classification process*. This classification schema extends strong classification by allowing transformations both on the source case and on the target case in order that they can match (in the same way as the unification of two terms in logic) in the following way: srce $\xrightarrow{T}$ T(srce) $\sqsubseteq$ G-sup $\sqsupseteq$ T'(srce) $\xleftarrow{T}$ target. T and T' are transformations such as a spatial modification based on a spatial inference rules or a generalization in the concept hierarchy (in the same way as strong classification). Three different kinds of spatial inference rules are available:

- Neighborhood rules : a relation can be transformed into a neighbor relation. Neighbor graphs have been described for RCC8 base relations in [19, 3],
- Composition rules : the relation between two regions can be inferred from the relation between the two regions and a third one. Composition rules have been described for distance relations [2] or for topological relations [4, 18],
- Specific inference rules relying on expert domain knowledge (an example is given below).

In order to illustrate the smooth classification process, let us consider the two bottom SOGs on figure 5 labeled by srce and target. The details of the matching process are given below.

The first step is the same as in strong classification process and consists in building the concept of which the individuals, i.e. srce and target, are instance

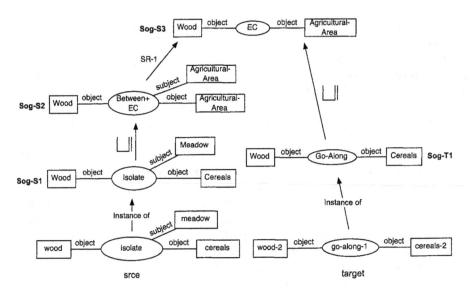

Fig. 5. A similarity path between `srce` and `target` based on smooth classification.

of. In the present case, the concept `Sog-S1` does not match the concept `Sog-T1`, i.e. there is no occurrence of `Sog-S1` in `Sog-T1`. Thus, a further generalization is applied (still in the same way as in the strong classification process), giving `Sog-S2`. In particular, the relation `Isolate` is generalized into `Between+EC`, and the concepts `Cereals` and `Meadow` are both generalized into `Agricultural-Area`. The `Sog-S2` concept does not match the `Sog-T1` concept, but there is a spatial inference rule, denoted in the following by SR-1, that can be applied to the vertex `Between+EC` (actually, a spatial inference rule is applied if needed and whenever its application conditions are fulfilled). In the present case, the SR-1 rule reads as follows : *if a region B is between a region A and a region C, where A is connected to B and C is connected to B, and if A and B are of the same type, then the regions A and B can be unified into a unique region denoted by A+B, and A+B is externally connected (EC) with C.* Here, C corresponds to the `Wood`, A and B to `Agricultural-Area`; the SR-1 inference rule applies, and the result of this application is `Sog-S3`. According to the relations existing between the spatial relations of example (see figure 6) the system finds a match between `Sog-S3` and `Sog-T1`, because `Go-Along` is a specialization of `EC`. Thus, a similarity path can be designed between the `srce` SOG and the `target` SOG, as summarized on figure 5.

The adaptation of the explanation associated with the source case to the target case can be done with respect to the similarity path. However, this is not a single copy with substitution as in the strong classification process. At present, this adaptation is left under control of the agronomist, the system being in charge to give the different generalizations and spatial inference rules applied.

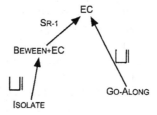

Fig. 6. A part of the relation hierarchy and the rule transformation used in the smooth classification process in figure 5.

Finally, it must be noticed that the classification and generalization operation rely on the subsumption test within the RACER system, while the application of spatial inference rules is controlled by an external module (implemented in JAVA).

5   Comments and Discussion

5.1   The Representation of Spatial Structure

The work presented here follows a set of works on the representation of spatial structures in object-based knowledge representation system and on qualitative spatial reasoning [14]. In these works, the recognition problem of spatial structures is treated in an object-based knowledge representation system, where patterns of structures are represented by classes and the structures to classify are represented by instances. A reification mechanism of relations allows to represent relations with attributes and facets, and to organize relations within a hierarchy. In the present work, instead of relying on predefined structure patterns, there is a collection of individual structures, namely E-SOGs, that are used as references for analyzing new spatial structures.

In addition, in the present work, the description logic system RACER has been used for the reasons that have been listed above: availability of the system, rich capabilities of representation and reasoning, follow-up and maintenance. Another possibility would be to use a platform based on conceptual graphs COGITANT[3] or CORESE[4] for example. This hypothesis is left open and forthcoming experiments have to be done in order to analyze the capabilities and the benefits that a conceptual graphs formalism could provide.

5.2   The SOG Matching and the Adaptation

The adaptation process in CBR is mainly dependent on the matching of SOGs. A number of problems have to be solved to provide an efficient and generic

---

[3] http://cogitant.sourceforge.net/
[4] http://www-sop.inria.fr/acacia/soft/corese.html

adaptation process. At present, the system returns one or more similarity paths leading from the retrieved case sources, the target case, and the corresponding explanations. The agronomist has in charge of accepting (validating) or rejecting the explanation proposed by the system, or building himself a new explanation on the base of the results retrieved by the system.

The adaptation of explanations can also be based on adaptation knowledge depending on the domain [10]. An adaptation knowledge acquisition process is planned to be carried out in a next future. More important, explanation are at present textual statements. They could also be described by graphs, namely explanation graphs, attached to SOGs, and managed in the same way as SOGs are. This research direction has still to be more deeply investigated [21].

Regarding the matching operation itself, there are two alternatives: a subsumption test controlled by the RACER classifier and a matching test controlled by an external module. More precisely, the matching of spatial structures relies on the comparison of the *composition* of structures. If the considered SOG is reduced to a single edge, i.e. a triplet (origin,relation,extremity), then the classification mechanism in RACER is used (actually these triplets are represented as defined concepts in RACER). If the considered SOG is composed of at least two triplets, then the external matching test has to be invoked. This external module has been designed especially for testing SOG matching and has been associated to RACER. The matching process of spatial structures presents the same characteristics as the matching of molecular structures introduced in [17].

5.3   Case-Based Reasoning in the ROSA System

Two fundamental modes of reasoning are used in the ROSA system: classification of concepts and relations, and case-based reasoning for the analysis of a new case. This work relies on a hierarchical approach of CBR and can be related to a number of other works [9, 8, 20]. In [9] the cases are indexed by a hierarchy of concepts and retrieval and adaptation are based on the classification of concepts. Some ideas presented in [9] regarding the indexing of concepts and the adaptation process could be used with profit in our own context. In [20], a case is described by an individual and the similarity between a source case and a target case is based on the search of the least common subsumer (LCS) of two concepts. The source cases retrieved are then classified according to a dissimilarity measure between the source and the target cases. In our approach, the cases are also represented by individuals, and the matching between a source case and a target case is based on the search of a graph playing a role similar to the LCS concept. In addition, transformation operations such as insertion, deletion and substitution, are used to build a similarity path.

6   Conclusion

In this paper, we have presented the knowledge-based system ROSA working on spatial and functional organizations of farms in agronomy. Actually, the overall approach presented in this paper is of general interest and is based on the representation and manipulation of graphs representing the knowledge units in a certain domain, agronomy in the present case. Thus, we have introduced the formalism of "spatial organization graphs" for modeling the space organization and the functioning of farms. We have shown how these graphs are represented and manipulated within the description logic system RACER. We have given details on the design of the solution of a new analysis problem, based on case-based reasoning combined with concept classification (in RACER) and spatial reasoning (involving spatial inference rules).

The ROSA system is currently under development, and there are a number of points that must be made precise and that must be worked further. One important point holds on the SOG matching and the available transformations for modifying a SOG in order that a matching can be found between a source and a target SOG, leading to the adaptation step in the CBR process. Questions such as the followings have still to be investigated for finding a correct answer: which graphs are comparable and how adaptation can be performed accordingly, i.e. how explanations can be adapted in the general case? The similarity path formalism is used as a guide for adaptation, but there are still progress to be made for adaptation knowledge acquisition and spatial inference rule application. A next experiment with agronomists is planned soon in order to collect adaptation knowledge units that will be used in a next version of the system.

References

1. Diego Calvanese, Giuseppe De Giacomo, Maurizio Lenzerini, and Daniele Nardi. Reasoning in expressive description logics. In Alan Robinson and Andrei Voronkov, editors, *Handbook of Automated Reasoning*, pages 1581–1634. Elsevier Science Publishers, 2001.

2. E. Clementini, P.D. Felice, and D. Hernández. Qualitative Representation of positional information. Technical Report FKI-208-95, Institut für Informatik, Technische Universität München, 1995.

3. A. G. Cohn, B. Bennett, J. Gooday, and N. M. Gotts. Representing and reasoning with qualitative spatial relations about regions. In *Spatial and Temporal Reasoning*, pages 97–134. Kluwer Academic Publishers, 1997.

4. M. J. Egenhofer. Reasoning about binary topological relations. In O. Gunther and H.-J. Schek, editors, *Proceedings SSD'91, Advance in Spatial Databases*, LNCS 525, pages 143–160. Berlin: Springer-Verlag, 1991.

5. J. Euzenat. Preserving modularity in XML encoding of description logics. In *Proceedings of the International Workshop on Description Logics (DL2001), Standford, USA*, pages 20–29, 2001.

6. V. Haarslev, C. Lutz, and R. Möller. A description logic with concrete domains and role-forming predicates. *Journal of Logic and Computation*, 9(3):351–384, 1999.

7. V. Haarslev and R. Möller. Racer system description. In R. Goré, A. Leitsch, and T. Nipkow, editors, *Proceedings of the International Joint Conference on Automated Reasoning, IJCAR'2001, Siena, Italy*, LNCS 2083, pages 701–706. Springer, 2001.

8. G. Kamp. Using description logics for knowledge intensive case-based reasoning. In B. Faltings and I. Smith, editors, *Third European Workshop on Case-Based Reasoning (EWCBR'96), Lausanne, Switzerland*, pages 204–218. Springer-Verlag, 1996.

9. J. Koehler. Planning from Second Principles. *Artificial Intelligence*, 87:145–186, 1996.

10. J. Kolodner. *Case-Based Reasoning*. Morgan Kaufmann Publishers, Inc., San Mateo, California, 1993.

11. S. Lardon, P. Maurel, and V. Piveteau, editors. *Représentations spatiales et développement territorial*. Hermès, 2001.

12. F. Le Ber, C. Brassac, and J.-L. Metzger. Analyse de l'interaction experts – informaticiens pour la modélisation de connaissances spatiales. In *IC'2002, Journées Francophones d'Ingénierie des Connaissances, Rouen*, pages 29–38. INSA Rouen, 2002.

13. F. Le Ber and A. Napoli. Design and comparison of lattices of topological relations based on Galois lattice theory. In *Proceedings of the 8th International Conference on Principles of Knowledge Representation and Reasoning (KR2002), Toulouse, France*, pages 37–46. Morgan Kaufmann, 2002.

14. F. Le Ber and A. Napoli. The design of an object-based system for representing and classifying spatial structures and relations. *Journal of Universal Computer Science*, 8(8):751–773, 2002. Special issue on Spatial and Temporal Reasoning.

15. J. Lieber and A. Napoli. Using Classification in Case-Based Planning. In W. Wahlster, editor, *Proceedings of the 12th European Conference on Artificial Intelligence (ECAI'96), Budapest, Hungary*, pages 132–136. John Wiley & Sons Ltd, Chichester, 1996.

16. J. Lieber and A. Napoli. Correct and Complete Retrieval for Case-Based Problem-Solving. In H. Prade, editor, *Proceedings of the 13th European Conference on Artificial Intelligence (ECAI'98), Brighton, UK*, pages 68–72. John Wiley & Sons Ltd, Chichester, 1998.

17. A. Napoli, C. Laurenço, and R. Ducournau. An object-based representation system for organic synthesis planning. *International Journal of Human-Computer Studies*, 41(1/2):5–32, 1994.

18. D. A. Randell and A. G. Cohn. Exploiting Lattices in a Theory of Space and Time. *Computers Math. Applic.*, 23(6-9):459–476, 1992.

19. D. A. Randell, Z. Cui, and A. G. Cohn. A Spatial Logic based on Regions and Connection. In *3rd International Conference on Principles of Knowledge Representation and Reasoning*, pages 165–176. Morgan Kaufmann Publishers, 1992.

20. S. Salotti and V. Ventos. Study and formalization of a case-based reasoning system using a description logic. In B. Smyth and P. Cunningham, editors, *Advances in Case-Based Reasoning, Proceedings of the 4th European Workshop (EWCBR-98), Dublin, Ireland*, LNCS 1488, pages 286–297. Springer, 1998.

21. R. Schank, C. Riesbeck, and A. Kass. *Inside Case-Based Explanation*. lLawrence Erlbaum Associates, Hillsdale, New Jersey, 1994.

Simple Conceptual Graphs Revisited: Hypergraphs and Conjunctive Types for Efficient Projection Algorithms

Jean-François Baget

INRIA Rhône-Alpes
baget@inrialpes.fr

Abstract. Simple Conceptual Graphs (SGs) form the cornerstone for the "Conceptual Graphs" family of languages. In this model, the subsumption operation is called projection; it is a labelled graphs homomorphism (a NP-hard problem). Designing efficient algorithms to compute projections between two SGs is thus of uttermost importance for the community building languages on top of this basic model.
This paper presents some such algorithms, inspired by those developped for Constraint Satisfaction Problems. In order to benefit from the optimization work done in this community, we have chosen to present an alternate version of SGs, differences being the definition of these graphs as hypergraphs and the use of conjunctive types.

1 Introduction

Introduced in [22], Simple Conceptual Graphs (or SGs) have evolved into a family of languages known as Conceptual Graphs (CGs). In the basic SG model, the main inference operator, *projection*, is basically a labelled graph homomorphism [10]. Intuitively, a projection from a SG H into a SG G means that all information encoded in H is already present in G. This operation is logically founded, since projection is sound and complete w.r.t. the first-order logics semantics Φ.

As a labelled graph homorphism, deciding whether a SG projects into another one is a NP-complete problem [10]; moreover, considering this *global* operation (instead of a sequence of local operations, as in [22]) allows to write more efficient algorithms. Indeed, this formulation is very similar to the *homomorphism theorem*, considered vital for database query optimization [1]. Designing more efficient projection algorithms is of uttermost importance to the GG community, not only for SGs reasonings, but for usual extensions of this basic model: reasonings in nested CGs can be expressed as projection of SGs [3], CG rules rely on enumerating SGs projections [20], a tableaux-like method for reasonings on full CGs uses SGs projection to cut branches of the exploration tree [16]...

The `BackTrack` algorithm (or BT) [13] has naturally been used in numerous CG applications to compute projections (e.g. in the platform CoGITaNT [12]). However, the CG community has not collectively tried to improve it. On the other hand, the Constraint Satisfaction Problem (CSP) community has been

A. de Moor, W. Lex, and B. Ganter (Eds.): ICCS 2003, LNAI 2746, pp. 229–242, 2003.

working for the last 20 years on various BT improvements. Although [18] have pointed out the strong connections between SG-PROJECTION and CSP, this work has mainly led to exhibit new polynomial subclasses for SG-PROJECTION. Our goal in this paper is to propose a translation of some generic algorithms developped in the CSP community to the SG-PROJECTION problem.

Though numerous algorithms have been proposed to solve CSPs, we considered the following criteria to decide which one(s) to adopt: 1) the algorithm must be *sound* and *complete* (it must find only, and all projections); 2) the algorithm must be *generic*, *i.e.* not developped for a particular subclass of the problem; 3) the algorithm must add as little *overhead cost* as possible. Let us now precise the third point. Many CSP algorithms rely on powerful filtering techniques to cope with exceptionally difficult instances of the problem (*i.e.* when the graphs involved are dense random graphs, corresponding to the *phase transition* [21]). However, when the graphs involved are sparse and well-structured graphs (and it seems to be the case for SGs written by human beings), these algorithms, inducing a high overhead cost, are not as efficient as those presented here.

The algorithms presented here are `BackMark` [11] and `Forward Checking` [14]. We present them in an unified way, using an original data structure that allow to write them with small code modification, and to execute them at little overhead cost. It was a surprise to see, in [8], how much more efficient was the `Forward Checking` algorithm when we considered SGs as *hypergraphs*, instead as their associated *bipartite graph* (as is usually done for SGs). We have then decided to present SGs as hypergraphs (following the proposal of [7], whose reasons were essentially to simplify definitions). Integrating a *type conjunction* mechanism (as done in [2, 9, 4]) also allows us to reduce the number of projections from a graph into another one, while keeping the meaningful ones.

This paper is organized in two distinct parts. In Section 2 (Syntax and Semantics), we recall main definitions and results about SGs, in our hypergraph formalism. Section 3 (Algorithms) is devoted to projection algorithms.

2   Syntax and Semantics

This section is devoted to a definition of SGs as hypergraphs. Syntax is given for the support (encoding ontological knowledge), and for SGs themselves (representing assertions). The notion of consequence is defined by model-theoretic semantics, and we show that projection is sound and complete with respect to these semantics. For space requirements, definitions are given without much examples. However, we discuss at the end of this section how our definitions relate to usual ones (as in [17], for example).

2.1   Syntax

The support encodes the vocabulary available for SGs labels: individual markers will be used to name the entities represented by nodes, and relation types, used to label hyperarcs, are ordered into a type hierarchy. SGs encode entities (the nodes) and relations between these entities.

Definition 1 (Support). *A* support *is a tuple $S = \langle \mathcal{M}, T_1, ..., T_k \rangle$ of pairwise disjoint partially ordered sets such that:*

- *$\mathcal{M}$, the set of* markers, *contains a distinguished element, the* generic marker *$*$, greater than the other pairwise non comparable* individual markers;
- *$T_1, ..., T_k$ are the sets of* types, *and a type $t \in T_i$ is said to be of* arity i.

We note $\leq$ the order relation on elements of the support, be it types or markers.

Definition 2 (Simple Conceptual Graphs). *A* simple conceptual graph, *defined over a support S, is a tuple $G = (V, U, mark, type)$ where V and U are two finite disjoint sets, and mark and type are two mappings.*

- *V is the set of* nodes, *and $mark : V \rightarrow \mathcal{M}$ labels each node by a marker of $\mathcal{M}$. Nodes labelled by $*$ are called* generic, *other are called* individual.
- *$U \subseteq V^+$ is a set of hyperarcs called* relations, *defined as non empty tuples of nodes, and $type : U \rightarrow (2^{T_1} \cup ... \cup 2^{T_k}) \setminus \{\emptyset\}$ labels each relation of size i by a non empty set of types of arity i.*

Though similar definitions of SGs as hypergraphs have already been proposed (e.g. in [7]), the lack of any type for nodes and the typing of relations by sets of types must be explained. Indeed, the type of an unary relation incident to a node can be seen as the type for this node (sect 2.5), and the set of types labelling a node can be understood as the conjunction of these types (sect. 2.2).

2.2   Model Theoretic Semantics

Motivations for expressing semantics in model theory are stated in [15]:
"A model-theoretic semantics for a language assumes that the language refers to a world, and describes the minimal conditions that a world must satisfy in order to assign an appropriate meaning for every expression in the language. A particular world is called an interpretation [...] The idea is to provide an abstract, mathematical account of the properties that any such interpretation must have, making as few assumptions as possible about its actual nature or intrinsic structure. [...] It is typically couched in the language of set theory simply because that is the normal language of mathematics."

Definition 3 (Interpretations, models). *An* interpretation *of a support S is a pair $(\mathcal{D}, \mathcal{I})$, where $\mathcal{D}$ is a set called the* domain, *and the interpretation function $\mathcal{I}$ associates to each individual marker in $\mathcal{M}$ a distinct element of $\mathcal{D}$, and to each type $t \in T_i$ a subset of $\mathcal{D}^i$ such that $t \leq t' \Rightarrow \mathcal{I}(t) \subseteq \mathcal{I}(t')$.*

Let G be a SG defined over S. An interpretation $(\mathcal{D}, \mathcal{I})$ of S is a model *of G if there exists a mapping $\sigma : V \rightarrow \mathcal{D}$ (called a* proof *that $(\mathcal{D}, \mathcal{I})$ is a model of G) such that:*

- *for each individual node x, $\sigma(x) = \mathcal{I}(mark(x))$;*
- *$\forall\, r = (x_1, ..., x_i) \in U(G)$, $\forall\, t \in type(r)$, $(\sigma(x_1), ..., \sigma(x_i)) \in \mathcal{I}(t)$.*

Note that a set of types is interpreted as the conjunction of these types. As usual in logics, the notion of consequence follows the notion of models:

Definition 4 (Consequence). *Let G and H be two SGs defined a support S. We say that H is a* consequence *of G if all models of G are also models of H.*

2.3 Projection

The inference mechanism (answering the question "is all information encoded in graph H already present in graph G?") was initially presented as a sequence of elementary, local operations trying to transform a SG H into a SG G [22]. These operations have been completed [17] to handle the case of disconnected graphs.

The reformulation of this mechanism as a labelled graph homomorphism called *projection* [10] allowed to write backtrack-based, more efficient algorithms.

Definition 5 (Projection). *Let G and H be two SGs defined over a support S. A projection from H into G is a mapping $\pi : V(H) \rightarrow V(G)$ such that:*

- *for each node $x \in V(H), mark(\pi(x)) \leq mark(x)$;*
- *for each relation $r = (x_1, \ldots, x_i) \in U(H), \forall t \in type(r)$, there exists a relation $r' = (\pi(x_1), \ldots, \pi(x_i)) \in (G)$ such that $\exists t' \in type(r'), t' \leq t$.*

2.4 Main Equivalence Result

It remains now to prove that projection is sound and complete with respect to the model-theoretic semantics defined above. The following theorems have been proven respectively in [23] and [18], where the semantics were given by translating the graphs in first-order logic formulas.

Theorem 1 (Soundness). *Let G and H be two SGs defined over a support S. If there exists a projection from H into G, then H is a consequence of G.*

Both [24] and [18] exhibited independently a counterexample to projection completeness w.r.t. the logical semantics. They proposed different solutions to this problem: [24] proposed to modify the SG H, putting it in *anti-normal form*, and [18] proposed to modify G, putting it in *normal form*. We present here the second solution, since a consequence of the first one is to lose structural information on the graph H, leading to less efficient projection algorithms.

Definition 6 (Normal Form). *A SG is said in* normal *form if no two distinct nodes share the same individual marker.*

To put a SG H into its normal form nf(H), we fusion into a single node all nodes sharing the same individual marker. Hyperarcs incident to the original nodes are made incident to the node resulting from this fusion. Should we obtain multiple relations (*i.e.* defined by the same tuple of nodes), we keep only one of those (our hypergraphs are not multigraphs, as usually defined for SGs), and label it by a type obtained by the union of the original types (*i.e.* their conjunction). Note that since H and nf(H) have the same semantics, Th. 1 remains true if we consider nf(H) instead of H.

Theorem 2 (Completeness). *Let G and H be two SGs defined over a support S. If H is a consequence of G, then there exists a projection from H into nf(G).*

Proof. Soundness ($\Rightarrow$) and completeness ($\Leftarrow$) rely on seeing an interpretation $(\mathcal{D}, \mathcal{I})$ of a support $\mathcal{S}$ as a SG $\mathcal{G}(\mathcal{D}, \mathcal{I})$ over $\mathcal{S}$, and a proof that $(\mathcal{D}, \mathcal{I})$ is a model as a projection.

- consider each element d of $\mathcal{D}$ as a node, whose marker is the individual marker m such that $\mathcal{I}(m) = d$ if defined, the generic marker $*$ otherwise;
- for every tuple $(d_1, \dots, d_i)$, consider the set of types $\{t_1, \dots, t_p\}$ such that, for $1 \leq j \leq p$, $(d_1, \dots, d_i) \in \mathcal{I}(t_j)$. If this set $\{t_1, \dots, t_p\}$ is non empty, then the graph contains a relation $(d_1, \dots, d_i)$ whose type is $\{t_1, \dots, t_p\}$.

Lemma 1. *Let G be a SG defined over $\mathcal{S}$, and $(\mathcal{D}, \mathcal{I})$ be an interpretation of $\mathcal{S}$. Then σ proves that $(\mathcal{D}, \mathcal{I})$ is a model of G iff σ is a projection from G into $\mathcal{G}(\mathcal{D}, \mathcal{I})$.*

($\Rightarrow$) Let us suppose there exists a projection from H into G. We have to prove that every model of G is a model of H. Let us take an arbitrary model $(\mathcal{D}, \mathcal{I})$ of G (this is proven by a mapping σ). Then (Lemma 1, $\Rightarrow$ part), σ is a projection from G into $\mathcal{G}(\mathcal{D}, \mathcal{I})$. As the composition of two projections is a projection, then $\pi \circ \sigma$ is a projection from H into $\mathcal{G}(\mathcal{D}, \mathcal{I})$, thus (Lemma 1, $\Leftarrow$ part) $(\mathcal{D}, \mathcal{I})$ is a model of H.

($\Leftarrow$) Let us now suppose that every model of G is a model of H. In particular, we consider the model $(\mathcal{D}, \mathcal{I})$ of G and H that is *isomorphic* to G, *i.e.* $\mathcal{I}$ establishes a bijection from $V(G)$ to $\mathcal{D}$, and $(d_1, \dots, d_i) \in \mathcal{I}(t), t \in T_i$ if and only if there is a type $t' \in \text{type}(\mathcal{I}^{-1}(d_1), \dots, \mathcal{I}^{-1}(d_i))$ such that $t' \leq t$. Note that such a bijection $\mathcal{I}$ is possible only since G is a SG in normal form. Then $\mathcal{I}$ is a proof that $(\mathcal{D}, \mathcal{I})$ is a model of G. Let us call σ the proof that $(\mathcal{D}, \mathcal{I})$ is a model of H. Check that $\sigma \circ \mathcal{I}^{-1}$ is a projection from H into G, verifying that $\mathcal{I}^{-1}$ is a projection, and conclude with Lemma 1. $\qquad\square$

2.5   Relationships with "Usual SGs"

Above definitions differ obviously from usual ones. Before pointing out the differences, we first rely on graphical representation to highlight their similarities.

Nodes of a SG are represented by rectangles. If $n \geq 2$ and $(x_1, \dots, x_n)$ is a relation, we draw it by an oval, write its type inside the oval, then for $1 \leq i \leq n$, we draw a line between this oval and the rectangle representing x_i and write the number i next to it. Inside the rectangle representing a node, we write the string "$T : M$" where T is the union of the types of unary relations incident to the node, and M is the individual marker labelling the node (we write nothing if the graph is generic). Moreover, to highlight that a set of types is interpreted as a conjunction of these types, we write $t_1 \sqcap \dots \sqcap t_p$ for the set $\{t_1, \dots, t_p\}$.

Up to our introduction of conjunctive types, this is exactly the drawing of a SG in a traditional sense, and these objects have exactly the same semantics. Indeed, SGs are usually defined as bipartite labelled multigraphs, where the two classes of nodes (concept nodes and relation nodes) correspond to the objects obtained when considering the *incidence bipartite* graphs of our hypergraphs.

Though conjunctive types may seem to add expressiveness to SGs, [2, 9, 4] have pointed out it is not the case. Indeed, SGs usually allow multiple relations (in our sense) being defined by the same tuple of nodes, and such a set of relations has the same interpretation as a single relation whose type is the conjunction of their types. And since a concept node type has the same interpretation as a unary relation type, is it possible to encode, *at a semantic level*, type conjunction in "usual SGs". However, since type conjunction is not integrated *at a syntactic level*, it is a problem when trying to fusion concept nodes having different types.

3   Algorithms

Before presenting the algorithms themselves, we recall that SG-PROJECTION is an NP-complete problem ([10], using a reduction to CLIQUE).

Theorem 3 (Complexity). *The problem* SG-PROJECTION *(given a support S, deciding whether or not a SG H projects into a SG G) is NP-complete.*

3.1   Number of Projections

Enumeration of projections from a SG H into a SG G benefits from our definitions. With usual SGs definitions, relations are nodes, that have to be projected.

Consider the following example: the support S contains only binary relation types $\{r, r_1, \ldots, r_k\}$ where $r_1, \ldots, r_k$ are pairwise non comparable and r is greater than all the other ones, and a unique unary relation type (or concept type), t. We build the graph G as a chain composed of n generic (concept) nodes, each one linked to the next by k relation nodes, respectively typed $r_1, \ldots, r_k$. Then consider the graph H, composed of a chain of nodes linked by relations typed r. Following the usual SGs definitions (where both concept nodes and relation nodes have to be projected), we have exacly $(n-1)^k$ projections from H into G. With our definitions, the graph G is a chain of n nodes, each one linked to the next by *one* relation node typed $r_1 \sqcap \ldots \sqcap r_k$, and there is exacly one projection from H into G, essentially capturing the meaning of the $(n-1)^k$ previous ones.

This potentially exponential gain in the number of projections can be important in CG models that rely on enumerating projections, such as rules [20].

3.2   BackTrack

The naïve algorithm searching one or all projections from a SG H into a SG G builds all mappings from $V(H)$ into $V(G)$, and tests whether one of them is a projection. This process can be seen as the building of a *seach tree*. Its nodes represent partial mappings of $V(H)$ into $V(G)$. The root represents the empty mapping. If a node N of the tree represents the partial mapping π, and x is the next unmapped node of $V(H)$, then the sons of N represent the mappings $\pi \cup \{(x, y_1)\}, \ldots, \pi \cup \{(x, y_p)\}$ where $y_1, \ldots, y_p$ are the nodes of G. Leaves of this search tree represent all mappings from $V(H)$ into $V(G)$. BackTrack [13] (or BT) explores this search tree with a depth-first traversal, but it does not wait to reach leaves to check whether a mapping is a projection. If a mapping from a subset of nodes of H is not a projection, then it cannot be extended to a projection (the node of the search tree corresponding to this partial mapping is called a *failure node*). Checking if we have a partial projection at each node of the tree (not just leaves) allows to cut sooner whole subtrees whose roots are failure nodes. Moreover, the sum of all these partial checks is no more costly than the unique global check made at the leaves by the naïve algorithm.

Data: A support S, two SGs H and G defined on S ($H \neq \emptyset$).
Result: All projections from H into nf(G).

$H \leftarrow$ Order(H);
level $\leftarrow$ 1;
up $\leftarrow$ FALSE;
while *(level $\neq$ 0)* **do**
    if *(level = $|V(H)| + 1$)* **then**
        Solution-Found(H);
        up $\leftarrow$ TRUE;
    else current-Node $\leftarrow V_H$[level] ;
    if *(up)* **then**
        if *(Other-Candidates(current-Node, G))* **then**
            up $\leftarrow$ FALSE;
            level $\leftarrow$ Next-Level(current-Node);
        else level $\leftarrow$ Previous-Level(current-Node) ;
    else
        if *(Find-Candidates(current-Node, G))* **then**
            level $\leftarrow$ Next-Level(current-Node);
        else
            up $\leftarrow$ TRUE;
            level $\leftarrow$ Previous-Level(current-Node);

The version of BT we present is an iterative one. As suggested in [19], it allows a better control of the stack of recursive calls, and is thus more convenient for enhancements presented later on. It will be used as the body of all BT improvements presented here: only the functions it calls should be rewritten. Let us briefly explain this algorithm. Following the writing of `Solution-Found`, it can compute one projection (when called, this function stops the execution, and returns the current mapping), or all projections (when called, it saves the current mapping, and all mappings will be read at the end of the program).

The variable `level` corresponds to the current level in the search tree, and thus identifies the node we have to examine (`current-Node`). The variable `up` will be set to `true` when the current mapping does not correspond to a projection, and a backtrack is required. Every node has a field `image` used to store its image by the curent partial projection. Let us now examine the different functions called by this algorithm. `Order` totally orders (consider an arbitrary order) nodes of the graph. We denote by $x_i = V(H)[i]$ the ith node, and by $\text{pre}_V(x_i) = \{x_1, \ldots, x_{i-1}\}$ the nodes placed before him in this table. In the same way, $\text{pre}_U(x_i) = \{(y_1, \ldots, y_k) \in U(H) | \exists 1 \leq j \leq k y_j = x_i$ and $\forall 1 \leq j \leq k, y_j \in \text{pre}_V(x_i) \cup \{x_i\}\}$. `Next-Level`($x_i$) (where x_i is at position i in the table) returns $i + 1$, and `Previous-Level`(x_i) returns $i - 1$. `Find-Candidates`(x_i) stores in the field `image` of x_i the first node $y \in V(G)$ that is a *candidate* for x_i: it means that mark(y) $\leq$ mark(x_i), and that $\forall r = (y_1, \ldots, y_k) \in \text{pre}_U(x_i), r' = (y_1.\text{image}, \ldots, y_k.\text{image}) \in U(G)$. We say that r is *supported* by r'. `Other-Candidates`(x_i) stores in the field `image` of x_i the next candidate unexplored since the last call to Find-Candidates(x_i). Both return **true** if they found a result, and **false** otherwise.

In the worst case, BackTrack explores the whole search tree, and has thus the same complexity as the naïve algorithm. However, in practical cases, it often cuts branches of this tree without needing to explore them to the end.

3.3   Types Comparisons

An immediate problem is the cost of type comparisons. Each time we look for a possible candidate for a node x, we compute type comparisons for all relations of $\text{pre}_U(x)$ incident to x. The number of operations performed can thus be expressed (in a simplified way) by $N \times k \times S$, where N is the number of candidates considered in the whole backtracking process, k the maximum size of pre_U sets, and S is the maximum cost of a type comparison. Since both N and S can be exponential (imagine types expressed in a description logics language), it is unwise to multiply these factors.

The solution is easy. We have at most $m_H \times m_G$ different types comparisons (where m_H and m_G respectively denote the number of relations in H and G). So each time one is computed, we can store its result in a table. BT will now run in $(N \times k) + (m_H \times m_G \times S)$, and the two exponential factors are no more multipliers for each other. In usual SG formalisms, S is polynomial (types hierarchies are given by their cover rellation): we can then consider $N \times k$ as the only significant factor, and forget the cost of type comparisons. Another advantage is that conjunctive types are compiled as a whole, and so it avoids to backtrack along multiple relations: conjunctive types are an algorithmic optimization.

3.4   BackMarking

We first extend to hypergraphs a datastructure proposed for BT in [6], not to reduce the size of the search tree, but to check candidates more efficiently.

If $\text{pre}_U(x) = r_0, \ldots, r_k$, and we suppose that r_0, if it exists, is a unary relation, then we provide (when calling Order(H)) the node x with k fields noted $\Delta_1(x), \ldots, \Delta_k(x)$. This data structure (the Δs) will be used to incrementally compute sets of candidates. When calling Find-Candidates(x), $\Delta_1(x)$ is built by taking all nodes y of G such that, if image(x) = y, r_1 is supported in G, then by removing from this set all nodes such that r_0 is not supported (if it exists). Then the following Δs are computed in the following way:

$$\Delta_{i+1}(x) = \{y \in \Delta_i(x) \mid y = \text{image}(x) \Rightarrow r_{i+1} \text{is supported in } G\}$$

Now Find-Candidates(x) computes this set, and Δ_k contains exactly all candidates for x. So the only work for Other-Candidates will be to iterate through this set. Let us consider the example in Fig. 1. It represents a step of the BT algorithm where images have been found for x_1 and x_2, and we are computing possible candidates for x. Assuming that all types are pairwise non comparable, check that $\Delta_1(x) = \{y_1, y_3\}$ (r_1 removed nothing, but r_0 removed y_2), and that $\Delta_2 = \{y_3\}$. The only possible candidate for x is y_3.

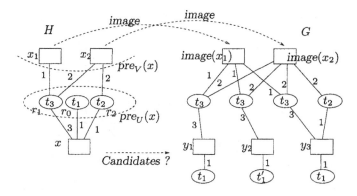

Fig. 1. Searching for possible candidates.

The complexity of the "natural" candidates research algorithm (Sect. 3.2) is in $\mathcal{O}(n_G^2 \times k)$, where n_G is the number of nodes of G and k the maximum arity of relations. But the worst-case complexity of the Δ based algorithm is $\mathcal{O}(n_G^3 \times k)$. There are two reasons to claim that this algorithm is still better: 1) we can implement BackMark, Forward-Checking, and BackJump on top of this algorithm at no additional overhead cost, and 2) an average case analysis shows that the natural algorithm still runs in $\mathcal{O}(n_G^2 \times k)$, while the Δ algorithm runs in $\mathcal{O}(n_G \times k \times R)$, where R is a factor decreasing with the density of G (see [6]).

Should we order the relations of $\text{pre}_U(x)$ in a particular way, our algorithm naturally evolves into BackMark [11]. Let x be a node of H. We order relations of $\text{pre}_U(x)$ in the following way (it can be done in linear time in the initialization phase, by the Order function): for each $r_j \in \text{pre}_U(x)$, we note $\text{last}(x, r_j)$ the greater node (as defined by Order) incident to r_j. We order relations of $\text{pre}_U(x)$ in such a way that if $\text{last}(x, r_i)$ is before $\text{last}(x, r_j)$, then r_i is before r_j.

Suppose now that, sometimes during the backtrack, we have built the different $\Delta_i(x)$, and that a failure forced to go up the search tree. What's happening when x becomes again the current node ? Let $(r_1, \ldots, r_k)$ be the non unary relations of $\text{pre}_U(x)$, ordered as indicated above. Then if the other arguments of $r_1, \ldots, r_p$ in $\text{pre}_V(x)$ did not change their images since the last candidates search, then $\Delta_1(x), \ldots, \Delta_p(x)$ do not need to be updated. When calling Find-Candidate, we look for the first $x_q \in \text{pre}_V(x)$ that changed since its last call, and build the Δ_i by beginning at Δ_q.

Note that this algorithm has been written at no overhead cost. To evaluate its interest, let us point out that the more "little backtracks" we encounter, the more we gain. And this is precisely the case on difficult instances of the problem.

3.5   Forward Checking(s)

The idea of Forward Checking (FC) [14] is to draw earlier the consequences when chosing a candidate. We present a first version,when all relations are binary,

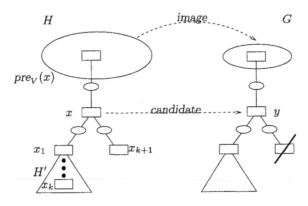

Fig. 2. Looking ahead of the current node.

that benefits greatly from our Δ data structure. Then, following [8], we point out that naive adaptations to hypergraphs are not efficient enough.

The Binary Case. Consider the example in Fig. 2, where all relations are binary. We have projected the subgraph H containing $pre_V(x)$, and have extended this partial projection to x (into y). Now it can take an exponentially long time to project the nodes $x_1, \ldots, x_k$. When at least we succeed, we remark that x_{k+1} had no possible candidate, and so this whole work was useless. To avoid this problem, we want the following property to be verified when considering candidates:
(FC): "If a candidate for x is y, then for each neighbour z of $pre_V(x) \cup \{x\}$, the current partial projection can be extended to a projection of z.

Rewriting the functions Find-Candidates and Other-Candidates, we obtain the algorithm FC, in its binary case, that respects the above property and is sound and complete for projection (note that our implementation, using Δs, naturally includes the BackMark algorithm, still at no overhead cost). Find-Candidates only puts an iterator at the beginning of the last Δ of x (that already contains all found candidates of x), then calls Other-Candidates. Other-Candidates advances the iterator, answer FALSE if the whole Δ has been explored, then calls FC-Propagates: if the result is TRUE (the property *FC* is verified), Other-Candidates also returns TRUE. Otherwise, it calls itself again to examine following candidates. So all checks are done in the function FC-Popagates, that looks ahead of the current node. This function considers all relations in $post_U(x)$, relations incident to x that are not in $pre_U(x)$. For each of these relations r, linking x to a node z, x contains a pointer to the associated $\Delta_i(z)$ (noted $\Delta'(x,r)$). This structure can be initialized in linear time during the call to Order. Then, when called, for each $r \in post_U(x)$, FC-Propagates builds the list $\Delta'(x,r)$, as indicated for the BackMark algorithm. If one of these Δ' becomes empty, then FC-Propagates stops and returns FALSE, otherwise it returns TRUE (after having correctly built all Δs for the neighbours of x).

We obtain this algorithm just by changing the order in which BackMark performs some operations. It has been experimentally proven very efficient.

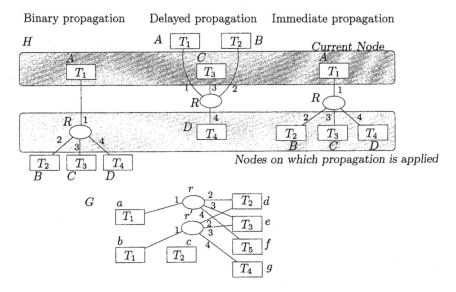

Fig. 3. When will propagation be called?

FC for Hypergraphs However, we are interested in *n*-ary relations, and the above algorithm only works in the binary case. A natural way is to consider the incidence bipartite of the hypergraph: nodes remain nodes, relations become nodes, and *x* is the *i*th node incident to *r* is translated into an edge labelled *i* between *x* and *r*. This is exactly the drawing of the hypergraph, viewed as a bipartite multigraph (it corresponds to the usual definition of SGs). Then it is sufficient to run the algorithm presented above on this associated binary graph. However, it cannot be considered as a true generalization of FC to hypergraphs: it propagates from (concept) node to (relation) node, then from (relation) node to (concept) node; it looks only half a step ahead. We will call **FC** this first "generalization" of the binary FC. In Fig. 3, when trying to project node *A* of *H* into node *a* of *G*, this *binary propagation* sees that there will be no problem to project the relation node *R*. However, looking a bit further should have been better...

To cope with this problem, it has been proposed to propagate at distance 2 in the incidence bipartite. This algorithm has been called **FC+**, but experiments showed it not as efficient as expected. The CSP community began then to be interested in hypergraphs themselves, and not in their incidence bipartite.

The first question to answer is *when to propagate* ? As illustrated in Fig. 3, we can either *delay* the propagation (*i.e.* check if a relation is supported as soon as all its arguments save one have been projected), or rely on an *immediate propagation* (*i.e.* check if a relation is supported as soon as one of its aruments has been projected). With *delayed propagation*, we only check if *R* is supported when *A* and *B* have already been projected. Suppose that *A* has been projected in *a*, *B* in *c*, and we are checking whether *C* can be projected in *e*. We deduce that *R* will not be supported, whatever the image of *D*. But we should have

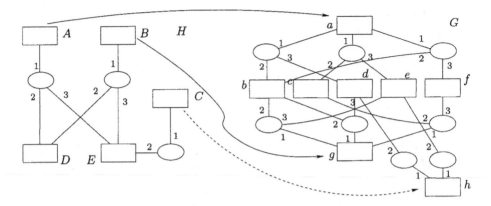

Fig. 4. Propagating along relations non incident to the current node.

remarked it sooner. With *immediate propagation*, as soon as we try to project A into a, we see that R will not be supported.

Immediate propagation always leads to smaller research trees. However, cost of propagation must be taken into account, and only experiments (done for CSPs) can show that cuts done in the research tree are worth that cost.

The next questions are *"where to propagate ?"*, and *"how many times ?"*. In the binary case, it is sufficient to propagate once along relations incident to the current node, but it is not the case in hypergraphs, as shown by Fig. 4. Suppose that A has been projected in a and B in g. When we immediately propagated from A, we deduced that (D, E) should only be projected in (b, d), (c, e) or (b, f). Candidates for D are $\{b, c\}$, and those for E are $\{d, e, f\}$. When we immediately propagated from B, we did not restrict these sets. The current node is now C, and we try to project it into h: propagating this choice, we remove f from the candidates of E. Let us check again the supports for the relation incident to A and the one incident to B (note they are not incident to C). The one incident to A does not remove any candidates, but the one incident to B proves that the only images for (D, E) are now (b, e) and (b, d): candidates for D and E are now respectively $\{b\}$ and $\{d, e\}$. Let us check again supports for the relation incident to A: candidates for D and E are now respectively $\{b\}$ and $\{d\}$.

So by propagating along relations non incident to the current node, we have removed more candidates, and more so by checking many times the supports for the same relation. It gives us many choices to implement an hypergraph generalization of FC (though all obtained algorithms are sound and complete, these choices define how much the research tree will be cut). 1) Do we only propagate along relations of $\mathrm{post}_U(x)$, relations incident to x that have an argument in $\mathrm{post}_V(x)$ (we call it *local propagation*), or do we propagate also along relations that have an argument in $\mathrm{pre}_V(x)$ and another in $\mathrm{post}_V(x)$ (we call it a *global propagation*). 2) Do we only propagate once (we call it *in one step*), or repeat propagations as long as one removes some candidates (we call it *in many step*). The following algorithms, implemented and experimented [8] in the CSP community, can be defined with the above criteria:

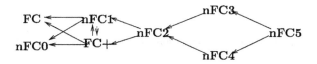

Fig. 5. Respective strengths of different n-ary Forward Checking.

- **FC** uses the binary FC on the incidence bipartite graph
- **FC+** uses a binary FC, modified to propagate at distance 2, on the incidence bipartite graph (it is equivalent to **nFC1**)
- **nFC0** uses delayed local propagation
- **nFC2** uses immediate local propagation, in one step
- **nFC3** uses immediate local propagation, in many step
- **nFC4** uses immediate global propagation, in one step
- **nFC5** uses immediate lobal propagation, in many step

We say that some propagation algorithm X is *stronger* than a propagation algorithm Y if, considering nodes of H and G in the same order, the research tree generated by X is a subtree of the one generated by Y. Note that, though it means that X generates fewer backtracks than Y, it does not mean that X is more efficient: the cost of propagation can be prohibitive. Fig. 5 compares the theoretical strengths of these different algorithms. Experiments in [8] show that, though there is no clear "best algorithm", there is really worst ones: **FC** and **FC+**. This is a justification for our hypergraphs point of view. Though **nFC5** is the best one when relations strongly constraint our choices (a sparse order on a big set of types, or sparse SGs), **nFC0** is the best when SGs are dense and few relation types are available. A good trade off is **nFC2**, whose efficiency is always satisfying, and can be implemented easily using our Δ data structure [5].

4   Conclusion

We have presented in this paper algorithms developped in the CSP community, that really improve algorithms used today for comparing SGs (by example, [12] uses an algorithm equivalent to **FC+**). This work is not a mere translation: the data structure we propose (the Δs) improves candidate search (in average case) and can be used to implement at the same time BackMark and **nFC2** without overhead cost. Moreover, our criteria describing the different FC generalizations to hypergraphs unify the different, unintuitive definitions given in [8].

This work has shown that considering SGs as hypergraphs greatly improved efficiency of projection. Experimental results rely on those done for CSPs [8]. Adding conjunctive types was needed for two reasons: to get rid of the problems encountered when fusioning nodes having different types, and to optimize backtracking process. Finally, though algorithms optimization is not a primary goal of our community, it should be interesting to follow the evolution of CSP algorithms.

References

1. S. Abiteboul, R. Hull, and V. Vianu. *Foundations of Databases.* Addison-Wesley, 1995.
2. F. Baader, R. Molitor, and S. Tobies. Tractable and Decidable Fragments of Conceptual Graphs. In *Proc. of ICCS'99*, pages 480–493. Springer, 1999.
3. J.-F. Baget. A Simulation of Co-Identity with Rules in Simple and Nested Graphs. In *Proc. of ICCS'99*, pages 442–455. Springer, 1999.
4. J.-F. Baget. Extending the CG Model by Simulations. Number 1867 in Lecture Notes in Artificial Intelligence, pages 277–291. Springer, 2000.
5. J.-F. Baget. *Repr senter des connaissances et raisonner avec des hypergraphes: de la projection la d rivation sous contraintes.* PhD thesis, Un. Montpellier II, 2001.
6. J.-F. Baget and Y. Tognetti. Backtracking through Biconnected Components of a Constraint Graph. In *Proceedings of IJCAI'01*, pages 291–296, Vol. 1, 2001.
7. Peter Becker and Joachim Hereth. A Simplified Data Model for CGs. http://tockit.sourceforge.net/papers/cgDataModel.pdf, 2001.
8. C. Bessière, P. Meseguer, E. C. Freuder, and X. Larrosa. On Forward Checking for Non-Binary Constraint Satisfaction. In *Proc. of CP'99*, pages 88–102. Springer, 1999.
9. T. H. Cao. *Foundations of Order-Sorted Fuzzy Set Logic Programming in Predicate Logic and Conceptual Graphs.* PhD thesis, Un. of Queensland, 1999.
10. M. Chein and M.-L. Mugnier. Conceptual Graphs: fundamental notions. *Revue d'Intelligence Artificielle*, 6(4):365–406, 1992.
11. J. Gaschnig. Performance measurement and analysis of certain search algorithms. Research report CMU–CS–79–124, Carnegie-Mellon University, 1979.
12. D. Genest and E. Salvat. A Platform Allowing Typed Nested Graphs: How CoGITo Became CoGITaNT. In *Proc. of ICCS'98*, pages 154–161. Springer, 1998.
13. S. W. Golomb and L. D. Baumert. Backtrack Programming. *Journal of the ACM*, 12:516–524, 1965.
14. R. M. Haralick and G. L. Elliott. Increasing Tree Search Efficiency for Constraint Satisfaction Problems. *Artificial Intelligence*, 14:263–314, 1980.
15. P. Hayes. RDF Model Theory. W3c working draft, 2001.
16. G. Kerdiles. Projection: A Unification Procedure for Tableaux in Conceptual Graphs. In *Proc. of TABLEAUX'97*, pages 138–152. Springer, 1997.
17. M.-L. Mugnier. Knowledge Representation and Reasonings Based on Graph Homomorphism. In *Proc. of ICCS'00*, pages 172–192. Springer, 2000.
18. M.-L. Mugnier and M. Chein. Repr senter des connaissances et raisonner avec des graphes. *Revue d'Intelligence Artificielle*, 10(1):7–56, 1996.
19. P. Prosser. Hybrid Algorithms for the Constraint Satisfaction Problem. *Computational Intelligence*, 9(3):268–299, 1993.
20. E. Salvat and M.-L. Mugnier. Sound and Complete Forward and Backward Chainings of Graphs Rules. In *Proc. of ICCS'96*. Springer, 1996.
21. B. Smith. Locating the Phase Transition in Binary Constraint Satisfaction Problems. *Artificial Intelligence*, 81:155–181, 1996.
22. J. F. Sowa. Conceptual Graphs for a Database Interface. *IBM Journal of Research and Development*, 20(4):6–57, 1976.
23. J. F. Sowa. *Conceptual Structures: Information Processing in Mind and Machine.* Addison-Wesley, Reading, MA, 1984.
24. M. Wermelinger. Conceptual Graphs anf First-Order Logic. In *Proc. of ICCS'95*. Springer, 1995.

Concept Graphs without Negations: Standard Models and Standard Graphs

Frithjof Dau

Technische Universität Darmstadt, Fachbereich Mathematik
Schloßgartenstr. 7, D-64289 Darmstadt
dau@mathematik.tu-darmstadt.de

Abstract. In this article, we provide different possibilities for doing reasoning on simple concept(ual) graphs without negations or nestings. First of all, we have on the graphs the usual semantical entailment relation $\models$, and we consider the restriction $\vdash$ of the calculus for concept graph with cuts, which has been introduced in [Da02], to the system of concept graphs without cuts. Secondly, we introduce a semantical entailment relation $\models$ as well as syntactical transformation rules $\vdash$ between models. Finally, we provide definitions for standard graphs and standard models so that we translate graphs to models and vice versa. Together with the relations $\models$ and $\vdash$ on the graphs and on the models, we show that both calculi are adequate and that reasoning can be carried over from graphs to models and vice versa.

1 Introduction

This paper is mainly based on two treatises: The dissertations of the author Dau ([Da02]) and of Prediger ([Pr98a]). In [Pr98a], Prediger developed a mathematical theory for simple concept graphs. In Predigers graphs, negation cannot be expressed, so these graphs correspond to the existential-conjunctive fragment of conceptual graphs or first order predicate logic (FOPL). Reasoning on these graphs can be performed in two different ways: First of all, Prediger introduced a sound and complete calculus which consists of fairly simple transformation rules. Secondly, she assigned to each graph a corresponding *standard model* (and, vice versa, to each model a corresponding *standard graph*). For graphs without generic markers, Predigers standard models encode exactly the same information as the respective graphs. Thus, for these graphs, the reasoning on graphs can be carried over to the models.

In [Da02], the author extended the syntax of concept graphs by adding the *cuts* of Peirce's existential graphs. Cuts are syntactical devices which serve to negate subgraphs. As negation can be expressed in concept graphs with cuts, these graphs correspond to full FOPL (see [Da02]). For these graphs, a sound and complete calculus is provided, too. This calculus is not a further development of Predigers calculus, but it is based on Peirce's calculus for existential graphs. In particular, some of its transformation rules are fairly complex. In contrast to concept graphs without negations, it turns out that the information which is

A. de Moor, W. Lex, and B. Ganter (Eds.): ICCS 2003, LNAI 2746, pp. 243–256, 2003.
© Springer-Verlag Berlin Heidelberg 2003

encoded by concept graphs with cuts cannot be encoded in standard models, so the notion of standard models has to be dropped.

In this article, we bring together the ideas of Prediger and Dau for concept graphs without negation (i. e. without cuts). First of all, on concept graphs without cuts, we introduce a restricted version of the calculus for concept graphs with cuts, where we have removed all rules where cuts are involved. Secondly, we extend Predigers notions of standard graphs and standard models (the differences to Predigers approach will be discussed in the next sections) such that even for graphs *with* generic markers, their standard models will encode exactly the same information as the respective graphs. On the models, we introduce a semantical entailment relation $\models$ as well as transformation rules. The latter can be seen as a kind of calculus which yields a relation $\vdash$ between models. It will turn out that the relations $\models, \vdash$ for graphs and for models and the notions of standard models and standard graphs fit perfectly together.

2   Basic Definitions

In this section, we provide the basic definitions for this paper. The first four definitions are directly adopted or slightly modified definitions which can be found in [Pr98a] as well as in [Da02]. Definition 5 is a restricted version –which suits for concept graphs without negation– of the calculus in [Da02]. We start with the definitions of the underlying alphabet (which is called the *support*, *taxonomy* or *ontology* by other authors) and of simple concept graphs.

Definition 1 (Alphabet).
An alphabet *is a triple* $\mathcal{A} := (\mathcal{G}, \mathcal{C}, \mathcal{R})$ *of disjoint sets* $\mathcal{G}, \mathcal{C}, \mathcal{R}$ *such that*

- $\mathcal{G}$ *is a finite set whose elements are called* object names,
- $(\mathcal{C}, \leq_C)$ *is a finite ordered set with a greatest element* $\top$ *whose elements are called* concept names, *and*
- $(\mathcal{R}, \leq_{\mathcal{R}})$ *is a family of finite ordered sets* $(\mathcal{R}_k, \leq_{\mathcal{R}_k})$, $k = 1, \ldots, n$ *(for an* $n \in \mathbb{N}$) *whose elements are called* relation names. *Let* $\doteq \in \mathcal{R}_2$ *be a special name which is called* identity.

On $\mathcal{G} \,\dot\cup\, \{*\}$ *we define an order* $\leq_{\mathcal{G}}$ *such that* $*$ *is the greatest element* $\mathcal{G} \,\dot\cup\, \{*\}$, *but all elements of* $\mathcal{G}$ *are incomparable.*

Definition 2 (Simple Concept Graphs).
A simple concept graph over $\mathcal{A}$ *is a structure* $\mathfrak{G} := (V, E, \nu, \kappa, \rho)$ *where*

- V *and* E *are pairwise disjoint, finite sets whose elements are called* vertices *and* edges,
- $\nu : E \to \bigcup_{k \in \mathbb{N}} V^k$ *is a mapping (we write* $|e| = k$ *for* $\nu(e) \in V^k$),
- $\kappa : V \cup E \to \mathcal{C} \cup \mathcal{R}$ *is a mapping such that* $\kappa(V) \subseteq \mathcal{C}$, $\kappa(E) \subseteq \mathcal{R}$, *and all* $e \in E$ *with* $|e| = k$ *satisfy* $\kappa(e) \in \mathcal{R}_k$, *and*
- $\rho : V \to \mathcal{G} \,\dot\cup\, \{*\}$ *is a mapping.*

The set of these graphs is denoted by CG^A. If the alphabet is fixed, we sometimes write CG. For the set E of edges, let $E^{id} := \{e \in E \mid \kappa(e) = \doteq\}$ and $E^{nonid} := \{e \in E \mid \kappa(e) \neq \doteq\}$. The elements of E^{id} are called identity-links. Finally set $V^* := \{v \in V \mid \rho(v) = *\}$ and $V^{\mathcal{G}} := \{v \in V \mid \rho(v) \in \mathcal{G}\}$. The nodes in V^* are called generic nodes.

These graphs correspond to the graphs in [Da02], where the cuts are removed. Although they are similar, there are some important differences between these graphs and the simple concept graphs as they are defined by Prediger in [Pr98a]:

First of all, in [Pr98a] Prediger assigned *sets* of objects instead of *single* objects to vertices (i.e. in [Pr98a] we have $\rho : V \to \mathfrak{P}(\mathcal{G}) \,\dot\cup\, \{*\}$ instead of $\rho : V \to \mathcal{G} \,\dot\cup\, \{*\}$). For concept graphs with cuts, it is not immediately clear what the meaning of a vertex is which is enclosed by a cut and which contains more than one object. For this reason, in [Da02] and thus in this article, ρ assigns single objects to vertices. The expressiveness of the graphs is not changed by this syntactical restriction.

Identity is in [Pr98a] expressed by an equivalence relation θ only on the set of generic vertices. In [Da02] and in this article, identity is expressed by identity links on the set of generic *and non-generic* vertices. Thus the concept graphs of this article have a slightly higher expressiveness than the concept graphs of [Pr98a]. This has to be taken into account in the definition of standard models and standard graphs, as well as in the calculus. To provide an example: Consider an alphabet with $\mathcal{C} := \{\top, A, B\}$ and $\mathcal{G} := \{a, b\}$, where A, B are incomparable concept names. In our approach, $\boxed{A\colon a}\!-\!\boxed{=}\!-\!\boxed{B\colon b}$ is a well-defined graph[1] which expresses a proposition which cannot represented in Predigers approach. This graph entails $\boxed{A\colon b}\!-\!\boxed{=}\!-\!\boxed{B\colon a}$. Obviously, this entailment is based on the unique meaning of identity. For this reason, we will have rules in our calculus which capture the role of the identity relation and which allow to derive the second graph from the first one. A derivation like this cannot be performed with projections[2] ([CM92]) or with the calculus presented in [Mu00].

The next two definitions describe the 'contextual models' which we consider. These definitions appear in [Pr98a] as well as in [Da02] and are based on the theory of Formal Concept Analysis (see [GW99]).

Definition 3 (Power Context Family).
A power context family $\vec{\mathbb{K}} := (\mathbb{K}_0, \mathbb{K}_1, \mathbb{K}_2, \ldots)$ is a family of formal contexts $\mathbb{K}_k := (G_k, M_k, I_k)$ that satisfies $G_0 \neq \emptyset$ and $G_k \subseteq (G_0)^k$ for each $k \in \mathbb{N}$. Then we write $\vec{\mathbb{K}} := (G_k, M_k, I_k)_{k \in \mathbb{N}_0}$. The elements of G_0 are the objects of $\vec{\mathbb{K}}$. A pair (A, B) with $A \subseteq G_k$ and $B \subseteq M_k$ is called a concept of $\mathbb{K}_k$, if and only

---

[1] In contrast to other authors, like Mugnier [Mu00], we allow that different object names may refer to the same object, i.e. we do not adopt the *unique name assumption*. The unique name assumption is needed when a graph shall be transformed into its normal-form. This graph cannot be transformed into a normal-form and is not a well-defined graph in the approach of Mugnier.

[2] Moreover, as projections rely on normal-forms for graphs, they have further restrictions. See [Mu00].

if $A = \{g \in G_k \mid gI_k n$ for all $b \in B\}$ and $B = \{m \in M_k \mid aI_k m$ for all $a \in A\}$.
$Ext(A, B) := A$ is called extension *of the concept (A, B), and $Int(A, B) := B$ is*
called intension *of the concept (A, B). The set of all concepts of a formal context*
$\mathbb{K}_k$ is denoted by $\mathfrak{B}(\mathbb{K}_k)$. The elements of $\bigcup_{k \in \mathbb{N}_0} \mathfrak{B}(\mathbb{K}_k)$ are called concepts,
and we set furthermore $\mathfrak{R}_{\vec{\mathbb{K}}} := \bigcup_{k \in \mathbb{N}} \mathfrak{B}(\mathbb{K}_k)$, and the elements of $\mathfrak{R}_{\vec{\mathbb{K}}}$ are called
relation-concepts.

To get a contextual structure over an alphabet, we have to interpret the object-, concept- and relation-names by objects, concepts and relation-concepts in a power context family. This is done in the following definition.

Definition 4 (Contextual Models).
For an alphabet $\mathcal{A} := (\mathcal{G}, \mathcal{C}, \mathcal{R})$ and a power context family $\vec{\mathbb{K}}$, we call the dis-
joint union $\lambda := \lambda_{\mathcal{G}} \,\dot{\cup}\, \lambda_{\mathcal{C}} \,\dot{\cup}\, \lambda_{\mathcal{R}}$ of the mappings $\lambda_{\mathcal{G}}: \mathcal{G} \to G_0$, $\lambda_{\mathcal{C}}: \mathcal{C} \to \mathfrak{B}(\mathbb{K}_0)$
and $\lambda_{\mathcal{R}}: \mathcal{R} \to \mathfrak{R}_{\vec{\mathbb{K}}}$ a $\vec{\mathbb{K}}$-interpretation of $\mathcal{A}$ if $\lambda_{\mathcal{C}}$ and $\lambda_{\mathcal{R}}$ are order-preserving,
and $\lambda_{\mathcal{C}}, \lambda_{\mathcal{R}}$ satisfy $\lambda_{\mathcal{C}}(\top) = \top$, $\lambda_{\mathcal{R}}(\mathcal{R}_k) \subseteq \mathfrak{B}(\mathbb{K}_k)$ for all $k = 1, \dots, n$, and
$(g_1, g_2) \in Ext(\lambda_{\mathcal{R}}(\doteq)) \Leftrightarrow g_1 = g_2$ for all $g_1, g_2 \in G_0$.[3] The pair $(\vec{\mathbb{K}}, \lambda)$ is called
contextual structure over $\mathcal{A}$ *or* contextual model over $\mathcal{A}$. *The set of these con-*
textual structures is denoted by $CS^{\mathcal{A}}$. If the alphabet is fixed, we sometimes write
CS.

The calculus for concept graphs with cuts consists of the following rules (see[Da02]): erasure, insertion, iteration, deiteration, double cuts, generalization, specialization, isomorphism, exchanging references, merging two vertices, splitting a vertex, $\top$-erasure, $\top$-insertion, identity-erasure and identity-insertion.

Only the double-cut-rule allows to derive a concept graph with cuts from a concept graph without cuts. The rules insertion and specialization can only be applied if we have a graph with cuts. As we consider concept graphs without cuts, we remove these three rules from the calculus and interpret the remaining rules as rules for the system of concept graphs without cuts. So we have the following definition (for examples, an explanation and a precise mathematical definition for the rules, we refer to [Da02]):

Definition 5 (Calculus for Simple Concept Graphs).
The calculus for concept graphs over the alphabet $\mathcal{A} := (\mathcal{G}, \mathcal{C}, \mathcal{R})$ consists of the
following rules:
Erasure, iteration, deiteration, generalization, isomorphism, exchanging ref-
erences, merging two vertices, splitting a vertex, $\top$-erasure, $\top$-insertion, identity-
erasure and identity-insertion.
If $\mathfrak{G}_a$, $\mathfrak{G}_b$ are concept graphs, and if there is a sequence $(\mathfrak{G}_1, \mathfrak{G}_2, \dots, \mathfrak{G}_n)$ with
$\mathfrak{G}_1 = \mathfrak{G}_a$, $\mathfrak{G}_b = \mathfrak{G}_n$ such that each $\mathfrak{G}_{i+1}$ is derived from $\mathfrak{G}_i$ by applying one of
the rules above, we say that $\mathfrak{G}_b$ can be derived from $\mathfrak{G}_a$ and write $\mathfrak{G}_a \vdash_{pos} \mathfrak{G}_b$.

---

[3] As Prediger does not consider identity links, the last condition does not appear in her definition of contextual structures.

3   Contextual Structures: Standard Models and Semantical Entailment

In this section, we will assign to a graph its *standard model* which encodes exactly the same information as the graph. This is based on [Wi97] and has already done by Prediger in Definition 4.2.5. of [Pr98a]. Remember that identity is in [Pr98a] expressed by an equivalence relation θ only on the set of generic vertices, so Prediger used the following approach: The set of objects of the standard model consists of all object names $G \in \mathcal{G}$ and of all equivalence classes of θ. But we can express identity between arbitrary, i. e. generic or non-generic, vertices, thus we have to extend this idea. We start by defining an equivalence relation $\theta_\mathfrak{G}$ on $V \dot\cup \mathcal{G}$, which is an appropriate generalization of Predigers θ.

Definition 6. *Let $\mathfrak{G} := (V, E, \nu, \kappa, \rho)$ be a concept graph over $\mathcal{A}$. We assume that V and $\mathcal{G}$ are disjoint. Let $\theta_\mathfrak{G}$ be the smallest equivalence relation on $V \dot\cup \mathcal{G}$ such that*

1. *if $\rho(v) = G \in \mathcal{G}$, then $v \theta_\mathfrak{G} G$, and*
2. *if $e \in E$ with $\nu(e) = (v_1, v_2)$ and $\kappa(e) \leq \doteq$, then $v_1 \theta_\mathfrak{G} v_2$.*

It is easy to see that two vertices which are equivalent must refer in each model to the same object, i. e. we have the following lemma:

Lemma 1. *Let $\mathfrak{G} := (V, E, \nu, \kappa, \rho)$ be a concept graph over $\mathcal{A}$. If $v_1, v_2 \in V$ with $v_1 \theta_\mathfrak{G} v_2$, then $ref(v_1) = ref(v_2)$ for each contextual structure $(\vec{\mathbb{K}}, \lambda)$ over $\mathcal{A}$ and each valuation $ref : V \to G_0$ with $(\vec{\mathbb{K}}, \lambda) \models \mathfrak{G}[ref]$.*

The opposite direction of the lemma holds as well, i. e. we could characterize $\theta_\mathfrak{G}$ by the condition in the lemma. This is not immediately clear, but it could easily be shown with the results of this paper.

Now we can assign to each concept graph an appropriate standard model which encodes exactly the same information as the graph.[4]

Definition 7 (Standard Model).
Let $\mathfrak{G} := (V, E, \nu, \kappa, \rho)$ be a concept graph over $\mathcal{A}$. We define the standard model of $\mathfrak{G}$ as follows:
For $\mathfrak{G} \neq \emptyset$ or $\mathcal{G} \neq \emptyset$,[5] we first define a power context family $\vec{\mathbb{K}}^\mathfrak{G}$ by

---

[4] This is possible because we consider only the existential-conjunctive fragment of concept graphs (in particular we do not consider negations or disjunctions of propositions), so we have to encode only the information a) whether objects have specific properties or whether objects are in a specific relation, b) whether objects *exist* with specific properties, and c) the conjunction of informations like these. These are the kinds of information which can be expressed in graphs (e.g. existential graphs or concept graphs) in an iconical way, i. e. we encode in standard models exactly the iconical features of concept graphs. For a deep discussion of this topic, we refer to [Sh99].

[5] As usual in logic, we only consider non-empty structures. For this reason, we have to treat the case $\mathfrak{G} = \emptyset = \mathcal{G}$ separately.

- $G_0^{\mathfrak{G}} := \{[k]\theta_{\mathfrak{G}} \mid k \in V \dot{\cup} \mathcal{G}\}$, and $G_i^{\mathfrak{G}} := (G_0^{\mathfrak{G}})^i$,
- $M_0^{\mathfrak{G}} := \mathcal{C}$, and $M_i^{\mathfrak{G}} := \mathcal{R}_i$ for $1 \leq i \leq n$.
- For $C \in \mathcal{C}$ and $g \in G_0$, we set
  $$g I_0^{\mathfrak{G}} C \iff C = \top \text{ or } \exists v \in V . g = [v]\theta_{\mathfrak{G}} \wedge \kappa(v) \leq C .$$
- For $R_i \in \mathcal{R}_i$ and $g_1, \ldots, g_i \in G_0$, we set $(g_1, \ldots, g_i) I_i^{\mathfrak{G}} R_i \iff$
  - $\exists e = (v_1, \ldots, v_i) \in E . g_1 = [v_1]\theta_{\mathfrak{G}} \wedge \ldots \wedge g_i = [v_i]\theta_{\mathfrak{G}} \wedge \kappa(e) \leq R_i$, or
  - $i = 2$ and $\doteq \leq R_i \wedge g_1 = g2 .$

The mappings $\lambda^{\mathfrak{G}}$ are defined canonically: $\lambda_{\mathcal{G}}^{\mathfrak{G}}(G) := [G]\theta_{\mathfrak{G}}$ for all $G \in \mathcal{G}$, $\lambda_{\mathcal{C}}^{\mathfrak{G}}(C) := \mu(C)$ for all $C \in \mathcal{C}$, and $\lambda_{\mathcal{R}}^{\mathfrak{G}}(R) := \mu(R)$ for all $R \in \mathcal{R}$.

If $\mathfrak{G} = \emptyset$ and $\mathcal{G} = \emptyset$, let g be an arbitrary element. We define $\vec{\mathbb{K}}^{\mathfrak{G}}$ as follows: $\vec{\mathbb{K}}_0 := (\{g\}, \{\top\}, \{(g, \top)\})$, $\vec{\mathbb{K}}_2 := (\{(g, g)\}, \{\doteq\}, \{((g, g), \doteq)\})$, and for $i \neq 0, 2$, let $\vec{\mathbb{K}}_i := (\emptyset, \emptyset, \emptyset)$. The mappings of $\lambda^{\mathfrak{G}}$ are defined canonically, i.e. $\lambda_{\mathcal{G}}^{\mathfrak{G}} := \emptyset$, $\lambda_{\mathcal{C}}^{\mathfrak{G}}(\top) := \mu(\top)$, and $\lambda_{\mathcal{R}}^{\mathfrak{G}}(\doteq) := \mu(\doteq)$. All remaining concept- or relation-names are mapped to the $\bot$-concept $(\emptyset'', \emptyset')$ of the respective formal context.

The contextual structure $(\vec{\mathbb{K}}^{\mathfrak{G}}, \lambda^{\mathfrak{G}})$ is called standard model of $\mathfrak{G}$ and is denoted by $\mathcal{M}^{\mathfrak{G}}$.[6]

It is easy to see that each graph holds in its standard model, i.e. we have:

Lemma 2. If $\mathfrak{G} := (V, E, \nu, \kappa, \rho)$ is a graph, then $(\vec{\mathbb{K}}^{\mathfrak{G}}, \lambda^{\mathfrak{G}})$ is a contextual structure over $\mathcal{A}$. For $ref^{\mathfrak{G}} := \{(v, [v]\theta_{\mathfrak{G}}) \mid v \in V\}$, we have $(\vec{\mathbb{K}}^{\mathfrak{G}}, \lambda^{\mathfrak{G}}) \models \mathfrak{G}[ref^{\mathfrak{G}}]$.

In the following, we provide some examples for simple concept graphs (over the alphabet $(\{a, b, c\}, \{A, B, C, \top\}, \{\doteq\})$ with incomparable concept names A, B, C and their standard models.

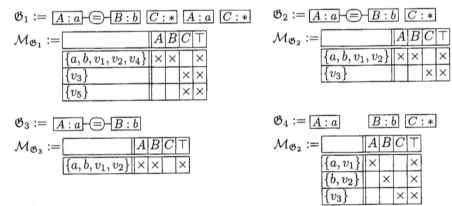

The well-known relation $\models$ on graphs can be understood as follows: $\mathfrak{G}_1 \models \mathfrak{G}_2$ holds iff $\mathfrak{G}_1$ contains the same or more information than $\mathfrak{G}_2$. This idea can be transferred to models as well. This yields the following definition:[7]

---

[6] We write $\mathcal{M}^{\mathfrak{G}}$ instead of $\mathcal{M}_{\mathfrak{G}}$, because the different contexts of $\vec{\mathbb{K}}^{\mathfrak{G}}$ and the mappings of $\lambda^{\mathfrak{G}}$ already have indices at the bottom.

[7] See also [Wi02], where Wille defines in a different way the *informational content* of a model.

Definition 8 (Semantical Entailment between Contextual Structures).

Let $\mathcal{M}^a := (\vec{\mathbb{K}}^a, \lambda^a)$ and $\mathcal{M}^b := (\vec{\mathbb{K}}^b, \lambda^b)$ be two $\mathcal{A}$-structures (with $\vec{\mathbb{K}}^x = (\mathbb{K}_0^x, \dots, \mathbb{K}_n^x)$ and $\mathbb{K}_i^x = (G_i^x, M_i^x, I_i^x)$, $i = 1, \dots, n$ and $x = a, b$). We set

$$\mathcal{M}^a \models \mathcal{M}^b :\Longleftrightarrow \text{ it exists a mapping } f : G_0^b \to G_0^a \text{ with:}$$

1. *For all $G \in \mathcal{G}$: $f(\lambda_{\mathcal{G}}^b(G)) = \lambda_{\mathcal{G}}^a(G)$ (i. e. f respects $\lambda_{\mathcal{G}}$.)*
2. *For all $g \in G_0^b$ and $C \in \mathcal{C}$: $g \in Ext(\lambda_{\mathcal{C}}^b(C)) \Longrightarrow f(g) \in Ext(\lambda_{\mathcal{C}}^a(C))$ (i. e. f respects $\lambda_{\mathcal{C}}$).*
3. *For all $\vec{g} \in G_i^b$ and $R \in \mathcal{R}_i$: $\vec{g} \in Ext(\lambda_{\mathcal{R}}^b(R)) \Longrightarrow f(\vec{g}) \in Ext(\lambda_{\mathcal{R}}^a(R))$ where $f(\vec{g}) = f(g_1, \dots, g_i) := (f(g_1), \dots, f(g_i))$ (i. e. f respects $\lambda_{\mathcal{R}}$).*

We will sometimes write $\mathcal{M}^a \models_f \mathcal{M}^b$ to denote the mapping f, too.

We want to make the following remarks on standard models:

1. The direction of f is not surprising: It corresponds to the direction of projections between conceptual graphs (see [CM92]). In fact, f can be roughly understood as projection between models (instead of between graphs). But note that projection between graphs is complete only under certain restrictions (e.g., on the normal form of one graph. But, as argued before Definition 3, not every graph can be transformed into a normal-form), but it will turn out that projection between models is complete without further restrictions. Thus, to evaluate whether a graph $\mathfrak{G}_a$ entails a graph $\mathfrak{G}_b$, a sound and complete approach is the following: First contruct the standard models $\mathcal{M}_{\mathfrak{G}_a}$ and $\mathcal{M}_{\mathfrak{G}_b}$, and then find out whether there is a 'projection' f from $\mathcal{M}_{\mathfrak{G}_b}$ to $\mathcal{M}_{\mathfrak{G}_a}$.

2. Please note that the models on the right are semantically equivalent, although they have a different number of objects. This is based on the fact that we cannot count in existential-conjunctive languages (without negation, we cannot express that we have *different* objects with the same properties).

$$\mathcal{M}_1 := \begin{array}{|c||c|c|} \hline \mathbb{K}_0^1 & P & \top \\ \hline\hline g & \times & \times \\ \hline \end{array} \qquad \mathcal{M}_2 := \begin{array}{|c||c|c|} \hline \mathbb{K}_0^2 & P & \top \\ \hline\hline g & \times & \times \\ \hline h & \times & \times \\ \hline \end{array}$$

3. In [Pr98a], standard models are compared as well, but Prediger compares only the restrictions of the incidence-relations to objects which are generated by non-generic nodes. E.g. in Predigers approach, the standard models of the graphs $\boxed{A : *}$ and $\boxed{B : *}$ are comparable, although they encode incomparable information. Thus Predigers approach is strictly weaker than semantical entailment between models.

The next theorem shows the main link between a graph and its standard model. As Prediger has no concept of semantical entailment between models, there is no corresponding theorem in [Pr98a].

Theorem 1 (Main Theorem for Graphs and Their Standard Models).

Let $\mathfrak{G} := (V, E, \nu, \kappa, \rho)$ be a graph, $\mathcal{M}^{\mathfrak{G}} := (\vec{\mathbb{K}}^{\mathfrak{G}}, \lambda^{\mathfrak{G}})$ be its standard model and $\mathcal{M} := (\vec{\mathbb{K}}, \lambda)$ be an arbitrary contextual structure. Then we have

$$\mathcal{M} \models \mathcal{M}^{\mathfrak{G}} \quad \Longleftrightarrow \quad \mathcal{M} \models \mathfrak{G}.$$

From the main theorem, we get the following corollary. The first equivalence of the corollary corresponds to Theorem 4.2.6. in [Pr98a]. But in [Pr98a], we find no result which can be compared to $\mathfrak{G}_1 \models \mathfrak{G}_2 \Leftrightarrow \mathcal{M}^{\mathfrak{G}_1} \models \mathcal{M}^{\mathfrak{G}_2}$. Again due to the lack of the concept of semantical entailment between models, Prediger has only proven an implication which is a weak variant of $\mathfrak{G}_1 \models \mathfrak{G}_2 \Rightarrow \mathcal{M}^{\mathfrak{G}_1} \models \mathcal{M}^{\mathfrak{G}_2}$

Corollary 1. *Let $\mathfrak{G}_1, \mathfrak{G}_2$ be two graphs. Then we have*

$$\mathfrak{G}_1 \models \mathfrak{G}_2 \quad \Longleftrightarrow \quad \mathcal{M}^{\mathfrak{G}_1} \models \mathfrak{G}_2 \quad \Longleftrightarrow \quad \mathcal{M}^{\mathfrak{G}_1} \models \mathcal{M}^{\mathfrak{G}_2}.$$

4   Standard Graphs

In the last section, we assigned to each graph a corresponding standard model which contains the same information as the graph. In this section, we do the same for the opposite direction: We assign to a model a standard graph which contains the same information. This is done with the following definition.

Definition 9 (Standard Graphs).
Let $\mathcal{M} := (\vec{\mathbb{K}}, \lambda)$ be a contextual structure. We define the standard graph *of $\mathcal{M}$ as follows:*

1. *For each $g \in G_0$, let $v_g := \boxed{\top : *}$ be a new vertex (i. e. we set $\kappa(v_g) := \top$ and $\rho(v_g) = *$).*
2. *For each $g \in \mathcal{G}$ with $\lambda_{\mathcal{G}}(G) = g \in G_0$, let $v_{g,G}$ be a new vertex and $e_{g,G}$ a new edge. We set $\kappa(v_{g,G}) := \top$, $\rho(v_{g,G}) := G$, $\nu(e) := (v_g, v_{g,G})$ and $\kappa(e) :=\doteq$ (i. e. we add the vertex and edge on the right).*

3. *For each $C \in \mathcal{C}\backslash\{\top\}$ with $g \in Ext(\lambda_{\mathcal{C}}(C))$, let $v_{g,C}$ be a new vertex and $e_{g,C}$ a new edge. We set $\kappa(v_{g,C}) := C$, $\rho(v_{g,C}) := *$, $\nu(e) := (v_g, v_{g,C})$ and $\kappa(e) :=\doteq$ (i. e. we add the vertex and edge on the right).*

4. *For each $R \in \mathcal{R}_i\backslash\{\doteq\}$ with $(g_1, \ldots, g_i) \in Ext(\lambda_{\mathcal{R}}(R))$, let $e_{g_1,\ldots,g_i,R}$ be a new vertex. We set $\kappa(e_{g_1,\ldots,g_i,R}) := R$ and $\nu(e) := (g_1, \ldots, g_i)$ (i. e. we add the edge on the right).*

We denote this graph by $\mathfrak{G}_{(\vec{\mathbb{K}},\lambda)}$ or $\mathfrak{G}_{\mathcal{M}}$.[8]

---

[8] $\mathfrak{G}_{\mathcal{M}}$ is given only up to isomorphism, but we have the implicit agreement that isomorphic graphs are identified.

In [Pr98a], Definition 4.2.15, we find a corresponding definition for Definition 9. But there is a crucial difference between Predigers approach and our approach: In [Pr98a], Prediger assigns a standard graph to a power context family instead to a contextual structure. Thus, she has first to define an alphabet which is derived from the power context family, then she defines the standard graph over this alphabet. Our approach is different: We fix an alphabet at the beginning and consider only graphs and structures over this fixed alphabet.

To get an impression of standard models and standard graphs, we provide a more extensive example. First we have to fix the alphabet. We set $\mathcal{A} :=$ $(\{a, b, c, d\}, \{A_1, A_2, B_1, B_2, C, E, \top\}, \{R_1, R_2, S, \doteq\})$, where R_1, R_2, S are dyadic relation names. The orderings on the concept- and relation-names shall be as follows:

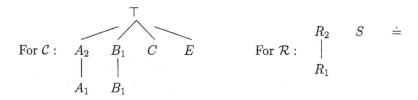

We consider the following graph over $\mathcal{A}$:

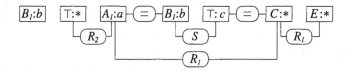

Below, we provide the standard model of this graph (the mappings λ_G, λ_C and λ_R are not explicit given, as they can easily be obtained from the power context family). We assume that the vertices of the graph are numbered from the left to the right, starting with 1, thus the i-th vertex is denoted by v_i.

$\mathbb{K}_0$	A_1	A_2	B_1	B_2	C	E	$\top$
$\{a, b, v_1, v_3, v_4\}$	×	×	×	×			×
$\{v_2\}$							×
$\{c, v_5, v_6\}$					×		×
$\{v_7\}$						×	×
$\{d\}$							×

$\mathbb{K}_2$	R_1	R_2	S	$\doteq$
$(\{v_2\}, \{a, b, v_1, v_3, v_4\})$		×		
$(\{a, b, v_1, v_3, v_4\}, \{c, v_5, v_6\})$	×	×	×	
$(\{c, v_5, v_6\}, \{v_7\})$	×	×		
$(\{v_2\}, \{v_2\})$				×
$(\{a, b, v_1, v_3, v_4\}, \{a, b, v_1, v_3, v_4\})$				×
$(\{c, v_5, v_6\}, \{c, v_5, v_6\})$				×
$(\{v_7\}, \{v_7\})$				×
$(\{d\}, \{d\})$				×

The standard graph of this model is given below. In the left, we sketch which vertices and edges are added by which step of Definition 9.

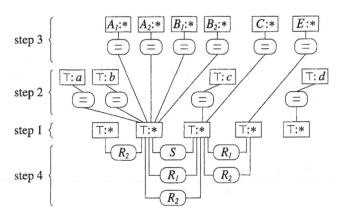

If we translate a model to a graph and then translate this graph back into a model, in general we do not get the same graph back, but at least a semantically equivalent graph:

Lemma 3 ($\mathcal{M}$ and $\mathcal{M}^{\mathfrak{G}_{\mathcal{M}}}$ are Semantically Equivalent).
Let $\mathcal{A} = (\mathcal{G}, \mathcal{C}, \mathcal{R})$ be an alphabet and let $\mathcal{M}$ be a contextual structure over $\mathcal{A}$.

1. *It holds $\mathcal{M} \models \mathcal{M}^{\mathfrak{G}_{\mathcal{M}}}$ and $\mathcal{M}^{\mathfrak{G}_{\mathcal{M}}} \models \mathcal{M}$.*
2. *If $\mathcal{M}$ satisfies furthermore $M_0 = \mathcal{C}$ and $M_i = \mathcal{R}_i$ for all $i \geq 1$, then $\mathcal{M}$ and $\mathcal{M}^{\mathfrak{G}_{\mathcal{M}}}$ are even isomorphic.*

From 2. of Lemma 3 we conclude that each model $\mathcal{M}$ with $M_0 = \mathcal{C}$ and $M_i = \mathcal{R}_i$ for all $i \geq 1$ is already isomorphic to a standard model of a graph. Together with 1., we see that each class in the quasiorder $(CS, \models)$ contains at least one standard model (but this is not uniquely determined: Each class contains infinitely many pairwise non-isomorphic standard models).

In the following, we will provide an corresponding result of the last lemma for graphs, that is for each graph $\mathfrak{G}$ we have that $\mathfrak{G}$ and $\mathfrak{G}_{\mathcal{M}^{\mathfrak{G}}}$ are equivalent. Note that in contrast to the last lemma, we state that $\mathfrak{G}$ and $\mathfrak{G}_{\mathcal{M}^{\mathfrak{G}}}$ are *syntactically* equivalent. As we know from [Da02] that the calculus for concept graphs with cuts is sound, we know that the restricted calculus $\vdash_{pos}$ we consider in this paper is sound, too. So from the next theorem we can immediately conclude that $\mathfrak{G}$ and $\mathfrak{G}_{\mathcal{M}^{\mathfrak{G}}}$ are semantically equivalent as well.

Theorem 2 ($\mathfrak{G}$ and $\mathfrak{G}_{\mathcal{M}^{\mathfrak{G}}}$ are Syntactically Equivalent).
Let $\mathfrak{G}$ be a concept graph. Then we have

$$\mathfrak{G} \vdash_{pos} \mathfrak{G}_{\mathcal{M}^{\mathfrak{G}}} \quad and \quad \mathfrak{G}_{\mathcal{M}^{\mathfrak{G}}} \vdash_{pos} \mathfrak{G} .$$

The class of all graphs over a given alphabet $\mathcal{A}$, together with the semantical entailment relation $\models$ is a quasiorder. The same holds for the class of all models. With the last theorem, we are now prepared to show that these quasiorders are isomorphic structures. More precisely, we have the following corollary:

Corollary 2 ($(CS, \models)$ and $(CG, \models)$ are isomorphic quasiorders).
The mappings $\mathcal{M} \mapsto \mathfrak{G}_{\mathcal{M}}$ and $\mathfrak{G} \mapsto \mathcal{M}^{\mathfrak{G}}$ are, up to equivalence, mutually inverse isomorphisms between the quasiordered sets $(CS, \models)$ and $(CG, \models)$.

Proof: As we know that $\vdash_{pos}$ is sound, the last theorem yields that $\mathfrak{G} \models \mathfrak{G}_{\mathcal{M}^{\mathfrak{G}}}$ and $\mathfrak{G}_{\mathcal{M}^{\mathfrak{G}}} \models \mathfrak{G}$ hold as well. We have furthermore

$$\mathcal{M}_1 \models \mathcal{M}_2 \overset{\text{L. 3}}{\Longleftrightarrow} \mathcal{M}^{\mathfrak{G}_{\mathcal{M}_1}} \models \mathcal{M}^{\mathfrak{G}_{\mathcal{M}_2}} \overset{\text{C. 1}}{\Longleftrightarrow} \mathfrak{G}_{\mathcal{M}_1} \models \mathfrak{G}_{\mathcal{M}_2} \ .$$

These results together with Lemma 3 and Corollary 1 yield this corollary. □

5   Transformation-Rules for Models

We still have to show that $\vdash_{pos}$ is complete. Although we have the last corollary, this cannot be derived from the results we have so far.

In order to prove the completeness of $\vdash_{pos}$, we will introduce four transformation rules for models:[9] *removing an element, doubling an element, exchanging attributes, and restricting the incidence relations.* This rules form a calculus for models, which will be denoted by $\vdash$. We will show that $\vdash_{pos}$ for graphs and $\vdash$ for models are complete. The main idea is the following: If we have two models $\mathcal{M}^a$ and $\mathcal{M}^b$ with $\mathcal{M}^a \models \mathcal{M}^b$, we will show that $\mathcal{M}^a$ can successively transformed to $\mathcal{M}^b$ with the rules for the models, and each transformation carries over to the standard graphs, i. e. we get simultaneous $\mathfrak{G}_{\mathcal{M}^a} \vdash_{pos} \mathfrak{G}_{\mathcal{M}^b}$. This is enough to prove the completeness of $\vdash_{pos}$ as well.

In the following, we will define the transformation rules for models and show that each rule is sound in the system of models, and that it carries over to the set of graphs, together with the calculus $\vdash_{pos}$.

Definition 10 (Removing an Element from a Contextual Structure).
Let $\mathcal{M} := (\vec{\mathbb{K}}, \lambda)$ be a contextual $\mathcal{A}$-structure and let $g \in G_0 \backslash \lambda_g[\mathcal{G}]$. We define a power context family $\vec{\mathbb{K}}'$ as follows:

1. *$G_0' = G_0 \backslash \{g\}$, $G_i' = G_i \cap (G_0')^i$ for all $i \geq 1$,*
2. *$M_i' = M_i$ for all i,*
3. *$I_i' = I_i \cap (G_i' \times M_i)$ for all i.*

For the contextual structure $\mathcal{M} := (\vec{\mathbb{K}}', \lambda)$, we say that $\mathcal{M}'$ is obtained from $\mathcal{M}$ by removing the element g.

Lemma 4 (Removing an Element).
Let $\mathcal{M} := (\vec{\mathbb{K}}, \lambda)$ be a model, let $\mathcal{M}' := (\vec{\mathbb{K}}', \lambda')$ be obtained from $\mathcal{M}$ by removing the element $g \in G_0$. Let $\mathcal{M}^a := (\vec{\mathbb{K}}^a, \lambda^a)$ be a model, let $f : G_0^a \rightarrow G_0$ with $g \notin f[G_0^a]$ and $\mathcal{M} \models_f \mathcal{M}_a$, and $id : G_0 \backslash \{g\} \rightarrow G_0$ the identity-mapping. Then we have $\mathcal{M}' \models_f \mathcal{M}^a$, $\mathcal{M} \models_{id} \mathcal{M}'$ and $\mathfrak{G}_{\mathcal{M}} \vdash_{pos} \mathfrak{G}_{\mathcal{M}'}$

---

[9] We have the implicit agreement that isomorphic models are identified. Isomorphism between power context families and between models is defined as usual.

Definition 11 (Doubling of an Element in a Contextual Structure).

Let $\mathcal{M} := (\vec{\mathbb{K}}, \lambda)$ be a contextual $\mathcal{A}$-structure and let $g, g' \in G_0$. For each tuple $\vec{h} = (g_1, \ldots, g_i) \in G_i$, we set $\vec{h}[g'/g] := (g_1', \ldots, g_i')$ with $g_j' := \begin{cases} g_j & g_j \neq g \\ g' & g_j = g \end{cases}$.

If $\mathcal{M}' := (\vec{\mathbb{K}}', \lambda')$ is a contextual structure over $\mathcal{A}$ such that there is a $g \in G_0$ with

1. *$G_0' = G_0 \dot{\cup} \{g'\}$ for a $g' \notin G_0$ and $G_i' = G_i \dot{\cup} \{\vec{h}[g'/g] \mid \vec{h} \in G_i\}$ for all $i \geq 1$,*
2. *$M_i' = M_i$ for all i,*
3. *$I_0' = I_0 \dot{\cup} \{(g', m) \mid g I_0 m\}$, and $I_i' = I_i \dot{\cup} \{(\vec{h}[g'/g], m) \mid \vec{h} I_i m\}$ for all $i \geq 1$,*
4. *$\lambda_{\mathcal{G}}'$ fulfills $\lambda_{\mathcal{G}}'(G) = \lambda_{\mathcal{G}}(G)$ for all $G \in \mathcal{G}$ with $\lambda_{\mathcal{G}}(G) \neq G$, and for all $G \in \mathcal{G}$ with $\lambda_{\mathcal{G}}(G) = G$ we have $\lambda_{\mathcal{G}}'(G) \in \{g, g'\}$,*
5. *$\lambda_{\mathcal{C}}'(C) = ((Int(\lambda_{\mathcal{C}}(C))^{I_0}, Int(\lambda_{\mathcal{C}}(C)))$ for all $C \in \mathcal{C}$, and*
6. *$\lambda_{\mathcal{R}}'(R) = ((Int(\lambda_{\mathcal{R}}(R))^{I_i'}, Int(\lambda_{\mathcal{R}}(R)))$ for all $R \in \mathcal{R}_i$,*

then we say that $\mathcal{M}'$ is obtained from $\mathcal{M}$ by doubling the element g.

Please note the following: If we double an element g in a model $\mathcal{M}$ such that different object names a, b refer to g (i.e. $\lambda_{\mathcal{G}}(a) = \lambda_{\mathcal{G}}(b) = g$), they information $\lambda_{\mathcal{G}}(a) = \lambda_{\mathcal{G}}(b)$ may get lost when g is doubled. So, if $\mathcal{M}'$ is obtained from $\mathcal{M}$ by doubling the element g, $\mathcal{M}'$ may contain less information than $\mathcal{M}$.

Furthermore we note that if $(\vec{\mathbb{K}}', \lambda'), (\vec{\mathbb{K}}'', \lambda'')$ are two contextual structures which are obtained from a contextual structure $(\vec{\mathbb{K}}, \lambda)$ by doubling the element $g \in G_0$, then they can (up to isomorphism) only differ between $\lambda_{\mathcal{G}}'$ and $\lambda_{\mathcal{G}}''$.

Lemma 5 (Doubling an Element).
Let $\mathcal{M} := (\vec{\mathbb{K}}, \lambda)$ be a model and let $\mathcal{M}' := (\vec{\mathbb{K}}', \lambda')$ be obtained from $\mathcal{M}$ by doubling the element $g \in G_0$. Then we have $\mathcal{M} \models \mathcal{M}'$ and $\mathfrak{G}_{\mathcal{M}} \vdash_{pos} \mathfrak{G}_{\mathcal{M}'}$. If we have a model $\mathcal{M}^a := (\vec{\mathbb{K}}^a, \lambda^a)$ with $\mathcal{M} \models \mathcal{M}_a$, then we can chose $\mathcal{M}'$ such that we have $\mathcal{M}' \models \mathcal{M}^a$.

Definition 12 (Restricting the Incidence Relations).
Let $\mathcal{M} := (\vec{\mathbb{K}}, \lambda)$ be a contextual $\mathcal{A}$-structure. If $\mathcal{M}' := (\vec{\mathbb{K}}', \lambda)$ is a contextual structure over $\mathcal{A}$ with $G_i' = G_i$ for all i, $M_i' = M_i$ for all i, and $I_i' \subseteq I_i$ for all i, then we say that $\mathcal{M}'$ is obtained from $\mathcal{M}$ by restricting the incidence relations.

Lemma 6 (Restricting the Incidence Relations).
If $\mathcal{M}' := (\vec{\mathbb{K}}', \lambda)$ is obtained from $\mathcal{M} := (\vec{\mathbb{K}}, \lambda)$ by restriction the incidence relations, we have $\mathcal{M} \models_{id} \mathcal{M}'$ and $\mathfrak{G}_{\mathcal{M}} \vdash_{pos} \mathfrak{G}_{\mathcal{M}'}$.

Definition 13 (Exchanging Attributes and Standardization).
Let $\mathcal{M} := (\vec{\mathbb{K}}, \lambda)$ be a contextual $\mathcal{A}$-structure. If $\mathcal{M}' := (\vec{\mathbb{K}}', \lambda')$ is a contextual $\mathcal{A}$-structure which satisfies

1. $G_i' := G_i$ for all i,
2. $gI_0\lambda_C(C) \Longleftrightarrow gI_0'\lambda_C'(C)$ for all $g \in G_0$ and $C \in \mathcal{C}$,
3. $\vec{g}I_i\lambda_R(R) \Longleftrightarrow \vec{g}I_i'\lambda_C'(R)$, for all $i \geq 1$, $\vec{g} \in G_i$ and $R \in \mathcal{R}_i$, and
4. $\lambda_G' := \lambda_G$,

then we say that $\mathcal{M}'$ is obtained from $\mathcal{M}$ by exchanging attributes of $\mathcal{M}$. If $\mathcal{M}'$ additionally satisfies $M_0' := \mathcal{C}$ and $M_i' := \mathcal{R}_i$ for all $i \geq 1$, then we say that $\mathcal{M}'$ is obtained from $\mathcal{M}$ by standardization of $\mathcal{M}$.

This rule is the only rule which does not weaken the informational content of a model. In particular, it can be carried out in both directions.

Lemma 7 (Exchanging Attributes and Standardization).
If $\mathcal{M}' := (\vec{\mathbb{K}}', \lambda')$ is obtained from $\mathcal{M} := (\vec{\mathbb{K}}, \lambda)$ by exchanging attributes, we have $\mathcal{M} \models_{id} \mathcal{M}'$, $\mathcal{M}' \models_{id} \mathcal{M}$, and $\mathfrak{G}_{\mathcal{M}} = \mathfrak{G}_{\mathcal{M}'}$ .Furthermore exists a standardization of $\mathcal{M}$ for each contextual structure $\mathcal{M}$.

The four rules form a calculus for models, i. e. we have the following definition:

Definition 14 (Calculus for Contextual Structures).
The calculus for contextual structures over the alphabet $\mathcal{A} := (\mathcal{G}, \mathcal{C}, \mathcal{R})$ consists of the following rules:
Removing an element, doubling an element, exchanging attributes, and restricting the incidence relations.
If $\mathcal{M}^a$, $\mathcal{M}^b$ are two models, and if there is a sequence $(\mathcal{M}^1, \mathcal{M}^2, \ldots, \mathcal{M}^n)$ with $\mathcal{M}^1 = \mathcal{M}^a$ and $\mathcal{M}^b = \mathcal{M}^n$ such that each $\mathcal{M}^{i+1}$ is derived from $\mathcal{M}^i$ by applying one of the rules of the calculus, we say that $\mathcal{M}^b$ can be derived from $\mathcal{M}^a$, which is denoted by $\mathcal{M}^a \vdash \mathcal{M}^b$.

Now we are prepared to show that the transformation rules for models are complete and respected by the construction of standard graphs.

Theorem 3 ($\vdash$ is Complete and Respected by Standard Graphs).
Let $\mathcal{M}^a := (\vec{\mathbb{K}}^a, \lambda^a)$, $\mathcal{M}^b := (\vec{\mathbb{K}}^b, \lambda^b)$ be two contextual structures such that $\mathcal{M}^a \models \mathcal{M}^b$. Then we have $\mathcal{M}^a \vdash \mathcal{M}^b$ and $\mathfrak{G}_{\mathcal{M}^a} \vdash_{pos} \mathfrak{G}_{\mathcal{M}^b}$.

This theorem yields that the calculus $\vdash_{pos}$ on the graphs is complete as well:

Corollary 3 (Both Calculi are Complete).
Let $\mathcal{M}_1, \mathcal{M}_2$ be models and $\mathfrak{G}_1, \mathfrak{G}_2$ be graphs. We have:
$$\mathcal{M}_1 \models \mathcal{M}_2 \quad \Longleftrightarrow \quad \mathcal{M}_1 \vdash \mathcal{M}_2 \quad and \quad \mathfrak{G}_1 \models \mathfrak{G}_2 \quad \Longleftrightarrow \quad \mathfrak{G}_1 \vdash_{pos} \mathfrak{G}_2 .$$

6 Conclusion

In Prediger ([Pr98a]), the notion of standard models is adaquate only for concept graphs without generic markers. For concept graphs with generic markers, a standard model of a graph may encode less information than the graph. Thus, in this case, the reasoning on concept graphs cannot be carried over to the models completely. This is a gap we have bridged with our notion of standard models, which extends Predigers approach. Moreover, reasoning on models can be carried out in two different ways: By the semantical entailment relation $\models$ on models (see Definition 8), and by transformation rules between models (see Section 5).

In [Da02], an adaquate calculus for concept graphs with cuts is provided. Thus, if we have two concept graphs without cuts $\mathfrak{G}_a$ and $\mathfrak{G}_b$ with $\mathfrak{G}_a \models \mathfrak{G}_b$, we have a proof for $\mathfrak{G}_a \vdash \mathfrak{G}_b$, that is: We have a sequence $(\mathfrak{G}_1, \mathfrak{G}_2, \ldots, \mathfrak{G}_n)$ with $\mathfrak{G}_1 = \mathfrak{G}_a$, $\mathfrak{G}_b = \mathfrak{G}_n$ such that each $\mathfrak{G}_{i+1}$ is derived from $\mathfrak{G}_i$ by applying one of the rules of the calculus. But now we have even more: From Corollary 3, we conclude that we can find a proof $(\mathfrak{G}_1, \mathfrak{G}_2, \ldots, \mathfrak{G}_n)$ such that *all* graphs $\mathfrak{G}_1, \ldots, \mathfrak{G}_n$ are concept graphs without cuts. This result cannot be directly derived from [Da02].

In this sense, this work is a completion of both [Da02]) and [Pr98a].

References

Da02.   F. Dau: *The Logic System of Concept Graphs with Negations and its Relationship to Predicate Logic*. PhD-Thesis, Darmstadt University of Technology, 2002. To appear in Springer Lecture Notes on Computer Science.

CM92.   M. Chein, M.-L. Mugnier: *Conceptual Graphs: Fundamental Notions*. Revue d'Intelligence Artificielle 6, 1992, 365–406.

Mu00.   M.-L. Mugnier: *Knowledge Representation and Reasonings Based on Graph Homomophism*. In: B. Ganter, G. W. Mineau (Eds.): Conceptual Structures: Logical, Linguistic and Computational Issues. LNAI 1867, Springer Verlag, Berlin–New York 2000, 172–192.

GW99.   B. Ganter, R. Wille: *Formal Concept Analysis: Mathematical Foundations*. Springer, Berlin–Heidelberg–New York 1999.

Pr98a.   S. Prediger: *Kontextuelle Urteilslogik mit Begriffsgraphen. Ein Beitrag zur Restrukturierung der mathematischen Logik*. Aachen, Shaker Verlag 1998.

Pr98b.   S. Prediger: *Simple Concept Graphs: A Logic Approach*. In: M. -L. Mugnier, M. Chein (Eds.): Conceptual Structures: Theory, Tools and Applications. LNAI 1453, Springer Verlag, Berlin–New York 1998, 225–239.

Sh99.   S. J. Shin: *The Iconic Logic of Peirce's Graphs*. Bradford Book, Massachusetts, 2002.

So84.   J. F. Sowa: *Conceptual Structures: Information Processing in Mind and Machine*. Addison Wesley Publishing Company Reading, 1984.

So00.   J. F. Sowa: *Knowledge Representation: Logical, Philosophical, and Computational Foundations*. Brooks Cole Publishing Co., Pacific Grove, CA, 2000.

Wi97.   R. Wille: *Conceptual Graphs and Formal Concept Analysis*. In: D. Lukose et al. (Eds.): Conceptual Structures: Fulfilling Peirce's Dream. LNAI 1257, Springer Verlag, Berlin–New York 1997, 290–303.

Wi02.   R. Wille: *Existential Concept Graphs of Power Context Families*. In: U. Priss, D. Corbett, G. Angelova (Eds.): Conceptual Structures: Integration and Interfaces. LNAI 2393, Springer Verlag, Berlin–New York 2002, 382–396.

A complete version of this paper, including all proofs, can be found at
`www.mathematik.tu-darmstadt.de/~dau/DauICCS03Complete.pdf`

Distributed Systems
with Simple Concept Graphs

Grit Malik*

Institut für Algebra
Fakultät Mathematik und Naturwissenschaften
Technische Universität Dresden
Germany
malik@math.tu-dresden.de

Abstract. In this article a theory of distributed systems consisting of formal contexts which was introduced by Barwise and Seligman is translated to distributed systems consisting of concept graphs. The aim is to construct for every distributed system a channel consisting of a concept graph representing the distributed system and a set concept graph morphisms representing the relation between the parts of the system and the whole system. To reach this aim two operations for constructing new concept graphs, the semiproduct and the quotient by an invariant, are introduced. These operations are the basis for constructing the concept graph of the channel which represents a distributed system as a whole.

1 Introduction

J. Barwise and J. Seligman introduced in [1] a theory of distributed systems. In their theory a distributed system is a very general concept. It consists of a family of formal contexts[1] representing the components of the distributed system and a set of infomorphisms. Infomorphisms are contravariant pairs of mappings between formal contexts which are supposed to describe how information flows between the formal contexts and thus inside the distributed system and they are supposed to describe how a formal context can be interpreted in another. In order to describe a distributed system as a whole object they introduced information channels covering the distributed system. Such a channel consists of a set of infomorphisms from the formal contexts of the system into a common formal context, the core of the channel. The core represents the distributed system as a whole and the channel infomorphisms represent the connection of the parts of the distributed system to the whole system. Further in the core information common to the parts of the distributed system is represented.

Here this theory is translated to distributed systems consisting of concept graphs instead of formal concepts and concept graph morphisms instead of infomorphisms. This translation is interesting because concept graphs which were

---

* supported by the DFG/Graduiertenkolleg 334
[1] Barwise and Seligman do not use the word formal context

A. de Moor, W. Lex, and B. Ganter (Eds.): ICCS 2003, LNAI 2746, pp. 257–270, 2003.

introduced by R. Wille (*cf.* [10]) are a mathematization of conceptual graphs. Conceptual Graphs introduced by J. Sowa (*cf.* [9]) are a tool for knowledge representation and graph representation of natural language. Thus naturally distributed systems can be described by conceptual graphs or concept graphs, as considered here.

Concept graph morphisms which correspond to infomorphisms were introduced in [4]. They consist of two contravariant functions between the edges and the vertices which are supposed to model the information flow as in the case of infomorphisms between formal contexts. Further they consist of three functions between the alphabets of the concept graphs. This allows to map completely different alphabets which is useful for instance if the distributed systems consists of descriptions of the same object by different people. Naturally they do not use a standardized language and thus have different names for the labels and references. The aim is to construct a concept graph which shows the common information of such descriptions and the relation of the descriptions to this common information.

In the literature other mappings between conceptual graphs are considered (*cf.* [2] and [5]). M. Chein and M.-L. Mugnier introduced projections in their formalization of conceptual graphs as a characterization of the inference of conceptual graphs. In particular inference of conceptual graphs requires that the conceptual graphs have the same alphabet. Thus projections or graph homomorphisms do not allow to map different alphabets which would be desirable in the case described above. Further they do not provide the different directions of the mappings between the edges and the mapping between the vertices, *i.e.*, they can not express the two directions of the information flow.

In order to follow the construction of a channel by Barwise and Seligman the article starts with some technical definitions. First the semiproduct of concept graphs and the quotient by an invariant are defined. The main result is the construction of a minimal cover of a distributed system by a channel, *i.e.*, of the limit of the distributed system. This result is stated in Theorem 1 in Section 4.

2   Concept Graphs and Concept Graph Morphisms

In this section the definitions of concept graphs by S. Prediger (*cf.* [6]) and of concept graph morphisms (*cf.* [4]) are recalled. At first, an alphabet of concept graphs is introduced.

Definition 1. *An alphabet*(C, G, R) *of concept graphs consists of finite ordered set* $(C, \leq_C)$ *of concept names, a finite set* G *of object names and a finite ordered set* $(R, \leq_R)$ *of relation names.* $(R, \leq_R)$ *is partitioned into the ordered sets* $(R_k, \leq_{R_k})$, $k = 1, \ldots, n$.

Now concept graphs are defined over such an alphabet.

Definition 2. *A (simple) concept graph over the alphabet* (C, G, R) *is a structure* $\mathfrak{G} := (V, E, \nu, \kappa, \rho)$ *such that*

- (V, E, ν) *is a finite directed multi-hypergraph, i.e., V and E are finite sets. The elements of V are called vertices, the elements of E are called edges and $\nu : E \to \bigcup_{k=1}^{n} V^k$, $n \geq 1$, is a function.*
- $\kappa : V \cup E \to \mathcal{C} \cup \mathcal{R}$ *is a function such that $\kappa(V) \subseteq \mathcal{C}$ and $\kappa(E) \subseteq \mathcal{R}$ and for all $e \in E$ with $\nu(e) = (v_1, \ldots, v_k)$ is $\kappa(e) \in \mathcal{R}_k$.*
- $\rho : V \to \mathfrak{P}(\mathcal{G}) \setminus \{\emptyset\}$ *is a function.*

For an edge $e \in E$ with $\nu(e) = (v_1, \ldots, v_k)$ let $|e| := k$ and $\nu(e)|_i := v_i$.

Figure 1 shows some examples of concept graphs. They show some properties of three chemical elements (*cf.* [8]), potassium in the first concept graph $\mathfrak{G}_1$, sodium in $\mathfrak{G}_2$, and lithium in $\mathfrak{G}_3$.

Instead of the definition of concept graph morphisms in [4] a slightly changed definition is given here. The different directions of the functions of a concept graph morphism, especially f_V and f_E, together with the conditions, especially the first condition, were introduced to formalize that information always flows in both directions. This condition says that a vertex v of the second concept graph is incident to the image of an edge $f_E(e)$ iff the image of the vertex $f_V(v)$ is incident to the edge e in the first concept graph. Conditions 2 and 3 say that images of labels of v and e are labels of the images of $f_V(v)$ and $f_E(e)$, respectively. In condition 4 it is only required that the references of the image of a vertex contains the image of its set of references which is denoted by $f_{\mathcal{G}}[\rho_2(v)]$. The direction determining function is the function between the edges.

Definition 3. *Let $\mathfrak{G}_1 := (V_1, E_1, \nu_1, \kappa_1, \rho_1)$ be a concept graph over $(\mathcal{C}_1, \mathcal{G}_1, \mathcal{R}_1)$ and $\mathfrak{G}_2 := (V_2, E_2, \nu_2, \kappa_2, \rho_2)$ a concept graph over $(\mathcal{C}_2, \mathcal{G}_2, \mathcal{R}_2)$. A concept graph morphism $f : \mathfrak{G}_1 \rightleftarrows \mathfrak{G}_2$ is a tuple $(f_E, f_V, f_{\mathcal{C}}, f_{\mathcal{G}}, f_{\mathcal{R}})$ of functions*

$$f_E : E_1 \to E_2$$
$$f_V : V_2 \to V_1$$
$$f_{\mathcal{C}} : \mathcal{C}_2 \to \mathcal{C}_1$$
$$f_{\mathcal{G}} : \mathcal{G}_2 \to \mathcal{G}_1$$
$$f_{\mathcal{R}} : \mathcal{R}_1 \to \mathcal{R}_2$$

such that the following conditions are satisfied:

1. $\exists_{i \leq |f_E(e)|} \, v = \nu_2(f_E(e))|_i \iff \exists_{j \leq |e|} \, f_V(v) = \nu_1(e)|_j$ *for all $e \in E_1, v \in V_2$,*
2. $\kappa_1(f_V(v)) = f_{\mathcal{C}}(\kappa_2(v))$ *for all $v \in V_2$,*
3. $\kappa_2(f_E(e)) = f_{\mathcal{R}}(\kappa_1(e))$ *for all $e \in E_1$, and*
4. $\rho_1(f_V(v)) \supseteq f_{\mathcal{G}}[\rho_2(v)]$ *for all $v \in V_2$.*

The composition of concept graph morphisms $f : \mathfrak{G}_1 \rightleftarrows \mathfrak{G}_2$ and $g : \mathfrak{G}_2 \rightleftarrows \mathfrak{G}_3$ is the concept graph morphism $g \circ f = (g_E \circ f_E, f_V \circ g_V, f_{\mathcal{C}} \circ g_{\mathcal{C}}, f_{\mathcal{G}} \circ g_{\mathcal{G}}, g_{\mathcal{R}} \circ f_{\mathcal{R}})$ from $\mathfrak{G}_1$ to $\mathfrak{G}_3$. Further $id_{\mathfrak{G}} = (id_E, id_V, id_{\mathcal{C}}, id_{\mathcal{G}}, id_{\mathcal{R}})$ is a concept graph morphism from $\mathfrak{G}$ to itself. Thus a preorder on concept graphs can be defined by $\mathfrak{G}_1 \preceq \mathfrak{G}_2$ iff there is a concept graph morphism from $\mathfrak{G}_1$ to $\mathfrak{G}_2$. This relation is indeed a preorder, *i.e.*, not antisymmetric because it is possible that there are

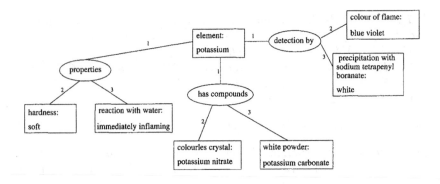

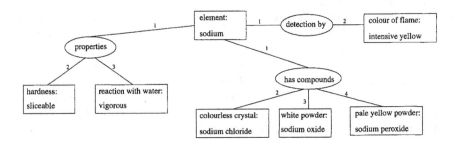

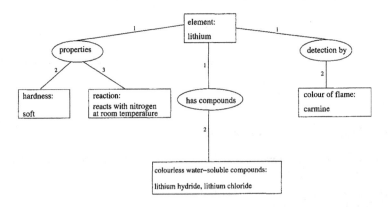

Fig. 1. Concept graphs representing properties of the chemical elements potassium (denoted by $\mathfrak{G}_1$), sodium ($\mathfrak{G}_2$), lithium ($\mathfrak{G}_3$)

concept graph morphisms in both direction although the concept graphs are not equivalent. This is possible even if the concept graph have the same alphabet because incomparable concept or relation names can be mapped. This relation is also not symmetric because there are concept graphs for which only concept graph morphisms in just one direction exist.

Possible concept graph morphisms between the concept graphs in Figure 1 are the following two in Figure 2 and Figure 3. These concept graph morphisms map properties of an element to similar properties of the other element. The vertices of the concept graphs are given in these figures by their concept names and their object names and the edges are given by the object names of their incident vertices.

$f_E^1 : E_1 \to E_2$

potassium - blue violet - white	$\mapsto$	sodium - intensiv yellow
potassium - p. nitrate - p. carbonate	$\mapsto$	sodium - s. chloride - s. oxide - s. peroxide
potassium - soft - imm. inflaming	$\mapsto$	sodium - sliceable - vigorous

$f_V^1 : V_2 \to V_1$

colour: intensiv yellow	$\mapsto$	colour: blue violet
hardness: sliceable	$\mapsto$	hardness: soft
reaction with H_2O: vigorous	$\mapsto$	reaction with H_2O: imm. inflaming
colourless crystal: s. chloride	$\mapsto$	colourless crystal: p. nitrate
white powder: s. oxide	$\mapsto$	white powder: p. carbonate
p.yellow powder: s. peroxide	$\mapsto$	white powder: p. carbonate
element: sodium	$\mapsto$	element: potassium

$f_C^1 : C_2 \to C_1$

colour	$\mapsto$	colour
hardness	$\mapsto$	hardness
reaction with H_2O	$\mapsto$	reaction with H_2O
colourless crystal	$\mapsto$	p. nitrate
white powder	$\mapsto$	white powder
p. yellow powder	$\mapsto$	white powder
element	$\mapsto$	element

$f_G^1 : G_2 \to G_1$

intensiv yellow	$\mapsto$	blue violet
sliceable	$\mapsto$	soft
vigorous	$\mapsto$	immediately inflaming
s. chloride	$\mapsto$	p. nitrate
s. oxide	$\mapsto$	p. carbonate
s. peroxide	$\mapsto$	p. carbonate
sodium	$\mapsto$	potassium

$f_R^1 : R_1 \to R_2$

detection by	$\mapsto$	detection by
properties	$\mapsto$	properties
compounds	$\mapsto$	compounds

Fig. 2. A concept graph morphism $f^1 : \mathfrak{G}_1 \rightleftarrows \mathfrak{G}_2$

3 Operations on Concept Graphs

In this section two operations producing new concept graphs out of given concept graphs are introduced. These operations are necessary to construct the concept graph which represents the whole distributed system, *i.e.*, the limit of the distributed system. Here the same operations which are used by Barwise and Seligman are introduced.

$f_E^2 : E_1 \to E_3$

potassium - blue violet - white	$\mapsto$ lithium - carmine
potassium - p. nitrate - p. carbonate	$\mapsto$ lithium - l. hydride, l. chloride
potassium - soft - imm. inflaming	$\mapsto$ lithium - soft - reacts w. nitrogen

$f_V^2 : V_3 \to V_1$

colour: carmine	$\mapsto$ colour: blue violet
hardness: soft	$\mapsto$ hardness: soft
reaction : reacts with nitrogen	$\mapsto$ reaction with H_2O: imm. infl.
colourl. water-sol. comp.: l. nitrate, l. chloride	$\mapsto$ colourless crystal: p. carbonate
element: lithium	$\mapsto$ element: potassium

$f_C^2 : C_3 \to C_1$

colour	$\mapsto$ colour
hardness	$\mapsto$ hardness
reaction	$\mapsto$ reaction with H_2O
colourless water-soluble comp.	$\mapsto$ colourless crystal
element	$\mapsto$ element

$f_\mathcal{G}^2 : \mathcal{G}_3 \to \mathcal{G}_1$

carmine	$\mapsto$ blue violet
soft	$\mapsto$ soft
reacts with nitrogen	$\mapsto$ immediately inflaming
l. hydride	$\mapsto$ p. nitrate
l. chloride	$\mapsto$ p. nitrate
lithium	$\mapsto$ potassium

$f_\mathcal{R}^2 : \mathcal{R}_1 \to \mathcal{R}_3$

detection by	$\mapsto$ detection by
properties	$\mapsto$ properties
compounds	$\mapsto$ compounds

Fig. 3. A concept graph morphism $f^2 : \mathfrak{G}_1 \rightleftarrows \mathfrak{G}_3$

3.1   The Semiproduct of Concept Graphs

The semiproduct of concept graphs corresponds to the semiproduct[2] of formal contexts. It combines the concept graphs of the distributed system into one. The semiproduct of formal contexts as well as an introduction into Formal Concept Analysis can be found in [3]. The vertices of the semiproduct of the concept graphs are pairs of vertices of the starting concept graphs and the set of the edges is the disjoint union of the edges of the starting concept graphs but the edges have a multiple of incident vertices. The semiproduct of the concept graphs of the distributed system will be the basis for the construction of the concept graph representing the system.

Definition 4. *Let* $\mathfrak{G}_i := (V_i, E_i, \nu_i, \kappa_i, \rho_i)$ *be concept graphs over* $(\mathcal{C}_i, \mathcal{G}_i, \mathcal{R}_i)$, $i = 1, 2$. *The semiproduct* $\mathfrak{G}_1 \mathbb{X} \mathfrak{G}_2$ *is the concept graph* $(V_1 \times V_2, E_1^\mathbb{X} \cup E_2^\mathbb{X}, \nu, \kappa, \rho)$ *over the alphabet* $(\mathcal{C}_1 \times \mathcal{C}_2, \mathcal{G}_1 \times \mathcal{G}_2, \mathcal{R}_1^\mathbb{X} \cup \mathcal{R}_2^\mathbb{X})$. $E_i^\mathbb{X}$ *is obtained from* E_i *by changing edges* e *with* k *vertices into edges* (e, i) *with* $k \cdot |V_j|$ *vertices* , $i \neq j$, *and* $\mathcal{R}_i^\mathbb{X}$, $i = 1, 2$, *is obtained from* $\mathcal{R}_i$ *by changing* k*-ary relation names* $r \in \mathcal{R}_i$ *into* $k \cdot |V_j|$*-ary relation names* $(r, i) \in \mathcal{R}_i^\mathbb{X}$. *The order on* $\mathcal{R}_i^\mathbb{X}$, $i = 1, 2$, *is obtained from the order on* $\mathcal{R}$ *by setting* $(r, i) \leq_{\mathcal{R}^\mathbb{X}} (s, i) \iff r \leq_\mathcal{R} s$.

---

[2] In [1] it is called sum.

The function $\nu : E_1^{\boxtimes} \cup E_2^{\boxtimes} \to \bigcup_{k=0}^{|V_1||V_2|} (V_1 \times V_2)^k$ *is defined by*

$$\nu(e, 1) = \begin{pmatrix} (v_1, w_1) & \cdots & (v_1, w_{|V_2|}) \\ \vdots & \ddots & \vdots \\ (v_k, w_1) & \cdots & (v_k, w_{|V_2|}) \end{pmatrix}$$

iff $\nu_1(e) = (v_1, \ldots, v_k)$. $\nu(e, 2)$ *is defined analogously.*

The function $\kappa : (V_1 \times V_2) \cup (E_1^{\boxtimes} \cup E_2^{\boxtimes}) \to (\mathcal{C}_1 \times \mathcal{C}_2) \cup (\mathcal{R}_1^{\boxtimes} \cup \mathcal{R}_2^{\boxtimes})$ *is defined by*

$$\kappa(v, w) = (\kappa_1(v), \kappa_2(w))$$

and

$$\kappa(e, i) = (\kappa_i(e), i).$$

$\rho : V_1 \times V_2 \to \mathcal{G}_1 \times \mathcal{G}_2$ *is defined by*

$$\rho(v, w) = \rho_1(v) \times \rho_2(w).$$

Figure 4 shows a small example of a semiproduct. Because the vertices are pairs of vertices of the factors in this figure the vertices are labeled by ordered pairs of concept names and ordered pairs of object names. Thats why in this picture the object names (*Li, white*) and (*carmine, K*) appear. *Li* and *carmine* are object names of the first concept graph and thus appear as the first component of the the pair. The object names *white* and *K* appear in the alphabet of the second concept graph and thus as the second component of the pair.

As seen in this figure the semiproduct can become very large. It has $|V_1| \cdot |V_2|$ vertices and $|E_1| + |E_2|$ edges. Mainly it will be of technical use.

The second aim was the construction of concept graph morphisms from the concept graphs of the system into the concept graph representing the system. Thus concept graph morphisms from the factors into the semiproduct are needed. These are the natural concept graph morphisms given in the following definition.

Definition 5. *For* $\mathfrak{G}_i$, $i = 1, 2$ *is* $\sigma^1 : \mathfrak{G}_1 \rightleftarrows \mathfrak{G}_1 \boxtimes \mathfrak{G}_2$ *with:*

$$\begin{array}{llll} \sigma_E^i : E_i \to E_1^{\boxtimes} \cup E_2^{\boxtimes} & \text{defined by} & e \mapsto (e, i), \\ \sigma_V^i : V_1 \times V_2 \to V_i & \text{defined by} & (v_1, v_2) \mapsto v_i, \\ \sigma_C^i : \mathcal{C}_1 \times \mathcal{C}_2 \to \mathcal{C}_i & \text{defined by} & (c_1, c_2) \mapsto c_i, \\ \sigma_{\mathcal{G}}^i : \mathcal{G}_1 \times \mathcal{G}_2 \to \mathcal{G}_i & \text{defined by} & (g_1, g_2) \mapsto g_i, \text{ and} \\ \sigma_{\mathcal{R}}^i : \mathcal{R}_i \to \mathcal{R}_1^{\boxtimes} \cup \mathcal{R}_2^{\boxtimes} & \text{defined by} & r \mapsto (r, i). \end{array}$$

the natural concept graph morphism from $\mathfrak{G}_1 \to \mathfrak{G}_1 \boxtimes \mathfrak{G}_2$.

3.2   Invariants and Quotients

A second method of constructing new concept graphs is the factorization by an invariant. In the case of formal contexts an invariant consists of a set of objects and a relation on the set of attributes such that the objects of the invariant

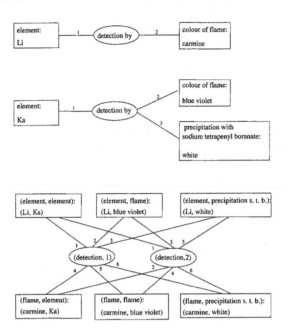

Fig. 4. Two concept graphs and their semiproduct

respect the relation on the attributes (*cf.* [1]). Analogously an invariant on a concept graph consists of a set of vertices, a relation on the edges, and a relation on the set of relation names such that the vertices of the invariant respect the relation on the edges. Further for all related edges the relation names which are assigned to the edges are related, too.

Definition 6. *Let* $\mathfrak{G} = (V, E, \nu, \kappa, \rho)$ *be a concept graph over* $(\mathcal{C}, \mathcal{G}, \mathcal{R})$. *An invariant[3]* $I := (W, \eta_E, \eta_R)$ *on* $\mathfrak{G}$ *consists of a set* $W \subseteq V$, *a relation* $\eta_E \subseteq E \times E$, *and a relation* $\eta_R \subseteq \mathcal{R} \times \mathcal{R}$ *such that holds*

1. $\exists_{i \leq |e_1|} \ v = \nu(e_1)|_i \iff \exists_{j \leq |e_2|} \ v = \nu(e_2)|_j$ *for every* $v \in W$, $(e_1, e_2) \in \eta_E$ *and*
2. $(e_1, e_2) \in \eta_E \implies (\kappa(e_1,), \kappa(e_2)) \in \eta_R$.

An example for an invariant is shown in Figure 5. This is an invariant on the semiproduct of the concept graphs in Figure 1. For readability reasons the relation names in the semiproduct are denoted by, e.g., (*detection by, K*) instead of (*detection by, 1*). For the edges this is omitted because the denotation is clear in this case. In the construction of the limit of a distributed system in the next section an invariant on the semiproduct of its concept graphs will be constructed out of the concept graph morphisms of the distributed system.

---

[3] The invariants correspond to the dual invariants in [1]. Barwise and Seligman introduce both, invariants and dual invariants.

W := { (element, element, element):(potassium, sodium, lithium), (hardness, hardness, hardness):(soft, sliceable, soft), (reaction with H_2O, reaction with H_2O, reaction):(immediately inflaming, vigorous, reacts with nitrogen at room temperature), (colourless crystal, colourless crystal, colourless water-soluble compound):(potassium nitrate, sodium chloride, lithium hydride), (colour of flame, precipitation with s. tetraphenyl borate, colour of flame):(blue violet, white, carmine), (colour of flame, colour of flame, colour of flame):(blue violet, intensive yellow, carmine) }

η_E := { (potassium - blue violet - white, lithium - carmine), (potassium - p. nitrate - p. carbonate, lithium - {l. hydride, l. chloride}), (potassium - soft - imm. inflaming, lithium - soft - reacts w. nitrogen), (potassium - blue violet - white, sodium - intensive yellow), (potassium - p. nitrate - p. carbonate, sodium - s. chloride - s. oxide - s. peroxide), (potassium - soft - imm. inflaming, sodium - sliceable - vigorous)}

η_R := { ((detection by, K), (detection by, Na)), ((detection by, K), (detection by, Li)), ((properties, K), (properties, Na)), ((properties, K),(properties, Li)), ((compounds, K),(compounds, Na)), ((compounds, K),(compounds, Li))}

Fig. 5. An invariant on the semiproduct of the concept graphs $\mathfrak{G}_1, \mathfrak{G}_2, \mathfrak{G}_3$

The quotient concept graph is built from the starting one by identifying edges which are related by the relation η_E. The vertices of the quotient are the vertices contained in W. In the alphabet only the set of relation names is replaced. In this set all relation names related by η_R are identified.

For the definition of the quotient concept graph the relations η_E and η_R of an invariant $I = (W, \eta_E, \eta_R)$ are assumed to be equivalence relations. If η_E and η_R are no equivalence relations at first E is restricted to the edges $e \in E$ such that there is a $d \in E$ with $(d, e) \in \eta_E$ or $(e, d) \in \eta_E$ and it is factorized by the smallest equivalence relation on the restricted set of edges containing η_E. The set of relation names is not restricted and again it is factorized by the smallest equivalence relation containing η_R. The quotient of the semiproduct of the concept graphs in Figure 1 by the invariant in Figure 5 is shown in Figure 6.

Definition 7. *Let* $I := (W, \eta_E, \eta_R)$ *be an invariant on a concept graph* $\mathfrak{G} = (V, E, \nu, \kappa, \rho)$ *over the alphabet* $(\mathcal{C}, \mathcal{G}, \mathcal{R})$. *The quotient*[4] $\mathfrak{G}/I$ *is the concept graph* $(W, E/\eta_E, \nu_I, \kappa_I, \rho_I)$ *over the alphabet* $(\mathcal{C}, \mathcal{G}, \mathcal{R}/\eta_R)$ *with*

1. $\exists_i \ v = \nu_I([e]_{\eta_E})_{|_i} \iff \exists_j \ v = \nu(e)_{|_j}$ *for every* $v \in W$, $[e]_{\eta_E} \in E/\eta_E$,
2. $\kappa_I([e]_{\eta_E}) = [\kappa(e)]_{\eta_R}$ *for every* $[e]_{\eta_E} \in E/\eta_E$,
3. $\kappa_I(v) = \kappa(v)$ *for every* $v \in W$, *and*
4. $\rho_I(v) = \rho(v)$ *for every* $v \in W$.

The order on $\mathcal{R}/\eta_R$ *is defined by* $[r]_{\eta_R} \leq_{\eta_R} [s]_{\eta_R} \iff \forall_{x \in [r]_{\eta_R}} \exists_{y \in [s]_{\eta_R}} x \leq_{\mathcal{R}} y$. *The indices of the vertices incident to an equivalence class* $[e]$ *are ordered at first by the minimality of the indices which the vertices have for the edges in* $[e]$, *at second by the frequency of occurrence of the indices, and at last arbitrarily.*

---

[4] As for invariants the quotients here correspond to the dual quotients in [1].

The size of the quotient depends on the size of the invariant. Always the resulting concept graph has less or equal vertices and edges then the starting concept graph. As in the case of the semiproduct a special concept graph morphism from the starting concept graph into the quotient concept graph is needed.

Definition 8. *Let I be an invariant on $\mathfrak{G}$. The canonical quotient concept graph morphism $\tau^I : \mathfrak{G} \rightleftarrows \mathfrak{G}/I$ consists of*

1. $\tau_E^I : E \to E/\eta_E$ *defined by* $e \mapsto [e]_{\eta_E}$,
2. *the inclusion function* $\tau_V^I : W \to V$,
3. *the identity function* $\tau_C^I : C \to C$,
4. *the identity function* $\tau_G^I : \mathcal{G} \to \mathcal{G}$, *and*
5. $\tau_R^I : \mathcal{R} \to \mathcal{R}/\eta_{\mathcal{R}}$ *defined by* $r \mapsto [r]_{\eta_{\mathcal{R}}}$.

4   Distributed Systems and Channels

In this section a minimal cover by a channel of a distributed system is constructed. This channel consists of a set of concept graph morphisms from the concept graphs of the distributed system into a common concept graph. This common concept graph is supposed to represent the whole distributed system and thus the common information of the parts of the distributed system. Further the relation of the components of the system to the whole system is represented by the concept graph morphisms of the channel.

Definition 9. *A distributed system $\mathfrak{D} := (\mathrm{cg}(\mathfrak{D}), \mathrm{cgm}(\mathfrak{D}))$ consists of a finite family $\mathrm{cg}(\mathfrak{D})$ of concept graphs and of a set $\mathrm{cgm}(\mathfrak{D})$ of concept graph morphisms between concept graphs of $\mathrm{cg}(\mathfrak{D})$, i.e., $\mathrm{cgm}(\mathfrak{D}) \subseteq \{f : \mathfrak{G}_1 \rightleftarrows \mathfrak{G}_2 \mid \mathfrak{G}_1, \mathfrak{G}_2 \in \mathrm{cg}(\mathfrak{D})\}$.*

It is possible that a concept graph occurs more then one time in a distributed system, i.e., $\mathfrak{G}_i = \mathfrak{G}_j$ for $i \neq j$. In the example the distributed system is $(\{\mathfrak{G}_1, \mathfrak{G}_2, \mathfrak{G}_3\}, \{f^1, f^2\})$ (*cf.* Figures 1, 2, and 3).

Definition 10. *A channel C is a set $\{\gamma_k : \mathfrak{G}_k \rightleftarrows \mathfrak{C} \mid k = 1, \ldots, n\}$ of concept graph morphisms into a common concept graph $\mathfrak{C}$. $\mathfrak{C}$ is called the core of C.*

As mentioned in the introduction channels are a model for representing a distributed system as a whole. The next definition gives the connection between distributed systems and channels.

Definition 11. *A channel $C := \{\gamma_k : \mathfrak{G}_k \rightleftarrows \mathfrak{C} \mid k = 1, \ldots, n\}$ covers a distributed system $\mathfrak{D}$ with $\mathrm{cg}(\mathfrak{D}) = \{\mathfrak{G}_k \mid k = 1, \ldots n\}$ if for every $i, j \leq n$ and every concept graph morphism $f : \mathfrak{G}_i \rightleftarrows \mathfrak{G}_j \in \mathrm{cgm}(\mathfrak{D})$ holds*

$$\gamma_i = \gamma_j \circ f.$$

C is a minimal cover of $\mathfrak{D}$ if for every other channel $C' = \{\delta_k : \mathfrak{G}_k \rightleftarrows \mathfrak{C}' \mid k = 1, \ldots, n\}$ covering $\mathfrak{D}$ there is a unique concept graph morphism h from $\mathfrak{C}$ to $\mathfrak{C}'$ with $h \circ \gamma_k = \delta_k$, $k = 1, \ldots, n$.

A minimal cover turns a distributed system into such a channel. Now the construction of a minimal cover of a distributed system is given. At first the concept graphs of $cg(\mathfrak{D})$ are combined in the semiproduct. The concept graph morphisms of $cgm(\mathfrak{D})$ are presented by an invariant on the semiproduct. The core of the minimal cover is the quotient by this invariant. Thus it represents the distributed system.

Definition 12. *Let $\mathfrak{D}$ be a distributed system with $cg(\mathfrak{D}) = \{\mathfrak{G}_k \mid k = 1, \ldots, n\}$. The limit $\lim \mathfrak{D}$ of $\mathfrak{D}$ is the channel constructed as follows.*

1. *The core $\mathfrak{C}$ of $\lim \mathfrak{D}$:*
   At first an invariant $I = (W, \eta_E, \eta_R)$ on $\mathop{\mathsf{X}}_{k=1}^{n} \mathfrak{G}_k$ is defined by
   - $W := \{(v_1, \ldots, v_n) \mid f_V(v_j) = v_i \text{ for all } f : \mathfrak{G}_i \rightleftarrows \mathfrak{G}_j \in cgm(\mathfrak{D})\}$,
   - $((d, i), (e, j)) \in \eta_E$ *iff there is an* $f : \mathfrak{G}_i \rightleftarrows \mathfrak{G}_j$ *with* $f_E(d) = (e)$, *and*
   - $((r, i), (s, j)) \in \eta_R$ *iff there is an* $f : \mathfrak{G}_i \rightleftarrows \mathfrak{G}_j$ *with* $f_R(r) = (s)$.
   The core $\mathfrak{C}$ is the quotient $\mathop{\mathsf{X}}_{k=1}^{n} \mathfrak{G}_k / I$.

2. *The concept graph morphisms $\gamma_k : \mathfrak{G}_k \rightleftarrows \mathfrak{C}$, $k = 1, \ldots, n$ are defined by*

$$\gamma_k := \tau^I \circ \sigma_k.$$

The next theorem shows that $\lim(\mathfrak{D})$ is the channel which represents the distributed system.

Theorem 1. *Let $\mathfrak{D}$ be a distributed system. Then $\lim \mathfrak{D}$ is a minimal cover of $\mathfrak{D}$ and it is unique up to isomorphism.*

In the example the core of the limit is the concept graph shown in Figure 6. This concept graph is the quotient of the semiproduct $\mathfrak{G}_1 \mathsf{X} \mathfrak{G}_2 \mathsf{X} \mathfrak{G}_3$ by the invariant shown in Figure 5. The vertices of this concept graph are the vertices of the invariant on the semiproduct. Their corresponding concept and object names are the same as in the semiproduct. The edges are equivalence classes of edges in the semiproduct and their corresponding relation names are equivalence classes of the relation names in the semiproduct. Thus the relation name [*properties*] is the set $\{(properties, K), (properties, Na), (properties, Li)\}$. A common information in this distributed system is for instance that all elements can be detected by the colour of the flame.

For practical applications one will surely not use the alphabet constructed in the limit but an isomorphic alphabet with appropriate concept, object, and relation names. This is possible because of Theorem 1.

The size of the channel depends on the size of the distributed system. If there are no concept graph morphisms in $cgm(\mathfrak{D})$ then the core of the channel is just the semiproduct of the concept graphs in $cg(\mathfrak{D})$ and the channel morphisms are the natural concept graph morphisms from the factors into the semiproduct. It is possible, too, that the limit of a distributed system has an empty set of vertices or the alphabet has an empty set of concept or object names. This can happen if there are no tuples of vertices (concept or object names) which

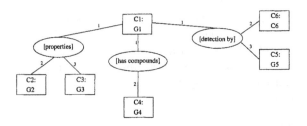

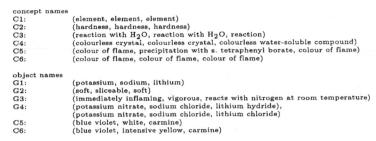

concept names
C1:   (element, element, element)
C2:   (hardness, hardness, hardness)
C3:   (reaction with H₂O, reaction with H₂O, reaction)
C4:   (colourless crystal, colourless crystal, colourless water-soluble compound)
C5:   (colour of flame, precipitation with s. tetraphenyl borate, colour of flame)
C6:   (colour of flame, colour of flame, colour of flame)

object names
G1:   (potassium, sodium, lithium)
G2:   (soft, sliceable, soft)
G3:   (immediately inflaming, vigorous, reacts with nitrogen at room temperature)
G4:   (potassium nitrate, sodium chloride, lithium hydride),
           (potassium nitrate, sodium chloride, lithium chloride)
C5:   (blue violet, white, carmine)
C6:   (blue violet, intensive yellow, carmine)

Fig. 6. The core of the channel covering the distributed system

respect all concept graph morphisms in cgm($\mathfrak{D}$), *i.e.*, for all tuples of vertices $(v_1, \ldots, v_n)$ exist two concept graph morphism $f : \mathfrak{G}_r \rightleftarrows \mathfrak{G}_s$ and $g : \mathfrak{G}_r \rightleftarrows \mathfrak{G}_t$ with $f_V(v_s) \neq g_V(v_t)$. If Ind($\mathfrak{D}$) denotes the set of all indices of concept graphs which are a domain of a concept graph morphism in cgm($\mathfrak{D}$) Generally the core of the channel has $\prod_{j \notin \mathrm{Ind}(\mathfrak{D})} V_j \cdot \min\{|V_i| \mid i \in \mathrm{Ind}(\mathfrak{D})\}$ vertices. Ind($\mathfrak{D}$) denotes here the set of all indices of concept graphs which are a domain of a concept graph morphism in cgm($\mathfrak{D}$)

5   Future Work

The next step is the examination of the contextual semantics of these constructions. In particular this would be interesting for the core of the limit of a distributed system.

In [1] Barwise and Seligman describe distributed systems further by local logics. On every component of the distributed system they have such a local logic which is supposed to describe the rules which are valid in this component. Infomorphisms are extended to logic infomorphisms, *i.e.*, to mappings additionally respecting the local logics. The core of the channel covering the system is provided with a local logic which is constructed out of the local logics of the concept graphs. The infomorphisms of the channel are logic infomorphisms, too. Further research has to find out if there is a corresponding theory based on concept graphs.

6   Appendix (Proof of Theorem 1)

Lemma 1. *Let* $f^1 : \mathfrak{G}_1 \rightleftarrows \mathfrak{G}_3$ *and* $f^2 : \mathfrak{G}_2 \rightleftarrows \mathfrak{G}_3$ *concept graph morphisms. Then there is a unique concept graph morphism* $f^1 \boxtimes f^2 : \mathfrak{G}_1 \boxtimes \mathfrak{G}_2 \rightleftarrows \mathfrak{G}_3$ *such that* $f^i = (f^1 \boxtimes f^2) \circ \sigma_i$, $i = 1, 2$.

Proof. $f^1 \boxtimes f^2$ consists of the functions

$$
\begin{array}{llll}
(f^1 \boxtimes f^2)_E : E_1^{\boxtimes} \cup E_2^{\boxtimes} \to E_3 & \text{defined by} & (e, i) \mapsto f_E^i(e),\ i = 1, 2, \\
(f^1 \boxtimes f^2)_V : V_3 \to V_1 \times V_2 & \text{defined by} & v \mapsto (f_V^1(v), f_V^2(v)), \\
(f^1 \boxtimes f^2)_C : C_3 \to C_1 \times C_2 & \text{defined by} & c \mapsto (f_C^1(c), f_C^2(c)), \\
(f^1 \boxtimes f^2)_{\mathcal{G}} : \mathcal{G}_3 \to \mathcal{G}_1 \times \mathcal{G}_2 & \text{defined by} & g \mapsto (f_{\mathcal{G}}^1(g), f_{\mathcal{G}}^2(g)),\ \text{and} \\
(f^1 \boxtimes f^2)_{\mathcal{R}} : \mathcal{R}_1^{\boxtimes} \cup \mathcal{R}_2^{\boxtimes} \to \mathcal{R}_3 & \text{defined by} & (r, i) \mapsto f_{\mathcal{R}}^i(r),\ i = 1, 2.
\end{array}
$$

It is easy to verify that $f^1 \boxtimes f^2$ defined in this way satisfies the required conditions. $\qquad\square$

Definition 13. *A concept graph morphism* $f : \mathfrak{G}_1 \rightleftarrows \mathfrak{G}_2$ *respects an invariant* $I = (W, \eta_E, \eta_{\mathcal{R}})$ *on* $\mathfrak{G}_1$ *if*

1. $f_V(v) \in W$ *for every* $v \in V_2$,
2. $f_E(d) = f_E(e)$ *for every* $(d, e) \in \eta_E$, *and*
3. $f_{\mathcal{R}}(r) = f_{\mathcal{R}}(s)$ *for every* $(r, s) \in \eta_{\mathcal{R}}$.

Lemma 2. *Let* $I = (W, \eta_E, \eta_{\mathcal{R}})$ *be an invariant on* $\mathfrak{G}_1$ *and* $f : \mathfrak{G}_1 \rightleftarrows \mathfrak{G}_2$ *an concept graph morphism that respects* I. *Then there is a unique concept graph morphism* $f^I : \mathfrak{G}/I \rightleftarrows \mathfrak{G}_2$ *such that* $f^I \circ \tau^I = f$ *holds.*

Proof. The concept graph morphism f^I consists of the functions

$$
\begin{array}{llll}
f_E^I : E_1/\eta_E \to E_2 & \text{defined by} & [e]_{\eta_E} \mapsto f_E(e), \\
f_V^I : V_2 \to W & \text{defined by} & v \mapsto f_V(v), \\
f_C^I : C_2 \to C_1 & \text{defined by} & c \mapsto f(c), \\
f_{\mathcal{G}}^I : \mathcal{G}_2 \to \mathcal{G}_1 & \text{defined by} & g \mapsto f(g),\ \text{and} \\
f_{\mathcal{R}}^I : \mathcal{R}_1/\eta_{\mathcal{R}} \to \mathcal{R}_2 & \text{defined by} & [r]_{\eta_{\mathcal{R}}} \mapsto f_{\mathcal{R}}(r).
\end{array}
$$

Again, it is easy to verify that f^I defined in this way satisfies the conditions. $\quad\square$

Proof of Theorem 1.

It is easy to show that C is a cover of $\mathfrak{D}$.

Let $C' := \{\delta_k : \mathfrak{G}_k \rightleftarrows \mathfrak{C}' \mid k = 1, \dots, n\}$ be another channel covering $\mathfrak{D}$, i.e., $\delta_i = \delta_j \circ f$ for every concept graph morphism $f : \mathfrak{G}_i \rightleftarrows \mathfrak{G}_j \in \mathrm{cgm}(\mathfrak{D})$. From Lemma 1 follows that there is a unique concept graph morphism $\delta : \boxtimes_{k=1}^n \mathfrak{G}_k \rightleftarrows \mathfrak{C}'$ with $\delta_k = \delta \circ \sigma_k$ for every $k = 1, \dots, n$.

Now, it is shown that δ respects the invariant I. Because $\delta(v)_i = f_V(\delta(v)_j)$ for every $f : \mathfrak{G}_i \rightleftarrows \mathfrak{G}_j \in \mathrm{cgm}(\mathfrak{D})$ it follows $\delta(v) \in W$ for every $v \in V_{\mathfrak{C}'}$. Further for every $((d, i), (e, j)) \in \eta_E$ there is an $f : \mathfrak{G}_i \rightleftarrows \mathfrak{G}_j \in \mathrm{cgm}(\mathfrak{D})$ with $\delta(d_i) = \delta_j(e)$. The condition for $\eta_{\mathcal{R}}$ is shown analogously.

Because of Lemma 2 there is a unique $\alpha : \aleph_{k=1}^{n} \mathfrak{G}_k / I \rightleftarrows \mathfrak{C}'$ with $\delta = \alpha \circ \tau^I$. Thus C is a minimal cover of $\mathfrak{D}$.

Let now $\tilde{C}$ be another minimal cover of $\mathfrak{D}$. From the minimality of C and $\tilde{C}$ follows that there is a unique concept graph morphism $\xi : \mathfrak{C} \rightleftarrows \tilde{\mathfrak{C}}$ and a unique concept graph morphism $\tilde{\xi} : \tilde{\mathfrak{C}} \rightleftarrows \mathfrak{C}$. Due to the fact that C and C' are minimal $id_{\mathfrak{C}}$ is the unique concept graph morphism from $\mathfrak{C}$ to $\mathfrak{C}$ (analogous for $\mathfrak{C}'$) and thus $\xi \circ \tilde{\xi} = id_{\mathfrak{C}'}$ and $\tilde{\xi} \circ \xi = id_{\mathfrak{C}}$. It follows that $\mathfrak{C}$ is isomorphic to $\mathfrak{C}'$.   □

References

1. J. Barwise, J. Seligman: Information Flow: The Logic of Distributed Systems. Cambridge University Press, Cambridge 1997.
2. M. Chein, M.-L. Mugnier: Conceptual Graphs: fundamental notions. Revue d'Intelligence Artificielle 6, 1992, 365-406.
3. B. Ganter, R. Wille: Formal Concept Analysis: Mathematical Foundations. Springer Verlag, Berlin Heidelberg New York 1999.
4. G. Malik: Information Transfer across Simple Concept Graphs. In: U. Priss, D. Corbett, G. Angelowa (eds.): Conceptual Structures: Integration and Interfaces. LNAI 2393. Springer, Berlin Heidelberg New York 2002, 48-61.
5. M.-L. Mugnier: Knowledge Representation and Reasoning Based on Graph Homomorphism. In B. Ganter, G. Mineau (eds.): Conceptual Structures: Logical, Linguistic, and Computational Issues. LNAI 1867. Springer Berlin Heidelberg New York 2000, 171-191.
6. S. Prediger: Simple Concept Graphs: a logic approach. In M.-L. Mugnier, M.Chein (eds.): Conceptual Structures: Theory, Tools and Applications. LNAI 1453. Springer, Berlin Heidelberg New York 1998, 225-239.
7. S. Prediger: Kontextuelle Urteilslogik mit Begriffsgraphen: Ein Beitrag zur Restrukturierung der mathematischen Logik. Dissertation, FB Mathematik, TU Darmstadt 1998. Verlag Shaker, Aachen 1998.
8. W. Schröter, K.-H. Lautenschläger, H. Bibrack, A. Schnabel: Chemie: Nachschlagebücher für Grundlagenfächer. Fachbuchverlag Leipzig, 1986.
9. J.F. Sowa: Conceptual Structures: information processing in mind and machine. Addison-Wesley, Reading 1984.
10. R. Wille: Conceptual Graphs and Formal Concept Analysis. In: D. Lukose, H. Delugach, M. Keeler, L. Searle, J. Sowa (eds.): Conceptual Structures: Fulfilling Peirce's Dream. LNAI 1257. Springer, Berlin Heidelberg New York 1997, 290-303.

Concept Graphs with Subdivision:
A Semantic Approach

Lars Schoolmann and Rudolf Wille

Technische Universität Darmstadt, Fachbereich Mathematik,
Schloßgartenstr. 7, D–64289 Darmstadt
wille@mathematik.tu-darmstadt.de

Abstract. The semantic approach defines a *concept graph with subdivision* as a mathematical structure derived from a triadic power context family. The aim of introducing concept graphs with subdivision is to represent modal information mathematically. Based on the notion of the *conceptual content* of a concept graph with subdivision, we can show that the concept graphs with subdivision of a triadic power context family form a complete lattice with respect to the *information order*. Finally, our approach is extended to *existential concept graphs with subdivision*.

1   Triadic Mathematization of Modal Information

A semantic approach to concept graphs and existential concept graphs has already been outlined in [Wi02]. That approach shall now be extended to *concept graphs with subdivision* which code modal information. A first attempt of establishing a mathematical theory of concept graphs with subdivision can be found in [Wi98] and [GW00]. The approach of this paper tries to simplify, to harmonize, and to expand definitions and results of the three mentioned papers. Again, *Triadic Concept Analysis* will be the basis for developing a useful semantics for coding modal information since already the basic notion of a triadic context enables the representation of modal information (cf. [DW00]). For recalling basic notions and results of Triadic Concept Analysis, the reader is referred to the Appendix and to the papers [LW95] and [WZ00]. Basics of Formal Concept Analysis are presented in [GW99].

First, an example shall show how a triadic mathematization and representation of modal information may be performed. The example is concerned with the two paintings in Fig. 1 which strongly resemble each other: Both show in their center the Madonna with a protective coat holding the infant Jesus in her arms; the Madonna is standing in front of a throne while five persons (members of the family of Jakob Meyer, mayor of Basel in the 16th century) are kneeling down on both sides of the Madonna; a nacked little boy stands in front who has been identified as John the Baptist. Today, the left painting is located in the castle museum of Darmstadt and therefore called the *Darmstädter Madonna*, the right painting is located in the Zwinger of Dresden and therefore called the *Dresdner Madonna* (cf. [BG98]).

A. de Moor, W. Lex, and B. Ganter (Eds.): ICCS 2003, LNAI 2746, pp. 271–281, 2003.

Fig. 1. The Darmstädter Madonna and the Dresdner Madonna

In the 19th century a strong controversy took place about the question: Which of the two paintings is the original? Finally, the art historians found out that the Darmstädter Madonna is the original painted 1526/28 by Hans Holbein d. J. in Basel and that the Dresdner Madonna is a free replica of Holbein's Madonna painted 1635/37 by Bartholomäus Sarburgh in Den Haag. What are the characteristics of the two paintings indicating that the Darmstädter Madonna is the original? Here only characteristics which are readable from the paintings shall be made explicit with the aim to test the *applicability of the triadic semantics* for representing modal information.

Fig. 2 depicts a *triadic context* $\mathbb{K}_0 := (G_0, M_0, B, Y_0)$ of which the object set G_0 consists of instances visible in the two paintings, the attribute set M_0 contains characteristics of those instances, and the modality set B is formed by the two modalities $< DarM >$ and $< DreM >$; the ternary relation Y_0, indicated by the crosses in Fig. 2, comprises all valid relationships of the form object-attribute-modality such as $(Jesus, child, < DarM >)$ which codes the elementary information that Jesus is a child in view of the "Darmstädter Madonna" as painting. Notice that there are the three *"conceptual objects"* $DarM$, $DreM$, and $DarMDreM$ which give rise to the three triadic concepts

$$\xi(DarM) := (\{DarM\}, \{painting\}, \{< DarM >\}),$$
$$\xi(DreM) := (\{DreM\}, \{painting\}, \{< DreM >\}),$$
$$\xi(DarMDreM) := (\{DarMDreM\}, \{comparison\}, \{< DarM >, < DreM >\}).$$

K₀	Darmstädter Madonna ⟨ DarM ⟩														Dresdner Madonna ⟨ DreM ⟩													
	Person	male	female	adult	child	biblical	Meyer family	part of throne	decoration	plant	space	visible	painting	comparison	Person	male	female	adult	child	biblical	Meyer family	part of throne	decoration	plant	space	visible	painting	comparison
Mary	×		×	×		×							×		×		×	×		×							×	
Jesus	×	×			×	×							×		×	×			×	×							×	
John	×	×			×	×							×		×	×			×	×							×	
Jakob Meyer	×	×		×			×						×		×	×		×			×						×	
1st Wife	×		×	×			×						×		×		×	×			×						×	
2nd Wife	×		×	×			×						×		×		×	×			×						×	
Daughter	×		×		×		×						×		×		×		×		×						×	
Son	×	×			×		×						×		×	×			×		×						×	
Niche								×					×									×					×	
Left Console								×					×									×					×	
Right Console								×					×									×					×	
Left Wall Front								×														×					×	
Right Wall Front								×														×					×	
Carpet									×				×										×				×	
Left Fig Spray										×			×											×			×	
Right Fig Spray										×			×											×			×	
Foreground											×	×													×	×		
Headroom											×	×													×	×		
DarM													×															
DreM																											×	
DarMDreM														×														×

Fig. 2. The triadic context K₀ representing information of the paintings in Fig. 1

The triadic concepts $\xi(DarM)$ and $\xi(DreM)$ represent the self-referential information that the Darmstädter Madonna and the Dresdner Madonna are paintings in view of the "Darmstädter Madonna" as painting and the "Dresdner Madonna" as painting, respectively. The triadic concept $\xi(DarMDreM)$ represents the information that the dyad [Darmstädter Madonna - Dresdner Madonna] is a comparison in the combined view of the "Darmstädter Madonna" as painting and the "Dresdner Madonna" as painting.

Since K₀ only indicates as difference between the paintings that the left and right wall fronts of the throne are visible in the Dresdner Madonna, but not in the Darmstädter Madonna, further differences are made explicit by the triadic context K₂ in Fig. 3. Those differences are graphically depicted in the form of a *"concept graph with subdivision"* in Fig. 4. The depicted differences already give strong arguments that the Darmstädter Madonna is a renaissance painting and the Dresdner Madonna is a mannerism painting because the architecture in the Dresdner Madonna is finer and gives more space so that the persons are less pressed together as in the Darmstädter Madonna (cf. [BG98], p.61ff.). Thus, the comparison of the two paintings yields good reasons to view the Darmstädter Madonna as the original.

For mathematically representing modal information in general, triadic power context families (with conceptual objects) are introduced as the basic structures of the triadic semantics: a *triadic power context family (with conceptual objects)* is defined as a sequence $\vec{\mathbb{K}} := (\mathbb{K}_0, \mathbb{K}_1, \mathbb{K}_2, \ldots)$ of triadic contexts

$\mathbb{K}_2$	⟨ DarM ⟩				⟨ DreM ⟩			
	higher	higher located	larger	finer	higher	higher located	larger	finer
(Niche, Niche)	×				×			
(Shell, Shell)		×	×			×	×	
(Left Console, Left Console)				×				×
(Right Console, Right Console)				×				×
(Foreground, Foreground)			×				×	
(Headroom, Headroom)			×				×	

Fig. 3. The triadic context $\mathbb{K}_2$ representing comparabilities between the paintings in Fig. 1

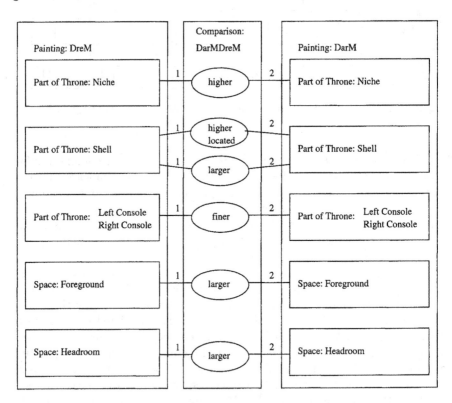

Fig. 4. Concept graph with subdivision representing information coded in $\mathbb{K}_2$

$\mathbb{K}_k := (G_k, M_k, B, Y_k)$ with $G_k \subseteq (G_0)^k$ for $k = 1, 2, \ldots$ and with a partial map ξ from G_0 into the set $\mathfrak{T}(\mathbb{K}_0)$ of all triadic concepts of $\mathbb{K}_0$; ξ is called the *subdivision map* of $\vec{\mathbb{K}}$ and the elements of *dom*ξ are called *conceptual objects*.

The sequence $(\mathbb{K}_0, \mathbb{K}_2)$ which codes modal information about the Darmstädter Madonna and the Dresdner Madonna is an example of a triadic power context family having $DarM$, $DreM$, and $DarMDreM$ as conceptual objects.

2 Concept Graphs with Subdivision

For introducing concept graphs with subdivision, we first mathematize their rhetoric structures as follows: A *relational graph with subdivision* is defined as a quadruple (V, E, ν, σ) for which V and E are disjoint sets, ν is a map from E into $\bigcup_{k=1,2,\dots} V^k$, and σ is a map from V into the power set $\mathfrak{P}(V \cup E)$ with $v \notin \sigma^n(v)$ for all $n \in \mathbb{N}$ and $v \in V$ (where $\sigma(X) := \bigcup_{w \in X \cap V} \sigma(w)$). The elements of V and E are called *vertices* and *edges*, respectively. Thus, ν maps every edge to a sequence of vertices. If $\nu(e) \in V^k$, the edge e is said to be $k - ary$, in symbols: $|e| = k$. Let $E^{(k)} := \{e \in E \mid |e| = k\}$ and let $E^{(0)} := V$; furthermore, let $V^\circ := V \setminus \bigcup_{v \in V} \sigma(v)$ and let $E^\circ := E \setminus \bigcup_{v \in V} \sigma(v)$.

An example of a relational graph with subdivision can be seen in Fig. 4 if only the rectangles, ovals, and connecting line segments with numbers are considered. The rectangles represent the vertices, the ovals represent the edges, and the line segments with their numbers indicate the sequence of vertices belonging to an edge. In our example, all edges are 2-ary where the vertex of the first component is on the left and the vertex of the second component is on the right. The subdivision is represented by three large vertices: one contains all vertices on the left, another contains all edges, and the third contains all vertices on the right. Examples in [Wi98] and [GW00] show that the subdivision of concept graphs need not to be a partition or a classification; it might be quite complex.

Now we are able to give a semantic definition of a concept graph with subdivision based on a triadic power context family $\vec{\mathbb{K}} := (\mathbb{K}_0, \mathbb{K}_1, \mathbb{K}_2, \dots)$ with $\mathbb{K}_k := (G_k, M_k, B, Y_k)$ for $k = 0, 1, 2, \dots$: A *concept graph with subdivision* of the triadic power context family $\vec{\mathbb{K}}$ is defined as a structure $\mathfrak{G} := (V, E, \nu, \sigma, \kappa, \rho)$ satisfying the following conditions:

1. (V, E, ν, σ) is a relational graph with subdivision;
2. $\kappa : V \cup E \to \bigcup_{k=0,1,2,\dots} \mathfrak{T}(\mathbb{K}_k)$ is a map with $\kappa(u) \in \mathfrak{T}(\mathbb{K}_k)$ for $k = 0, 1, 2, \dots$ and $u \in E^{(k)}$;
3. $\rho : V \to \mathfrak{P}(G_0) \setminus \{\emptyset\}$ is a map with $|\rho(v)| = 1$ if $\sigma(v) \neq \emptyset$; furthermore, $\rho(v) \subseteq Ext(\kappa(v))$ for $v \in V^\circ$ and $\rho(e) := \rho(v_1) \times \cdots \times \rho(v_k) \subseteq Ext(\kappa(e))$ for $e \in E^\circ$ with $\nu(e) = (v_1, \dots, v_k)$;
4. for $v \in \sigma(w) \cap V$, there is a unique conceptual object $g \in \rho(w) \cap dom\xi$ with $\rho(v) \subseteq (Int(\kappa(v)) \times Mod(\xi(g)))^{(1)}$;
5. for $u_0 \in \sigma(w_0) \cap E$ with $\nu(u_0) = (u_1, \dots, u_k)$, there is, if $u_i \in \sigma(w_i)$, a unique conceptual object $h_i \in \rho(w_i) \cap dom\xi$ with $\rho(w_i) \subseteq (Int(\kappa(w_i)) \times Mod(\xi(h_i)))^{(1)}$ $(i = 0, 1, \dots, k)$ and furthermore $Mod(\xi(h_i)) \subseteq Mod(\xi(h_0))$.

It is convenient to consider the map ρ not only on vertices, but also on edges: For $e \in E$ with $\nu(e) = (v_1, \dots, v_k)$, let $\rho(e) := \rho(v_1) \times \cdots \times \rho(v_k)$ (as already indicated in 3.).

Fig. 4 represents a concept graph with subdivision $\mathfrak{G} := (V, E, \nu, \sigma, \kappa, \rho)$ of the power context family $(\mathbb{K}_0, \mathbb{K}_2)$ of Section 1 which can be proved as follows: Condition 1 has already been confirmed above. Condition 2 and 3 can be easily checked within the triadic contexts of Fig. 2 and Fig. 3. Now, let us consider the case $v \in \sigma(w)$ of Condition 4 where, in Fig. 4, w is represented by the large rectangle on the left and v by the uppermost rectangle in that large rectangle. Then the desired conceptual object g is $DreM$ for which we obtain indeed

$$
\begin{aligned}
\rho(v) &= \{Niche\} \\
&\subseteq (\{part\,of\,throne, visible\} \times \{< DreM >\})^{(1)} \\
&= (Int(\kappa(v)) \times Mod(\xi(DreM)))^{(1)}.
\end{aligned}
$$

The proof for the other choices of v can be analogously performed. To demonstrate the proof in the case $u_0 \in \sigma(w)$ of Condition 5, we choose w_0 to be represented by the large rectangle in the middle of Fig. 4 and u_0 by the uppermost oval in that large rectangle. Then the desired conceptual object h_0 is $DarMDreM$ for which we obtain

$$
\begin{aligned}
\rho(e) &= \{(Niche, Niche)\} \\
&\subseteq (\{higher\} \times \{< DarM >, < DreM >\})^{(1)} \\
&= (Int(\kappa(e)) \times Mod(\xi(DarMDreM)))^{(1)}.
\end{aligned}
$$

The proof for the other choices of u_0 can be analogously performed. The rest of Condition 5 is confirmed by

$$
\begin{aligned}
Mod(\xi(DarM)) &= \{< DarM >\} \\
&\subseteq \{< DarM >, < DreM >\} = Mod(\xi(DarMDreM)),
\end{aligned}
$$

$$
\begin{aligned}
Mod(\xi(DreM)) &= \{< DreM >\} \\
&\subseteq \{< DarM >, < DreM >\} = Mod(\xi(DarMDreM)).
\end{aligned}
$$

Next we try to answer the question: *What is the information given by a concept graph with subdivision?* Our answer generalizes the approach described in [Wi02]. The coded information in a concept graph with subdivision is constituted by the elementary information of the form: an object g is an element of the extension of a triadic concept. In correspondence with the definition of a concept graph with subdivision, such extension should be constructed as the (1)-derivation of the direct product of the intension of a triadic concept $\mathfrak{b}$ not related to the subdivision and the modus of a triadic concept $\mathfrak{c}$ related to the subdivision (i.e., $\mathfrak{c} = \xi(h)$ for some $h \in G_0$), or $\mathfrak{c}$ should be equal to $\mathfrak{b}$ (recall that $Ext(\mathfrak{b}) = (Int(\mathfrak{b}) \times Mod(\mathfrak{b}))^{(1)}$). This means that, in our triadic semantics, the *conceptual informations units* are triples

$$
(g, \mathfrak{b}, \mathfrak{c}) \text{ with } g \in (Int(\mathfrak{b}) \times Mod(\mathfrak{c}))^{(1)} \text{ or } (g, \mathfrak{b}, \mathfrak{b}) \text{ with } g \in Ext(\mathfrak{b}).
$$

Such triples are called *triadic conceptual instances*; the set of all triadic conceptual instances of the context $\mathbb{K}_k$ of a triadic power context family $\vec{\mathbb{K}}$ is denoted by $\mathfrak{T}^{inst}(\mathbb{K}_k)$ $(k = 0, 1, 2, \ldots)$.

Our answer to the posed question is: The information given by a concept graph with subdivision consists of the triadic conceptual instances coded in that graph and further triadic conceptual instances which can be inferred from those instances by background knowledge coded in the underlying triadic power context family $\vec{\mathbb{K}} := (\mathbb{K}_0, \mathbb{K}_1, \mathbb{K}_2, \ldots)$ (cf. [Br94]). The inferences which we propose are between sets of objects and between pairs consisting of two sets of triadic concepts. The connection between these two types of inferences is given by the triadic conceptual instances coded in the considered concept graph with subdivision. The following definitions make this explicit: Let $k = 0, 1, 2, \ldots$; for $A, B \subseteq G_k$,

$$A \to B \text{ is an inference in } \mathbb{K}_k \text{ if } A^{(1)} \subseteq B^{(1)}$$

and, for $\mathfrak{C}, \mathfrak{E} \subseteq \mathfrak{T}(\mathbb{K}_k)$ and $\mathfrak{D}, \mathfrak{F} \subseteq \mathfrak{T}(\mathbb{K}_k) \cup im\xi$,

$$(\mathfrak{C}, \mathfrak{D}) \to (\mathfrak{E}, \mathfrak{F}) \text{ is an inference in } \mathbb{K}_k \text{ if}$$

$$\left(\bigcup_{c \in \mathfrak{C}} Int(c) \times \bigcup_{\mathfrak{d} \in \mathfrak{D}} Mod(\mathfrak{d}) \right)^{(1)} \subseteq \left(\bigcup_{e \in \mathfrak{E}} Int(e) \times \bigcup_{f \in \mathfrak{F}} Mod(f) \right)^{(1)}.$$

The inferences $A \to B$ and $(\mathfrak{C}, \mathfrak{D}) \to (\mathfrak{E}, \mathfrak{F})$ give rise to a closure system $\mathcal{C}(\mathbb{K}_k)$ on $\mathfrak{T}^{inst}(\mathbb{K}_k)$ consisting of all subsets Y of $\mathfrak{T}^{inst}(\mathbb{K}_k)$ which have the following property:

(P_k) If $A \times \mathfrak{C} \times \mathfrak{D} \subseteq Y$ and if $A \to B$ and $(\mathfrak{C}, \mathfrak{D}) \to (\mathfrak{E}, \mathfrak{F})$ are inferences in $\mathbb{K}_k$ then $B \times \mathfrak{E} \times \mathfrak{F} \subseteq Y$.

The *k-ary conceptual content* $C_k(\mathfrak{G})$ of a concept graph with subdivision $\mathfrak{G} := (V, E, \nu, \sigma, \kappa, \rho)$ of a triadic power context family $\vec{\mathbb{K}}$ is defined as the closure of

$$\{(g, \kappa(u), \kappa(u)) \mid u \in E^{(k)} \cap (V^\circ \cup E^\circ) \text{ and } g \in \rho(u)\} \cup$$
$$\{(g, \kappa(u), \xi(h)) \mid \exists w \in V : h \in \rho(w) \cap dom\xi, u \in E^{(k)} \cap \sigma(w), \text{ and } g \in \rho(u)\}$$

with respect to the closure system $\mathcal{C}(\mathbb{K}_k)$ $(k = 0, 1, 2, \ldots)$. Then

$$C(\mathfrak{G}) := (C_0(\mathfrak{G}), C_1(\mathfrak{G}), C_2(\mathfrak{G}), \ldots)$$

is called the *conceptual content* of the concept graph $\mathfrak{G}$.

For concept graphs with subdivision, $\mathfrak{G}_1 := (V_1, E_1, \nu_1, \sigma_1, \kappa_1, \rho_1)$ is defined to be *less informative* (*more general*) than $\mathfrak{G}_2 := (V_2, E_2, \nu_2, \sigma_2, \kappa_2, \rho_2)$ (in symbols: $\mathfrak{G}_1 \lesssim \mathfrak{G}_2$) if $C_k(\mathfrak{G}_1) \subseteq C_k(\mathfrak{G}_2)$ for $k = 0, 1, 2, \ldots$; $\mathfrak{G}_1$ and $\mathfrak{G}_2$ are called *equivalent* (in symbols: $\mathfrak{G}_1 \sim \mathfrak{G}_2$) if $\mathfrak{G}_1 \lesssim \mathfrak{G}_2$ and $\mathfrak{G}_2 \lesssim \mathfrak{G}_1$ (i.e., $C_k(\mathfrak{G}_1) = C_k(\mathfrak{G}_2)$ for $k = 0, 1, 2, \ldots$). The set of all equivalence classes of concept graphs with subdivision of a triadic power context family $\vec{\mathbb{K}}$ together with the order induced by the quasi-order $\lesssim$ is an ordered set denoted by $\widetilde{\Gamma}(\vec{\mathbb{K}})$. For this ordered set one can prove the following structural result:

Proposition 1. *Let* $\vec{\mathbb{K}} := (\mathbb{K}_0, \mathbb{K}_1, \mathbb{K}_2, \ldots)$ *be a triadic power context family and let* $\underline{\mathcal{C}}(\vec{\mathbb{K}}) := \times_{k=0,1,2,\ldots} \underline{\mathcal{C}}(\mathbb{K}_k)$ *with* $\underline{\mathcal{C}}(\mathbb{K}_k) := (\mathcal{C}(\mathbb{K}_k), \subseteq)$ *for* $k = 0, 1, 2, \ldots$, *and let* $\mathfrak{o}_3 := (G_0, M_0, (G_0 \times M_0)^{(3)})$. *Then the map* $\iota : \widetilde{\Gamma}(\vec{\mathbb{K}}) \to \underline{\mathcal{C}}(\vec{\mathbb{K}})$ *with*

$\iota\widetilde{\mathfrak{G}} := C(\mathfrak{G})$ *for each equivalence class* $\widetilde{\mathfrak{G}}$ *represented by a concept graph* $\mathfrak{G}$ *of* $\vec{\mathbb{K}}$
is an isomorphism from the ordered set $\widetilde{\Gamma}(\vec{\mathbb{K}})$ *onto the ordered subset of* $\underline{C}(\vec{\mathbb{K}})$
consisting of all those sequences $(Z_0, Z_1, Z_2, \ldots)$ *for which*

- *there exists a generating subset* $\underline{Z}_0$ *of the closure* Z_0 *such that* $(g, \mathfrak{c}, \mathfrak{d}) \in$
  $\underline{Z}_0$ *with* $\mathfrak{c} \neq \mathfrak{d}$ *implies that there exists an* $h \in dom\xi$ *with* $\xi(h) = \mathfrak{d}$ *and*
  $(h, \mathfrak{o}_3, \mathfrak{o}_3) \in \underline{Z}_0$, *and*
- *for* $k = 1, 2, \ldots$ *there exists a generating subset* $\underline{Z}_k$ *of the closure of* Z_k *such*
  that $((g_1, \ldots, g_k), \mathfrak{c}, \mathfrak{d}) \in \underline{Z}_k$ *always implies* $(g_i, \mathfrak{o}_3, \mathfrak{o}_3) \in \underline{Z}_0$ *for* $i = 1, \ldots, k$
  and if $\mathfrak{c} \neq \mathfrak{d}$ *there exists* $h \in dom\xi$ *with* $\xi(h) = \mathfrak{d}$ *and* $(h, \mathfrak{o}_3, \mathfrak{o}_3) \in \underline{Z}_0$.

Proof: Clearly, ι is an injective and order-preserving map. In order to show
that it is an isomorphism, we construct a concept graph with subdivision for
each sequence $(Z_0, Z_1, Z_2, \ldots)$ described in the proposition:

$$V := \underline{Z}_0$$
$$E^{(k)} := \underline{Z}_k \ (k = 1, 2, \ldots)$$
$$\nu((g_1, \ldots, g_k), \mathfrak{c}, \mathfrak{d}) := ((g_1, \mathfrak{o}_3, \mathfrak{o}_3), \ldots, (g_k, \mathfrak{o}_3, \mathfrak{o}_3))$$
$$\kappa(g, \mathfrak{c}, \mathfrak{d}) := \mathfrak{c}$$
$$\rho(g, \mathfrak{c}, \mathfrak{d}) := \{g\}$$
$$\sigma(h, \mathfrak{o}_3, \mathfrak{o}_3) := \{(g, \mathfrak{c}, \mathfrak{d}) \in V \cup E \mid \mathfrak{c} \neq \mathfrak{d} \text{ and } \xi(h) = \mathfrak{d}\}$$
$$\sigma(g, \mathfrak{c}, \mathfrak{d}) := \emptyset \text{ if } \mathfrak{c} \neq \mathfrak{o}_3 \text{ or } \mathfrak{d} \neq \mathfrak{o}_3$$

Then $\mathfrak{G} := (V, E, \nu, \sigma, \kappa, \rho)$ is a concept graph with subdivision (w.l.o.g. one can
assume that $V \cap E = \emptyset$). It is easy to see that $C_k(\mathfrak{G}) = Z_k$ for $k = 0, 1, 2, \ldots$.
Thus, we have constructed the inverse of ι which, obviously, is order-preserving.
$\square$

Corollary 1. $\widetilde{\Gamma}(\vec{\mathbb{K}})$ *is always a complete lattice.*

3   Existential Concept Graphs with Subdivision

For defining existential concept graphs with subdivision we generalize the construction of a free X-extension of a power context family introduced in [Wi02]
to the triadic case. For a set X of variables, an X-*interpretation* into a set G_0
with $G_0 \cap X = \emptyset$ is defined as a map $\chi : G_0 \cup X \to G_0$ with $\chi(g) = g$ for all
$g \in G_0$; the set of all X-interpretations into G_0 is denoted by $B(X, G_0)$.

Let $\vec{\mathbb{K}} := (\mathbb{K}_0, \mathbb{K}_1, \mathbb{K}_2, \ldots)$ be a triadic power context family with $\mathbb{K}_k :=$
(G_k, M_k, B, Y_k) for $k = 0, 1, 2, \ldots$ and let X be a set of variables with $G_0 \cap X = \emptyset$.
The triadic contexts $\mathbb{K}_0, \mathbb{K}_1, \mathbb{K}_2, \ldots$ can be *freely extented by* X by the following
construction:

- let $\mathbb{K}_0[X] := (G_0[X], M_0[X], B[X], Y_0[X])$ with $G_0[X] := G_0 \cup X$, $M_0[X] :=$
  M_0, $B[X] := B$, and $Y_0[X] := Y_0 \cup \{(x, m, b) \mid x \in X \text{ and } \{(m, b)\}^{(1)} \neq \emptyset\}$;
- let $\mathbb{K}_k[X] := (G_k[X], M_k[X], B[X], Y_k[X])$ $(k = 1, 2, \ldots)$ with $G_k[X] :=$
  $\{(u_1, \ldots, u_k) \in G_0[X]^k \mid \exists \chi \in B(X, G_0) : (\chi(u_1), \ldots, \chi(u_k)) \in G_k\}$,
  $M_k[X] := M_k$, $B[X] := B$, and $((u_1, \ldots, u_k), m, b) \in Y_k[X] :\Longleftrightarrow \exists \chi \in$
  $B(X, G_0) : ((\chi(u_1), \ldots, \chi(u_k)), m, b) \in Y_k$.

The triadic power context family $\vec{\mathbb{K}}[X] := (\mathbb{K}_0[X], \mathbb{K}_1[X], \mathbb{K}_2[X], \ldots)$ is called the *free X-extension* of the triadic power context family $\vec{\mathbb{K}}$. For defining existential concept graphs with subdivision, we need a map which naturally assigns the triadic concepts of $\vec{\mathbb{K}}[X]$ to triadic concepts of $\vec{\mathbb{K}}$: For each $k = 0, 1, 2, \ldots$, we define the map $\pi_k^X : \mathfrak{T}(\mathbb{K}_k[X]) \to \mathfrak{T}(\mathbb{K}_k)$ by

$$\pi_k^X(A_1, A_2, A_3) := (A_1 \cap G_k, (A_1 \cap G_k)^{(1,2,A_3)}, ((A_1 \cap G_k) \times ((A_1 \cap G_k)^{(1,2,A_3)})^{(3)})$$

for $(A_1, A_2, A_3) \in \mathfrak{T}(\mathbb{K}_k[X])$. An *existential concept graph with subdivision* of a power context family $\vec{\mathbb{K}}$ is defined as a concept graph with subdivision $\mathfrak{G} := (V, E, \nu, \sigma, \kappa, \rho)$ of a free X-extension $\vec{\mathbb{K}}[X]$ for which an X-interpretation χ into G_0 exists such that $\mathfrak{G}^\chi := (V, E, \nu, \sigma, \kappa^\chi, \rho^\chi)$ with $\kappa^\chi(u) := \pi_k^X(\kappa(u))$ and $\rho^\chi(v) := \chi(\rho(v))$ is a concept graph with subdivision of $\vec{\mathbb{K}}$; $\mathfrak{G}$ is also said to be an existential concept graph with subdivision of $\vec{\mathbb{K}}$ over X. The *conceptual content* of an existential concept graph with subdivision $\mathfrak{G}_X$ of a triadic power context family $\vec{\mathbb{K}}$ is defined as the conceptual content of $\mathfrak{G}_X$ understood as a concept graph of the free X-extension $\vec{\mathbb{K}}[X]$.

For existential concept graphs with subdivision, $\mathfrak{G}_1 := (V_1, E_1, \nu_1, \sigma_1, \kappa_1, \rho_1)$ is defined to be *less informative (more general)* than $\mathfrak{G}_2 := (V_2, E_2, \nu_2, \sigma_2, \kappa_2, \rho_2)$ (in symbols: $\mathfrak{G}_1 \lesssim \mathfrak{G}_2$) if $C_k(\mathfrak{G}_1) \subseteq C_k(\mathfrak{G}_2)$ for $k = 0, 1, 2, \ldots$; $\mathfrak{G}_1$ and $\mathfrak{G}_2$ are called *equivalent* (in symbols: $\mathfrak{G}_1 \sim \mathfrak{G}_2$) if $\mathfrak{G}_1 \lesssim \mathfrak{G}_2$ and $\mathfrak{G}_2 \lesssim \mathfrak{G}_1$ (i.e., $C_k(\mathfrak{G}_1) = C_k(\mathfrak{G}_2)$ for $k = 0, 1, 2, \ldots$). The set of all equivalence classes of existential concept graphs with subdivision of a triadic power context family $\vec{\mathbb{K}}$ over a set X of variables together with the order induced by the quasi-order $\lesssim$ is an ordered set denoted by $\widetilde{\Gamma}(\vec{\mathbb{K}}; X)$. For this ordered set one can prove the following structural result:

Proposition 2. *Let $\vec{\mathbb{K}} := (\mathbb{K}_0, \mathbb{K}_1, \mathbb{K}_2, \ldots)$ be a triadic power context family, let X be a set with $X \cap G_0 = \emptyset$, and let $\mathfrak{o}_3^X := (G_0[X], M_0[X], (G_0[X] \times M_0[X])^{(3)[X]})$. The map $\iota_X : \widetilde{\Gamma}(\vec{\mathbb{K}}; X) \to \underline{\mathcal{C}}(\vec{\mathbb{K}}[X])$ with $\iota_X \mathfrak{G}_X := C(\mathfrak{G}_X)$ for each equivalence class $\mathfrak{G}_X$ represented by an existential concept graph with subdivision $\mathfrak{G}_X$ of $\vec{\mathbb{K}}$ is an isomorphism from the ordered set $\widetilde{\Gamma}(\vec{\mathbb{K}}; X)$ onto the ordered subset of $\underline{\mathcal{C}}(\vec{\mathbb{K}}[X])$ consisting of all those sequences $(Z_0, Z_1, Z_2, \ldots)$ for which*

- *there exists a generating subset $\underline{Z}_0$ of the closure Z_0 such that $(g, \mathfrak{c}, \mathfrak{d}) \in \underline{Z}_0$ with $\mathfrak{c} \neq \mathfrak{d}$ implies that there exists an $h \in \mathrm{dom}\xi$ with $\xi(h) = \mathfrak{d}$ and $(h, \mathfrak{o}_3^X, \mathfrak{o}_3^X) \in \underline{Z}_0$,*
- *for $k = 1, 2, \ldots$ there exists a generating subset $\underline{Z}_k$ of the closure of Z_k such that $((g_1, \ldots, g_k), \mathfrak{c}, \mathfrak{d}) \in \underline{Z}_k$ always implies $(g_i, \mathfrak{o}_3^X, \mathfrak{o}_3^X) \in \underline{Z}_0$ for $i = 1, \ldots, k$ and if $\mathfrak{c} \neq \mathfrak{d}$ there exists $h \in \mathrm{dom}\xi$ with $\xi(h) = \mathfrak{d}$ and $(h, \mathfrak{o}_3^X, \mathfrak{o}_3^X) \in \underline{Z}_0$, and*
- *there is an X-interpretation $\chi \in B(X, G_0)$ such that $(u, \mathfrak{c}, \mathfrak{d}) \in \underline{Z}_0$ always implies $(\chi(u), \pi_0^X(\mathfrak{c}), \pi_0^X(\mathfrak{d})) \in \mathfrak{T}^{inst}(\mathbb{K}_0)$ and $((u_1, \ldots, u_k), \mathfrak{c}, \mathfrak{d}) \in \underline{Z}_k$ ($k = 1, 2, \ldots$) always implies $((\chi(u_1), \ldots, \chi(u_k)), \pi_k^X(\mathfrak{c}), \pi_k^X(\mathfrak{d})) \in \mathfrak{T}^{inst}(\mathbb{K}_k)$.*

The proof of Proposition 2 is analogous to the proof of Proposition 1. It is still an open problem to characterize more specifically the ordered set $\widetilde{\Gamma}(\vec{\mathbb{K}}; X)$ of the equivalence classes of existential concept graphs with subdivision of a triadic power context family over a given set of variables. For a comprehensive theory of concept graphs with subdivision, an instructive solution of this problem would be very desirable.

4 Appendix: Basics of Triadic Concept Analysis

Triadic Concept Analysis has been introduced as an extension of Formal Concept Analysis in [LW95] and [Wi95] (see also [Bi98]). It is mathematically based on the notion of a triadic context defined as a structure (G, M, B, Y) where G, M, and B are sets and Y is a ternary relation between G, M, and B, i.e. $Y \subseteq G \times M \times B$; the elements of G, M, and B are called (*formal*) *objects*, *attributes*, and *modalities*, respectively, and $(g, m, b) \in Y$ is read: the object g *has* the attribute m *in* the modality b.

A *triadic concept* of a triadic context (G, M, B, Y) is defined (analogously to a dyadic concept in Formal Concept Analysis) as a triple (A_1, A_2, A_3) with $A_1 \subseteq G$, $A_2 \subseteq M$, and $A_3 \subseteq B$ such that the triple (A_1, A_2, A_3) is maximal with respect to component-wise set inclusion in satisfying $A_1 \times A_2 \times A_3 \subseteq Y$. If (G, M, B, Y) is described by a three-dimensional cross table, this means that, under suitable permutations of rows, columns, and layers of the cross table, the triadic concept (A_1, A_2, A_3) is represented by a maximal rectangular box full of crosses. For a particular triadic concept $\mathfrak{c} := (A_1, A_2, A_3)$, the components A_1, A_2, and A_3 are called the *extent*, the *intent*, and the *modus* of $\mathfrak{c}$, respectively; they are denoted by $Ext(\mathfrak{c})$, $Int(\mathfrak{c})$, and $Mod(\mathfrak{c})$.

For the description of derivation operators, it is convenient to denote the underlying context alternatively by $\mathbb{K} := (K_1, K_2, K_3, Y)$. For $\{i, j, k\} = \{1, 2, 3\}$ with $j < k$ and for $X \subseteq K_i$ and $Z \subseteq K_j \times K_k$, the (i)-*derivation operators* are defined by

$$X \mapsto X^{(i)} := \{(a_j, a_k) \in K_j \times K_k \mid a_i, a_j, a_k \text{ are related by } Y \text{ for all } a_i \in X\},$$
$$Z \mapsto Z^{(i)} := \{a_i \in K_i \mid a_i, a_j, a_k \text{ are related by } Y \text{ for all } (a_j, a_k) \in Z\}.$$

It can be easily seen that a triple (A_1, A_2, A_3) with $A_i \subseteq K_i$ for $i = 1, 2, 3$ is a triadic concept of $\mathbb{K}$ if and only if $A_i = (A_j \times A_k)^{(i)}$ for $\{i, j, k\} = \{1, 2, 3\}$ with $j < k$. For constructing a triadic concept (A_1, A_2, A_3) out of two sets $X_i \subseteq K_i$ and $Z_k \subseteq K_k$, the following derivation operator is useful: $X_i \mapsto$

$$X_i^{(i,j,Z_k)} := \{a_j \in K_j \mid a_i, a_j, a_k \text{ are related by } Y \text{ for all } a_i \in X_i, a_k \in Z_k\}.$$

The construction produces first $A_j := X_i^{(i,j,Z_k)}$, then $A_i := A_j^{(j,i,Z_k)}$, and finally $A_k := (A_i \times A_j)^{(k)}$ if $i < j$, or $A_k := (A_j \times A_i)^{(k)}$ if $j < i$.

The set $\mathfrak{T}(\mathbb{K})$ of all triadic concepts of the triadic context $\mathbb{K} := (K_1, K_2, K_3)$ is structured by set inclusion considered in each of the three components of the triadic concepts. For each $i \in \{1, 2, 3\}$, one obtains a quasi-order $\lesssim_i$ and its corresponding equivalence relations $\sim_i$ defined by

$$(A_1, A_2, A_3) \lesssim_i (B_1, B_2, B_3) :\Longleftrightarrow A_i \subseteq B_i \quad \text{and}$$
$$(A_1, A_2, A_3) \sim_i (B_1, B_2, B_3) :\Longleftrightarrow A_i = B_i \quad (i = 1, 2, 3).$$

The relational structure $\underline{\mathfrak{T}}(\mathbb{K}) := (\mathfrak{T}(\mathbb{K}), \lesssim_1, \lesssim_2, \lesssim_3)$ is called the *concept trilattice* of the triadic context $\mathbb{K}$. The *ik-joins* $(i, k \in \{1, 2, 3\}$ with $i \neq k)$ are the trilattice operations ∇_{ik} on $\underline{\mathfrak{T}}(\mathbb{K})$ which are defined as follows: For a pair $(\mathfrak{X}_i, \mathfrak{Z}_k)$ of sets of triadic concepts and for $X_i := \bigcup\{A_i \mid (A_1, A_2, A_3) \in \mathfrak{X}_i\}$ and $Z_k := \bigcup\{A_k \mid (A_1, A_2, A_3) \in \mathfrak{Z}_k\}$,

$$\nabla_{ik}(\mathfrak{X}_i, \mathfrak{Z}_k) := (B_1, B_2, B_3)$$

with $B_j := X_i^{(i,j,Z_k)}$, $B_i := B_j^{(j,i,Z_k)}$, and $B_k := (B_i \times B_j)^{(k)}$ if $i < j$, or $B_k := (B_j \times B_i)^{(k)}$ if $j < i$. The algebraic theory of concept trilattices with the *ik*-operations ∇_{ik} is intensively studied in [Bi98].

References

[BG98]   O. Bätschmann, P. Griener: *Hans Holbein d. J. Die Darmstädter Madonna.* Fischer Taschenbuch Verlag, Frankfurt 1998.

[Bi98]   K. Biedermann: A foundation of the theory of trilattices. Dissertation, TU Darmstadt 1998. Shaker Verlag, Aachen 1998.

[Br94]   R. B. Brandom: *Making it explicit. Reasoning, representing, and discursive commitment.* Havard University Press, Cambridge 1994.

[DW00]   F. Dau, R. Wille: On the modal understanding of triadic contexts. In: R. Decker, W. Gaul (eds.): *Classification and information processing at the turn of the millennium.* Spinger, Heidelberg 2000, 83–94.

[GW99]   B. Ganter, R. Wille: *Formal Concept Analysis: mathematical foundations.* Springer, Heidelberg 1999; German version: Springer, Heidelberg 1996.

[GW00]   B. Groh, R. Wille: Lattices of triadic concept graphs. In: B. Ganter, G. W. Mineau (eds.): *Conceptual structures: logical, linguistic and computational issues.* LNAI **1867**. Springer, Heidelberg 2000, 332-341.

[LW95]   F. Lehmann, R. Wille: A triadic approach to formal concept analysis. In: G. Ellis, R. Levinson, W. Rich, J. F. Sowa (eds.): *Conceptual structures: applications, implementations, and theory.* LNAI **954**. Springer, Heidelberg 1995, 32–43.

[Wi95]   R. Wille. The basic theorem of triadic concept analysis. *Order* **12** (1995), 149–158.

[Wi98]   R. Wille: Triadic concept graphs. In: M. L. Mugnier, M. Chein (eds.): *Conceptual structures: theory, tools and applications.* LNAI **1453**. Springer, Heidelberg 1998, 194-208.

[Wi02]   R. Wille: Existential concept graphs of power context families. In: U. Priss, D. Corbett, G. Angelova (eds.): *Conceptual structures: integration and interfaces.* LNAI **2393**. Springer, Heidelberg 2002, 382–395.

[WZ00]   R. Wille, M. Zickwolff: Grundlagen einer triadischen Begriffsanalyse. In: G. Stumme, R. Wille (eds.): *Begriffliche Wissensverarbeitung: Methoden und Anwendungen.* Springer, Heidelberg 2000, 125–150.

A Generic Scheme for the Design
of Efficient On-Line Algorithms for Lattices

Petko Valtchev[1], Mohamed Rouane Hacene[1], and Rokia Missaoui[2]

[1] DIRO, Université de Montréal, C.P. 6128, Succ. "Centre-Ville",
Montréal, Québec, Canada, H3C 3J7
[2] Département d'informatique et d'ingénierie, UQO, C.P. 1250, succursale B
Gatineau, Québec, Canada, J8X 3X7

Abstract. A major issue with large dynamic datasets is the processing of small changes in the input through correspondingly small re-
arrangements of the output. This was the motivation behind the design of *incremental* or *on-line* algorithms for lattice maintenance, whose
work amounts to a gradual construction of the final lattice by repeatedly adding rows/columns to the data table. As an attempt to put the
incremental trend on strong theoretical grounds, we present a generic
algorithmic scheme that is based on a detailed analysis of the lattice
transformation triggered by a row/column addition and of the underlying sub-structure. For each task from the scheme we suggest an efficient
implementation strategy and put a lower bound on its worst-case complexity. Moreover, an instanciation of the incremental scheme is presented
which is as complex as the best batch algorithm.

1 Introduction

Formal concept analysis (FCA) [5] studies the lattice structures built on top
of binary relations (called concept lattices or *Galois* lattices as in [1]). As a
matter of fact, the underlying algorithmic techniques are increasingly used in the
resolution of practical problems from software engineering [6], data mining [7]
and information retrieval [3].

Our study investigates the new algorithmic problems related to the analysis of
volatile data sets. As a particular case, on-line or incremental lattice algorithms,
as described in [8, 3], basically maintain lattice structures upon the insertion of
a new row/column into the binary table. Thus, given a binary relation $\mathcal{K}$ and its
corresponding lattice $\mathcal{L}$, and a new row/column o, the lattice $\mathcal{L}^+$ corresponding
to the augmented relation $\mathcal{K}^+ = \mathcal{K} \cup \{o\}$ is computed. Most of the existing on-
line algorithms have been designed with practical concerns in mind, e.g., efficient
handling of large but sparse binary tables [8] and therefore prove inefficient
whenever data sets get denser [9].

Here, we explore the suborder of $\mathcal{L}^+$ made up of all new nodes with respect
to $\mathcal{L}$ and use an isomorphic suborder of $\mathcal{L}$ (the *generators* of the new nodes) that
works as a guideline for the completion of $\mathcal{L}$ to $\mathcal{L}^+$. Structural properties of the
latter suborder underly the design of a generic completion scheme, i.e., a sequence

A. de Moor, W. Lex, and B. Ganter (Eds.): ICCS 2003, LNAI 2746, pp. 282–295, 2003.

of steps that can be separately examined for efficient implementations. As a first offspring of the scheme, we describe a novel on-line algorithm that relies both on insights on the generator suborder and on some cardinality-based reasoning while bringing down the overall cost of lattice construction by subsequent completions to the current lower bound for batch construction.

The paper starts by recalling some basic FCA results (Section 2) and fundamentals of lattice construction (Section 3). The structure of the generator/new suborders in the initial/target lattice, respectively, is then examined (Section 4). Next, a generic scheme for lattice completion is sketched and for each task of the scheme implementation, directions are discussed (Section 5). Finally, the paper presents an effective algorithm for lattice maintenance and clarifies its worst-case complexity (Section 6).

2 Formal Concept Analysis Background

FCA [5] studies the partially ordered structure, known under the names of *Galois lattice* [1] or *concept lattice*, which is induced by a binary relation over a pair of sets O (*objects*) and A (*attributes*).

Definition 1. *A formal context is a triple $\mathcal{K} = (O, A, I)$ where O and A are sets and I is a binary (incidence) relation, i.e., $I \subseteq O \times A$.*

Within a context (see Figure 1 on the left), objects are denoted by numbers and attribute by small letters. Two functions, f and g, summarize the context-related links between objects and attributes.

Definition 2. *The function f maps a set of objects into the set of common attributes, whereas g^1 is the dual for attribute sets:*

- $f : \mathcal{P}(O) \to \mathcal{P}(A)$, $f(X) = X' = \{a \in A | \forall o \in X, oIa\}$
- $g : \mathcal{P}(A) \to \mathcal{P}(O)$, $g(Y) = Y' = \{o \in O | \forall a \in Y, oIa\}$

For example, $f(14) = fgh^2$. Furthermore, the compound operators $g \circ f(X)$ and $f \circ g(Y)$ are closure operators over $\mathcal{P}(O)$ and $\mathcal{P}(A)$ respectively. Thus, each of them induces a family of *closed* subsets, called $\mathcal{C}^o$ and $\mathcal{C}^a$ respectively, with f and g as bijective mappings between both families. A couple (X, Y), of mutually corresponding closed subsets is called a *(formal) concept*.

Definition 3. *A formal concept is a couple (X, Y) where $X \in \mathcal{P}(O)$, $Y \in \mathcal{P}(A)$, $X = Y'$ and $Y = X'$. X is called the extent and Y the intent of the concept.*

Thus, $(178, bcd)$ is a concept, but $(16, efh)$ is not. Moreover, the set $\mathcal{C}_\mathcal{K}$ of all concepts of the context $\mathcal{K} = (O, A, I)$ is partially ordered by intent/extent inclusion: $(X_1, Y_1) \leq_\mathcal{K} (X_2, Y_2) \Leftrightarrow X_1 \subseteq X_2 (Y_2 \subseteq Y_1)$.

---

[1] Hereafter, both f and g are denoted by $'$.
[2] We use a separator-free form for sets, e.g. 127 stands for $\{1, 2, 7\}$ and $g(abc) = 127$ w.r.t. the table $\mathcal{K}$ in figure 1, on the left, and ab for $\{a, b\}$.

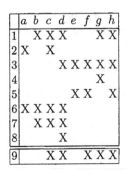

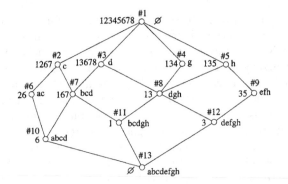

Fig. 1. Left: Binary table $\mathcal{K} = (O = \{1, 2, ..., 8\}, A = \{a, b, ..., h\}, R)$ and the object 9.
Right: The Hasse diagram of the lattice derived from $\mathcal{K}$.

Theorem 1. *The partial order* $\mathcal{L} = \langle \mathcal{C}_\mathcal{K}, \leq_\mathcal{K} \rangle$ *is a complete lattice with joins and meets as follows:*

- $\bigvee_{i=1}^{k}(X_i, Y_i) = ((\bigcup_{i=1}^{k} X_i)'', \bigcap_{i=1}^{k} Y_i),$
- $\bigwedge_{i=1}^{k}(X_i, Y_i) = (\bigcap_{i=1}^{k} X_i, (\bigcup_{i=1}^{k} Y_i)'').$

The Hasse diagram of the lattice $\mathcal{L}$ drawn from $K = (\{1, 2, ..., 8\}, A, R)$ is shown on the right-hand side of Figure 1 where intents and extents are drawn on both sides of a node representing a concept. For example, the join and the meet of $c_{\#6} = (26, ac)$ and $c_{\#3} = (13678, d)$ are $(12345678, \emptyset)$ and $(6, abcd)$ respectively.

3   Constructing the Lattice Efficiently

A variety of efficient algorithms exists for constructing the concept set or the entire concept lattice of a context (see [9] for a detailed study). As we are interested in incremental algorithms as opposed to batch ones, we consider those two groups separately.

3.1   Batch Approaches

The construction of a Galois lattice may be carried out with two different levels of output structuring. Indeed, one may look only for the set $\mathcal{C}$ of all concepts of a given context $\mathcal{K}$ without any hierarchical organization:

Problem COMPUTE-CONCEPTS

   Given : a context $\mathcal{K} = (O, A, I)$,
   Find : the set $\mathcal{C}$ of all concepts from $\mathcal{K}$.

An early FCA algorithm has been suggested by Ganter [4] based on a particular order among concepts that helps avoid computing a given concept more than once.

However, of greater interest to us are algorithms that not only discover $\mathcal{C}$, but also infer the lattice order $\leq$, i.e., construct the entire lattice $\mathcal{L}$. This more complex problem may be formalized as follows:

Problem COMPUTE-LATTICE

> **Given** : a context $\mathcal{K} = (O, A, I)$,
> **Find** : the lattice $\mathcal{L} = \langle \mathcal{C}, \leq \rangle$ corresponding to $\mathcal{K}$.

Batch algorithms for the COMPUTE-LATTICE problem have been proposed first by Bordat [2] and later on by Nourine and Raynaud [10]. The former algorithm relies on structural properties of the precedence relation in $\mathcal{L}$ to generate the concepts in an appropriate order. Thus, from each concept the algorithm generates its upper covers which means that a concept will be generated a number of times that corresponds to the number of its lower covers. Recently, Nourine and Raynaud suggested an efficient procedure for constructing a family of *open* sets and showed how it may be used to construct the lattice (see Section 5.4).

There is a known difficulty in estimating the complexity of lattice construction algorithms uniquely with respect to the size of the input data. Actually, there is no known bound (other than the trivial one, i.e., the number of all subsets of O or A) of the number of concepts depending on the dimensions of the binary relation, i.e., the size of the object set, of the attribute set, or of the binary relation. Even worse, it has been recently proven that the problem of estimating the size of $\mathcal{L}$ from $\mathcal{K}$ is #P-complete. For the above reasons, it is admitted to include the size of the result, i.e., the number of the concepts, in the complexity estimation. Thus, with $|\mathcal{L}|$ as a factor, the worst-case complexity expression of the classical algorithms solving COMPUTE-CONCEPT is $O((k + m)lkm)$, where $l = |\mathcal{L}|$, $k = |O|$, and $m = |A|$. The algorithm of Bordat can be assessed to be of complexity $O((k+m)l|I|)$ where the size of the binary relation (i.e., the number of positive entries in $\mathcal{K}$) is taken into account. Finally, the work of Nourine and Raynaud has helped reduce the complexity order of the problem to $O((k+m)lk)$.

3.2 Incremental Approaches

On-line or incremental algorithms do not actually construct the lattice, but rather maintain its integrity upon the insertion of a new object/attribute into the context:

Problem COMPUTE-LATTICE-INC

Given : a context $\mathcal{K} = (O, A, I)$ with its lattice $\mathcal{L}$ and an object o,
Find : the lattice $\mathcal{L}^+$ corresponding to $\mathcal{K}^+ = (O \cup \{o\}, A, I \cup \{o\} \times o')$.

Obviously, the problem COMPUTE-LATTICE may be polynomially reduced to COMPUTE-LATTICE-INC by iterating COMPUTE-LATTICE-INC on the entire set $O(A)$. In other words, an (extended) incremental method can construct the lattice $\mathcal{L}$ starting from a single object o_1 and gradually incorporating any new object o_i (on its arrival) into the lattice $\mathcal{L}_{i-1}$ (over a context $\mathcal{K} = (\{o_1, ..., o_{i-1}\}, A, I)$), each time carrying out a set of structural updates.

Godin et al. [8] suggested an incremental procedure which locally modifies the lattice structure (insertion of new concepts, completion of existing ones, deletion of redundant links, etc.) while keeping large parts of the lattice untouched. The basic approach follows a fundamental property of the *Galois connection* established by f and g on $(\mathcal{P}(O), \mathcal{P}(A))$: both families $\mathcal{C}^o$ and $\mathcal{C}^a$ are closed under intersection [1]. Thus, the whole insertion process is aimed at the integration into $\mathcal{L}_{i-1}$ of all concepts whose intents correspond to intersections of $\{o_i\}'$ with intents from $\mathcal{C}_{i-1}^a$, which are not themselves in $\mathcal{C}_{i-1}^a$. These additional concepts (further called *new* concepts in $\mathbf{N}^+(o)$), are inserted into the lattice at a particular place, i.e., each new concept is preceded by a specific counterpart from the initial lattice, called its *generator* (the set of generators is denoted $\mathbf{G}(o)$). Two other categories of concepts in $\mathcal{L} = \mathcal{L}_{i-1}$ are distinguished: *modified* ($\mathbf{M}(o)$) concepts correspond to intersections of $\{o_i\}'$ with members of $\mathcal{C}_{i-1}^a$ that already exist in $\mathcal{C}_{i-1}^a$, while the remaining set of concepts in the initial lattice are called *old* or *unchanged*. In the final lattice $\mathcal{L}^+ = \mathcal{L}_i$, the old concepts preserve all their characteristics, i.e., intent, extent as well as upper and lower covers. Generators do not experience changes in their information content, i.e., intent and extent, but a new concept is added to their upper covers. In a modified concept, the extent is augmented by the new object o while in the set of its lower covers, any generator is replaced by the corresponding new concept. In the next sections, we shall stick to this intuitive terminology, but we shall put it on a formal ground while distinguishing the sets of concepts in the initial lattice ($\mathbf{M}(o)$ and $\mathbf{G}(o)$) from their counterparts in the final one ($\mathbf{M}(o)^+$ and $\mathbf{G}(o)^+$, respectively).

Example 1 (Insertion of object 9). Assume $\mathcal{L}$ is the lattice induced by the object set 12345678 (see Figure 1 on the right) and consider 9 as the new object. The set of unchanged concepts has two elements, $\{c_{\#6}, c_{\#10}\}$, where as the set of modified and generators are $\mathbf{M}(o) = \{c_{\#1}, c_{\#2}, c_{\#3}, c_{\#4}, c_{\#5}, c_{\#8}\}$ and $\mathbf{G}(o) = \{c_{\#7}, c_{\#9}, c_{\#11}, c_{\#12}, c_{\#13}\}$ respectively. The result of the whole operation is the lattice $\mathcal{L}$ in Figure 2. Thus, the set of the new concept intents is: $\{cd, fh, cdgh, dfgh, cdfgh\}$.

Another incremental algorithm for lattice construction has been suggested by Carpineto and Romano [3].

In a recent paper [11], we generalized the incremental approach of Godin et al.. For this purpose, we applied some structural results from the lattice assembly framework defined in [14]. In particular, we showed that the incremental problem COMPUTE-LATTICE-INC is a special case of the more general lattice assembly problem ASSEMBLY-LATTICE. More recently, we have presented a theoretical framework that clarifies the restructuring involved in the resolution of

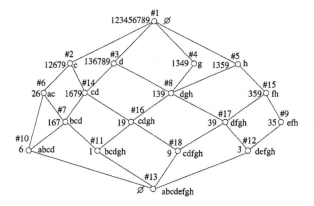

Fig. 2. The Hasse diagram of the concept (Galois) lattice derived from $\mathcal{K}$ with $O = \{1, 2, 3, ..., 9\}$.

COMPUTE-LATTICE-INC [13] and further enables the design of procedures that explore only a part of the lattice $\mathcal{L}$ (see Section 6).

In the next section, we recall the basic results from our framework.

4 Theoretical Foundations

For space limitation reasons, only key definitions and results that help the understanding of the more topical developments are provided in this section.

First, a set of mappings is given linking the lattices $\mathcal{L}$ and $\mathcal{L}^{+3}$. The mapping σ sends a concept from $\mathcal{L}$ to the concept in $\mathcal{L}^+$ with the same intent whereas γ works other way round, but respects extent preservation (modulo o). The mappings χ and χ^+ send a concept in $\mathcal{L}$ to the maximal element of its class $[]_{\mathcal{Q}}$ in $\mathcal{L}$ and $\mathcal{L}^+$, respectively.

Definition 1 *Assume the following mappings:*

- $\gamma : \mathcal{C}^+ \to \mathcal{C}$ *with* $\gamma(X, Y) = (X_1, X_1')$, *where* $X_1 = X - \{o\}$,
- $\sigma : \mathcal{C} \to \mathcal{C}^+$ *with* $\sigma(X, Y) = (Y', Y)$ *where* Y' *is computed in* $\mathcal{K}^+$,
- $\chi : \mathcal{C} \to \mathcal{C}$ *with* $\chi(X, Y) = (Y_1', Y_1'')$, *where* $Y_1 = Y \cap \{o\}'$,
- $\chi^+ : \mathcal{C} \to \mathcal{C}^+$ *with* $\chi^+(X, Y) = (Y_1', Y_1)$, *where* $Y_1 = Y \cap \{o\}'$ *(' over* $\mathcal{K}^+$*).*

The above mappings are depicted in Figure 3. Observe that σ is a *join-preserving* order embedding, whereas γ is a *meet-preserving* function with $\gamma \circ \sigma = id_{\mathcal{C}}$. Moreover, both mappings underly the necessary definitions (skipped here) for the sets $\mathbf{G}(o)$ and $\mathbf{M}(o)$ in $\mathcal{L}$ and their counterparts $\mathbf{G}^+(o)$ and $\mathbf{M}^+(o)$ in $\mathcal{L}^+$ to replace the intuitive descriptions we used so far.

A first key result states that $\mathbf{G}(o)$ and $\mathbf{M}(o)$ are exactly the maximal concepts in the equivalence classes induced by the function $\mathcal{Q} : \mathcal{C} \to 2^A$ defined as $\mathcal{Q}(c) =$

---

[3] In the following, the correspondence operator $'$ is computed in the respective context of the application co-domain (i.e. $\mathcal{K}$ or $\mathcal{K}^+$).

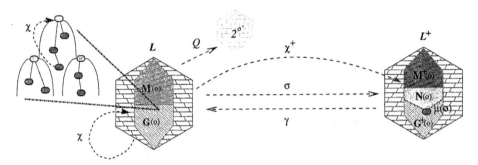

Fig. 3. The lattices $\mathcal{L}$, $\mathcal{L}^+$ and 2^A related by the mappings χ, χ^+, σ, γ and $\mathcal{Q}$.

$Y \cap \{o\}'$ where $c = (X, Y)$. Moreover, the suborder of $\mathcal{L}$ made up of $\mathbf{G}(o)$ and $\mathbf{M}(o)$ is isomorphic, via χ^+, to $\uparrow \nu(o)$, i.e., the prime filter of $\mathcal{L}^+$ generated by the minimal concept including o. Consequently, $(\mathbf{G}(o) \cup \mathbf{M}(o), \leq)$ is a meet-semi-lattice.

Finally, the precedence order in $\mathcal{L}^+$ evolves from the precedence in $\mathcal{L}$ as follows. Given a new concept c, its generator $\sigma(c)$ is a lower cover of c while the possible other lower covers of c ($Cov^l(c)$) lay in $\mathbf{N}^+(o)$. The upper covers of c are the concepts from $\mathbf{M}^+(o) \cup \mathbf{N}^+(o)$, that correspond, via σ, to the upper covers of the generator $\sigma(c)$ in the semi-lattice $(\mathbf{G}(o) \cup \mathbf{M}(o), \leq)$. The latter set may be extracted from the set of actual upper covers of $\sigma(c)$ in $\mathcal{L}$, $Cov^l(\sigma(c))$, by considering the maxima of their respective classes for $\mathcal{Q}$, i.e., the values of χ on $Cov^l(sigma(c))$, and keeping only the minimal values of those values. With a modified concept c in $\mathbf{M}^+(o)$, its lower covers in $\mathcal{L}^+$ differ from the lower covers of $\gamma(c)$ in $\mathcal{L}$ by (i) the (possible) inclusion of concepts from $\mathbf{N}^+(o)$, and (ii) the removal of all members of $\mathbf{G}^+(o)$. These facts are summarized as follows:

Property 1. *The relation $\prec^+$ is obtained from $\prec$ as follows:*

$$\prec^+ = \quad \{(\sigma(\gamma(c)), c) \mid c \in \mathbf{N}^+(o)\}$$
$$\cup \{(c, \bar{c}) \mid c \in \mathbf{N}^+(o), \ \bar{c} \in \text{Min}(\{\chi(\hat{c}) \mid \gamma(c) \prec \hat{c}\})\}$$
$$\cup \{(c_1, c_2) \mid (\gamma(c_1), \gamma(c_2)) \in (\prec - \mathbf{G}(o) \times \mathbf{M}(o))\}$$

5   A Generic Scheme for Incremental Lattice Construction

The structural results from the previous paragraphs underlie a generic procedure that, given an object o, transforms $\mathcal{L}$ into $\mathcal{L}^+$.

5.1   Principles of the Method

A generic procedure solving COMPUTE-LATTICE-INC may be sketched out of the following main tasks: (i) partition of the concepts in $\mathcal{L}$ into classes (by computing intent intersections), (ii) detection of maxima for every class $[]_{\mathcal{Q}}$ and test of its status, i.e., modified or generator, (*iii*) update of modified concepts, (*iv*) creation of new elements and computation of their intent and extent, (*v*) computation of lower and upper covers for each new element, and (*vi*) elimination of obsolete links for each generator. These tasks, when executed in the previously indicated order, complete a data structure representing the lattice $\mathcal{L}$ into a structure representing $\mathcal{L}^+$ as shown in Algorithm 1 hereafter.

1: **procedure** COMPUTE-LATTICE-INC(**In/Out:** $\mathcal{L} = \langle \mathcal{C}, \leq \rangle$ a lattice; **In:** o an object)
2:
3: **for all** c in $\mathcal{C}$ **do**
4:     Put c in its class in $\mathcal{L}_{/\mathcal{Q}}$ w.r.t. $\mathcal{Q}(c)$
5: **for all** $[]_{\mathcal{Q}}$ in $\mathcal{L}_{/\mathcal{Q}}$ **do**
6:     Find $c = \max([]_{\mathcal{Q}})$
7:     **if** $Intent(c) \subseteq o'$ **then**
8:         Put c in $\mathbf{M}(o)$
9:     **else**
10:         Put c in $\mathbf{G}(o)$
11: **for all** c in $\mathbf{M}(o)$ **do**
12:     $Extent(c) \leftarrow Extent(c) \cup \{o\}$
13: **for all** c in $\mathbf{G}(o)$ **do**
14:     $\hat{c} \leftarrow$ NEW-CONCEPT($Extent(c) \cup \{o\}', \mathcal{Q}(c)$)
15:     Put $\hat{c}$ in $\mathbf{N}(o)$
16: **for all** $\hat{c}$ in $\mathbf{N}(o)$ **do**
17:     Connect $\hat{c}$ as an upper cover of its generator c
18:     COMPUTE-UPPER-COVERS($\hat{c}, c$)
19: **for all** c in $\mathbf{G}(o)$ **do**
20:     **for all** $\bar{c}$ in $Cov^u(c) \cap \mathbf{M}(o)$ **do**
21:         Disconnect c and $\bar{c}$

Algorithm 1: Generic scheme for the insertion of a new object into a concept (Galois) lattice.

The above procedure is an algorithmic scheme that generalizes the existing incremental algorithms in the sense of specifying the full scope of the work to be done and the order of the tasks to be carried out. However, the exact way a particular algorithm might instantiate the scheme deserves a further clarification. On one hand, some of the tasks might remain implicit in a particular method. Thus, the task (*i*) is not explicitly described in most of the methods from the literature, except in some recent work on lattice-based association rule mining [13, 12]. However, all incremental methods do compute the values of the function $\mathcal{Q}$ for every concept in $\mathcal{L}$, as a preliminary step in the detection of class maxima. On the other hand, there is a large space for combining subtasks into larger steps, as major existing algorithms actually do. For example, the algorithms in [8, 3] perform all the sub-tasks simultaneously, whereas Algorithm 7 in [13] separates the problem into two stages: tasks (*i – iii*) are first carried out, followed by tasks $(iv-vi)$. In the next paragraphs, we discuss various realizations of the above subtasks.

5.2 Partitioning of C into Classes $[]_Q$

All incremental algorithms explore the lattice, most of the time in a top-down breadth-first traversal of the lattice graph. Classes are usually not directly manipulated. Instead, at each lattice node, the status of the corresponding concept within its class is considered. Classes are explicitly considered in the methods described in [13, 12], which, although designed for a simpler problem, i.e., update of $(C^a, \subseteq)$ and C^a, respectively, can be easily extended to first-class methods for COMPUTE-LATTICE-INC. Both methods apply advanced techniques in order to avoid the traversal of the entire lattice when looking for class maxima. The method in [13] skips the entire class induced by the empty intersection, i.e., $Q^{-1}(\emptyset)$. Except for small and very dense contexts where it can even be void, $Q^{-1}(\emptyset)$ is by far the largest class, and skipping it should result in substantial performance gains. An alternative strategy consists to explore class convexity (see Property 2 below) in order to only partially examine each class [12]. For this purpose, a bottom-up (partial) traversal of the lattice is implemented: whenever a non-maximal member of a class is examined, the method "jumps" straight to the maximum of that class.

5.3 Detection of Class Maxima

Top-down breadth-first traversal of the lattice eases the direct computation of each class maxima, i.e., without constructing the class explicitly. The whole traversal may be summarized as a gradual computation of the functions Q. Thus, it is enough to detect each concept c that produces a particular intersection $Int = Intent(c) \cap o'$, for the first time. For this task, the method of Godin *et al.* relies on a global memory for intersections that have already been met. This approach could be efficiently implemented with a trie structure which helps speed-up the lookups for a particular intersection (see Algorithms 3 and 4 in [13]). However, we suggest here another technique, based exclusively on locally available information about a lattice node. The technique takes advantage of the convexity of the classes $[]_Q$:

Property 2. *All classes $[]_Q$ in $\mathcal{L}$, are convex sets:*

$$\forall c, \bar{c}, \underline{c} \in C, \underline{c} \le c \le \bar{c} \text{ and } [\bar{c}]_Q = [\underline{c}]_Q \Rightarrow [\bar{c}]_Q = [c]_Q.$$

In short, for a non-maximal element c, there is always an upper cover of c, say $\bar{c}$, which is in $[c]_Q$. Thus, the status of c in $[c]_Q$ can be established by only looking at its upper covers. Moreover, as Q is a monotonous function ($c \le \bar{c}$ entails $Q(\bar{c}) \subseteq Q(c)$), one set inclusion can be tested on set sizes.

5.4   Computation of the Upper Covers of a New Concept

Given a generator c, "connecting" the new concept $\hat{c} = \chi^+(c)$ in the lattice requires the upper and lower covers of $\hat{c}$. A top-down breadth-first traversal of $\mathcal{L}$ allows the focus to be limited on upper covers while the work on lower covers is done for free. Moreover, at the time $\hat{c}$ is created, all its upper covers in $\mathcal{L}^+$ are already processed so they are available for lookup and link creation. In [8], a straightforward technique for upper cover computation is presented which amounts to looking for all successors of c that are not preceded by another successor. A more sophisticated technique as in [10] uses a property of the set difference between extents of two concepts (sometimes called the *face* between the concepts in the literature). The property states that a concept c precedes another concept $\bar{c}$ in the lattice, *iff* for any object $\bar{o}$ in the set difference $Extent(\bar{c}) - Extent(c)$, the closure of the set $\{\bar{o}\} \cup Extent(c)$ is $Extent(\bar{c})$:

Property 3. *For any* $c = (X, Y), \bar{c} = (\bar{X}, \bar{Y}) \in \mathcal{L}$, $c \prec \bar{c}$ *iff* $\bar{X} - X = \{\bar{o} \in O | (\{\bar{o}\} \cup X)'' = \bar{X}\}$.

This is easily checked through intersections of concept intents and a subsequent comparison of set cardinalities. To detect all upper covers of a concept $c = (X, Y)$, one needs to check the closures of $\{\bar{o}\} \cup X$ for every $\bar{o} \in O - X$ and select successors of c that satisfy the above property. This leads to a complexity of $k(k+m)$ per concept, where k comes from the factor $O - X$ and m is the cost of set-theoretic operations on intents.

To further cut the complexity of the task, we suggest a method that should at least improve the practical performances. It can be summarized as follows (see [14] for details). First, instead of considering *all* the potential successors of a new concept c, we select a subset of them, $Candidates = \{\chi^+(\bar{c}) \mid \bar{c} \in Cov^u(\gamma(c))\}$, i.e., the images by χ^+ of all upper covers of the generator $\gamma(c)$. $Candidates$ is a (not necessarily strict) subset of $\uparrow c - \{c\}$, whereby the convexity of the classes $[]_Q$ and the monotonicity of Q, insure the inclusion of all upper covers of $Cov^u(c) = \min(\uparrow c - \{c\})$ in the former set. Since the concepts in $Cov^u(c)$ coincide with the minima of $Candidates$, the former set can be computed through a direct application of a basic property of formal concepts stating that extent faces between c and the members of $Cov^u(c)$ are pairwise disjoint.

Property 4. *For any* $c = (X, Y) \in \mathcal{L}$, *and* $\bar{c}_1 = (\bar{X}_1, \bar{Y}_1), \bar{c}_2 = (\bar{X}_2, \bar{Y}_2) \in Cov^u(c)$, $\bar{X}_1 \cap \bar{X}_2 = X$.

For any $\hat{c} = (\hat{X}_1, \hat{Y}_1)$ from $Candidates - Cov^u(c)$ there is an upper cover $\bar{c} = (\bar{X}, \bar{Y})$ such that $\bar{c} \leq \hat{c}$ whence $\hat{X} \cap \bar{X} = \bar{X} \supseteq X$, where X is the extent of c. The elements of $Candidates - Cov^u(c)$ can therefore be filtered by a set of inclusion tests on $Candidates$. To do this efficiently and avoid testing of all possible couples, a buffer of attributes can be used to cumulate all the faces of valid upper covers of c that are met so far. Provided that candidates are listed in an order compatible with $\leq$ (so that smaller candidates are met before larger ones), a simple intersection with the buffer is enough to test whether a candidate is un upper cover or not. This above filtering strategy eliminates non-minimal candidates while also discarding copies of the same concept (as several upper covers of c may belong to the same class). Finally, the computation of χ^+ which is essential for the upward detection of class maxima is straightforward: while modified concepts in $\mathcal{L}$ take their own σ values for χ^+ (same intent), generators take the respective new concept, and unchanged concepts simply "inherit" the appropriate value from an upper cover that belongs to the same class $[]_\varrho$.

To assess the cost of the operation, one may observe that $|Cov^u(\gamma(c))|$ operations are needed, which is at most $d(\mathcal{L})$, i.e., the (outer) degree of the lattice taken as an oriented graph. Moreover, the operations of extent intersection and union, with ordered sets of objects in concept extents takes linear time in the size of the arguments, i.e., no more than $k = |O|$. Only a fixed number of such operations are executed per member of *Candidates*, so the total cost is in the order of $O(kd(\mathcal{L}))$. Although the complexity order remains comparable to $O(k^2)$, the factor $d(\mathcal{L})$ will be most of the time strictly smaller than k, and, in sparse datasets, the difference could be significant.

5.5   Obsolete Link Elimination

Any modified $\hat{c}$ which is an immediate successor of a generator $\bar{c}$ in $\mathcal{L}$ should be disconnected from $\bar{c}$ in $\mathcal{L}^+$ since $\chi^+(\hat{c})$ is necessarily an upper cover of the corresponding new element $c = \chi^+(\bar{c})$:

Property 5. *For any $\bar{c} \in \mathbf{G}(o)$, $\hat{c} \in \mathbf{M}(o) : \bar{c} \prec \hat{c} \Rightarrow \hat{c} \in \min(\{\chi^+(\hat{c}) \mid \hat{c} \in Cov^u(\bar{c})\})$.*

As the set $Cov^u(\bar{c})$ is required in the computation of $Cov^u(c)$, there is no additional cost in eliminating $\hat{c}$ from the list of the upper covers of $\bar{c}$. This is done during the computation of *Candidates*. Conversely, deleting $\bar{c}$ from the list of the lower covers of $\hat{c}$ (if such list is used), is done free of extra effort, i.e., by replacing $\bar{c}$ with $c = \chi^+(\bar{c})$.

6   An Efficient Instantiation of the Scheme

The algorithm takes a lattice and a new object[4] and outputs the updated lattice using the same data structure $\mathcal{L}$ to represent both the initial and the resulting lattices. The values of $\mathcal{Q}$ and χ^+ are supposed to be stored in a generic structure allowing indexing on concept identifiers (structure *ChiPlus*).

First, the concept set is sorted to a linear extension of the order $\leq$ required for the top-down traversal of $\mathcal{L}$ (primitive SORT on line 3). The overall loop (lines 4 to 20) examines every concept c in $\mathcal{L}$ and establishes its status in $[c]_\mathcal{Q}$ by comparing $|\mathcal{Q}(c)|$ to the maximal $|\mathcal{Q}(\bar{c})|$ where $\bar{c}$ is an upper cover of c (line 6). To this end, the variable *new-max* is used. Initialized with the upper cover maximizing $|\mathcal{Q}|$ (line 5), *new-max* eventually points to the concept in $\mathcal{L}^+$ whose intent equals $\mathcal{Q}(c)$, i.e., $\chi^+(c)$. Class maxima are further divided into modified

---

[4] The set A is assumed to be known from the beginning, i.e., $\{o\}' \subseteq A$.

and generators (line 7). A modified concept c (lines 8 to 10) has its extent updated. Then, such a c is set as its own value for χ^+, $\chi^+(c) = c$ (via *new-max*). Generators, first, give rise to a new concept (line 12). Then, the values of χ^+ for their upper covers are picked up (in the *Candidates* list, line 13) to be further filtered for minimal concepts (MIN-CLOSED, line 14). Minima are connected to the new concept and those of them which are modified in $\mathcal{L}$ are disconnected from the generator c (lines 15 to 17). Finally, the correct maximum of the class $[c]_\mathcal{Q}$ in $\mathcal{L}^+$, i.e., $\chi^+(c)$ is set (line 18) and the new concept is added to the lattice (line 19). At the end of the loop, the value of χ^+ is stored for further use.

```
 1: procedure ADD-OBJECT(In/Out: L = ⟨C, ≤⟩ a lattice; In: o an object)
 2:
 3:   SORT(C)
 4:   for all c in C do
 5:     new-max ← argmax({|Q(c̄)| | c̄ ∈ Covᵘ(c)})
 6:     if |Q(c)| ≠ |Q(new-max)| then
 7:       if |Q(c)| = |Intent(c)| then
 8:         Extent(c) ← Extent(c) ∪ {o}        {c is modified}
 9:         M(o) ← M(o) ∪ {c}
10:         new-max ← c
11:       else
12:         ĉ ← NEW-CONCEPT(Extent(c) ∪ {o}', Q(c))    {c is generator}
13:         Candidates ← {ChiPlus(c̄) | c̄ ∈ Covᵘ(c)}
14:         for all c̄ in MIN-CLOSED(Candidates) do
15:           NEW-LINK(ĉ, c̄)
16:           if c̄ ∈ M(o) then
17:             DROP-LINK(c, c̄)
18:         new-max ← ĉ
19:         L ← L ∪ {ĉ}
20:     ChiPlus(c) ← new-max
```

Algorithm 2: Insertion of a new object into a Galois lattice.

Example 2. Consider the same situation as in Example 1. The trace of the algorithm is given in the following table which provides the intent intersection and the χ^+ image for each concept. Concepts in $\mathcal{L}^+$ are underlined to avoid confusion with their counterparts in $\mathcal{L}$).

c	$Q(c)$	$\chi^+(c)$	Cat.	c	$Q(c)$	$\chi^+(c)$	Cat.	c	$Q(c)$	$\chi^+(c)$	Cat.
$c_{\#1}$	$\emptyset$	$\underline{c}_{\#1}$	mod.	$c_{\#2}$	c	$\underline{c}_{\#2}$	mod.	$c_{\#3}$	d	$\underline{c}_{\#3}$	mod.
$c_{\#4}$	g	$\underline{c}_{\#4}$	mod.	$c_{\#5}$	h	$\underline{c}_{\#5}$	mod.	$c_{\#6}$	c	$\underline{c}_{\#2}$	old
$c_{\#7}$	cd	$\underline{c}_{\#14}$	gen.	$c_{\#8}$	dgh	$\underline{c}_{\#8}$	mod.	$c_{\#9}$	fh	$\underline{c}_{\#15}$	gen.
$c_{\#10}$	cd	$\underline{c}_{\#14}$	old	$c_{\#11}$	$cdgh$	$\underline{c}_{\#16}$	gen.	$c_{\#12}$	$dfgh$	$\underline{c}_{\#17}$	gen.
$c_{\#13}$	$cdfgh$	$\underline{c}_{\#18}$	gen.								

To illustrate the way our algorithm proceeds, consider the processing of concept $c_{\#12} = (3, defgh)$. The value of $Q(c_{\#12})$ is $dfgh$ whereas *Candidates* contains the images by χ^+ of the upper covers of $c_{\#12}$, i.e., $c_{\#8}$ and $c_{\#9}$: *Candidates*= $\{\underline{c}_{\#8} = (139, dgh), \underline{c}_{\#15} = (359, fh)\}$. Obviously, neither of the intents is as big as $\mathcal{Q}(c_{\#12})$, so $c_{\#12}$ is a maximum, more precisely a generator. The new concept, $\underline{c}_{\#17}$ is $(39, dfgh)$ and its upper covers are both concepts in *Candidates* (since both are incomparable). Finally, as $\underline{c}_{\#8}$ is in $\mathbf{M}(o)$, its link to $c_{\#12}$ is removed.

6.1   Complexity Issues

Let $\Delta(l) = |\mathcal{C}^+| - |\mathcal{C}|$ and let us split the cost of a single object addition into two factors: the cost of the traversal of $\mathcal{L}$ (lines $3 - 7$ and 20 of Algorithm 2) and the cost of the restructuring of $\mathcal{L}$, i.e., the processing of class maxima (lines $8 - 19$). First, as sorting concepts to a linear extension of $\leq$ only requires comparison of intent sizes, which are bound by m, it can be done in $O(l)$. Moreover, the proper traversal takes $O(l)$ concept examinations which are all in $O(k + m)$. Thus, the first factor is in $O(l(k + m))$. The second factor is further split into modified and generator costs whereby the first cost is linear in the size of $\mathbf{M}(o)$ (since lines $8 - 10$ may be executed in constant time even with sorted extents) and therefore could be ignored. The generator-related cost has a factor $\Delta(l)$ whereas the remaining factor is the cost of creating and properly connecting a single new concept. The dominant component of the latter is the cost of the lattice order update (lines $14 - 17$) which is in $O(k^2)$ as we mentioned earlier. Consequently, the global restructuring overhead is in $O(\Delta(l)k^2)$. This leads to a worst case complexity of $O(\Delta(l)k^2 + l(k+m))$ for a single insertion, which is a lower bound for the complexity of COMPUTE-LATTICE-INC (see also [11]).

The assessment of the entire lattice construction via incremental updates is delicate since it requires summing on all k insertions whereas the cost of steps 1 to $k - 1$ depends on parameters of the intermediate structures. Once again, we sum on the above high-level complexity factors separately. Thus, the total cost of the k lattice traversals is bound by k times the cost of the most expensive traversal (the last one), i.e., it is in $O(kl(k + m))$. The total cost of lattice restructuring is in turn bound by the number of all new concepts (the sum of $\Delta(l_i)$) times the maximal cost of a new concept processing. The first factor is exactly $l = |\mathcal{C}^+|$ since each concept in the final lattice is created exactly once which means the restructuring factor of the construction is in $O(l(k + m)k)$, thus leading to a global complexity in the same class $O(l(k + m)k)$. The above figures indicate that the complexity of COMPUTE-LATTICE, whenever reduced to a series of COMPUTE-LATTICE-INC, remains in the same class as the best known lower bound for batch methods [10].

7   Conclusion

The present study is motivated by the need for both efficient and theoretically-grounded algorithms for incremental lattice construction. In this paper, we complete our own characterization of the substructure that should be integrated into the initial lattice upon each insertion of an object/attribute into the context.

Moreover, we show how the relevant structural properties support the design of an effective maintenance methods which, unlike previous algorithms, avoid redundant computations. As guidelines for such design, we provide a generic algorithmic scheme that states the limits of the minimal work that needs to be done in the restructuring. A concrete method that instantiates the scheme is proposed whose worst-case complexity is $O(ml + \Delta(l)k^2)$, i.e., a function which puts a new and smaller upper bound for the cost of the problem COMPUTE-LATTICE-INC. Surprisingly enough, when applied as a batch method for lattice construction, the new algorithm shows the best known theoretical complexity, $O((k + m)lk)$, which is only achieved by one algorithm. As a next stage of our study, we are currently examining the pragmatic benefits of the scheme, i.e., the practical performances of specific scheme instantiations.

References

[1] M. Barbut and B. Monjardet. *Ordre et Classification: Algèbre et combinatoire.* Hachette, 1970.

[2] J.-P. Bordat. Calcul pratique du treillis de Galois d'une correspondance. *Mathématiques et Sciences Humaines*, 96:31–47, 1986.

[3] C. Carpineto and G. Romano. A Lattice Conceptual Clustering System and Its Application to Browsing Retrieval. *Machine Learning*, 24(2):95–122, 1996.

[4] B. Ganter. Two basic algorithms in concept analysis. preprint 831, Technische Hochschule, Darmstadt, 1984.

[5] B. Ganter and R. Wille. *Formal Concept Analysis, Mathematical Foundations.* Springer-Verlag, 1999.

[6] R. Godin and H. Mili. Building and maintaining analysis-level class hierarchies using Galois lattices. In *Proceedings of OOPSLA'93, Washington (DC), USA*, special issue of ACM SIGPLAN Notices, 28(10), pages 394–410, 1993.

[7] R. Godin and R. Missaoui. An Incremental Concept Formation Approach for Learning from Databases. *Theoretical Computer Science*, 133:378–419, 1994.

[8] R. Godin, R. Missaoui, and H. Alaoui. Incremental concept formation algorithms based on galois (concept) lattices. *Computational Intelligence*, 11(2):246–267, 1995.

[9] S. Kuznetsov and S. Ob'edkov. Algorithms for the Construction of the Set of All Concept and Their Line Diagram. preprint MATH-AL-05-2000, Technische Universität, Dresden, June 2000.

[10] L. Nourine and O. Raynaud. A Fast Algorithm for Building Lattices. *Information Processing Letters*, 71:199–204, 1999.

[11] P. Valtchev and R. Missaoui. Building concept (Galois) lattices from parts: generalizing the incremental methods. In H. Delugach and G. Stumme, editors, *Proceedings, ICCS-01*, volume 2120 of *Lecture Notes in Computer Science*, pages 290–303, Stanford (CA), USA, 2001. Springer-Verlag.

[12] P. Valtchev and R. Missaoui. A Framework for Incremental Generation of Frequent Closed Itemsets. *Discrete Applied Mathematics*, submitted.

[13] P. Valtchev, R. Missaoui, R. Godin, and M. Meridji. Generating Frequent Itemsets Incrementally: Two Novel Approaches Based On Galois Lattice Theory. *Journal of Experimental & Theoretical Artificial Intelligence*, 14(2-3):115–142, 2002.

[14] P. Valtchev, R. Missaoui, and P. Lebrun. A partition-based approach towards building Galois (concept) lattices. *Discrete Mathematics*, 256(3):801–829, 2002.

Towards Building Active Knowledge Systems with Conceptual Graphs

Harry S. Delugach

Computer Science Dept., Univ. of Alabama in Huntsville,
Huntsville, AL 35899 U.S.A.
delugach@cs.uah.edu

Abstract. This paper outlines a vision for using conceptual graphs to build active knowledge systems that have the capability to solve practical and complex problems. A key ingredient in an active knowledge system is its ability to interact (not just interface) with the real world. Basic features of such systems go beyond logic to include support for data mining, intelligent agents, temporal actors, active sensors, a system for knowledge interchange and finally, support for knowledge-in-the-large.

1 Introduction

I am a system builder. I am interested in solving real-world problems through the use of automated knowledge-based systems. This paper outlines a framework for building systems *active knowledge systems*: systems that have the capability to solve practical and complex problems. A key ingredient in an active knowledge system is its ability to inter*act* (not just inter*face*) with the real (and changeable) world. If a system can do that, it becomes more of an intelligent system, whose full potential we are just beginning to understand. This paper is intended to encourage and promote effective techniques for building these systems.

It is my claim that conceptual graphs have the potential – much of it unexplored – to support active knowledge systems that will explicitly model processes, and interact with the real world. The power of these capabilities becomes apparent when serious developers begin building large-scale and robust systems for general use. Several systems based on conceptual graphs (CGs) have recently become available e.g., PROLOG+CG [8], Notio [14], WebKB [10], CoGITaNT [6] and CharGer [4] [5]. Researchers are beginning to formulate practical guidelines as to how such systems should be designed and implemented [9]. In addition, interest in the Semantic Web [2] and support for distributed communities [12] [13] make it essential to find ways for systems to involve themselves in the real world.

While these existing conceptual graph systems each have their own clear strengths, I have found that these systems lack some features that give a system rich semantics. Because such systems generally support the declarative core of conceptual graphs, they aim to represent knowledge mostly about *things*; only a few (e.g., [12], [5]) are interested in semantic modeling of *procedures* and *activities*. The CG core, based on first-order predicate calculus, can certainly support useful results, especially within a closed, monotonic (i.e., unchanging) environment; however, to be more generally useful, a system must have additional features that lie outside of the core.

A. de Moor, W. Lex, and B. Ganter (Eds.): ICCS 2003, LNAI 2746, pp. 296–308, 2003.
© Springer-Verlag Berlin Heidelberg 2003

My interest in building useful systems places me at the juxtaposition of several areas of research. I am not a philosopher who studies the limits of interpretability, yet I require formal ways to interpret the results of the systems I want to build. I am not a theorist who studies the soundness or completeness of conceptual graph systems, yet I require a system to be sound and complete with respect to some portion of the real world I am modeling. It is these struggles (and others) which have convinced me to propose something more.

2 Practical Active Knowledge Systems

In this paper, I propose using conceptual graphs to support a full-fledged active knowledge system. Many of the features described here can, of course, be employed with other approaches besides conceptual graphs; they are gathered here in order to suggest the capabilities that a system ought to have. The notion of an active knowledge system is not new – a number of researchers have been exploring agents in distributed systems, most notably the Semantic Web [2] and the DAML effort (www.daml.org).

In this paper, I argue that useful knowledge (e.g., in the sense of the Semantic Web [2] or Interspace [13]) is coupled implicitly with activity or procedures that are rooted in the real-world. Every piece of knowledge encoded in a formal system must reflect some actual knowledge that real people have, if the system is to be practical.

I use the term *practical system* to mean a knowledge-based system that has the following properties. Though these descriptions are somewhat subjective, I propose them as a starting point for discussion.

A practical system is a system that:

- Addresses one or more needs that can be articulated in the real world,
- Produces results that are meaningful and useful to people in the real world in addressing their needs,
- Can be easily (re-)configured to provide meaningful and useful results in various domains,
- Can incorporate and accommodate new information,
- Can reason about processes and activities, as well as their steps and goals,
- Can recognize nonsense,
- Has the capability to explain how it operates, including where its knowledge comes from, and how it reached its conclusions,
- Is scalable to human-sized problems.

It would be a great leap forward for us to identify a single feature or technique that would unify all these characteristics, but no single feature or technique is going to suddenly make a system useful and intelligent. A collection of useful features and techniques can be very powerful, however, which is why this paper proposes a unifying framework to support a collection of useful features. Conceptual graphs, because of their demonstrated semantic modeling capabilities [15], [16] and their intrinsic ability to support reasoning about semantics [3] are a reasonable candidate for a representation.

The rest of this paper describes a few of the issues to consider in building practical systems.

2.1 Logic Support

The basis for most of our current systems is first-order logic. This is essential for a workable system, as some have argued for many years, starting with [15]; there is widespread agreement that logic is an important foundation for any knowledge-based system. For symbolic reasoning, we need logic, and if logical formulae are to be passed among different systems, then we must have a complete, consistent and practical standard for the interchange of logical expressions. One important point is that practicality is not a feature to be added to a standard after it is established; a standard should already include those best parts of a practical system and attempt to generalize them. As an ISO working group [1] considers a common logic standard, we may finally have a standard with which system builders can exchange knowledge between systems, as well as gauge their systems' logical consistency and completeness. I support current efforts to work out the details so that a standard can be agreed upon.

While nearly everyone agrees that logic is necessary, there is some disagreement as to whether it is sufficient. I believe that intelligence is composed of many elements, only one of which is logic. I further argue in this paper that, with logic as a basis, additional capabilities can support useful and practical intelligent systems based on conceptual graphs.

2.2 Support for Denotational Semantics

A logic system provides predicates and functions, which allow the representation of relationships between elements in a system. The rules of logic ensure that the truth of these predicates is preserved; e.g., if I assert that **dog(Spot)** is true and then perform any number of logical transformations, a (consistent) system should never derive the expression **not dog(Spot)**. This is a necessary property of any system that is meant to preserve truth. Most knowledge-based systems therefore operate as in Fig. 1. Elements in the model represent predicates, functions, etc.

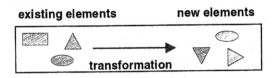

Fig. 1. Knowledge Based System Model.

There is a larger question that often goes unanswered (at least in a formal way): what do the predicates mean? Logicians remind us that a predicate is assumed to be a primitive relation between symbols; a logical system's only obligation is to preserve those relationships throughout its logical operations. The symbols themselves, however, are arbitrary (at least as far as the logic operations are concerned). This presents a difficulty for a system builder, whose symbols are not arbitrary at all! Symbols used in a knowledge base stand for something, usually in the real world.

Figure 2 depicts a model as it relates to the real world. Things in the real world are represented in a knowledge-based system (KBS) through a mapping **E** (for "encoding") that allows their representations to be transformed within the system. Once transformed, new representations can arise within the system. These new representations' symbols can then be mapped back to the real world through a mapping **D** (for "decoding") that captures their interpretation. This model is intentionally simplistic, ignoring some important philosophical issues, in order to provide a basis for the following discussion.

It is crucial for a knowledge system to preserve truth internally, and of course the system builder requires the same of any system they build. Furthermore, the system builder needs for the system to preserve the symbols' meanings as they exist in the real world. The mapping D in Fig. 2 represents this need.

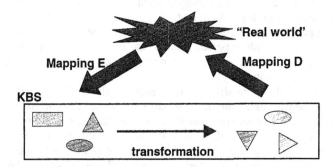

Fig. 2. Knowledge Based System Model's Relationship To Reality.

To be fair, logicians freely acknowledge this limitation, reminding us that the power of logic is that symbols and predicates can come from anywhere. No matter where the symbols come from, our logic system will preserve them and their relationships through a series of well-founded transformation steps. Of course, this is an appropriate use of logic and a clearly powerful feature for a knowledge system. One of my purposes in this paper is to describe clearly what are the limitations of logic and what additional features a practical knowledge system needs.

Figure 3 adds an essential characteristic of any useful knowledge model – namely that the real mapping to/from symbols depends upon a human's interpretation. This dependency was clearly articulated over a century ago by Charles S. Peirce (e.g., see [17, Lecture Two]) and reinforced by others. Indeed, much of the difficulty in

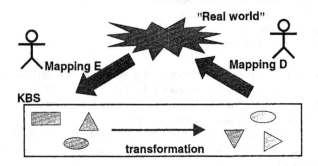

Fig. 3. Human Role In Interpretation.

building effective knowledge systems stems from our inability to formalize these mappings. In any knowledge-based system (whether formal or informal), there exist boundaries where formality can no longer apply. Even for a formal system, these boundaries constitute formality's "end-of-the-world" beyond which lies the *terra incognita* of interpretations.

We often overlook (or take for granted) this significant boundary. We have become so accustomed to using symbols for the actual things they represent that we forget the symbols themselves have no inherent meaning. English speakers see **cat** and immediately envision a small fluffy animal with a tail that is often treated as a human pet. One way to remind ourselves of this boundary is to pick up any document written in a foreign language that we do not know. When we see a sentence in such a language, we not only do not know what the symbols represent, we do not even know what is being written about! That is why foreign travelers spend a lot of time looking at the pictures accompanying written text.

I do not propose that we will ever be able to formalize interpretations from inside our models all the way into a person's mind; the best we can attain is the ability to formalize enough of the interpretation process so that we can share them among multiple persons and multiple systems.

One step toward that formalization is already in place. Conceptual graph theory provides a mechanism for getting at least part way toward this interpretation through *individual markers*. Markers are identifiers for real things, so that we can establish some facts about individuals in the real world, most notably what type they are. This part of conceptual graph theory therefore does attempt to show a formal model of the real world (albeit a shallow one) as follows: *the world consists of individuals, each of which can be identified using a unique individual marker.* Essentially this is a naming approach, whereby every thing is given a name. A system therefore contains an associative lookup between unique identifiers and particular things.

Of course this approach does not solve the problem of what symbols stand for; it merely defers the question (or restates it): namely, how do we construct the map that associates individual markers with actual real-world individuals? It also does not directly address the question of how to relate things (such as aggregates) that may not be denoted by a single typed concept. Constructing the map constitutes a *procedure*; it is not a purely logical or structural feature.

The real-world things to which a model refers are called its *denotation*. I will use that notion of how a knowledge system relates to the real world by describing a set of features to support denotational semantics. Denotational semantics can be supported by several technologies, such as:

- Data mining with augmented semantics
- Intelligent agent operations
- Temporal reasoning
- Sensor operation and interpretation

Figure 4 shows the "heaven" of a conceptual graph sheet of assertion, with the real world below it. Within the plane of the assertion sheet, the rules of logic hold; outside of the plane of assertion are found links to the real world. In a practical system, there must be actuators to cause information to flow along these links. The links shown in Fig. 4 are not meant as a definitive structure. They are meant to illustrate some examples of the range in behavior we would expect from a practical system that is linked to the real world.

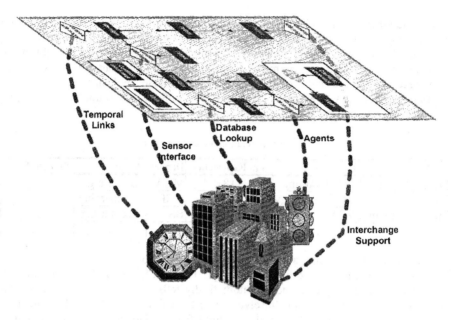

Fig. 4. Active Knowledge System Framework.

The sheet of assertion contains our knowledge based system model, with appropriate concepts, relations, contexts and (most importantly) actors. These actors provide "hooks" to the real world, giving our system its eyes, ears, fingers, and feet (and even a sense of "smell"!). Different kinds of actors provide different kinds of senses and connection to the world; they are briefly described in the following sections.

Within the sheet of assertion, traditional rules of logic can apply; however, since actors will necessarily react to a changing world, the contents of the sheet will also change. Though it is possible to build a system without actors (and actually solve some problems with it), such a system would only be able to model those parts of the world that do not change.

2.3 Data Mining with Augmented Semantics

There is much useful information currently stored around the world in conventional databases. A key ingredient in a practical system is its ability to acquire information from these databases. While a database itself possesses insufficient semantics for real knowledge based reasoning, there is much to be gained from the data if we use a semantically driven data mining process. A knowledge engineer can model the database's semantics by building conceptual graph models in a knowledge system, while populating the models with the actual identities of concepts that are retrieved from a conventional (relational) database as needed.

Here is a simple illustration. Suppose a database contains company records and has an employee relation as in Table 1. Because we humans can read and understand the column (field) labels, and because the first few entries' values seem to match our understanding, we tend to think that the relation table has meaning, but it really has very little inherent semantics. For example, each row appears to capture information about a different person, each of whom is employed by the organization that is being modeled by the database. As another example, "Yrs Experience" might mean experience in business environments, or experience in that particular position, or experience at that particular organization. And of course, replacing the (apparently meaningful) names with arbitrary identifiers like "bq598" would not alter the formal characteristics of database accesses at all.

Table 1. A relational database table.

Name	Position	Yrs Experience	Degree	Major	Percent Stock
Karen Jones	VP Marketing	18	MBA	Marketing	3
Kevin Smith	VP Technology	12	MSE	Engineering	4
Keith Williams	VP Finance	15	BS	Accounting	3
...	...	...	...	...	...

We can easily construct a conceptual graph with actors that will retrieve values from the database and insert them into a graph; in fact, CharGer already does this [4]. The fact that it can be constructed automatically tells us that there is no particular meaning in either the values themselves, or in the column labels (i.e., field names). Figure 5 shows how data values are acquired from a database and inserted into a knowledge base. The **<lookup>** actor is a procedural element, which performs the association between attribute names and a lookup key.

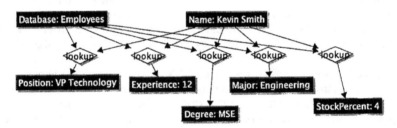

Fig. 5. Database Information Without Semantics.

We have shown how limited are the semantics the database possesses by itself. In order to show semantics, we want to know what all those values mean. A richer representation would be found in Fig. 6 which shows the conceptual relationships (beyond the "relations" in a conventional database) that exist in the real world between the values and a particular employee. Now we are getting closer to what an intelligent system needs to operate.

Additional semantics can be provided to Fig. 6 by adding additional graph elements and by including a type or relation hierarchy.

2.4 Intelligent Agent Operations

Researchers are already working on conceptual graph models of intelligent agents (e.g., [11] [7]). An intelligent agent is an autonomous entity, capable of controlling its

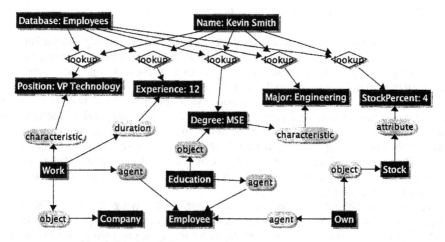

Fig. 6. Database Information with Augmented Semantics.

own communication, deciding which other agents should be contacted, and evaluating the quality of the knowledge gained. Agents are well suited for supporting a knowledge system, particularly if they can use the knowledge system's semantics to formulate queries and communicate with other agents.

One might argue that conceptual graph actors are already intelligent agents, since they interact with the outside world. Unfortunately, an actor is too limited (in general): its interactions with the outside are carefully pre-planned and constrained. The **<lookup>** actor in the previous section is a good example. **<lookup>** is "hard-wired" to a specific database by the graph itself and has a very limited function (e.g., it is not clear what the **<lookup>** actor should do if it cannot find a value to match its lookup key).

An intelligent agent is an autonomous entity that can interact with other knowledge based systems and/or their agents, making decisions along the way. A knowledge-based system will need to create new agents that are sent off into the real work to collect information that will be sent back to its creator. How would this work?

The active knowledge system first creates an intelligent agent, and may start it off with a "briefcase" of knowledge to work with. The knowledge system may also create an interface (or use a previously created one) through which the agent can communicate with the system. At a minimum, an agent must be able to report back its results. It should also be able to report back preliminary findings and have its search refined via interaction with the knowledge base. Of course, the agent could be supplied with all the knowledge it needs beforehand, but an adaptive agent cannot know all the content it needs in advance.

An agent is much more "intelligent" than an actor, which might merely be a functional relation. Whereas an actor takes a set of referents and returns one or more new referents, an agent has the ability to evaluate a set of graphs and return one or more graphs, as well as having the capability to alter existing graphs in the set it is given.

2.5   Temporal Reasoning (What Time Is It?)

By temporal support, I mean access to clocks and other time-keeping mechanisms so that the system will maintain some notion of what time it is in the real world. Just as most computer operating systems have access to a wall-clock time, so should the knowledge system have access to the world's actual time.

The notion of real-world time need not always mean the local wall clock. One advantage to having temporal support is that we can use it for models operating in time frames other than the one we all share in the real world. For example, a simulation can be supported by a knowledge system that operates on "simulator" time. Past scenarios or event records can be re-played using time values corresponding to when things actually happened. Demographic or geological models can use a much coarser granularity of time.

Temporal reasoning encompasses more than just having a clock. Temporal concepts include reasoning about duration, intervals, and relationships between them. Unlike pure logic, which does not care the order in which expressions are evaluated, real world events often have constraints on which ones must precede others. An active knowledge system can use its temporal reasoning abilities to help solve problems dealing with processes and time.

2.6   Sensor Operation and Interpretation

Regardless of how sophisticated we make our models, they remain closed off from things in the real world. A closed system relies completely on the decisions of people and other systems for any knowledge it acquires. These external decisions restrict a system's potential to autonomously gather new information.

I consider here a general notion of a sensor: the eyes and ears (fingers, toes, etc.) of a system which can continually update locations, states, and processes that can only be statically represented by concepts in the representation.

The temporal support in the previous section is a simple example of a sensor that merely reads a clock. Other sensors can be attached to entities in the real world so that we may get a "reading" on them – their location, state, etc. In this way, some concepts' referent values (or the concepts themselves) become changeable, reflecting the real world, which the model represents.

Sensors are only placed on those entities and attributes that we choose to track. These choices already constrain a system's ability to effectively use the sensor's information, especially if the structure of the environment changes. This is a persistent problem in current (passive) monitoring systems. For example, there is a feature in human beings (*selective attention*), whereby we are able to focus our senses more sharply on some parts of our surroundings and reduce the focus on others. This is not just a matter of efficiency in applying the resources of our sensory apparatus; it is also a matter of selecting priorities about what is important to see. Active knowledge systems need support for a kind of "selective attention" in the same way.

3 Other Practical Considerations

There are some other considerations that are not necessarily part of a system's interaction with the outside world, but are nonetheless essential to a practical usable system. These considerations are:

- Interchange mechanisms and standards
- Support for knowledge-in-the-large
- Explanation facilities

3.1 Interchange Mechanisms and Standards

Since constructing a knowledge model is both labor-intensive and time-consuming, we would like to re-use our models on other projects and also share our models with other systems. There are many issues involved in both re-use and sharing that must be addressed, or else each developer will find himself building little isolated solar systems of knowledge instead of contributing to a galaxy of inter-related knowledge.

One key issue is that of an agreed-upon interchange format (see sec. 2.1). It is obviously inefficient for each pair of system developers to get together and create a one-time interchange scheme for their systems. It seems clear that formal standards are the best solution to the interchange problem, and such standards are being developed. A standard does not solve all the problems associated with exchanging information; it merely makes a solution possible.

A larger problem when hooking systems to the real world is that when exchanging knowledge, the interpretive mapping from symbol to actual thing may be lost. One solution is to specify some procedure that performs the mapping, and have the capability to exchange such procedures as well. This means that whenever a particular symbol appears in the model, there is an associated procedure that can be invoked to get values for a set of attributes associated with that symbol.

3.2 Support for Knowledge-in-the-Large

The previous sections dealt with specific technologies that provide a range of different capabilities for an active system. Regardless of the capability being considered, however each such feature must be practical for very large models that may be tightly coupled to the real world through actors. All of the technologies mentioned here must be scalable if a system is to solve practical problems.

Knowledge-in-the-large does not just refer to the sheer number of things. If we simply had to perform linear searches through long lists of things, there are well-known ways to process these lists efficiently. The real problems in scalability fall into two categories:

- Computational complexity: looking for partial matches, mining for new relationships, etc.
- Concurrency complexity: large numbers of agents all operating simultaneously on large numbers of graphs

- Conceptual complexity: the explanation of a result may turn out to be difficult or impossible for a person to understand, regardless of how reliable is the process by which the results are obtained

The first of these problems has received considerable attention in conceptual graphs, most notably the efforts of Levinson [9]. The second problem is only now beginning to be understood in terms of what is expected of agents; their computational aspects are not yet being addressed. The third problem will be discussed in a moment.

3.3 Explanation Facilities

A practical system must have a way of explaining itself to its human users. It must be able to explain its own internal knowledge, its lines of reasoning in answering queries, and even its own internal operations. Most current systems require an in-depth explanation as to what they do; these explanations often must come from the programmers themselves or from a detailed examination of the design or source code, since the intelligent system itself does not have access to its own functioning.

In order for a system to explain itself, it must have direct and explicit access to all of its internal inference and reasoning procedures. "Hard wiring" these procedures (usually through program code) has the effect of removing them from the actual knowledge model; yet, these reasoning mechanisms are what gives the system its power!

3.4 Limitations on Our Understanding

The problem of conceptual complexity is an interesting one whose issues are ultimately influenced by the limits of our ability as human beings to understand our world. In our daily lives, our economic lives and our political decisions, we encounter situations where our intuition can lead us to an understanding that is counter to what logic or actual observation tells us. Though automated reasoning machines may be able to gain our confidence and trust, we still may not really be able to comprehend what they tell us. This makes it even more important for us to design and build knowledge-based systems with great care, since they may one day surpass our ability to understand them.

4 Conclusion

Can this become real? The support features described here may seem ambitious – can we really build them all? And do we really need them all? To be sure, they may go beyond what any particular system builder would need. Instead I am proposing that architectures and algorithms be developed for all these features, so that the system builder need not start from the beginning to develop them.

Though the list may seem to include unnecessary features, they may help a system builder to see beyond their immediate needs and thereby design new capabilities that might not have seemed possible without a pre-existing architecture.

This paper has been a brief attempt at illuminating the landscape of intelligent systems' possibilities, in order to help others as they begin to explore it. I look forward to seeing these features begin to take shape in practical knowledge-based development so that active knowledge systems can begin to help us solve more difficult and interesting problems.

Acknowledgments

I thank Bernhard Ganter, Aldo de Moor and the reviewers for their careful reading as well as their thoughtful and helpful comments on drafts of this paper. I am also grateful to Dan Rochowiak for our many stimulating conversations on these topics.

References

[1]   "Metadata Standards", ISO/IEC JTC1 SC32 WG2, http://metadata-stds.org/

[2]   Tim Berners-Lee, James Hendler, and Ora Lassila, "The Semantic Web," in *Scientific American*, vol. 284, 2001.

[3]   Dan Corbett, *Reasoning and Unification over Conceptual Graphs*. New York: Kluwer Academic, 2003.

[4]   Harry Delugach, "CharGer: Some Lessons Learned and New Directions," in *Working with Conceptual Structures: Contributions to ICCS 2000*, G. Stumme, Ed. Aachen, Germany: Shaker Verlag, 2000, pp. 306-309.

[5]   "CharGer - A Conceptual Graph Editor", Univ. of Alabama in Huntsville, http://www.cs.uah.edu/~delugach/CharGer

[6]   "GoGITaNT", LIRMM - Montpellier, http://cogitant.sourceforge.net/index.html

[7]   Lois W. Harper and Harry S. Delugach, "Using Conceptual Graphs to Capture Semantics of Agent Communication," in *Conceptual Structures for Knowledge Creation and Communication, Lecture Notes in Artificial Intelligence*, B. Ganter, A. d. Moor, and W. Lex, Eds. Berlin: Springer-Verlag, 2003.

[8]   Adil Kabbaj and Martin Janta-Polcynzki, "From Prolog++ to Prolog+CG: A CG Object-Oriented Logic Programming Language," in *Conceptual Structures: Logical, Linguistic and Computational Issues*, vol. 1867, *Lecture Notes in Artificial Intelligence*, B. Ganter and G. W. Mineau, Eds. Berlin: Springer-Verlag, 2000, pp. 540-554.

[9]   Robert Levinson, "Symmetry and the Computation of Conceptual Structures," in *Conceptual Structures: Logical, Linguistic and Computational Issues*, vol. 1867, *Lecture Notes in Artificial Intelligence*, B. Ganter and G. W. Mineau, Eds. Berlin: Springer-Verlag, 2000, pp. 496-509.

[10] Philippe Martin, "The WebKB set of tools: a common scheme for shared WWW Annotations, shared knowledge bases and information retrieval," in *Conceptual Structures: Fulfilling Peirce's Dream*, vol. 1257, *Lecture Notes in Artificial Intelligence*, D. Lukose, H. S. Delugach, M. Keeler, L. Searle, and J. F. Sowa, Eds.: Springer-Verlag, 1997, pp. 585-588.

[11] Guy W. Mineau, "A first step toward the Knowledge Web: Interoperability Issues among Conceptual Graph Based Software Agents," in *Conceptual Structures: Integration and Interfaces*, vol. 2363, *Lecture Notes in Artificial Intelligence*, U. Priss, D. Corbett, and G. Angelova, Eds. Berlin: Springer-Verlag, 2002, pp. 250-260.

[12] Aldo de Moor, "Applying Conceptual Graph Theory to the User-Driven Specification of Network Information Systems," in *Conceptual Structures: Fulfilling Peirce's Dream*, vol. 1257, *Lecture Notes in Artificial Intelligence*, D. Lukose, H. S. Delugach, M. Keeler, L. Searle, and J. F. Sowa, Eds.: Springer-Verlag, 1997, pp. 536-550.

[13] Bruce R. Schatz, "The Interspace: Concept Navigation Across Distributed Communities," *IEEE Computer*, vol. 35, pp. 54-62, 2002.

[14] Finnegan Southey and James G. Linders, "NOTIO - A Java API for Developing CG Tools," in *Conceptual Structures: Standards and Practices*, vol. 1640, *Lecture Notes in Artificial Intelligence*, W. Tepfenhart and W. Cyre, Eds. Berlin: Springer-Verlag, 1999, pp. 262-271.

[15] J.F. Sowa, *Conceptual Structures: Information Processing in Mind and Machine*. Reading, Mass.: Addison-Wesley, 1984.

[16] J.F. Sowa, *Knowledge Representation: Logical, Philosophical, and Computational Foundations*: Brooks/Cole, 2000.

[17] Patricia Ann Turrisi, *Pragmatism as a Principle and Method of Right Thinking, by Charles Sanders Peirce: The 1903 Harvard Lectures on Pragmatism*. Albany, NY, U.S.A.: State University of New York Press, 1997.

Formalizing Botanical Taxonomies

Uta Priss

School of Computing, Napier University
u.priss@napier.ac.uk

Abstract. Because botanical taxonomies are prototypical classifications it would seem that it should be easy to formalize them as concept lattices or type hierarchies. On closer inspection, however, one discovers that such a formalization is quite challenging to obtain. First, botanical taxonomies consist of several interrelated hierarchies, such as a specimen-based plant typology, a name hierarchy and a ranking system. Depending on the circumstances each of these can be primary or secondary for the formation of the taxonomy. Second, botanical taxonomies must comply with botanical nomenclature which follows a complicated rule system and is historically grown. Third, it may be difficult to design a formalization that is both mathematically appropriate and has a semantics which matches a taxonomist's intuition. The process of formalizing botanical taxonomies with formal concept analysis methods highlights such problems and can serve as a foundation for solutions.

1 Introduction

Biological taxonomies are often regarded as prototypical examples of classifications. First, it is usually assumed that taxonomic classes are defined using precise characteristics in the form of biological criteria that can be measured or in the form of specimen sets that are precisely grouped. Second, classes are usually non-overlapping with clear boundaries. Characteristics are usually non-gradual, i.e., they are either true or not true for an object that is to be classified. There is no grey-zone where a characteristic might apply only to a certain degree. Third, the classification usually forms a tree-hierarchy.

Given this assumed nature of biological taxonomies, it should be straightforward to formalize them as concept lattices in the sense of formal concept analysis. On closer inspection, however, one discovers that even biological taxonomies are more difficult to formalize than it might appear on first sight. This is demonstrated using the example of botanical taxonomies in this paper.

Botanical taxonomies are historically grown objects, which do not necessarily follow a simple construction principle. There are a variety of influences and contributing systems. For example, there are several separate hierarchies that contribute to the formation of a botanical taxonomy. These are a specimen-based plant typology (also called a "classification"), a name hierarchy and a ranking system. Each of the three is described in section 2. The notion of "taxonomy" usually refers to a combination of a classification with a name hierarchy. A taxonomy consists of "taxa" (singular: "taxon").

Section 3 provides a short introduction to the basic terms of formal concept analysis. But it is assumed in this paper that the reader is already somewhat familiar with formal

A. de Moor, W. Lex, and B. Ganter (Eds.): ICCS 2003, LNAI 2746, pp. 309–322, 2003.

concept analysis. If not, the reader is referred to Ganter & Wille (1999). Section 4 provides a short description of the Prometheus Project (Pullan et al., 2000), which is based on an object-relational modeling of taxonomies. Section 5 discusses some of the advantages and challenges involved in formalizing taxonomies with formal concept analysis. Section 6 considers the dynamic nature of botanical taxonomies and the notion of "closure systems". Section 7 provides the details of the formalization of taxonomies with formal concept analysis and Section 8 discusses some problems involved in applying a ranking system to a combination of multiple taxonomies.

2   Three Hierarchies in Botanical Taxonomies

A specimen-based plant typology (or classification) is created by taxonomists who group the specimens that they are interested in according to characteristics that can be identified for these specimens. This corresponds to a formal context in formal concept analysis which has specimens as objects and characteristics as attributes. (Although the ranking system must be included in such a formal context and it is actually not practical to use characteristics as attributes. See below for details.) One problem is that because of the size of modern taxonomies and the fact that usually many taxonomists contribute to a classification, it is difficult to oversee a single classification. Other problems are mentioned by Pullan et al. (2000) who state that classifications must cope with "historical data, newly described taxa, new revisions and conflicting opinions".

The second component of taxonomies is a name hierarchy. Name hierarchies are based on a strict nomenclature, "the International Code of Botanical Nomenclature" (Greuter et al., 1994)), which is called "the Code" in the rest of this paper. Unfortunately, because both name hierarchies and the nomenclature are historically grown and the Code is a very complicated rule system, there is no simple mapping from any specimen-based plant typology to any name hierarchy. Names are only valid within the context in which they are published. That means that a single name can refer to different specimens in different taxonomies. For example, the family of "lilies (Liliaceae)" includes "asparagus" in some taxonomies, whereas in others it does not. Furthermore, a single specimen may have different names in different taxonomies. Names above "genus" are formed differently from names below or at genus level. Names depend on rank. For example, the same specimen "Eriogonum umbellatum var. subaridum" at the variety rank, is called "Eriogonum umbellatum subsp. ferrisii" at the subspecies rank and "Eriogonum biumbellatum" at the species rank (Reveal, 1997). There are also rules of historical authorship. Usually the first published name for a taxon should be used. But there are cases where publications are "forgotten" and then rediscovered (Reveal, 1997). In this case, the Code may specify exceptions to the normal rules and state explicitly which names are to be used in preference.

The ranking system contains the ranks "variety", "species", "genus", "family" and so on. The same ranking system is used in principle for all taxonomies. That means that no taxonomist can invent new ranks. But not all taxonomies must have all ranks. For example, the rank of "subspecies" may not always be used. As mentioned above, names depend on ranks. But even classifications depend on ranks. The lowest common supertype of two specimens can occur at different ranks in different classifications.

Furthermore, the same specimen set can be used at different ranks. For example, if a family has only one genus, then the specimen set of that family and of that genus are identical. But the taxon for this family and this genus are different and will have different names assigned. Thus a classification cannot be generated from just the specimen sets. The rank information is essential. While it is relatively easy to define ranks in a tree hierarchy, assigning ranks to a multi-hierarchy or lattice is more difficult because there could be paths of different length from the top element to any taxon. Thus if different taxonomies are combined, the ranking system becomes problematic. But because they are used both in the classification and in the naming hierarchy, ranks cannot be ignored.

3   Formal Concept Analysis

This section provides a short introduction to the basic notions of formal concept analysis which are relevant for taxonomies. Formal concept analysis (Ganter & Wille, 1999) provides a formal framework and vocabulary suited for classification systems, taxonomies, ontologies and hierarchies. Its main unit of study are concepts within a conceptual hierarchy. In the case of type hierarchies, concepts are called "types"; in classification systems they are called "classes"; in ontologies they are called either "concepts" or "types" or "classes"; and in taxonomies they are called "taxa". Concepts are to be distinguished from objects and attributes, which are usually formal representations of elements of some domain. Objects can also be called elements, individuals, tokens, instances or specimens. They describe the extensional aspects of concepts. Attributes can also be called features, characteristics, characters or defining elements. They describe the intensional aspects of concepts. Both extensional and intensional elements can be used separately or together to describe or define concepts. Alternatively, concepts can be defined by establishing relations with other concepts. Thus there are multiple ways of describing, characterizing or comparing concepts. Formal concept analysis provides a formal framework in which all of these can be combined and applied.

There are two primary views of the data in formal concept analysis: formal contexts and concept lattices. Formal contexts are represented as binary relational tables. A formal context (O, A, I) consists of a set O of objects, a set A of attributes and a binary relation I between them, i.e., $I \subseteq O \times A$. Instances of the relation are either denoted by oIa or by $(o, a) \in I$ for an object $o \in O$ and an attribute $a \in A$. They are both read as "object o has the attribute a". Two mappings are defined: $I(o) := \{a \in A | (o, a) \in I\}$ provides for each object the set of its attributes and $I(a) := \{o \in O | (o, a) \in I\}$ provides for each attribute the set of objects that have that attribute. If one applies this mapping repeatedly, one will find that after two iterations there are no further changes. That is if one starts with a set of objects, takes all their attributes and then looks up all other objects that also have these attributes one can stop at exactly that point. Further applications of $I()$ will not add further objects or attributes. Formally, $\forall_{o \in O, a \in A} : I(I(I(o))) = I(o)$ and $I(I(I(a))) = I(a)$.

Concepts are pairs of sets of objects and sets of attributes of a formal context that do not change if $I()$ is applied. Formally a pair (O_1, A_1) with $O_1 \subseteq O$ and $A_1 \subseteq A$ is a concept if $I(O_1) = A_1$ and $O_1 = I(A_1)$. The set of objects of a concept c is called its *extension* and denoted by $ext(c)$. The set of attributes of a concept is called its *intension* and denoted by $int(c)$. An ordering relation is defined on a set of concepts as follows: a concept c_1 is a subconcept of another concept c_2 if $ext(c_1) \subseteq ext(c_2)$. This is equivalent to $int(c_2) \subseteq int(c_1)$. For example, "Bellis (bellflower)" is a subconcept of "Asteraceae" if all instances of Bellis are also instances of Asteraceae or, equivalently, all attributes of Asteraceae also apply to Bellis. This conceptual ordering is denoted by $\leq$. The set of concepts with its ordering, $(C, \leq)$, is not only a partially ordered set but even a (concept) lattice, which means that for each set of concepts there exists a unique infimum (largest common subconcept) and a unique supremum (largest common superconcept). Infimum and supremum of concepts are denoted by $c_1 \wedge c_2$ and $c_1 \vee c_2$, respectively.

There are many applications for formal concept analysis. To convert a traditional tree-hierarchy into a concept lattice, one must formally add a bottom concept which has an empty extension and has all the attributes of the underlying formal context in its intension. In some applications only either sets of objects or sets of attributes are of interest. It is sufficient to either only specify the extensions of a concept lattice or only the intensions to generate the lattice. The intensions (or extensions, respectively) can then be formally labeled, such as A_1, A_2, and so on, but have no further meaning for the domain. It should be pointed out that concepts, suprema, infima, the conceptual ordering and so on all depend on the formal context in which they are defined.

4   The Prometheus Project

The Prometheus project (Pullan et al., 2000) provides an object-oriented formalism for taxonomies that is implemented in an object database system (Prometheus, 2002). This implementation improves on prior attempts at implementing taxonomies which according to Berendsohn (1997) often suffer from "a frequent error of over-simplification of taxonomic data by non-taxonomist database designers" (quoted after Pullan et al. (2000)). A first main achievement of Prometheus is a clear separation of the specimen-based classification and the nomenclature-based name hierarchy. Apparently prior implementations of taxonomic databases failed to separate these two thus making it impossible to combine more than one taxonomy in a single database. Prometheus recognizes the value of combining several different taxonomies and purposefully designs the separation of classification and name hierarchy to achieve this combination. It also provides an impressive graphical interface that facilitates simultaneous browsing through multiple taxonomies (Graham, 2001).

A second main achievement of Prometheus is a concentration on the specimens as the main drivers for the classification. Biological characteristics are not as unambiguous as one might naively assume (Pullan et al, 2000). In fact characteristics, such as "hairy leaves", are very imprecise. Furthermore, they are not usually checked for all specimens in the database but only for those within a certain smaller subset. Some taxonomists might even assign characteristics based on a prototypical image that they have of the specimens in a set, which may not be true for all actual specimens. Thus biological characteristics are not attributes in the sense of formal concept analysis. The Prometheus project defines the "circumscription" not in the traditional manner as referring to the characteristics of a taxon but instead as the set of specimens of a taxon. Thus in formal concept analysis terms, the taxonomy is entirely extensionally defined. This ensures that both taxa and circumscriptions are precisely defined and that different taxonomies can be combined and compared based on the specimen sets. The potential problem of having to incorporate too much data (if all specimens of the world were included in a taxonomy) is overcome by focusing on "type specimens", i.e., such specimens that actually contribute to the distinctions made in the classification. In formal concept analysis terms this corresponds to using a "purified" formal context.

5 Advantages and Challenges Involved in Formalizing Taxonomies with Formal Concept Analysis

There are some shortcomings of the object-oriented modeling as used in the Prometheus Project which a formal concept analysis modeling can clarify. It is not discussed in this paper whether or not a formal concept analysis-based implementation would improve the performance of the database. That may or may not be the case. But it is claimed that a formal concept analysis-based representation simplifies the terminology and the conceptual understanding of the problems involved in modeling taxonomies. There is, however, at least one definite practical advantage of using formal concept analysis modeling instead of object-oriented modeling, which is that the Prometheus software had to be specifically designed for its purpose. This required a significant amount of resources. Most likely Prometheus cannot easily be used for other types of hierarchical data without major redesign. On the other hand, formal concept analysis software provides general multi-purpose tools for dealing with hierarchical data. Each new project requires some specific adjustments but the required time and effort is much smaller than designing a single-purpose object-oriented implementation.

The remainder of this section discusses some of the differences between object-oriented and formal concept analysis modeling. It also discusses the challenges that formal concept analysis models must overcome to be acceptable to application domain experts. For a start, domain experts often do not see a need for mathematical formalization at all. In the case of taxonomists, they already have a complicated semi-formal system in their Code. One advantage of the Prometheus model is that on the surface it is simpler. It matches fairly closely to the taxonomic terminology without introducing formal constructs that do not have an obvious reason for existence from a taxonomist's viewpoint. An example for such formal constructs is the bottom concept with an empty set of specimens that a formal concept analysis modeling adds to every tree-hierarchy.

Another formal construct is what Berendsohn (1997) calls "potential taxa" which can be automatically generated higher up in the hierarchy, Pullan et al. (2002) express discomfort with these, which they call "informationless" taxa. They state that Prometheus should not extrapolate "far beyond what can be supported by the original sources". From a formal concept analysis viewpoint automatically generated information does not pose a problem as long as users are aware of what is input to the system and what is generated. But supporting such awareness and ensuring that users feel comfortable with the system can be difficult. Thus a formal concept analysis model is more complicated on the surface than an object-oriented formalism because it contains these structures which Pullan et al. call "informationless". But it is claimed in this paper that the "deeper level" structures, i.e., underlying mathematical formalisms and formal semantics, of a formal concept analysis modeling are actually simpler.

There is no detailed mathematical formalism with formal semantics provided for Prometheus. But it can be foreseen that such a formalism would be complicated. An advantage of a simpler underlying mathematical formalism is that it is easier to be checked for consistency, completeness and semantics. An extreme example of a simple surface model which completely lacks an underlying formalism is the Code itself. Due to its lack of mathematical formalism it is ambiguous as it is. It would be very difficult to model it precisely and to check it for consistency.

It has already been mentioned that formal concept analysis allows to reuse its software for different hierarchy-related projects instead of having to write ad-hoc solutions for each project. A similar argument can be made on the level of mathematical structures. Formal concept analysis builds on complex but well known mathematical structures, such as lattices. If it can be shown that taxonomies can be represented as lattices in the sense of formal concept analysis, then it follows that numerous statements about formal properties can automatically be inferred about taxonomies. Formalizations only have a purpose if they support such inferences. There is a chance that inferentially derived knowledge about taxonomies contradicts domain knowledge. While such discrepancies could be caused by inappropriate use of formalizations, they can also point to problems within the domain knowledge. Philosophers, such as Peirce or Brandom (1994) emphasize the use of logic (formalisms and inferences) for the purpose of "making it explicit". This paper claims that formal concept analysis can serve as an inferential tool that explicates structures in taxonomies. For example, it will be shown below that the notion of "rank" causes problems at least in the case of a combination of multiple taxonomies. Maybe taxonomists should review their use of "rank".

Another advantage of formal concept analysis is the explicit use of "context". A consequence of a lack of considering context, is a confusion between member-of and subset-of relations. For example, CYC (1997) explains the differences between what they call the isa and the genls relation and the resulting problems with transitivity. From a formal concept analysis viewpoint, a concept is a subconcept of another concept if the extension of the first concept is contained in the extension of the second concept. The subconcept relation among concepts thus corresponds to a subset relation on the set of extensions. The subconcept relation is transitive. The member-of or element-of relation holds between objects and concepts. If an object is an element of a concept than it is also an element of all its superconcepts. But an object is always at the left-hand side of the member-of relation. Nothing can be an element of or a subset of an object. At least not within the same context. Concepts are never elements of other concepts within the same context. But a concept in one context can correspond to an object in another context which is then an element of other concepts in that second context.

For example, "Asteraceae" is the name of the family of asters. In a context which has plant specimens as objects and contains a concept "Asteraceae", certain specimens can be elements of "Asteraceae". In a different context, "Asteraceae" can denote an object, which can then be an element of a concept called "family". Thus the statements "this specimen is an element of Asteraceae" and "Asteraceae is a (or is an element of) family" can both be true, but not in the same context! The false statement "this specimen is a family" is not implied because inferences across contexts are not allowed without further constraints.

For Prometheus the distinction between element-of and subset-of is important because a form of "element-of" is used in classifications to describe the assignment of specimens to taxa. Two forms of "subset-of" are used in name hierarchies: first, there is the "placement" of a taxon within a higher-ranked taxon. Second, there is the "type definition" which for each taxon points to a taxon of lower rank or to a specimen set. The reason for the need of type definitions is that the name of a higher level taxon is formed based on one prototypical specimen or lower level taxon. For example, the family "Asteraceae" is named after the prototypical genus "Aster". Going up in the hierarchy, "Asteraceae" provides the basis for "Asterales", which in turn provides the name for "Asteridae". But the notion of "type definition" can lead to a confusion of "element-of" and "subtype-of" because either specimens or subtaxa are used. A solution would be to formalize name hierarchies without referring to specimens at all, but instead use names as basic elements, ignoring whether they are names of specimens or names of taxa.

6   Dynamic Taxonomies and Taxonomy Merging

Several further aspects of taxonomies should be mentioned before the formalization is explained in detail. Taxonomists have an implied dynamical view of classification. Because new specimens are discovered, existing classifications are revised, and new scientific knowledge is gained through new methods (such as genetic analysis), taxonomic knowledge is continuously changing. A certain static view of taxonomies is imposed by distinguishing between published taxonomies and taxonomies that are "in-progress". Published taxonomies never change because if revisions of a taxonomy are published, then this is viewed as a new separate taxonomy. On the other hand, working taxonomists have their own personal taxonomies "in-progress" which they can be changing on a daily basis. But these are not shared beyond an immediate circle of colleagues. As long as a taxonomy is changing it is private and not published. Name hierarchies are attached to published taxonomies. Taxonomies that are in-progress normally consist of classifications, not name hierarchies. In Prometheus, the names can be automatically calculated to avoid naming errors on the taxonomist's part.

A second aspect to be mentioned is the notion of a "closure system", which may be the minimal formal structure that should be considered for combining taxonomies and classifications. Viewing a specimen-based classification as a closure system means that if sets of specimens are intersected these intersections are also part of the classification. For example, a gardener might use both a taxonomy to obtain information on plant families and a plant catalog, which provides information on plant habitat and hardiness. A plant catalog is not a taxonomy in the sense of the Code, but it can be considered what happens if it is combined with a taxonomy. The plant catalog may provide sets of specimens, such as "spring flowering plants", which can be intersected with specimen sets from a taxonomy to form sets such as "asters that flower in spring". These are meaningful sets which are easily obtained from the combination. Thus it is reasonable to demand that intersections are calculated when classifications or taxonomies are combined. Of course, only those intersections are added to the merged classification that provide new information. Since roses flower in summer there will not be a new class "summer-flowering roses". Instead the specimen set for "roses" will be contained in the specimen set for "summer-flowering plants". Thus if classifications are merged that are very similar, not many new interesting intersections may occur.

Unions of specimen sets, however, are usually not considered at all. For example, a set of plants that are either asters or spring-flowering is an odd combination. While unions of specimen sets need not be included in a classification, it is important to know which is the smallest set that contains a union. This is called a *closure set*. For example, while a union of asters and roses may not be a meaningful set, it is important to know which is the smallest set that does contain both asters and roses because that is the point where the upwards paths from aster and rose meet in a classification. In the case of asters and spring flowering plants the closure set is probably the set of all plants, which is not very interesting but by default contained in a classification. A *closure system* thus provides intersections and closure sets for all sets of specimens and is an intuitive formalization of classifications. Tree hierarchies are automatically closure systems if the empty set is formally added as the shared bottom node of the tree. Formal concept lattices are closure systems with additional structure.

7 The Formalization of (Botanical) Taxonomies

7.1 Specimens, Taxa, Ranks

In what follows S denotes a set of specimens; T denotes a set of taxa; and R denotes a set of ranks. It should be noted that all sets are finite and $S \cap R = \emptyset$, $S \cap T = \emptyset$, $T \cap R = \emptyset$.

7.2 Closure Systems

Let $P(S)$ denote the power set (or set of subsets) of S and $\overline{S}$ a subset of $P(S)$. According to standard order theory, a set $\overline{S}$ forms a *closure system* if

$$S \in \overline{S} \text{ and } (\overline{S}_1 \subseteq \overline{S} \Longrightarrow \bigcap \overline{S}_1 \in \overline{S}) \tag{1}$$

Also according to standard order theory it follows that if $\overline{S}$ is a closure system then $\overline{S}$ with $\subseteq$ forms a complete (concept) lattice

$$(\overline{S}, \subseteq) \text{ with formal context } (S, \overline{S}, \in) \text{ and } \top := S \text{ and } \bot := \bigcap \overline{S} \tag{2}$$

7.3 Tree Hierarchy

Tree hierarchies are closure systems if a bottom node is formally added. They are defined as follows:
A set $\overline{S}$ with $\overline{S} \subseteq P(S)$ forms a *tree hierarchy with bottom element* if

$$S \in \overline{S} \text{ and } (\forall_{a,b \in \overline{S}} : a \nsubseteq b \text{ and } b \nsubseteq a \Rightarrow a \cap b = \emptyset) \tag{3}$$

If a tree has at least two incomparable nodes and $\emptyset \in \overline{S}$ then $\overline{S}$ is a closure system and $(\overline{S}, \subseteq)$ is a complete lattice with top element $\top := S$ and bottom element $\bot := \emptyset$.

It is straightforward to interpret a specimen-based plant typology as a concept lattice in the sense of formal concept analysis. But this is not yet sufficient for the treatment of specimen classifications because as discussed before these depend on rank. Two different taxa can have the same specimen set but different ranks. This is always the case, if a taxon at one rank has only one immediate subtaxon at a lower rank.

7.4 Rank Order and Rank

The ranking systems considered by taxonomists are always linear orders, thus only linear orders are considered here. Using specimens as objects and ranks as attributes would not yield a useful concept lattice because the property of "being at a certain rank" is not inherited. For example, members of "species" are not also members of "genus". But the property of "being at most at a certain rank" is inherited. Thus ranking systems can be formalized by using "at-least-ranks" as objects and "at-most-ranks" as attributes.

A partially ordered set $(R, \preceq)$ is called a *rank order* if $\preceq$ is a linear order. The elements of R are called *ranks*. With $R^{\geq} := R$ and $R^{\leq} := R$, a rank order can be modeled as a concept lattice

$$(R, \preceq) \text{ with formal context } (R^{\geq}, R^{\leq}, \vdash) \text{ and } r_1 \vdash r_2 \iff r_1 \preceq r_2 \qquad (4)$$

For example, "at-least-species below at-most-genus" if and only if "species below genus".

7.5 (Botanical) Taxonomies

The following two assumptions are necessary for the definition of a (botanical) taxonomy:

1) The set of sets of specimens $(\overline{S}, \subseteq)$ forms a closure system.
2) A map $rank : \overline{S} \to R$ assigns to each set of specimens in $\overline{S}$ exactly one rank with the condition

$$\forall_{S_1, S_2} \in \overline{S} : S_1 \subset S_2 \Rightarrow rank(S_1) \prec rank(S_2) \qquad (5)$$

The assigned rank is the maximum rank at which a taxonomist wants to see this specimen set uniquely describing a taxon. A specimen set can describe different adjacent ranks if the taxonomist sees no need to divide the set into subsets at the lower rank. To formally distinguish these sets, each set of specimens is unioned with its rank, i.e., $S_1 \cup rank(S_1)$. If $S_1 \subset S_2$, $rank(S_1) \prec r \prec rank(S_2)$ and there exists no S_3 with $rank(S_3) = r$, then both $S_2 \cup rank(S_2)$ and $S_2 \cup r$ are formed. The resulting set of specimen sets unioned with ranks is denoted by $\overline{S}^R$. The map $rank$ is extended to $\overline{S}^R$ by mapping each set to its rank. It follows that if the top rank is assigned to any specimen set, then $rank$ on $\overline{S}^R$ is surjective.

A *(botanical) taxonomy* is now defined as a concept lattice $(T, \leq)$ with the following formal context:

$(S \cup R, \overline{S}^R, I_{SR})$ with $I_{SR} = \{(s_1, S_1) | s_1 \in S_1\} \cup \{(r_1, S_1) | r_1 \vdash rank(S_1)\}$.

Because of $S \cap R = \emptyset$, this is a disjoint union and can be represented as $I_{SR} = I_{SR}|_S \cup I_{SR}|_R$. The elements of $(T, \leq)$ are denoted by $t \in T$ and are called *taxa*. The map $rank$ can be extended to all taxa as follows: $rank : T \to R$ with $rank(t) = max\{r \in R^{\geq} | r \in ext(t)\} = max_r ext(t)|_R$.

A taxon is uniquely characterized by its rank and the set of specimens it refers to. Figure 1 shows two examples of taxonomies. The terms in boxes represent names from a name hierarchy which are assigned to the taxonomies. From a formal concept analysis viewpoint, there is no difference between the dashed and the solid lines. From a taxonomist's viewpoint, the dashed lines are pointers from the ranks to the taxa at those ranks. In figure 1, "carrot" in the left taxonomy and "apple" in the right taxonomy represent examples of specimen sets that occur at two different ranks. On $\overline{S}$, these are mapped to rank 2; in $\overline{S}^R$ these correspond to {carrot, 1}, {carrot, 2}, {apple, 1}, {apple, 2}.

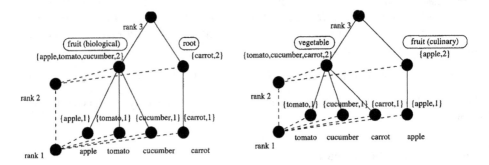

Fig. 1. Two taxonomies

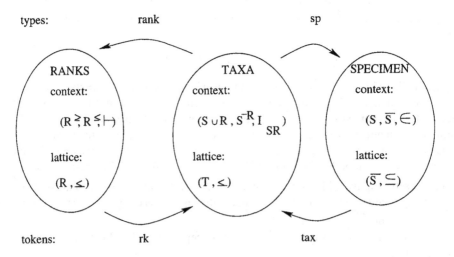

Fig. 2. A specimen/rank information channel

7.6   A Specimen/Rank Information Channel

A taxonomy as defined in the previous section can also be interpreted as an information channel (in the sense of Barwise & Seligman (1997)) between a specimen-based closure system $(S, \overline{S}, \in)$ and a rank order $(R^{\geq}, R^{\leq}, \vdash)$ (cf. figure 2). This corresponds to an intuitive notion that the specimen classification and the rank order are independent of each other but both contribute to the taxonomy.

The infomorphisms are constructed as follows:

a) With $sp : \overline{S}^R \to \overline{S}$ and $tax : S \to S$ as identity map it follows that
$s_1 \in sp(S_1) \Longleftrightarrow tax(s_1) \, I_{SR} \, S_1$.

b) With $rk : R^{\geq} \to R$ as an identity map and $rank : \overline{S}^R \to R^{\leq}$ as defined above it follows that
$r_1 \vdash rank(S_1) \Longleftrightarrow rk(r_1) \, I_{SR} \, S_1$ because of the definition of I_{SR}.

The map sp is extended to all taxa in the following manner: $sp : T \to \overline{S}$ with $sp(t) := \{s \in S | s \in ext(t)\} = ext(t)|_S$. This is well defined because $sp(t)$ is an element of $\overline{S}$ due to the fact that $I_{SR}|_S$ is equivalent to "$\in$" in $(S, \overline{S}, \in)$.

7.7   Properties of Taxonomies

The following 7 properties of taxonomies can be proved. These properties correspond to intuitive notions that taxonomists have about their taxonomies and demonstrate that the semantics of the formalization as derived so far does not contradict traditional taxonomic knowledge.

-   1) For each specimen exists at least one taxon so that $s \in ext(t)$.
-   2) For each specimen exists exactly one minimum taxon $tax(s)$ in $(T, \leq)$ so that
    $s \in ext(t)$, i.e., $tax(s) := \bigwedge\{t \in T | s \in ext(t)\}$.
    Proof: $(T, \leq)$ is a lattice.
-   3) For each rank exists at least one taxon so that $r \in ext(t)$.
    Proof: Due to the set of objects: $S \cup R^{\geq}$.
-   4) $t_1 \leq t_2 \Longleftrightarrow sp(t_1) \subseteq sp(t_2)$ and $rank(t_1) \preceq rank(t_2)$.
    Proof: $sp(t_1) \subseteq sp(t_2)$ and $rank(t_1) \preceq rank(t_2) \Longleftrightarrow$
    $ext(t_1)|_S \subseteq ext(t_2)|_S$ and $max_r ext(t_1)|_R \preceq max_r ext(t_2)|_R \Longleftrightarrow$
    $ext(t_1)|_S \subseteq ext(t_2)|_S$ and $ext(t_1)|_R \subseteq ext(t_2)|_R \Longleftrightarrow$
    $ext(t_1) \subseteq ext(t_2) \Longleftrightarrow t_1 \leq t_2$
-   5) $t_1 = t_2 \Longleftrightarrow sp(t_1) = sp(t_2)$ and $rank(t_1) = rank(t_2)$
    Because $\leq$ is ordering and $t_1 \leq t_2$ and $t_2 \leq t_1 \Longleftrightarrow$
    $sp(t_1) \subseteq sp(t_2)$ and $rank(t_1) \preceq rank(t_2)$ and
    $sp(t_2) \subseteq sp(t_1)$ and $rank(t_2) \preceq rank(t_1)$
-   6) $t_1 < t_2 \Longleftrightarrow sp(t_1) \subset sp(t_2)$ and $rank(t_1) \preceq rank(t_2)$ or
    $sp(t_1) \subseteq sp(t_2)$ and $rank(t_1) \prec rank(t_2)$
    Because: $t_1 \neq t_2 \Longleftrightarrow sp(t_1) \neq sp(t_2)$ or $rank(t_1) \neq rank(t_2)$
    $t_1 \leq t_2$ and $t_1 \neq t_2 \Longleftrightarrow sp(t_1) \subseteq sp(t_2)$ and $rank(t_1) \preceq rank(t_2)$ and
    $(sp(t_1) \neq sp(t_2)$ or $rank(t_1) \neq rank(t_2))$
-   7) $t_1 \wedge t_2 = t \Longleftrightarrow sp(t_1) \wedge sp(t_2) = sp(t)$ and $rank(t_1) \wedge rank(t_2) = rank(t)$.
    $t_1 \vee t_2 = t \Longleftrightarrow sp(t_1) \vee sp(t_2) = sp(t)$ and $rank(t_1) \vee rank(t_2) = rank(t)$.

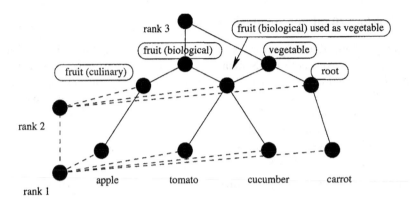

Fig. 3. Merging taxonomies

8   Rank and Multiple Classifications

Botanical taxonomies are usually modeled as tree-hierarchies. But if multiple hierarchies are combined, then the resulting structure is a multi-classification, which is not usually a tree-hierarchy. Intersections between specimen sets from different classifications emerge as new classes. The example in figures 1 and 3 shows a biological classification in which "tomato", "cucumber" and "apple" are in a class called "fruit (biological)". In a culinary classification, "tomato" and "cucumber" are grouped under "vegetable", and "apple" under "fruit (culinary)". In the merged classification new classes arise, such as "fruit (biological) used as vegetable" which contains "tomato" and "cucumber". This new class emerges because it adds new information. Because all elements under "fruit (culinary)" are also under "fruit (biological)" there is no new class "fruit (biological) used as fruit (culinary)".

A problem with multi-classifications is the meaning of "rank". In a tree-hierarchy it is fairly easy to establish ranks because there is only one path from each specimen set to the top of the tree. In a multi-hierarchy there are possibly several paths of different length from a single taxon to the top of the classification. It is thus not possible to start at the bottom and assign ranks by following all paths that lead upwards. An obvious idea is to somehow combine the original ranks of the original tree-like classifications into a shared rank of the multi-classification. But it is a question which ranks to assign to the emerging classes. In the example, "fruit (culinary)", "fruit (biological)" and "vegetable" are all of rank 2. What rank should then be assigned to "fruit (biological) used as vegetable"? It is not possible to choose an already existing rank that is smaller than 2 because "tomato" and "cucumber" are at rank 1. Another possibility would be to invent new ranks for the new emerging classes (such as rank 1.5) but botanical taxonomists are not allowed to invent new ranks.

Thus the only remaining possibility is for a class that emerges as an intersection of classes x_1, x_2, and so on to assign the maximum rank that is shared by all these classes. In the example above, that means that "fruit (biological) used as vegetable" receives rank 2 even though that is the same rank as its parent classes. In this approach, the main function of using ranks is to ensure that every class can be uniquely identified by its specimens and its rank. This approach is consistent with the definitions and propositions of section 7. But as the example in figure 4 shows it is not in general correct to assume that if a set of specimens of one class is contained in the set of specimens of another class that then the second class should be of higher or equal rank than the first class. That means that $sp(t_1) \subset sp(t_2) \not\Rightarrow rank(t_1) \preceq rank(t_2)$ and thus while equation (5) is true for the original specimen sets of a classification it cannot be generalized to all taxa.

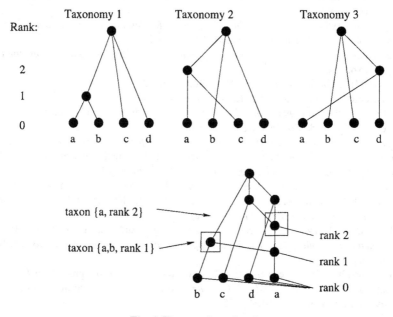

Fig. 4. The question of rank

9   Conclusion

This paper provides the foundation for a mathematical formalization of botanical taxonomies. The formalization shows that the notion of "rank" is problematic with respect to merged taxonomies and non-tree-like taxonomies. In these cases, ranks only serve to distinguish taxa but not to rank-order them. If "rank" cannot be generalized to these cases, then maybe taxonomists should review their use of ranks. Maybe ranks should be applied to taxonomies as a meta-level property. But then they could not be used to define taxa.

For future research, it would be interesting to compare the process of botanical classification to similar historically grown classification tasks. Library cataloging employs a complex rule based system similar to the Code used in botanical taxonomies. A formalization of library cataloging rules might provide insights into botanical classification and vice versa.

Acknowledgments

I wish to thank my colleagues Jessie Kennedy, Andrew Cumming, Gordon Russell and Cedric Raguenaud for stimulating discussions on this topic.

References

1. Barwise, J.; Seligman, J. (1997). *Information Flow. The Logic of Distributed Systems*. Cambridge University Press.
2. Brandom, R. (1994). *Making It Explicit: Reasoning, Representing, and Discursive Commitment*. Cambridge, MA: Harvard University Press, 1994.
3. Berendsohn, W. G. (1997). *A taxonomic information model for botanic databases: the IOPI Model*. Taxon, 44, p 207-212.
4. CYC (1997). *A note about isa and genls*. Online available at http://www.cyc.com/cyc-2-1/footnotes.html#IsaGenls
5. Ganter, B.; Wille, R. (1999a). *Formal Concept Analysis*. Mathematical Foundations. Berlin-Heidelberg-New York: Springer, Berlin-Heidelberg.
6. Graham, M.; Kennedy J. (2001). *Combining linking & focusing techniques for a multiple hierarchy visualisation*. 5th International Conference on Information Visualisation - IV2001, University of London, London, IEEE Computer Society Press, p 425-432.
7. Greuter, W.; Barrie, F. R.; Burdet, H. M.; Chaloner, W. G.; Demoulin, V.; Hawksworth, D. L.; Jorgensen, P. M.; Nicholson, D. H.; Silva, P. C.; Trehane, P.; McNeill, J. (1994). *International Code of Botanical Nomenclature (Tokyo Code)*. Regnum Veg. 131.
8. Pullan, M. R.; Watson, M. F.; Kennedy, J. B.; Raguenaud, C.; Hyam, R. (2000). *The Prometheus Taxonomic Model: a practical approach to representing multiple classifications*. Taxon, 49, p 55-75.
9. Prometheus (2002). On-line available at http://www.dcs.napier.ac.uk/ prometheus/
10. Reveal, J. (1997) *One and Only One Correct Name?*. On-line available at http://www.inform.umd.edu/PBIO/pb250/onename.html

Conceptual Graphs Self-tutoring System

Albena Strupchanska[1], Milena Yankova[1], and Svetla Boytcheva[2]

[1] Bulgarian Academy of Sciences, Linguistic Modelling Lab, CLPP,
25A Acad. G. Bonchev Str., 1113 Sofia, Bulgaria,
{albena,myankova}@lml.bas.bg
[2] Department of Information Technologies, FMI, Sofia University,
5 J. Bauchier Blvd., 1164 Sofia, Bulgaria,
svetla@fmi.uni-sofia.bg

Abstract. The present paper describes a self-tutoring system CG-EST (Conceptual Graphs Elearning Self-Tutoring system). The purpose of this system is to generate questions to the user from given text lessons, to process user answers and to check their correctness. CG-EST includes modules for: parsing and producing logical forms (LF); generating conceptual graphs (CG) based on LF; generating questions based on LF; defining the scope of correct answer using CG operations; matching the user's and correct answers. This approach assists the automatic eLearning systems being developed for various domains with some requirements: all sentences in the lessons have to use Controlled English (CE); to each lesson a simple concept type hierarchy and a synonyms set have to be attached. However these restrictions are useful for semantic interpretation of Natural Language (NL) texts. Thus CG-EST presents a solution for self-service eLearning that is ideal for training in a specific field.

1 Introduction

Nowadays the eLearning industry faces an immense development. In order to satisfy the increasing user interest and demands of eLearning systems it is necessary to ease the process of development of such kind of systems. One possible step towards is to automate the knowledge acquisition and the testing of a user comprehension. Many different theories for cognitive aspects of body of knowledge/knowledge acquisition exists nowadays. They distinguish different levels of knowledge with different granularity. We shall mention only two of them as our research is closer to these paradigms. Benjamin Bloom's famous "Taxonomy of Educational Objectives" [1] identifies six levels: knowledge (rote memory skills - facts, terms, procedures, classification systems), comprehension (the ability to translate, paraphrase, interpret or extrapolate material), application (the capacity to transfer knowledge from one setting to another), analysis (the ability to discover and differentiate the component parts of a larger whole), synthesis (the ability to weave component parts into a coherent whole) and evaluation (the ability to judge the value or use of information using a set of standards). Another theory claims that research in educational psychology identifies three

A. de Moor, W. Lex, and B. Ganter (Eds.): ICCS 2003, LNAI 2746, pp. 323–336, 2003.

major types of knowledge: rote (i.e., memorization), declarative (i.e., knowledge of concepts) and procedural (i.e., knowledge of a physical or intellectual process, method, or skill). Knowledge does not automatically transfer between declarative and procedural knowledge [16]. The procedural type of knowledge is more complicated to be automatically examined since it requires deep semantic inferences so we focus only on the rote and the declarative ones. Usually the learning process includes set of lessons presented as texts and some drills that test user knowledge comprehension.

The presented system CG-EST (Conceptual Graphs Elearning Self-Tutoring system) translates sentences from the given text lessons into CG and automatically generates questions from them. User comprehension is tested by processing user answers and checking their correctness. Scope of correctness is presented as *minimal required* and *maximal expected* CG sets. Once the answer is translated into CG, CG-EST tries to match it to the so defined scope. The answer is correct only if it is a subset of *maximal* and superset of *minimal* CG sets.

Automatic extraction of formal knowledge specifications from free-text still looks too hard, effort consuming and time expensive. Automatic knowledge acquisition requires preliminary defined description of each (important) word expected in the input text. In addition an upper layer type hierarchy and some semantic primitives are required. That's why the task of automatic knowledge acquisition from free text is shifted to extraction from restricted NL input only. For our purposes we define "controlled English" as "complete declarative sentences in English". This ease the automatic processing of the lesson texts in CG-EST as well as the user answers.

This paper is structured as follows. Section 2 overviews briefly some related research. Section 3 describes the architecture of CG-EST in general. Section 4 presents the preprocessing module for generation of CG from the lessons. Section 5 focuses on self-tutoring module for questions generation and on defining the scope of correct answer by using of CG operations. Section 6 presents answering module and illustrates in more details the matching algorithm between user answer and the scope of correct ones. Section 7 contains the conclusion and sketches future work.

2   Related Work

A lot of research is carried out concerning the usage of CG for the semantic parsing of NL sentences and storing the semantic information in CG. The first algorithm was proposed by Sowa [17–19] and it was further applied to several prototypes for different languages as Italian [20], French, German and English [14, 15].

The prototypes for acquisition of CG from text de facto process controlled English. The restrictions for most systems can be summarized as: *(i)* limited vocabulary of the NL input, *(ii)* processing of phrases and simple sentences, *(iii)* often missing syntactic analysis as a separate module. For example CG Mars Lander [7, 10] skips unknown words in the input sentences exemplified in [7], which is a flexible strategy to define a controlled language. The creators of the workbench WebKB developed so called Formalized English that provides acquisition of different ontological structures from text statements [11]. Another prototype CGExtract [2] is restricted in a similar manner since it deals with restricted vocabulary, grammar rules and restricted set of semantic primitives.

Several prototypes have been recently developed for generating CG from text in CE. CGExtract [2] constructs Knowledge Base (KB) of CG corresponding to the input text using the NL analysis and Understanding (NLU) module Parasite [13], an initial type hierarchy and a relevant lexicon. Parasite processes coreferences in extended discourse in a few sentences. The extraction of CG is automatic but the filling of lexicon and the acquitting, encoding and testing of the meaning postulates is time consuming. On the other hand CGExtract checks for semantic consistency of an extracted graph to already created CG KB, which is an attempt for achieving a better functionality. Another prototype [12] extracts CG from English texts, where the syntactic structure is analyzed by Conexor syntactic parser and represented as a "grammatical CG". The system uses concepts and relations type hierarchy, WordNet, and manually created rules for transformation of "grammatical CG" to CG.

Zhang and Yu present in [21] another method for generating CG from domain specific sentences using a machine-learning based approach. Each sentence in the text is parsed by a link parser. This approach uses a vocabulary of words, types of links that can be attached to them and labels indicating how they are attached in a particular sentence.

Methods for automatic generation of questions are presented in [9]. Kunichica et al. extracts syntactic and semantic information of sentences using Definite Clause Grammar, sematic categories of nouns and verbs with corresponding interrogative pronouns. They also define a correspondence between roles of prepositional phrases and interrogative pronouns. Their approach for asking questions about the content of a sentence is to determine the sematic role which will be questioned, to find appropriate interrogative pronoun for it using mentioned above correspondences, to replace query part with interrogative pronoun and to translate the sentence into interrogative one. In order to generate more complicated questions synonyms and antonyms dictionaries are used for replacement of words in sentences. Using modifiers of different instance or different words is another way of generating questions with different contents from the original sentences. Methods for questioning about contents represented by plural sentences using a relative pronoun and questioning about relationship of time and space are also described. All methods generate approximately 93% semantically correct sentences due to the rough classification of semantic categories. Moreover, the words in parsed sentences should be classified according to the defined categories in advance.

3   CG-EST System

CG-EST is an intelligent eLearning system for self-tutoring in specific scientific domains. CG-EST has two main modules: preprocessing module and self-tutoring module (fig. 1). The main purpose of the preprocessing module is to generate a KB of the system from given by the tutor *(i)* lessons in NL, *(ii)* type hierarchy and *(iii)* terminology. The self-tutoring module trains the user in a chosen domain by presenting some lessons and asking questions related to them to check his/her comprehension. The innovative strategy implemented in this module is an automatic questions generation from a dynamically extracted KB. In addition the self-tutoring module applies another intelligent technique that handles checking process of user answers in CE using dynamically generated scope of the correct answer. The preprocessing module is activated once but the self-tutoring module could be used many times by different users.

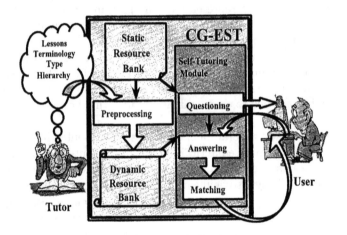

Fig. 1. Architecture of CG-EST system

4   Preprocessing Module for KB Generation

This section describes the preprocessing module (fig. 2). Its purpose is to generate a KB of CG from lessons in CE. At the beginning a lexical (morphological) analysis of the input is needed to automate the process. Then a syntactic analysis is needed. At this stage we use the system GATE 2.1 beta 1 (GATE - General Architecture for Text Engineering) [6] which provides several NLP modules. The most important for this work are the Lexical Analysis modules. They turn a text into a sequence of sentences, each of them is a sequence of lexical items (tokens). Each token is looked up in the dictionary to determine its possible features. Also part-of-speech types are assigned here. The syntactic analysis translates sentences into LF using parsing rules. LF and CG are two different forms used

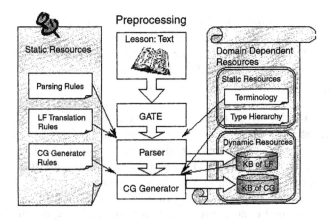

Fig. 2. CG-EST: Preprocessing Module

for formal representation of the syntactic and the sematic structure of sentences. LF consist of lemma-predicates corresponding to the words and θ-predicates corresponding to the thematic roles in sentences. CG have concepts and relations between them. The concepts correspond to the words and the relations correspond to the thematic roles in a sentence. Thus LF and CG have very similar structure except for different granularity of lemma-predicates in LF and concepts in CG.

4.1   Logical Forms

A specially developed left-recursive, top-down depth-first parser, implemented in Sicstus Prolog, is used for translation of sentences in CE to LF. This parser uses grammar rules and rules for translation into LF from our static resource bank. In LF we represent all words as predicates (*lemma-predicates*) with predicate symbol (the corresponding base form (lemma) of the word) and two arguments, where the first one is a variable and the other is a Part-Of-Speech (POS) tag attached to the word. POS tag indicates which word-form of the lemma is used in the raw text, this is necessary for the further analysis. For example the word "*trades*" will be represented in LF as $trade(X,'VBZ')$, where $'VBZ'$ is tag for "verb - 3rd person, singular, present". For thematic roles we add predicates with predicate symbol θ and three arguments $\theta(X, agnt, Y)$. The constant that represents the thematic role is placed as a second argument. The rest of the arguments are bound with the lemma-predicates or constants, concerning this thematic role (see example 1). These predicates are called θ-*predicates*. All proper names are represented as constants that occur as arguments of the corresponding thematic roles.

Example 1:
Text in NL: *Primary market trades newly issued stocks and bonds.*
LF: *logical_form(primary_market(_451,' NN')&trades(_425,' VBZ')&*
θ(_425, agnt, _451)&θ(_425, with, _1569&_1570)&newly(_1169,' RB')&
θ(_1569&_1570, attr, _1169)&issued(_1367,' RB')&θ(_1569&_1570, attr, _1367)&
stocks(_1569,' NNS')&bonds(_1570,' NNS')).

The adverbial relations in thematic roles are not specified (e.g. location, instrument, time etc.), because this task is extremely difficult and there is no need of such kind of "deep" semantic (dependency and context) analysis. Therefore all adverbial relations are marked with prepositions as they are met in the text.

An important resource given by the knowledge engineer (tutor) with the lesson text is a terminology bank. The terminology bank is a list of words and phrases used as terms in the given text, which should be represented as single concepts. During the parsing process the compound terms from the terminology bank are recognized in the text and the consisting words are concatenated with "_" in a single predicate name. For instance *"primary market"* becomes *"primary_market"*. This allows correct interpretation of terms as concepts in the next step of generating CG form LF. Once the lesson text is parsed, all LF are stored into the KB of LF for further processing. All LF in this KB are in conjunctive normal form (CNF).

4.2   Conceptual Graphs Generator

The CG generator module consists of a preprocessor and a translator submodules. The preprocessor instantiates the arguments of θ-predicates with the lemma-predicates and removes all lemma-predicates from the LF. So after this stage LF contains only θ-predicates which could be almost directly translated to relations. The name of the CG relation is taken from the second argument of a θ-predicate and the other arguments are translated as concepts corresponding to it. The translator uses some rules for translating the θ-predicates into relations and concepts. Most of θ-predicates are translated as relations while others just add some features to already created concepts (see example 2).

Example 2:
This part of LF:
θ([year,' NN'], count, [one,' CD'])&θ([security,' NNS'], maturing, [year,' NN'])
will be translated as the following CG:

```
[SECURITY]->(MATURING)->[YEAR: @1].
```

Some rules for translation of the POS types assigned by the Lexical Analysis modules to concept referents are also used during the concept creation.

The CG generator creates mostly simple graphs and the only kind of compound graphs are type definitions graphs. They are not created just by looking at the sentence but also by checking the type hierarchy (see example 3). So the

CG generator module takes as input the LF KB, applies some rules and translates the LF KB into CG KB.

Example 3:
The sentence:
"Security is a financial instrument which represents debt of corporation"
will be translated as the following CG:

```
[SECURITY]->(DEF)->[Definition: [REPRESENT]-
                    (AGNT)-> [FINANCIAL_INSTRUMENT]
                    (OBJ)-> [DEBT]->(POSS)->[CORPORATION] ] ,
```

if we have a clause in the type hierarchy *isa(security, financial_instrument)*.

Using string similarities it is relatively easy to automatic construct "naive" type hierarchy e.g. *primary_market* is a subtype of *market*. For more complex cases as in the example above more powerful and complex methods for generation are needed. Considering that a type hierarchy corresponding to a lesson contains just a few concepts, it would be ineffective to apply any methods. So we chose the type hierarchy to be an input for the CG-EST system. A tutor gives the type hierarchy as clauses that are used in the CG generator and the subsequent modules.

5   Self-tutoring Module

CG-EST self-tutoring module automatically generates questions and defines the scope of correct answer. It aims to assist the process of checking user comprehension of material presented as lessons in NL.

5.1   Questions Generation

The main goal of this module is not just to invert one of the sentences given in the lesson text to questions, but to generate a more complicated question by making some inference. Then it is clear that if the user answer is correct, s/he understands the lesson, not just repeats explicitly said statements in the text.

The module for question generation functions by performing the following steps (see fig. 3 and example 4):

- step 1: Module selects one logical form L_i from the generated KB of LF (lets denote as $\sum$);
- step 2: Module finds the set:
  $S = \{L_j | L_j \in \sum \wedge \exists \; positive \; literal \; l \in L_i \wedge \exists \; substitution \; \sigma(l\sigma \in L_j)\}$
  This set contains all correct statements related to the question, which will be given to the user;
- step 3: Then we found the set C of all positive literals that represent concepts and belong to the LFs from S. So we skip all literals for thematic roles, time, place etc. $C = \{c | \exists L_i \in S \wedge \exists \; substitution \; \sigma(c\sigma \in L_i)\}$

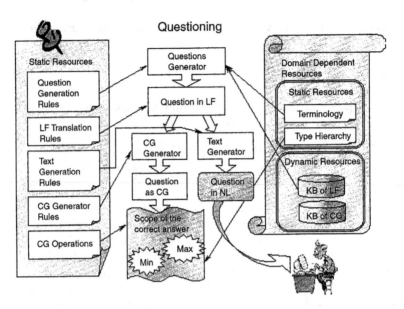

Fig. 3. CG-EST: Questions Generation Module

- step 4: Module chooses a non empty subset $A \subseteq C$. The subset A contains those concepts, which will be questioned;
- step 5: The set S is reduced to the set S' by skipping those LF that do not contain chosen concepts in A, i.e:
  $$S' = \{L_p \in S | \exists c \in A \wedge \exists \text{ substitution } \sigma(c\sigma \in L_p)\}$$
- step 6: Question logical form Q is produced by replacing variables in the terms representing concepts from A with the variable "UNIV" in S' and taking common parts of LF from S' by applying unification algorithm. Further an additional step reduces Q to Q' by skipping all "singleton" literals i.e literals that do not contain variables for which exists literal from Q that contains any of them. This step is necessary to generate coherent, correct and sensitive questions;
  $$Q' = Q \setminus \{l \in Q | X = \text{arg\_set}(l), \forall x \in X : \neg \exists l_i \in Q(l_i \neq l \wedge x \in \text{arg\_set}(l_i))\}$$
- step 7: Text Generation module generates a question in NL from question's logical form Q' by using backward operation for reconstruction of the sentence from its LF. This operation is simple because we do not have discourse and all sentences are universally quantified due to the specific domain.

The algorithm guarantees that we will generate only correct questions because we don't use any kind of generalization, specialization of type hierarchy inference. **Example 4:**

Let the given Lesson from which KB of LF and KB of CG are produced be:

(4.1) *Financial market trades financial instruments. Security is a financial instrument which represents debt of corporation. Bond is a security which represents debt of corporation. A stock is a security that represents an equity interest in a business. The owner of a stock is a stockholder. The stockholder has the right to vote in stockholder meetings. Stockholder meetings take place regularly at the end of the fiscal year. Primary market is a financial market that operates with newly issued financial instruments. Primary market trades newly issued securities and provides new investments. Secondary market is a financial market that operates with already issued financial instruments. Secondary market trades already issued securities.*

For instance let it be chosen LF (see Example 1) of the sentence:

(4.2) *Primary market trades newly issued stocks and bonds.*

Then the set of related sentences to (4.2) that is found after logical inference will contain LF of the following sentences:

(4.3.1) *Financial market trades financial instruments.*

(4.3.2) *Security is a financial instrument which represents debt of corporation.*

(4.3.3) *Bond is a security which represents debt of corporation.*

(4.3.4) *A stock is a security that represents an equity interest in a business.*

(4.3.5) *Primary market trades newly issued stocks and bonds.*

(4.3.6) *Primary market trades newly issued securities and provides new investments.*

(4.3.7) *Secondary market trades already issued securities.*

(4.3.8) *Primary market is a financial market that operates with newly issued financial instruments.*

(4.3.9) *Secondary market is a financial market that operates with already issued financial instruments.*

(4.3.10) *Financial market trades securities.*

(4.3.11) *Financial market trades bonds.*

(4.3.12) *Financial market trades stocks.*

(4.3.13) *Secondary market trades already issued bonds.*

(4.3.14) *Secondary market trades already issued stocks.*

The set of common concepts of (4.3) is:

(4.4) *[already_issued, bond, financial_instrument, financial_market, newly_issued, primary_market, secondary_market, security, stock]*

Let the chosen subset of (4.4) also contains the relation "trade securities":

(4.5) *[financial_market, primary_market, secondary_market]*

Then the reduced set of LF (4.3) contains LF of the following sentences:

(4.6) *Secondary market trades already issued securities. Primary market trades newly issued securities and provides new investments. Financial market trades securities.*

To produce LF of a question all concepts from (4.5) are replaced by "UNIV" and common parts are taken from LF :

(4.7) $logical\_form(trades(\_425,'VBZ')\&\theta(\_425, agnt, UNIV)\&$
$\theta(\_425\, obj, \_1569)\&securities(\_1569,'NNS')$.

The text generation module produces: (4.8) *Who does trade securities?*

5.2   Scope of Correct Answer

One approach for automatic check-up of answers is to assign scores to them. In [8] the answer is matched against three types of patterns: QA patterns (manually prepared), Qtargets (sematic type of determined answer) and Qargs (additional information related to Qtargets), and a score is calculated according to the matching. The standard approach for checking user answer is to compare it with a given beforehand correct answer. Both approaches need some manual work in advance. Our approach is to check the answer correctness by using an automatic generated scope. To have the user freedom in answering and according to [3] we chose to define correct answer scope as two sets: a *minimal required* and a *maximal expected* answer set. The approach adopted in [3] checks user answer both syntactically and semantically, it also gives a useful feedback to the user. On the other hand a lot of manually prepared meaning postulates are needed.

Since the knowledge representation model in CG-EST is CG, our approach consists of translating the question into CG (query graph) and applying CG operations to determine the scope of a user answer. The LF of the question obtained by the previous module is processed by the CG generator and a query CG is created (see example 5).

Example 5: Query graph for (4.8)

```
[UNIV]<-(AGNT)<-[TRADE]->(OBJ)->[SECURITY: {*}].
```

The projection operation projects one graph into another if the latter is a specialization of the former one. So by projecting the query graph to the knowledge base, CG-EST extracts all CG that fulfill the query graph (see example 6). These graphs also contain minimal information that should be included in the answer. Thus the resulting graphs are considered as the minimal required answer.

Example 6: Minimal answer graphs for the question (4.8)

```
[PRIMARY_MARKET]<-(AGNT)<-[TRADE]->(OBJ)->[BOND: {*}].
[PRIMARY_MARKET]<-(AGNT)<-[TRADE]->(OBJ)->[STOCK: {*}].
[PRIMARY_MARKET]<-(AGNT)<-[TRADE]->(OBJ)->[SECURITY: {*}].
[SECONDARY_MARKET]<-(AGNT)<-[TRADE]->(OBJ)->[SECURITY: {*}].
```

In addition these CG are processed as in [4] to obtain the corresponding pairs (query concept/ KB concept). For each retrieved KB concept, all CG from minimal required answer are obtained and a special maximal join operation is executed. "Special" here means that we perform maximal join on batch of CG until only one CG is obtained. The algorithm for the operation consists of two steps:

- step 1: arbitrary selection of two graphs, maximal join carried out on them and replacement of these two graphs by the result of the join in order to receive new CG batch;
- step 2: performing step 1 over the CG batch until the CG batch consists of one CG.

Thus, by applying the algorithm we reduce the set of minimal answer graphs.

Example 7: Reduced minimal answer graphs for the question (4.8)

```
(7.1) [PRIMARY_MARKET]<-(AGNT)<-[TRADE]-(OBJ)-[BOND: {*}]
                                      -(OBJ)-[STOCK: {*}].
(7.2) [SECONDARY_MARKET]<-(AGNT)<-[TRADE]->(OBJ)->[SECURITY: {*}].
```

The maximal answer set has to contain all true statements related to the question. These are not only explicitly said in the lesson text facts but are also facts that can be extracted with some inferences. A set containing all these truth utterances, is already created by the question generation module in step 2 (see example 4, especially (4.3)) . The representation of facts in this set is in LF and the CG generator from the preprocessing module is used again to translate them into CG and create maximal answer CG set.

6 Answering Module

This module expects user answer in a few full declarative sentences in English as an input. Answering module performs similar techniques as the preprocessing module but instead of generating KB of LF and CG, it produces user answer in LF and user answer as CG (fig. 4).

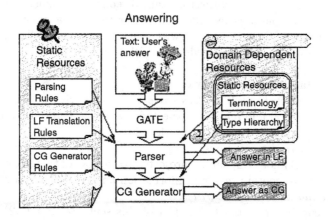

Fig. 4. CG-EST: User Answer Processing Module

6.1 Matching Algorithm

The matching algorithm realizes the actual checking of the user utterance correctness against the scope (fig.5). The main goal of this step is to verify the user answer and to give him/her detailed feedback. The feedback given by CG-EST for truth to the user is elaborated taking into account the feedbacks used in [5].

The correct user answer has to include the minimal required answer and to be subset of the maximal expected answer. Since the knowledge representation model in CG-EST is CG, our approach consists of determining the user answer scope as a set of CG and applying CG operations to define its correctness. There are two steps provided for matching the answer into the scope of correctness. Both steps use the matching rules as described below.

The first step validates if the user answer contains minimal required information. As the only operation that preserves truth is the projection, it is used to match each CG from the minimal answer set to a user answer CG set. If the match succeeds, this step is followed by the second one. Otherwise, the algorithm tries to project the user answer to the minimal answer CG set. If the projection succeeds then the answer is incomplete, see example 8, especially (8.2), because the minimal required information is partially expressed. If the projections mentioned above are impossible, the answer is considered as absolutely wrong, see example 8, especially (8.1).

The second step aims to prove the truth of additional statements made by the user. They should be related to the question and be expressed in the lesson. This means that the user answer CG set should be a subset of the maximal answer CG set. The algorithm checks if every CG from the user answer set can be projected to a CG from the maximal answer set. If the projection succeeds the user answer is absolutely correct, see example 8, especially (8.4). If not - there are some utterances that are false and the answer is only partially correct as we have already proved that user answer contained required statements in minimal answer set, see example 8, especially (8.3). After the user answer is processed all information about the correctness (wrong, incomplete, partially correct and correct) is given to the user as a feedback in NL form.

Example 8: Matching user answer
Let's ask the question (4.8) *Who does trade securities?* with minimal required answer set (example 7) and maximal expected answer set translation of (4.3) to CG. The following answers are given by the user:
(8.1) *Stockholder trades securities.*
Step 1: Neither (7.1) nor (7.2) can be projected to the CG of this sentence. Thus the answer does not contain required information and is *absolutely wrong*.
(8.2) *Primary market trades securities.*
Step 1: (7.1) can be projected to the CG of this sentence, but (7.2) can not. Thus the answer contains only part of required information and it is *incomplete*.
(8.3) *Financial market trades securities. However secondary market trades newly issued securities.*
Step 1: Both (7.1) and (7.2) can be projected to the CG of the first sentence. Therefore the answer contains required information and the process continues.
Step 2: The CG of the first sentence can be projected to the CG of (4.3.10), but the CG of the second sentence can not be projected to any of the CG from the maximal expected answer CG set. This means that the answer contains additional wrong statements and it is only *partially correct*.

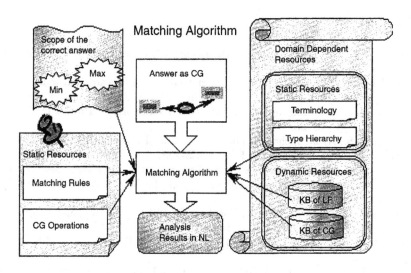

Fig. 5. CG-EST: Matching Algorithm

(8.4) *Financial market trades securities.*
Step 1: Both (7.1) and (7.2) can be projected to the CG of this sentence. Therefore the answer contains required information and the process continues.
Step 2: The CG of this sentence can be projected to the CG of (4.3.10), so the answer is *correct*.

7   Conclusion and Further Work

This paper presents an attempt for building a "more intelligent" self-tutoring approach to support eLearning paradigm. The main innovations of our approach are: *(i)* dynamically extraction of a CG KB from the text lessons; *(ii)* automatic generation of complicate questions from the KB by making some inference; *(iii)* distinguishing correct, partially correct, incomplete and wrong answers using formal CG processing. The system helps a tutor to easily create self testing materials for a user and assists the user in studying specific material. Unfortunately there is not known any other similar system to compare our results with it.

It is clear that such kind of systems can process only text in CE because in such a way we use more "formalized" sentences in English. The presented approach shows how only with some poor semantic information (type hierarchy and terms) KB of CG can be automatically constructed. CG-EST uses the results of GATE, therefore CG-EST performance depends on GATE's performance and the built-in GATE data corpus. However CG-EST can be improved by adding more rules in Static resource bank.

References

1. Bloom, B.S. and Krathwohl, D.R., Taxonomy of Educational Objectives, Addison Wesley, NY, 1984.
2. Boytcheva, S., P. Dobrev and G. Angelova, CGExtract: Toward Extraction of Conceptual Graphs from Controlled English, In Contr. ICCS-2001, pp. 89-102
3. Boytcheva, S., ILP Techniques for Free-Text Processing, In Proc. AIMSA-2002, LNAI 2443, pp. 101-110
4. Boytcheva, S., A. Strupchanska and G. Angelova, Processing Negation in NL Interfaces to Knowledge Bases, In Proc. ICCS-2002, LNAI 2393, pp.137-150
5. Boytcheva, S., O. Kalaydjiev, A. Strupchanska and G. Angelova, Bewveen Language Correcness and Domain Answering on the World Wide Web, In Proc. RANLP-2001, pp. 40-46
6. Cunningaham, H., D. Mayard, K. Boncheva, V. Tab-lan, C. Ursu and M. Dimitrov (2002) "The GATE User Guide". http://gate.ac.uk/.
7. Fuchs, G. and R. Levinson, The CG Mars Lander, In Proc. ICCS-97, LNAI 1257, pp. 611-614.
8. Hovy, E., Gerber, L., Hermjakob, U., Junk, M., and Lin, C.-Y., Question Answering in Webclopedia. In Proc. TREC-9, NIST, Gaithersburg, MD
9. Kunichika, H., Katayama, T., Hrashima, T., and Takeuchi, A., Automated Question Generation Methods for Intelligent English Learning Systems and its Evaluation,In Proc. ICCE 2002, New Zeeland.
10. Levinson, R. Symmetry and the Computation of Conceptual Structures. Proc. ICCS-2000, Darmstadt, Germany, LNAI 1867, pp. 496-509.
11. Martin Ph.,P. Eklund, Large-Scale Cooperatively-Build KDs,ICCS-2001, LNAI 2120, pp. 231-244
12. Nicolas St., Mineau G. and B. Moulin, Extracting Conceptual Structures from English Texts Using a Lexical Ontology and a Grammatical Parser, In Contr. ICCS-2002, pp. 15-28
13. Parasite, http://www.co.umist.ac.uk/~allan/index.html. Parasite is free for academic applications.
14. Rassinoux, A.-M., Baud, R. H. and J.-R. Scherrer. A Multilingual Analyser of Medical Texts, In Proc. ICCS'94, LNAI 835, pp. 84-96.
15. Schroeder, M. Knowledge Based Analysis of Radiology Reports using Conceptual Graphs, In Proc. 7th Annual Workshop: Conceptual Structures: Theory and Implementation, July 1992, LNAI 754.
16. Smilkstein, Rita, Acquiring Knowledge and Using It, Journal Gamut, Published by Seattle Community College District (Washington) 1993, pp. 16-17,41-43
17. Sowa, J. Conceptual Structures: Information Processing in Mind and Machine, Addison-Wesley, Reading, MA, 1984.
18. Sowa, J. and E. Way. Implementing a Semantic Interpreter using Conceptual Graphs, IBM Journal R&D, Vol. 30 (1), 1986, pp. 57-69.
19. Sowa, J. Using a Lexicon of Canonical Graphs in a Semantic Interpreter. Relational Models of the Lexicon, Cambridge University Press, 1988, pp. 113-137.
20. Velardi, P.,M. Pazienza and M. De'Giovanetti. Conceptual Graphs for the Analysis and Generation of Sentences. In: IBM J. Res. and Develop. Vol. 32 (2),March 1988, pp. 251-267.
21. Zhong L., Y. Yu, Learning o Generate CGs from Domain Specific Sentences, ICCS-2001, LNAI 2120, pp. 44-57

Cognitive Map of Conceptual Graphs: A Graphical Model to Help for Decision

Ghania Aissaoui, David Genest, and Stéphane Loiseau

LERIA – Université d'Angers
2, Boulevard Lavoisier - 49045 Angers cedex 1 - France
{aissaoui,genest,loiseau}@info.univ-angers.fr

Abstract. We propose a new model for decision support systems. A decision support system helps a user to make a choice between alternatives to reach a goal. The cognitive map model enables a user to visualize the influences of the alternatives, and to compute the propagation of influence on the goal. Like cognitive maps, our model offers a graphical representation of influences between notions (alternatives and goals). The distinctive feature of our model is that on a unique support, each notion is precisely defined by conceptual graphs. One of the major advantages of our model is the combination of operations of cognitive maps and operations of conceptual graphs. Firstly, the definition of a notion and the projection provides a solution to compute semantically linked notions. Secondly, original propagations can be computed from such semantically linked notions.

Keywords. cognitive map, decision support system, influence.

1 Introduction

To present some information, human being uses graphical representations. Such representations as geographical map, diagram, or inheritance hierarchy between classes, provide a way to synthesize information in an easily interpreted and exploited form. The graphical media is a good one to apprehend the information globally before paying attention to interesting parts of it.

This article focuses on tools for decision support systems. A decision support system uses knowledge and provides mechanisms to help the user. A decision is the choice of a user among different alternatives to reach the goal that he fixes. Some systems, as expert systems, determine which alternative must be chosen in function of the goal. Some systems give more initiatives to the user, they simply present him different alternatives and provide tools to help him to make his own decision. The main idea of such systems is to help the user to take into account the semantic bonds existing between the different represented notions: by knowing its consequences, a user can take his decision. However, to make this help efficient, the user must understand the represented knowledge, he must access easily to the alternatives and he must know their consequences. Our contribution takes place in this kind of approach, in which the interaction is based on maps.

The interest to represent knowledge with maps is not new in decision support systems. The cognitive map model [Chaib-draa et al, 1998] has been studied in

A. de Moor, W. Lex, and B. Ganter (Eds.): ICCS 2003, LNAI 2746, pp. 337–350, 2003.

cognitive or political sciences. It provides a solution to represent knowledge graphically [Louça, 2000]. Tolman [Tolman, 1948] was the first one to use the expression "cognitive map"; it represents a space made of alternatives for an animal. Later, some works, like [Axelrod, 1976] use cognitive maps to represent graphically believes of a person on a particular domain. The goal is to capture influence relations, called bonds, between different notions: a notion is either favorable or not favorable to an other notion. Today, cognitive maps are mainly used in three main approaches. First, cognitive maps are used to structure ideas [Huff et al, 1992]. Cognitive cartography clarifies a confuse idea because it structures the idea in a representation. Second, cognitive maps are often used as a medium to communicate between people [Eden, 1998]. A cognitive map helps decision-makers to share their ideas and becomes a tool to communicate. Third, cognitive maps are used in decision support systems [Huff et al, 1992]: the cognitive map is a model which aim is to represent the way by which a person can find a solution for a given problem. Our work is about this third approach. A cognitive map contains two kinds of information: on the one hand nodes called *states* represent notions, and on the other hand directed edges between those nodes represent influence links, called *bonds* generally positive or negative. Fig. 1 presents an example of a cognitive map. The right part of the map can be interpreted as followed: "disobeying a stop sign increases the risk of fatal accident". The influence is represented by a bond labelled by +. The map also represents that "The presence of policemen decreases infractions due to non respect of stops". The cognitive map model also defines an inference mechanism that consists into the *propagation* of influences. This inference process is the basis of the decision support process. In this way, a user confronted to a choice among many alternatives (represented by different states) to reach a goal (a state), can use the propagation mechanism to know the influence of each alternative on the goal.

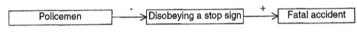

Fig. 1. A cognitive map

A first drawback of cognitive maps is their too important flexibility because a state can be represented by any linguistic label with no definition, what causes ambiguity: for instance different users can use the same map with a different vocabulary. A second drawback of the model is the lack of structure, that implies that implicit links between states (that are not only influence) cannot be expressed. So, the choice of alternatives is only made by the user who has no help. It is particularly difficult with large maps, with which the user may be lost. A solution to face those problems is to represent more precisely each state.

Different models provide a solution to represent knowledge precisely, and to define query operations to search sets of knowledge that are logically connected. Among those models, we selected the conceptual graph model [Sowa, 1984] [Chein et al, 1992] first because it is a graphical knowledge model, and so it is homogenous with the visual aim of cognitive maps and second, it has a logical semantics. A conceptual graph is a graph composed of *concept* vertices representing entities and *relation* vertices representing relation between entities. A graph is defined on a structure called *support* that to specifies and organizes in hierarchy the basic vocabulary used for concepts and relations. The support corresponds to a representation of the ontology of the domain. A formal operation, called *projection*, provides a way to search logical links between graphs.

A main weakness of cognitive maps is their lack of clear definition for their states; use of conceptual graphs overcomes this weakness. Some works such as [Povolina, 1999] underline the interest to use conceptual graphs for decision support. A distinctive feature of our work is that our model combines cognitive maps and conceptual graphs. More precisely, the main idea of the model which is presented in this article under the name of *cognitive map of conceptual graphs* is that each state of a cognitive map is defined by a conceptual graph. On the one hand, the use of a conceptual graph, associated to each state of the map, provides a way to define precisely each state referring the definition to an ontology that is the support. On the other hand, the model offers the possibility to select the set of states of a map that are semantically linked with a given state, called a *query state*. The result set, called *collection* is automatically built using the projection operation: this collection contains the states of the map that are defined by a graph such that there is a projection from the conceptual graph defining the query state to it. This possibility can be combined with the power of propagation, issued of cognitive maps, that describes and computes influences between states. For instance, it is possible to visualise the influence of a query state, through all the states of its collection, on other states.

The article is composed of three parts. In the first part, we describe the cognitive map of conceptual graphs model. In the second part, we define the operations that provide the reasoning and we describe the benefits of such operations for decision support systems. The third part is a short presentation of the system that implements the cognitive maps of conceptual graph model; an example illustrates its use.

2   The Model of Cognitive Map of Conceptual Graphs

In this part, we briefly recall the conceptual graph model we used and that is issued of [Chein et al, 1992] [Mugnier, 2000]. Then we define our new model of cognitive map of conceptual graphs and we compare the objects of this new model with objects of the conceptual graph model and the cognitive map model.

2.1   Conceptual Graph

A conceptual graph has a meaning only relatively to a support.

Definition: support

A *support* S is a quintuplet $(T_C, T_R, \sigma, I, \tau)$ such that :

- T_C, *the set of concept types*, is a set partially ordered by a "a kind of" relation (denoted $\leq$) having a maximum element denoted T.

- T_R, *the set of relation types*, is a partially ordered set, partitioned in sub-sets of relations types of same arity. $T_R = T_{Ri1} \cup \ldots \cup T_{Rip}$, where T_{Rij} is the set of relations types of arity i_j, $i_j \,] \, 0$. Every T_{Rij} has a maximum element (denoted T_{ij}).

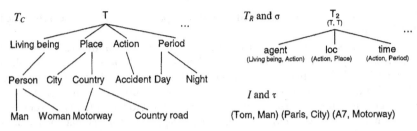

Fig. 2. A support

- σ is the application that associates to each relation type a *signature*, it is the maximal type of each of its arguments. More precisely, to each $t_r \in T_{Rij}$, σ associates a tuple $\sigma(t_r) \in (T_c)^{ij}$. We denote $\sigma_i(t_r)$ the ith argument of $\sigma(t_r)$.
- *I* is the *set of individual markers*. An individual marker represents an individual of the knowledge base. We dispose of a generic marker denoted ✻ that represents a non specified individual. $I \cup \{✻\}$ is provided by the following order : ✻ is greater than any individual marker, and individual markers are pairwise incomparable.
- τ is the *conformity* relation. It is an application of *I* in T_c, that associates a concept type to each individual marker.

The support described in Fig. 2 defines concept types such as *City* (a kind of *Place*), relation types such as *agent* (can be used between a *Living being* and an *Action*) and individual markers such as *Tom* (who is a *Man*). A conceptual graph expresses a fact using the vocabulary defined in the support. It is a bipartite graph, it includes two kinds of nodes : the concept one's and the relations one's. Those nodes are labelled to express knowledge.

Definition: conceptual graph

A *conceptual graph* $G = (C_G, R_G, E_G, lab_G)$ defined on a support S, is a bipartite non oriented multigraph, where :

- C_G is the set of *concept vertices* of the graph and R_G the set of *relation vertices* ($C_G \cap R_G = \varnothing$). We denote N_G the set of vertices ($N_G = C_G \cup R_G$).
- E_G is the set of *edges*. Every edge of a conceptual graph G has an extremity in C_G and the other in R_G
- lab_G is the application that associates a label to each vertex of N_G and to each edge of E_G:
  - if $r \in R_G$, $lab_G(r) \in T_R$.
  - if $c \in C_G$, $lab_G(c) \in T_C \times (I \cup \{✻\})$. The label of a concept vertex c is a couple $(type(c), marker(c))$. There is a partial order on the labels of concept vertices, which is the product of the order of T_C and the order on $I \cup \{✻\}$: having two labels $e = (t, m)$ and $e' = (t', m')$, $e \leq e'$ if and only if $t \leq t'$ and $m \leq m'$.
  - if $e \in E_G$, $lab_G(e) \in \mathbf{N}$.
- lab_G verifies the constraints fixed by σ and τ :
  - The set of adjacent edges of each relation vertex r is totally ordered, that is represented by labelling the edges from 1 to the degree of r. We denote $G_i(r)$ the ith neighbour of r in G.
  - For each $r \in R_G$, $type(G_i(r)) \leq \sigma_i(type(r))$.
  - For each $c \in C_G$, if $marker(c) \in I$ then $type(c) = \tau(marker(c))$.

Fig. 3. A conceptual graph

Fig. 3 presents a conceptual graph defined on the support described in Fig. 2. This graph can be interpreted by "fatal accident" (accident in which a person dies). As the figure shows, the graphical nature of conceptual graphs makes easy to interpret them, to modify them or to create new one's. This facility to create, is reinforced by the explicit separation of different kinds of knowledge, and more precisely the definition of an object distinct of graphs, the support, that helps during the creation of the graphs, because the labels of the vertices must be chosen in this support. The conceptual graph model is associated to a first order logic (FOL) semantics: the relation and concepts types correspond to predicates, the individuals correspond to constants, the generic markers represent variables, the links "kind of" are represented by implications and graphs by simple assertions [Sowa, 1984] [Chein et al, 1992].

2.2 Cognitive Map of Conceptual Graphs

A cognitive map of conceptual graphs provides a way to represent influence relations between different notions, called states, each of these states being defined by a conceptual graph. In the same way that a conceptual graph is related to a support, a cognitive map of conceptual graphs must be defined relatively to a unique support.

Definition: state
A *state* defined on a support S is a pair (i,G) where i is a *title* describing the state and $G=(C_G,R_G,E_G,lab_G)$ a conceptual graph defined on S. We say that G *defines* the state.

Definition: cognitive map of conceptual graphs
A *cognitive map of conceptual graphs* defined on a support S is an oriented graph $X = (E_x, L_x, lab_x)$ where :
- E_x is the set of nodes.
- L_x is the set of the directed edges, called *bonds* of the map. $L_x \subseteq E_x \times E_x$
- lab_x is a labelling function that to each element l of L_x associates a label $lab_x(l) \in \{+, -\}$ and to each element e of E_x associates a state defined on S : $lab_x(e) = (i_e, G_e)$ where G_e is a conceptual graph defined on S

For convenience, we often call "state" not only the couple (i, G) but also only the title i or the corresponding node. Fig. 4 presents an example of a cognitive map of conceptual graphs: on a wet road, the risk of fatal accident is more important.

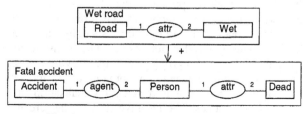

Fig. 4. A cognitive map of conceptual graphs

2.3 Comparison with Conceptual Graph and Cognitive Map

The graphical representation of a cognitive map of conceptual graphs evokes a nested graph [Chein et al, 1998]. Let us recall that a nested graph is a conceptual graph in which the label of a concept vertex can contain not only a type and an individual but also a conceptual graph. So, a cognitive map of conceptual graphs can be seen as a nested graph: each state is represented by a concept vertex, in which should be included its definition (a conceptual graph), and each bond is represented as a specific relation + or -. But, on the one hand, the (nested) conceptual graphs provides no specific semantics to relation vertex, as influence; and on the other hand, the title of states are independent of the support for flexibility reasons (a simple cognitive map is also a cognitive map of conceptual graphs used with an empty support and empty graphs). In cognitive map of conceptual graphs, the states are connected with influence bonds that have a meaning and so a specific usage that is not the same as the relation in conceptual graph; next part will describe this use.

The model proposed in this paper is different from the model of cognitive map because it increases the representation of states with conceptual graphs. To define a cognitive map of conceptual graphs, a support must be provided. But the price to pay to construct the support is rewarded: the lack of structure that we criticize on simple cognitive maps, disappears thanks to conceptual graphs. The states are clearly defined by conceptual graphs which labels belong to the unique support. In a simple cognitive map, a state labelled as "alcohol" can be interpreted from different ways: is it the driver that has drunk ? or is it the pedestrian that is drunk and was implicated in the accident ? or is it an other thing ? In the new model, we clearly see what is the idea of the designer because he does not only give the title to a state but also its definition by a conceptual graph (that has a logic semantics); for instance in Fig. 5, its semantic is that the driver is drunk. So the ambiguity that could exist on the meaning of this state disappears. The conceptual graph model has a logical semantics so it provides a logical semantics to our states definitions and not only a graphic one.

Fig. 5. A conceptual graph defining the state "alcohol"

The introduction of conceptual graph structure in our model and the definition of a support on which are based our definition of states, also increase the utility of a solution based on cognitive maps not only as a decision support system but also as a media of communication between people. Everyone can construct its map, and maps can be compared if the users use the same support.

As for models which it is issued, one major point of our model of cognitive map of conceptual graphs is its graphical approach. Fig. 6 shows a cognitive map of conceptual graphs in which conceptual graphs have been hidden. The display of titles only is the first level of the map. If the user selects a state, he can access to a second level in which the conceptual graph that defines the state is displayed. We call this process the zoom on states. This zoom provides a simple way to access to conceptual graphs, without surcharging the map. One important point of this model is to keep the main advantage of cognitive map, that is a general visualisation of the influence network between the different states. The negative (red on the screen, here labelled with the "-" symbol) and positive bonds (green on the screen, here labelled with the "+" symbol) provides an overview vision of the map. It is also possible to focus our interest only on certain parts of the map, or to access to definitions of some states.

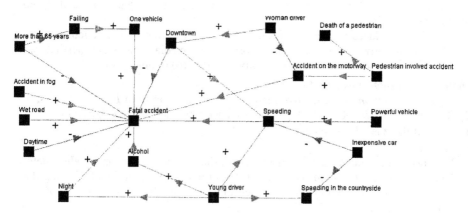

Fig. 6. Overview of a cognitive map of conceptual graphs

3 Reasoning

In addition to the semantics associated to each state by a conceptual graph, one of the distinctive feature of this new model is the reasoning it allows. Indeed, this model defines a mode of reasoning integrating inference techniques of both the conceptual graph model and the cognitive map model: logical deduction from conceptual graphs and propagation of influences, a kind of fuzzy reasoning, from cognitive maps . We will focus on the way to see our model as a decision-support tool allowing some operations that are impossible with simple cognitive maps. First, we describe the mechanisms allowing to propagate the influences between states of the map. In the second time, we describe a *query state*, which is a state defined as a conceptual graph. Such a state gives access to the states of the map which relate to the subject represented in the query. Let us note that a query state can be either a state of a cognitive map of conceptual graphs or a state only defined to execute queries. Thus, from a query state, we can find, thanks to projection, the collection of states that are semantically bound to it, called the *collection of the state*, and see, for each of these states, the influences they allow through the propagation of influence bonds. Our method thus allows, starting from only one state, to calculate influences, from the states (or to the states) which are bound to it by their definition.

3.1 Influence and Propagation in Cognitive Maps

Using the map on Fig. 1, we can intuitively deduce that the presence of policemen decreases the risk of accident. We now define an inference mechanism leading to such deductions. This process of inference is the base of the decision support, it is presented for example in [Chaib-draa et al., 1998] [Chaib-draa, 2001] and formalized in the following.

Definition: influence path

An *influence path* H on X is a succession of states h_1, h_2, ..., h_n of E_X such that there exists a path between these nodes, i.e., for any $i \in [1, n-1]$, $(h_i, h_{i+1}) \in L_X$. h_1 is called the *origin* of the path, h_n the *destination*, and the other states are called the *intermediate* states.

The effect produced by the origin state on the destination state is a function of the positive and negative bonds between the origin, intermediate states and destination. This effect can be positive (noted +), negative (-), null (0) or ambiguous (?).

Definition: influence function

An *influence function* of a cognitive map X is an application from $E_X \times E_X$ to $\{+, -, 0, ?\}$. The *elementary influence* I_1 of X is an influence function such that $I_1(e_i, e_j) =$
- $lab_X(l)$ if there is a $l = (e_i, e_j)$ in L_X ;
- 0 if not.

Two operations are used to measure the influence of a state on another state: *addition* corresponds to the idea of sum of the influences of several paths having the same destination and *multiplication* corresponds to the idea of transitivity of the influences in a path.

Definition: addition

Addition in $\{+, -, 0, ?\}$ is noted $\oplus$ and defined by the following table:

$\oplus$	+	-	0	?
+	+	?	+	?
-	?	-	-	?
0	+	-	0	?
?	?	?	?	?

We say there is an ambiguity when there is at least a conflict, i.e. when we have to sum a positive influence and a negative influence.

Definition: multiplication

Multiplication in $\{+, -, 0, ?\}$ is noted $\otimes$ and defined by the following table:

$\otimes$	+	-	0	?
+	+	-	0	?
-	-	+	0	?
0	0	0	0	?
?	?	?	?	?

In an intuitive way, the multiplication is based on the following principle: "a friend of an enemy is an enemy" and "an enemy of an enemy is a friend". With these two operations, it is possible to define the *total influence* of a state e_i on a state e_j as the sum of the indirect effects of the paths from e_i to e_j.

Definition: total influence

Total influence I_T of X is an influence function of X such that:

$$I_T(e_i, e_j) = \bigoplus_{H \in H_{i,j}, i \in [1, card(H)-1]} \bigotimes I_1(h_i, h_{i+1})$$ (where $H_{i,j}$ is the set of paths from e_i to e_j)

Thus the calculation of the total influence is the basic operation of the model, it can be used to determine the influence of a state on another. Fig. 7 can be interpreted by "Women often drive in downtown, and they often drive small cars. Fatal accident are more frequent at night, and with small cars, and they are less frequent in downtown". As an example, let us calculate the total influence of "night" (e_5) on "fatal accident" (e_1): There is one path between these two states, and this path contains only one bond: $I_T(e_5, e_1) = I_1(e_5, e_1) = +$. One can deduce that there are more fatal accidents at night. Now let us calculate the total influence of "woman driver" (e_3) on "fatal accident"

(e_1): There are two paths between these states: $I_T(e_3, e_1) = (I_I(e_3, e_2) \otimes I_I(e_2, e_1)) \oplus (I_I(e_3, e_4) \otimes I_I(e_4, e_1)) = (+ \otimes -) \oplus (+ \otimes +) = ?$. The influence is thus ambiguous, which means that there are a positive influence and a negative influence.

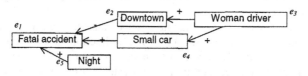

Fig. 7. A cognitive map

Total influence represents the nature of the influence bond between two states. We propose to extend this mechanism to visually represent the influences of a particular state, chosen by the user and called *choice state*, on each state of the map. Total influences $I_T(a, e_i)$ are calculated for each state e_i of the map, and, in order to present in a synthetic way this propagation, we propose to define a colouring of the map.

Definition: propagation colouring

A *propagation colouring* from a state a of a cognitive map of conceptual graphs $X = (E_X, L_X, lab_X)$ is an application *propagation* : $E_X \rightarrow \{+, -, 0, ?, *\}$ such that *propagation*(a) = * and *propagation*(e) = $I_T(a, e)$ if $e \neq a$.

In order to display a map, we can associate each element $\{+, -, 0, ?, *\}$ a real colour, for example blue (*, selected state), green (+, positive influence), red (-, negative influence), grey (0, null influence), orange (?, ambiguous influence). An example will be given in Fig. 10. In a symmetrical way, one can be interested by knowing the states that influence a given state b, called *target state*.

Definition: reception colouring

A *reception colouring* related to a state b of a cognitive map of conceptual graphs $X = (E_X, L_X, lab_X)$ is an application *reception* : $E_X \rightarrow \{+, -, 0, ?, *\}$ such that *reception*(b) = * and *reception*(e) = $I_T(e, b)$ if $e \neq b$.

3.2   Collection of States

The colouring mechanisms presented above relate to every state of the map. In a context of decision support, a user often prefers to select one or more choice states e_1, e_2, ..., e_m, corresponding to the alternatives he considers and a target state e_j, corresponding to the objective. Then the system has to determine the influence of these choices on the target: the total influence of each choice state on the target state can then be used to inform the user. If $I_T(e_x, e_j) = +$, the choice state e_x has a positive effect on the target state, and the choice has to be taken into account. If the obtained value is -, the effect is negative, and the choice is rejected. If the value is 0 or ?, no advice can be given. For example, in order to locate the causes of fatal accidents to target a prevention campaign, by using the cognitive map of Fig. 7, one can choose "Fatal accident" as target and "Night" and "Woman driver" as choices. The calculation of the total influences allows to deduce that conductors who drives nightly can be targeted in the campaign ($I_T(e_5, e_1) = +$), but the system does not give any advice on women drivers ($I_r(e_3, e_1) = ?$). In order to offer such a mechanism, we have to handle set of states.

Definition: collection of states
A *collection of states* F of a cognitive map of conceptual graphs $X = (E_X, L_X, lab_X)$ is a subset of the states set of $X : F \subseteq E_X$.

A collection of states allows to handle a set of states. Such a collection will be used, for example, to visualize how states of a collection of choice states have an influence on a target state. However, a manual selection of many states to build a collection can be tiresome. In addition to a waste of time, there is another issue. Indeed, if the user of a map is not the designer of the map, he can not understand the significance of the name of some states, and he has to zoom on some states to know if they belong to his search. Moreover, he can forget some interesting states. The author of the map himself may forget some states he created. He also may have used a title of a state, not precise enough, and forget this state during its selection. For example, if a state is entitled "pile-up" and thus must be selected in a collection of states about accidents, a user can forget this state because the word "accident" does not appear in the name of the state.

One of the issues of the cognitive map model is that the links can only represent influence: it is impossible to connect each state representing various alternatives of a given subject, in order to find them easily. For example, if there was a link between all the states representing the various types of accidents, it would be easy to build a collection of states containing them. Definition of states with conceptual graphs provide implicit links between states that can be used. That is why we propose a new operation, allowing to automatically build a collection that contains all specializations of a given state, using the conceptual graph definition of each state.

Definition: query state
A *query state* r defined on a support S is a state defined on S that is designed to compute its collection.

A query state is a couple (i, G) that will be used to select the states of a map (its collection) that are defined by a graph which "answers" G. In order to execute such a query, we use the projection operation [Sowa, 1984].

Definition: projection
A *projection* from $G = (C_G, R_G, E_G, lab_G)$ to $H = (C_H, R_H, E_H, lab_H)$ is an application Π: $N_G \rightarrow N_H$, which :
- preserves adjacency and labels of edges : for each edge rc in E_G, $\Pi(r)\Pi(c)$ is an edge of E_H and $lab_G(rc) = lab_H(\Pi(r)\Pi(c))$;
- may decrease labels of vertices : for each n in N_G, $lab_H(\Pi(n)) \leq lab_G(n)$.

Definition: collection of a query state
The *collection of a query state* $r = (i, G)$ related to a support S in a cognitive map of conceptual graphs $X = (E_X, L_X, lab_X)$ related to the same support, is the set of nodes $e \in E_X$ labelled by (i_e, G_e) such that there exists a projection from G to G_e.

Fig. 8 shows an example of use of a query state. By creating the query state labelled by "accident in the countryside", it is possible to simply query a map and retrieve the states representing all the "types of accidents in the countryside". For example, there is a projection from the graph defining the query state to the graph

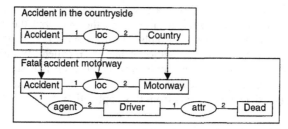

Fig. 8. "Fatal accident motorway" is a state of the collection of the query state "Accident in the countryside"

defining the state "Fatal accident motorway" of the map. Let us note that there is such a projection because the support defines *Motorway* ≤ *Countryside*. The state "Fatal accident motorway" thus belongs to the collection of the query state "Accident in the countryside". In a more general way, to choose a query state labelled by a graph *G* is equivalent to choose every state that is labelled by a graph which represents a fact more specialized than *G*. These states are selected, but the user does not have to explicitly select each of them, thanks to the projection. Notice that, if we only consider conceptual graphs in normal form (i.e. a given individual marker appears only once in a graph) we can provide a FOL semantics to the collection of a query state: using the support, the logical semantics of the definition of the query state is implied by the logical semantics of the definition of any state of the collection : It is due to the FOL semantics of projection [Sowa, 1984] [Chein et al, 1998].

We have used a collection to represent the various alternatives (choices) taken into account by a user to achieve a goal (target). But it is possible to use a collection as a target. Thus, if the user selects one choice state and a collection of target states, he can simply visualize the influences of one state on a set of states. The most general mode of use consists in defining a collection of choice states and a collection of target states. In this case, the system calculates each influence between a state of the collection of choices and a state of the collection of targets.

In conclusion, the first asset of this reasoning is to take into account collections of states as choices or targets, without having to select them explicitly. The second asset is that a query state allows to represent, as a graph, a concept which is perhaps not exactly present in the map, and to use this graph to find all specializations of this state among the states of the map. That brings practical advantages since it is a saving of time and the best way of not forgetting an interesting state. Finally, query states also allow a user to search if some ideas are present in the map, by querying states of the map.

4   Application

We developed a tool called "*3CG*" [Mahier, 2002] in Java. This tool is based upon *Swing* for the GUI and *Notio* [Linders et al., 1999] for conceptual graphs. 3CG is built around a zone displaying a map. This tool allows to build a cognitive map of conceptual graphs but also to use and query an existing map using the above described mechanisms. Fig. 9 shows a cognitive map of conceptual graphs in 3CG. Conceptual graphs are not displayed directly in the map, but the zoom mechanism allows to see, in a new window, the graph defining a state. It is available by a double-click on the state. We built an application using 3CG, called "Warning" that informs users about the causes of road accidents. The application is based upon a cognitive map, defined in 3CG, which relates to the field of the road safety. The examples presented in this article are excerpts from this application.

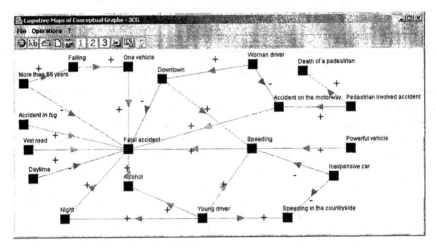

Fig. 9. A cognitive map of conceptual graphs in 3CG

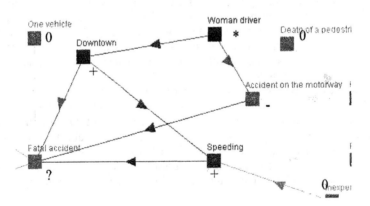

Fig. 10. Propagation

In our system, we can ask to see the propagation colouring from a selected state: the influences that starts from this state appear, and each state of the map is coloured using the propagation colouring (green or "+" for positive influence, red or "-" for negative influences, orange or "?" for ambiguous influence and grey or "0" for null influence). In Fig. 10, we see such a propagation from the state "Woman driver": we can see that among the concerned states, "Downtown" and "Speeding" are positively influenced, despite "Accident on the motorway" is negatively influenced and that the influence on "Fatal accident" is ambiguous. That allows to conclude, for instance, that women have less accident on motorway, they drive often in town and they often speed. The colour system provides an elegant solution to make evidence on the total influences arriving on states, because all the concerned states by this propagation have their colour changing depending on the influence they receipt.

The user can construct a query state using a conceptual graph editor. From such a state, the system can compute and display the collection of the query state, so it provides a way to question the map and facilitates the research of states in big maps. Once a collection is made (by projection), the user can use it as choice or target and can start the propagation of influences. For instance, to know in which types of accidents the women are implicated, a user can select the state "Woman driver" as choice, to make a query state defined with a conceptual graph made of only the vertex "accident" and choose for target the collection that is computed from this query state. The result is provided on Fig. 11. The collection of the query state contain all the states of the map that are defined using a vertex "accident" or a more specific one. They are denoted by "target" on the map of Fig. 11. The system presents, on a window titled "Results", the influences that are computed between the state choice "Woman driver" denoted by "choice", and the state of the collection as targets; we can, for instance, notice that women have less accident on motorways.

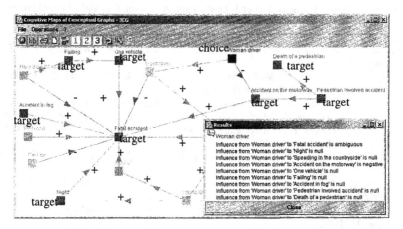

Fig. 11. Influence from "Woman driver" to "Accident"

5   Conclusion

The model presented in this article is original for two reasons. First, it provides a method to represent knowledge including cognitive maps model [Tolman, 1948] and conceptual graphs model [Sowa, 1984]. On one hand, influence relations between conceptual graphs can be represented and exploited, on the other hand, the support on which states of a cognitive map are defined is formal (each state is defined by a conceptual graph constructed on the support). Second, clear mechanisms to reason are proposed. The projection, issued from conceptual graph model, is combined with the propagation, issued from cognitive maps. This combination provides a powerful mean to make reasoning: it is possible to define a query state (or to take a state of the map) that enables to compute a collection of semantically linked states (the states that are associated to it by projection); so it is possible to take into account those states of the collection to test the propagation. To conclude, our model proposes a graphical representation clearly defined, with a clear semantic for the designer, simple to interpret for the user and some reasoning operations that helps him to exploit its cognitive map of conceptual graphs to make his choice during a session.

In large maps, it is frequent to obtain ambiguity during a propagation because there are many ways to connect a state to an other, and they may be contradictory. An extension of our work should provide a solution in spite of ambiguous solutions. A solution is to look for some works made in cognitive maps, as [Kosko, 1986] that proposes to label each bond not only by "+" or "-" but also by a number that represents the force of the influence. We prefer another solution which consists to replace "+" and "-" labels by linguistic labels such as "very positive", "very little negative".

References

[Axelrod, 1976] R. Axelrod. *Structure of decision : the cognitive maps of political elites*, 1976. Princeton University Press.

[Chaib-draa et al, 1998] B. Chaib-draa et J. Desharnais. A relational model of cognitive maps. *International Journal Of Human-Computer Science studies*, vol. 49, p. 191-200, 1998.

[Chaib-draa, 2001] B. Chaib-draa. Causal maps : Theory, implementation and practical applications in multiagent environments. *IEEE Transaction On Knowledge and Data Engineering*, 2001.

[Chein et al, 1992] M. Chein and M.-L. Mugnier. Conceptual Graphs: Fundamental Notions. *Revue d'Intelligence Artificielle*, vol. 6 (4), p. 365-406, 1992.

[Chein et al, 1998] M. Chein, M.-L. Mugnier and G. Simonet. *Nested Graphs: A Graph-based Knowledge Representation Model with FOL Semantics,* Proceedings of the 6th International Conference "Principles of Knowledge Representation and Reasoning" (KR'98), p. 524-534. Morgan Kaufmann Publishers, 1998.

[Eden, 1998] C. Eden. Cognitive mapping. *European Journal of Operational Research*, vol. 36, 1988.

[Huff et al, 1992] A. Huff and M. Fiol. Maps for managers : where are we ? where do we go from here? *Journal of Management Studies*, vol. 29, 1992.

[Kosko, 1986] B. Kosko. Fuzzy cognitive maps. *International Journal of Man-Machines Studies*, vol. 25, 1986.

[Linders et al, 1999] J. Linders and F. Southey. Notio, a Java API for Developing CG Tools. *Proceedings of ICCS'99*. LNAI 1640, p. 262-271. Springer, 1999.

[Louça, 2000] J.A. Louça. Cartographie cognitive, réflexion stratégique et interaction distribuée. PhD Thesis, Paris IX-Dauphine, 2000.

[Mahier, 2002] E. Mahier. *Cartes cognitives de graphes conceptuels – un modèle graphique d'aide à la prise de décision*. Mémoire de DEA, Université d'Angers 2002.

[Mugnier, 2000] M.-L. Mugnier. Knowledge Representation and Reasonings Based on Graph Homomorphism. *Proceedings of ICCS'00*, LNAI 1867, p. 172-192. Springer, 2000.

[Povolina et al, 1999] S. Polovina and A. Vile. A Gateway Too Far? Experiences on Integrating Accounting Information Systems with Strategic Knowledge, *Journal of Financial Information Systems*.

[Sowa, 1984] J. F. Sowa. *Conceptual Structures : Information Processing in Mind and Machine*. Addison Wesley, 1984.

[Tolman, 1948] E. Tolman. Cognitive maps in rats and men. *Psychological Review*, vol. 55, 1948.

Mining Concise Representations
of Frequent Multidimensional Patterns

Alain Casali, Rosine Cicchetti, and Lotfi Lakhal

Laboratoire d'Informatique Fondamentale de Marseille (LIF), CNRS UMR 6166
Université de la Méditerranée, Case 901, 163 Avenue de Luminy
13288 Marseille Cedex 9, France
{casali,cicchetti,lakhal}@iut.univ-aix.fr

Abstract. In this paper, we consider an instance of the constraint pattern problem: the frequent multidimensional patterns. We propose the first concise representation of frequent multidimensional patterns which on one hand is not based on the powerset lattice framework and on the other hand differs from the representations by closed patterns and non derivable patterns. From such a representation, we show that any frequent multidimensional pattern along with its conjunction, disjunction and negation frequencies can be obtained using inclusion-exclusion identities.

1 Introduction

Discovering constrained multidimensional patterns is a very important issue for multidimensional data mining because it makes it possible to solve various problems such as mining multidimensional association rules, multidimensional constrained gradients, classification rules and correlation rules. It is also the fundamental step when computing full, iceberg or range cubes materialized for OLAP [Han].

Adapting to this new multidimensional context, approaches and algorithms successfully used when mining binary databases is possible but not really relevant. However such adaptations have been frequently proposed for the extraction of quantitative association rules [SA96] and for classification [Han]. We believe that a precise semantics is required for characterizing the search space and solving multidimensional data mining problems such as computing datacubes, version spaces, quotient cubes (succinct summaries of datacubes) and template multidimensional associations. Such a semantics can be captured through an algebraic structure, the cube lattice [CCL03], provided with a similar expression power than the power set lattice when it is used for binary database mining.

In this paper, within the groundwork of the cube lattice, we consider an instance of the constraint pattern problem: the frequent multidimensional patterns. We propose the first concise representation of frequent multidimensional patterns which on one hand is not based on the powerset lattice framework and on the other hand differs from the representations by close patterns [PBTL99,PHM00]

A. de Moor, W. Lex, and B. Ganter (Eds.): ICCS 2003, LNAI 2746, pp. 351–361, 2003.

[ZH02,STB$^+$02,VMGM02] and non derivable patterns [CG02]. From such a representation, we show that any frequent multidimensional pattern along with its conjunction, disjunction and negation frequencies can be obtained using inclusion-exclusion identities [Nar82].

Organization of the Article:
In section 2, we detail the structure of the cube lattice. Assuming that we are dealing with frequent multidimensional patterns, our concise representation of the solution space is defined in section 3. Further research work is summarized as a conclusion.

2   Background: The Cube Lattice Framework

Throughout the paper, we make the following assumptions and use the introduced notations. Let r be a relation over the schema $\mathcal{R}$. Attributes of $\mathcal{R}$ are divided in two sets (i) $\mathcal{D}$ the set of dimensions, also called categorical or nominal attributes, which correspond to analysis criteria for OLAP, classification or concept learning [Mit97] and (ii) $\mathcal{M}$ the set of measures (for OLAP) or class attributes. Moreover, attributes of $\mathcal{D}$ are totally ordered (the underlying order is denoted by $<_d$) and $\forall A \in \mathcal{D}, \text{Dim}(A)$ stands for the projection of r over A.

The multidimensional space of the categorical database relation r groups all the valid combinations built up by considering the value sets of attributes in $\mathcal{D}$, which are enriched with the symbolic value ALL. The latter, introduced in [GCB$^+$97] when defining the operator Cube-By, is a generalization of all the possible values of any dimension. Thus $\forall A \in \mathcal{D}, \forall a \in \text{Dim}(A), \{a\} \subset$ ALL.

The multidimensional space of r is noted and defined as follows: $\text{Space}(r) = \{\times_{A \in \mathcal{D}}(\text{Dim}(A) \cup \text{ALL})\} \cup \{<\emptyset, \ldots, \emptyset>\}$ where $\times$ symbolizes the Cartesian product, and $<\emptyset, \ldots, \emptyset>$ stands for the combination of empty values. Any combination belonging to the multidimensional space is a tuple and represents a multidimensional pattern.

Example 1. - Table 1 presents the categorical database relation used all along the paper to illustrate the introduced concepts. In this relation, A, B, C are dimensions and M is a measure.
The following tuples $t_1 = <a_1, b_1, \text{ALL}>, t_2 = <a_1, b_1, c_1>, t_3 = <a_1, b_2, c_1>$, $t_4 = <a_1, \text{ALL}, c_1>$ and $t_5 = <\text{ALL}, b_1, \text{ALL}>$ are elements of $\text{Space}(r)$.

The multidimensional space of r is structured by the generalization/specialization order between tuples. This order is originally introduced by T. Mitchell [Mit97] in the context of machine learning. Let u, v be two tuples of the multidimensional space of r:

$$u \geq_g v \Leftrightarrow \begin{cases} \forall A \in \mathcal{D}, v[A] \subseteq u[A] \\ \text{or } v = <\emptyset, \ldots, \emptyset> \end{cases}$$

If $u[A]$ and $v[A] \neq$ ALL then $u[A]$ and $v[A]$ correspond to singletons just encompassing a value of $\text{Dim}(A)$. If $u \geq_g v$, we say that u is more general than

Table 1. Relation example r

RowId	A	B	C	M
1	a_1	b_1	c_1	3
2	a_1	b_1	c_2	2
3	a_1	b_2	c_1	2
4	a_1	b_2	c_2	2
5	a_2	b_1	c_1	1
6	a_3	b_1	c_1	1

v in $Space(r)$. The covering relation of $\geq_g$ is noted $\succ_g$ and defined as follows: $\forall t, t' \in Space(r), t \succ_g t' \Leftrightarrow t >_g t'$ and $\forall t'' \in Space(r), t \geq_g t'' \succ_g t' \Rightarrow t = t''$. In the multidimensional space of our relation example, we have: $t_5 \geq_g t_2$, i.e. t_5 is more general than t_2 and t_2 is more specific than t_5 and $t_1 \succ_g t_2$. Moreover any tuple generalizes the tuple $<\emptyset, \emptyset, \emptyset>$ and specializes the tuple $<$ALL, ALL, ALL$>$.

The two basic operators provided for tuple construction are: Sum (denoted by +) and Product (noted •). The Semi-Product operator (noted ⊙) is a constrained product.

The Sum of two tuples yields the most specific tuple which generalizes the two operands. Let u and v be two tuples in $Space(r)$.

$$t = u + v \Leftrightarrow \forall A \in \mathcal{D}, t[A] = \begin{cases} u[A] \text{ if } u[A] = v[A] \\ \text{ALL elsewhere.} \end{cases}$$

We say that t is the Sum of the tuples u and v. In our example of $Space(r)$, we have $t_2 + t_3 = t_4$. This means that t_4 is built up from the tuples t_2 and t_3.

The Product of two tuples yields the most general tuple which specializes the two operands. If it exists, for these two tuples, a dimension A having distinct and real world values (i.e. existing in the original relation), then the only tuple specializing them is the tuple $<\emptyset, \ldots, \emptyset>$ (apart from it, the tuple sets which can be used to construct them are disjoined). Let u and v be two tuples in $Space(r)$. We define the tuple z as follows: $\forall A \in \mathcal{D}, z[A] = u[A] \cap v[A]$. Then,

$$t = u \bullet v \Leftrightarrow \begin{cases} t = z \text{ if } \nexists A \in \mathcal{D} \mid z[A] = \{\emptyset\} \\ <\emptyset, \ldots, \emptyset> \text{ elsewhere.} \end{cases}$$

We say that t is the Product of the tuples u and v. In our example of $Space(r)$, we have $t_1 \bullet t_4 = t_2$. This means that t_1 and t_4 generalize t_2 and t_2 participates to the construction of t_1 and t_4 (directly or not). The tuples t_1 and t_3 have no common point apart from the tuple of empty values.

In some cases, it is interesting to know the attributes for which the values associated with a tuple t are different from the value ALL. This is why the function Attribute is introduced. Let t be a tuple of $Space(r)$, we have Attribute$(t) = \{A \in \mathcal{D} \mid t[A] \neq$ ALL$\}$. In our example of $Space(r)$, we have Attribute$(t_1) = \{A, B\}$.

The Semi-Product operator is a constrained product operator useful for candidate generation in a levelwise approach [AMS+96,MT97]. Provided with multidimensional patterns at the level i, we only generate candidates of level $i+1$, if they exist (else the constrained product yields $<\emptyset, \ldots, \emptyset>$). Moreover, each tuple is generated only once. Let u and v be two tuples of $\text{Space}(r)$, $X = \text{Attribute}(u)$ and $Y = \text{Attribute}(v)$.

$$t = u \odot v \Leftrightarrow \begin{cases} t = u \bullet v \text{ if } X \backslash \max_{<_d}(X) = Y \backslash \max_{<_d}(Y) \\ \text{and } \max_{<_d}(X) <_d \max_{<_d}(Y) \\ <\emptyset, \ldots, \emptyset> \text{ elsewhere.} \end{cases}$$

where $<_d$ is the total order over $\mathcal{D}$. In our example, we have $t_1 \odot t_4 = t_2$ and $t_4 \odot t_1 = <\emptyset, \ldots, \emptyset>$.

By providing the multidimensional space of r with the generalization order between tuples and using the above-defined operators Sum and Product, we define an algebraic structure which is called cube lattice. Such a structure provides a sound foundation for multidimensional data mining issues.

Theorem 1. *Let r be a categorical database relation over $\mathcal{D} \cup \mathcal{M}$. The ordered set $\text{CL}(r) = \langle \text{Space}(r), \geq_g \rangle$ is a complete, graded, atomistic and coatomistic lattice, called cube lattice in which Meet ($\bigwedge$) and Join ($\bigvee$) elements are given by:*

1. $\forall \, T \subseteq \text{CL}(r)$, $\bigwedge T = +_{t \in T} \, t$
2. $\forall \, T \subseteq \text{CL}(r)$, $\bigvee T = \bullet_{t \in T} \, t$

Example 2. - Figure 1 exemplifies the cube lattice of our relation example (Cf. Table 1) over the attributes A and B.. In this diagram, the edges represent the generalization or specialization links between tuples.

Through the two following propositions, we characterize the order-embedding from the cube lattice towards the powerset lattice of the whole set of attribute values and analyze the number of elements (for a given level or in general) of the cube lattice. Each value is prefixed by the name of the concerned attribute.

Proposition 1. *Let $\mathcal{L}(r)$ be the powerset lattice of attribute value set, i.e. the lattice $\langle \mathcal{P}(\bigcup_{A \in \mathcal{D}} A.a, \forall a \in \text{Dim}(A)), \subseteq \rangle$[1]. Then it exists an order-embedding:*

$$\Phi : \text{CL}(r) \to \mathcal{L}(r)$$

$$t \mapsto \begin{cases} \bigcup_{A \in \mathcal{D}} A.a, \forall a \in \text{Dim}(A) \text{ if } t = <\emptyset, \ldots, \emptyset> \\ \{A.t[A] \mid \forall A \in \text{Attribute}(t)\} \text{ elsewhere.} \end{cases}$$

We denote by $\mathcal{A}t(\text{CL}(r))$ the atoms of the cube lattice (i.e. $\{t \in \text{CL}(r) \mid |\Phi(t)| = 1\}$).

---

[1] $\mathcal{P}(X)$ is the powerset of X.

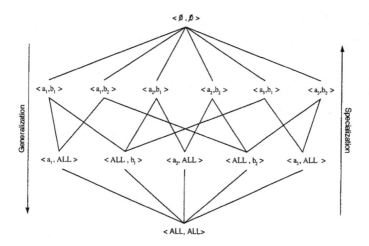

Fig. 1. Hasse diagram of the cube lattice for the projection of r over AB

3   Concise Representations of Frequent Multidimensional Patterns

In this section, we propose a concise representation of the frequent cube lattices. From such a representation, any frequent multidimensional pattern along with its frequency can be derived by using inclusion-exclusion identities adapted to the cube lattice framework.

3.1   Frequency Measures and Inclusion-Exclusion Identities

Let us consider $t \in CL(r)$ (r is a relation over $\mathcal{D} \cup \{M\}$), we define three weight measures, which are compatible with the weight functions defined in [STB+02], for t: (i) its frequency (noted $Freq(t)$), (ii) the frequency of its disjunction (noted $Freq(\vee t)$) and (iii) the frequency of its negation (noted $Freq(\neg t)$).

$$Freq(t) = \frac{\sum\limits_{t' \in r} t'[M] \mid t \geq_g t'}{\sum\limits_{t' \in r} t'[M]} \tag{1}$$

$$Freq(\vee t) = \frac{\sum\limits_{t' \in r} t'[M] \mid t + t' \neq <ALL, \ldots, ALL>}{\sum\limits_{t' \in r} t'[M]} \tag{2}$$

$$Freq(\neg t) = \frac{\sum\limits_{t' \in r} t'[M] \mid t + t' = <ALL, \ldots, ALL>}{\sum\limits_{t' \in r} t'[M]} \tag{3}$$

Example 3. In the cube lattice of our relation example, we have:
$\text{Freq}(<a_1, b_1, \text{ALL}>) = 5/11$, $\text{Freq}(\vee <a_1, b_1, \text{ALL}>) = 1$ and
$\text{Freq}(\neg <a_1, b_1, \text{ALL}>) = 0$.

The inclusion-exclusion identities make it possible to state, for a tuple t, the relationship between the frequency, the frequency of the disjunction and the frequency of the negation, as follows:

$$\text{Freq}(\vee t) = \sum_{\substack{t' \geq_g t \\ \perp \neq t'}} (-1)^{(|\Phi(t')|-1)} \text{Freq}(t'). \tag{4}$$

$$\text{Freq}(t) = \sum_{\substack{t' \geq_g t \\ \perp \neq t'}} (-1)^{(|\Phi(t')|-1)} \text{Freq}(\vee t'). \tag{5}$$

$$\text{Freq}(\neg t) = 1 - \text{Freq}(\vee t) \text{ (from De Morgan Law)} \tag{6}$$

where Φ is the order-embedding (Cf. proposition 1) and $\perp$ stands for the tuple $<\text{ALL}, \ldots, \text{ALL}>$.

Example 4. In the cube lattice of our relation example, we have:

1. $\text{Freq}(\vee <a_1, b_1, \text{ALL}>) = \text{Freq}(<a_1, \text{ALL}, \text{ALL}>) + \text{Freq}(<\text{ALL}, b_1, \text{ALL}>)$
   $-\text{Freq}(<a_1, b_1, \text{ALL}>) = 9/11 + 7/11 - 5/11 = 1$
2. $\text{Freq}(<a_1, b_1, \text{ALL}>) = \text{Freq}(<a_1, \text{ALL}, \text{ALL}>) + \text{Freq}(<\text{ALL}, b_1, \text{ALL}>)$
   $-\text{Freq}(\vee <a_1, b_1, \text{ALL}>) = 9/11 + 7/11 - 1 = 5/11$
3. $\text{Freq}(\neg <a_1, b_1, \text{ALL}>) = 1 - \text{Freq}(\vee <a_1, b_1, \text{ALL}>) = 0$.

Computing the frequency of the disjunction for a tuple can be performed along with computing its frequency and thus the execution time of levelwise algorithms is not altered.

Provided with the frequency of the disjunction for the tuples, a concise representation of frequent tuples can be defined and the computation of the negation frequency is straightforward.

3.2   Frequent Essential Tuples

Definition 1. Let $t \in CL(r)$ be a tuple, and minfreq a given threshold. We say that t is frequent if and only if $\text{Freq}(t) \geq \text{minfreq}$. If $t \neq <\text{ALL}, \ldots, \text{ALL}>$, then t is said essential if and only if

$$\text{Freq}(\vee t) \neq \max_{DLB(t)} (\text{Freq}(\vee t')) \tag{7}$$

where $DLB(t) = \{t' \in CL(r) \mid t' \succ_g t\}$. Let us note Essential(r) the set of essential tuples and FCL(r) the set of frequent tuples.

Example 5. In the cube lattice of our relation example (Cf. Table 1), $<a_1, b_1, \text{ALL}>$ is essential because $\text{Freq}(\vee <a_1, b_1, \text{ALL}>) \neq \text{Freq}(\vee <a_1, \text{ALL}, \text{ALL}>)$ and $\text{Freq}(\vee <a_1, b_1, \text{ALL}>) \neq \text{Freq}(\vee <\text{ALL}, b_1, \text{ALL}>)$.

Lemma 1. *Let us consider the twofold constraint: "t is frequent" (C_1) and "t is essential" (C_2). Such a constraint conjunction is antimonotone according to the order $\geq_g$ (i.e. if t is either an essential or frequent tuple and $t' \geq_g t$, then t' is also essential or frequent).*

The three following formulas show firstly how to compute the frequency of the disjunction from the set of essential tuples and secondly how to optimize the inclusion-exclusion identity for finding efficiently the frequency of a frequent tuple. Formula 8 should be used if t is essential, formula 9 elsewhere.

Lemma 2.

$$\forall t \in CL(r), Freq(\lor t) = \max_{t' \in Essential(r)} (\{Freq(\lor t') \mid t' \geq_g t\}).$$

Lemma 3. $\forall t \in FCL(r)$, *let* $u = Argmax(\{Freq \lor t' \mid t' \geq_g t \text{ and } t' \in Essential(r)\})$, *then we have:*

$$Freq(t) = \sum_{\substack{t' \geq_g t \\ \perp \neq t'}} (-1)^{|\Phi(t')|-1} \begin{cases} Freq(\lor u) & \text{if } u \geq_g t' \\ Freq(\lor t') & \text{if not} \end{cases} \tag{8}$$

$$\Leftrightarrow Freq(t) = \sum_{\substack{t' \geq_g t \\ \perp \neq t' \\ t' \not\geq_g u}} (-1)^{|\Phi(t')|-1} Freq(\lor t')$$

$$+ (\sum_{p=0}^{q=|\Phi(t)|-|\Phi(u)|} \binom{p}{q} (-1)^{|\Phi(u)|+p-1}) Freq(\lor u) \tag{9}$$

Definition 2. *RC(r) is a concise representation for FCL(r) if and only if (i) $RC(r) \subseteq FCL(r)$ and (ii) $\forall t \in FCL(r), RC(r) \models Freq(t)$.*

Unfortunately, $Essential(r)$ cannot be a concise representation of $FCL(r)$, this is why we add the set of maximal frequent tuples $(S_{C_1}^+)$ to obtain a concise representation. Thus a tuple is frequent if and only if it generalizes a tuple of $S_{C_1}^+$.

Theorem 2. $S_{C_1}^+ \cup \{t \in CL(r) \mid t \text{ is a frequent essential tuple}\}$ *is a concise representation of the frequent cube lattice.*

Claim:

1. In a ROLAP context, when the function count is used, we have: $Datacube(r) = \{(t, freq(t)) \mid t \in CL(r) \text{ and } Freq(t) > 0\}$. Thus frequent essential tuples and $S_{C_1}^+$ provide a concise representation of iceberg datacubes [BR99,Han].
2. The framework of frequent essential tuples can be used in the context of frequent pattern mining and the set of frequent essential patterns is a concise representation of frequent patterns. It is closely related to the representation by key patterns [BTP+00,STB+02] since essential patterns are key patterns for the disjunction.

3.3 GLAE Algorithm

In order to yield the frequent essential tuples, we propose a levelwise algorithm with maximal frequent $(S_{C_1}^+)$ pruning. Our algorithm includes the function Max_Set_Algorithm which discovers maximal frequent multidimensional patterns. It could be enforced by modifying the Max-Miner algorithm [Bay98].

Algorithm 1 GLAE Algorithm

1: $S_{C_1}^+ := \text{Max\_Set\_Algorithm}(r, C_1)$
2: $L_1 = \{t \in \mathcal{A}t(CL(r)) \mid \exists u \in S_{C_1}^+ : t \geq_g u\}$
3: $i := 1$
4: **while** $L_1 \neq \emptyset$ **do**
5:     $C_{i+1} := \{v = t \odot t' \mid t, t' \in L_i, v \neq <\emptyset, \ldots, \emptyset>, \exists u \in S_{C_1}^+ : v \geq_g u \text{ and } \forall w \in DLB(v), w \in L_i\}$
6:     Scan the database to find the frequencies of the disjunction $\forall t \in C_{i+1}$
7:     $L_{i+1} := C_{i+1} \backslash \{t \in C_{i+1} \mid \exists t' \in DLB(t) : \text{Freq}(\vee t) = \text{Freq}(\vee t')\}$
8:     $i := i + 1$
9: **end while**
10: **return** $\cup_{j=1..i} L_j$

Example 6. The concise representation of the relation illustrated in Table 1 for the antimonotone constraint "$\text{Freq}(t) \geq 2/11$" is the following: the set of frequent essential tuples is exemplified in Table 2 and the set $S_{C_1}^+$ is given in Table 3.

1. We want to know if $<a_1, b_2, ALL>$ is frequent and if it is, what is the frequency of its disjunction and negation.
   (a) Since $<a_1, b_2, ALL> \geq_g <a_1, b_2, c_1>$, the former tuple is frequent. We use lemma 3 to retrieve its frequency:
   $\text{Freq}(<a_1, b_2, ALL>) = \text{Freq}(<a_1, ALL, ALL>) + \text{Freq}(<ALL, b_2, ALL>) - \text{Freq}(\vee <a_1, b_2, ALL> = 9/11 + 4/11 - 9/11 = 4/11$
   (b) In order to retrieve the frequency of the disjunction, we apply lemma 2:
   $\text{Freq}(\vee <a_1, b_2, ALL>) = \text{Freq}(\vee <a_1, ALL, ALL> = 9/11.$
   (c) We use De Morgan law to retrieve the frequency of the negation:
   $\text{Freq}(\neg <a_1, b_2, ALL>) = 1 - \text{Freq}(\vee <a_1, b_2, ALL>) = 2/11.$
2. We want to retrieve the frequencies of the tuple $<a_1, b_2, c_1>$.
   (a) Since $<a_1, b_2, c_1> \in S_{C_1}^+$, it is frequent. We use lemma 3 to compute its frequency:
   $\text{Freq}(<a_1, b_2, c_1>) = \text{Freq}(<a_1, ALL, ALL>) + \text{Freq}(<ALL, b_2, ALL>) + \text{Freq}(<ALL, ALL, c_1>) - (\text{Freq}(\vee <a_1, b_2, ALL>) + \text{Freq}(\vee <a_1, ALL, c_1>) + \text{Freq}(\vee <ALL, b_2, c_1>)) + \text{Freq}(\vee <a_1, b_2, c_1>) = 9/11 + 4/11 + 7/11 - (9/11 + 1 + 9/11) + 1 = 2/11$
   (b) We use lemma 2 to yield the frequency of the disjunction:
   $\text{Freq}(\vee <a_1, b_2, c_1>) = \text{Freq}(\vee <a_1, ALL, c_1> = 1.$

Table 2. Essential tuples for "$Freq(t) \geq 2/11$"

rank	tuple t	$Freq(\lor t)$
1	$<a_1, ALL, ALL>$	9/11
1	$<ALL, b_1, ALL>$	7/11
1	$<ALL, b_2, ALL>$	4/11
1	$<ALL, ALL, c_1>$	7/11
1	$<ALL, ALL, c_2>$	4/11
2	$<a_1, b_1, ALL>$	1
2	$<a_1, ALL, c_1>$	1
2	$<ALL, b_1, c_1>$	9/11
2	$<ALL, b_2, c_1>$	9/11
2	$<ALL, b_2, c_2>$	6/11

Table 3. S^+ for $C_1 =$ "$sum(M) \geq 2/11$"

tuple
$<a_1, b_1, c_1>$
$<a_1, b_1, c_2>$
$<a_1, b_2, c_1>$
$<a_1, b_2, c_2>$

(c) We use De Morgan law to retrieve the frequency of the negation:
$Freq(\neg<a_1, b_2, c_1>) = 1 - Freq(\lor<a_1, b_2, c_1>) = 0$.

3. We want to know if $<a_2, b_1, c_1>$ is frequent. Since $\nexists t \in S^+_{C_1} | <a_2, b_1, c_1> \geq_g t$, $<a_2, b_1, c_1>$ is not frequent.

Conclusion

In this paper, we use the cube lattice framework for solving an instance of the constraint multidimensional pattern problem. We propose a new concise representation: the frequent essential tuples and an algorithm preserving the original complexity of a levelwise algorithm in a binary context.

A work in progress addresses the definition of a closure operator on the cube lattice. Our aim is to show that the set of closed tuples provided with the generalization order is a coatomistic lattice which seems to be isomorphic to the Galois (concept) lattice [GW99] of the binary relation representing the original categorical database relation. If such an isomorphism exists, it is possible to compare the number of frequent essential tuples with the number of frequent closed tuples. This comparison is of great interest because the frequent closed patterns [PBTL99,PHM00,ZH02,STB+02,VMGM02] are currently the smallest concise representation of frequent patterns.

References

[AMS⁺96] Rakesh Agrawal, Heikki Mannila, Ramakrishnan Srikant, Hannu Toivonen, and A. Inkeri Verkamo. Fast Discovery of Association Rules. In *Advances in Knowledge Discovery and Data Mining*, pages 307–328, 1996.

[Bay98] R. J. Bayardo, Jr. Efficiently Mining Long Patterns from Databases. In *Proceedings of the International Conference on Management of Data, SIGMOD*, pages 85–93, 1998.

[BR99] Kevin Beyer and Raghu Ramakrishnan. Bottom-Up Computation of Sparse and Iceberg CUBEs. In *Proceedings of the International Conference on Management of Data, SIGMOD*, pages 359–370, 1999.

[BTP⁺00] Yves Bastide, Rafik Taouil, Nicolas Pasquier, Gerd Stumme, and Lotfi Lakhal. Mining Frequent Patterns with Counting Inference. In *SIGKDD Explorations*, volume 2(2), pages 66–75, 2000.

[CCL03] Alain Casali, Rosine Cicchetti, and Lotfi Lakhal. Cube Lattices: a Framework for Multidimensional Data Mining. In *Proceedings of the 3rd SIAM International Conference on Data Mining, SDM*, 2003.

[CG02] Toon Calders and Bart Goethals. Mining all non-derivable frequent itemsets. In *In Proceedings of the 6th European Conference on Principles of Data Mining and Knowledge Discovery, PKDD* , pages 74–85, 2002.

[GCB⁺97] Jim Gray, Surajit Chaudhuri, Adam Bosworth, Andrew Layman, Don Reichart, Murali Venkatrao, Frank Pellow, and Hamid Pirahesh. Data cube: A relational aggregation operator generalizing group-by, cross-tab, and subtotals. In *Data Mining and Knowledge Discovery*, volume 1(1), pages 29–53, 1997.

[GW99] Bernhard Ganter and Rudolf Wille. *Fomal Concept Analysis: Mathematical Foundations*. Springer, 1999.

[Han] Jiawei Han. Data mining, data warehousing, and knowledge discovery in databases. In *http://db.cs.sfu.ca/sections/publication/kdd/kdd.html*.

[Mit97] Tom M. Mitchell. *Machine learning*. MacGraw-Hill Series in Computer Science, 1997.

[MT97] Heikki Mannila and Hannu Toivonen. Levelwise Search and Borders of Theories in Knowledge Discovery. In *Data Mining and Knowledge Discovery*, volume 1(3), pages 241–258, 1997.

[Nar82] Hiroshi Narushima. Principle of Inclusion-Exclusion on Partially Oder Sets. In *Discrete Math*, volume 42, pages 243–250, 1982.

[PBTL99] Nicolas Pasquier, Yves Bastide, Rafik Taouil, and Lotfi Lakhal. Discovering frequent closed itemsets for association rules. In *Proceedings of the 7th International Conference on Database Theory, ICDT*, pages 398–416, 1999.

[PHM00] Jian Pei, Jiawei Han, and Runying Mao. CLOSET: An Efficient Algorithm for Mining Frequent Closed Itemsets. In *Workshop on Research Issues in Data Mining and Knowledge Discovery, DMKD*, pages 21–30, 2000.

[SA96] Ramakrishnan Srikant and Rakesh Agrawal. Mining quantitative association rules in large relational tables. In *Proceedings of the International Conference on Management of Data, SIGMOD*, pages 1–12, 1996.

[STB⁺02] Gerd Stumme, Rafik Taouil, Yves Bastide, Nicolas Pasquier, and Lotfi Lakhal. Computing Iceberg Concept Lattices with Titanic. In *Data and Knowledge Engineering*, volume 42(2), pages 189–222, 2002.

[VMGM02] Petko Valtchev, Rokia Missaoui, Robert Godin, and Mohamed Meridji. Generating Frequent Itemsets Incrementally : Two Novel Approaches based on Galois Lattice Theory. In *Experimental and Theoretical Artificial Intelligence (JETAI): Special Issue on Concept Lattice-based theory, methods and tools for Knowledge Discovery in Databases*, volume 14(2/3), pages 115–142, 2002.

[ZH02] Mohammed J. Zaki and Ching-Jui Hsio. CHARM: An Efficient Algorithm for Closed Itemset Mining . In *Proceedings of the 2nd SIAM International Conference on Data mining*, 2002.

Sesei: A CG-Based Filter
for Internet Search Engines

Stéphane Nicolas, Bernard Moulin, and Guy W. Mineau

Université Laval, Québec G1R 1J4, Québec, Canada,
{stephane.nicolas, bernard.moulin, guy.mineau}@ift.ulaval.ca

Abstract. The web faces a major contradiction: it has been created to present information to human beings but the increase of available information sources on the web makes it more difficult for people to find the most relevant documents when browsing through numerous pages of URLs returned by search engines. Classic indexing and retrieval techniques have an important limitation: the retrieved documents do not necessarily match the meaning of the user's query. In order to address this issue, we designed a system capable of matching conceptual structures extracted from both the user's query and documents retrieved by a classic search engine like Google. This article presents our fully functional application and provides details about its behavior. The application provides a new framework for broader experiments in the manipulation of conceptual structures extracted from natural language documents and is built upon research and tools developped by the CG community.

1   Introduction

The web faces a major contradiction: it has been created to present information to human beings but the increase of available information sources on the web makes it impossible for people to find the most relevant documents when browsing through numerous pages of URLs returned by search engines. Although, classic indexing and retrieval techniques have proved their efficiency, they have an important limitation: they offer neither a good recall (finding all relevant documents) nor a good precision (finding only relevant documents). The reason is that the retrieved documents do not necessarily match the meaning of the user's query. In order to address this issue, we designed a system capable of matching conceptual structures extracted from both the user's query and documents retrieved by a classic search engine like Google.

In our literature review, we identified two main practices to extract conceptual structures from texts:

1. The use of a set of canonical conceptual graphs representing the acceptable semantic uses of concepts (mainly verb concepts). These graphs define a set of relations between concepts that have to be matched with the syntagms obtained through syntactic parsing[10, 8].

A. de Moor, W. Lex, and B. Ganter (Eds.): ICCS 2003, LNAI 2746, pp. 362–377, 2003.

2. The use of a set of transformation rules. These rules rely upon the identification of certain templates in grammatical and syntactic structures. The rules bind semantic representations to these templates[11, 1].

These two approaches are very different: the former aims at creating conceptual structures through a top-down process guided by semantics, whereas the latter uses a bottom-up process guided by syntax and grammar. The main drawback of the first method results from the large amount of canonical graphs needed to interpret a reasonable number of natural language sentences. This is why it has been usually limited to specific domains in which a relatively small set of frames is sufficient to express a large part of the knowledge.

On the other hand, the use of transformation rules seems to contradict a fundamental principle of modern linguistics [2]: the syntactic structure should be derived from the semantic structure and not the other way around. This assertion is supported by the existence of meaningless but valid grammatical structures [3] such as 'Colorless green ideas sleep furiously'.

Obviously, even though the second technique allows to convert a large number of natural language sentences into conceptual structures, we cannot expect them to be valid. Moreover, several semantic structures may share the same syntactic representation (e.g. 'I travel *to* visit a friend' can express both a goal and a reason), and they cannot be distinguished by applying a syntax-driven process. Thus, we cannot assert that it is possible to infer the whole semantic structure of a sentence from its syntactic structure, though we think it is reasonable to expect to get a valuable part of it in most cases.

In our previous research [6], we proposed a mechanism that extracts conceptual graphs from natural language sentences. This technique belongs to the second familly and uses a set of rules to transform syntactic structures into conceptual structures. This article describes an application (called sesei) that we developped around this technique. Figure 1 provides an overview of sesei's workflow. The goal of our system is to use both linguistic and semantic techniques to enhance the precision of a classic search engine like Google. The main steps of our approach are: 1) to build a semantic representation of a query sentence (called query graph) expressed in natural English, 2) to build a semantic representation (called resource graphs) of documents in natural English retrieved on the Internet by a classic search engine and 3) finally, to evaluate the semantic similarity of these structures in order to improve precision over the set of documents retrieved by conventional means. We used the conceptual graph (CG) formalism to represent the conceptual structures of the natural language sentences because CGs offer a knowledge representation close to natural language and processable by computers.

Traditionally, natural language processing is a time consuming task for a computer since it involves a huge amount of data and a complex set of operations. However, we wanted our system to offer a reasonable processing time to retrieve documents. Thus, we added a preliminary stage that filters the documents and highlights the most valuable sentences according to our transformation process. The ontology-based metric used in this filter is described in Section 2.

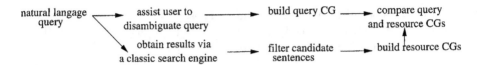

Fig. 1. Overview of sesei's workflow.

Our application uses a method [6] for conceptual graph extraction from natural language documents. Our approach is driven by a syntactic analysis. Section 3 briefly describes this extraction process.

Once the system has built the query graph and resource graphs for the most valuable resource sentences, it compares these semantic structures using a semantic distance to evaluate the similarity between the documents and the query. The semantic score of a resource graph is calculated using the notion of common shared semantic information as described in Section 4.

Finally, Section 5 gives an overview of the results obtained using sesei. It provides examples illustrating different aspects of the system.

2   Preliminary Selection of Candidate Sentences

The pre-selection step aims at identifying the sentences of a resource document that can produce valuable conceptual structures according to our extraction process. We call such sentences 'candidate sentences'. The main idea is to highlight the sentences that are composed of words semantically close to the words used in the query.

The first step of the preselection process is to segment the text into sentences. We chose a simple segmentation technique: we consider that a sentence ends with a period followed by (at least) a space character. Although this approach tends to produce many sentences, it offers a fairly good coverage (76%). Moreover, we added a size constraint on sentences: contains at most 60 words. This constraint is fairly reasonable for English texts (a candidate sentence length of 30 words maximum is a good guide line in English).

In the HTML documents found on the Internet, we do analyse neither graphical structures such as images, nor presentation features such as lists, menus, tables, frames, etc. Although these structures often embed a significant part of a document content, their interpretation is a complex problem that is beyond the scope of our work. We simply extract the textual content of the documents and remove the html tags except some special tags such as <BR>, </P> and <P> because they help to detect sentence boundaries. As we will see in the following sections, processing the textual content of documents seems sufficient for filtering purposes.

The second step is to score each sentence extracted from the resource document in order to isolate potentially useful sentences. We built an ontological similarity measure to evaluate the relevance of a sentence. As will be detailed below, the idea is to build a list of words in the semantic neighborhood of the words of the query. Then, we rate every word contained in this list according to a score function. Finally, each sentence of a resource document is scored according to the value of the words it contains.

This measure uses Wordnet as a lexical ontology. Wordnet offers a vast set of concept types and their representation in English. The concept types are organised into a hypernymic ontology and the words that represent the concept types are labeled using grammatical information such as their part of speech (POS) and derived forms (noun plurals, verb conjugations, comparative and superlative forms of adjectives). Wordnet emphasizes polysemy: a single word may have multiple meanings (called sense).

To disambiguate the words of the query, we ask the user to select the senses of the words she/he used in the query via a GUI. This step is the last operation that involves the user until the end of the document retrieval process. We provide the user with a list of the different senses for each word of the query and a small definition (in natural language) for each sense. Thus, the user has to select the desired senses. In Wordnet, senses are ranked by order of familiarity, so we can select the first sense by default to assist the user in this process.

Using Wordnet, we extract the list of words that represent a generalisation (hypernym), a specialisation (hyponym) or an equivalent (synonym) of the concept type associated with each of the words of the query. The list also contains all the derived forms of these words. For example, the word *cure* would produce the list of words: $\lambda_{cure} = \{$cure, cures, remedy, remedies, curative, curatives, elixir, elixirs, ...$\}$. In order to avoid too general and too specific terms, and to give reasonable boundaries to our search space, we restricted the list to the first five levels of generalisation and specialisation. Hence, the following definition:

Definition 1. *Let qs be a query sentence, q be a word in qs, rs be a resource sentence. Let wn(x) be a function that binds a word x to a concept type in Wordnet, $\delta(a,b)$ be the shortest specialisation/generalisation path length from concept type a to concept type b in Wordnet. Then we have:*

$$\forall \, q, \forall \, w \in \lambda_q, \delta(wn(w), wn(q)) \in [M, N]$$
$$where \; M = -min(\delta(\bot, wn(q)), 5) \; and \; N = min(\delta(\top, wn(q)), 5)$$

Thus, we produce a score for each word in the list. The score of a word w is a value in $[0,1]$ that estimates how relevant w is when specialising a word q of a query. Formula 1 details the score function. The function only uses the degree of hypernymy: words are rated regardless of their derived form, their meaning or the word of the query they are related to. In this measure, the score of a word linearly depends on the number of nodes of the ontology that relate its concept type to the concept types of the words of the query. Indeed the evaluation of the semantic distance between two concept types is dependent on Wordnet: Wordnet indicates specialisation links between concept types but not the degree of this specialisation. This is the reason why we consider all these links as equivalent. Adopting the same notation as in Definition 1, we can define the score of a word contained in λ_q by the formula

$$score(w,q) = \begin{cases} 1 & \text{if } wn(w) \text{ is a hyponym of } wn(q) \\ 1 & \text{if } wn(w) \text{ is a synonym of } wn(q) \quad (1) \\ 1 - \delta(wn(w), wn(q))/N & \text{if } wn(w) \text{ is a hypernym of } wn(q) \end{cases}$$

In order to emphasize the importance of a word w in the query, we allow the user to give a weight to each word of the query: $weight(w)$. The words of the query have a default weight based on their part of speech. The default values are 1.0 for verbs, 0.8 for nouns, 0.2 for adverbs and 0.16 for adjectives. The default value for nouns has been set empirically. The values for adjectives and adverbs have been set to specific values because we wanted to emphasize their roles as qualifiers: adjectives qualify nouns and adverbs qualify verbs. Qualification can be seen as a kind of specialisation and we adjusted the weight of the qualifiers so that qualification and specialisation are scored equally. Figure 2 illustrates this principle. Considering the noun phrase *canadian company* as part of the query, the noun phrases *canadian institution* and *company* would be ranked equally in a resource document because we consider these phrases as equally relevant generalisations of the query noun phrase.

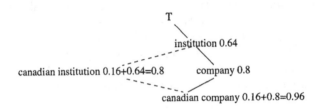

Fig. 2. Specialisation or qualification of a noun are scored equally.

Now, we can define the score function that evaluates the score of a word in the union of all λ_q: $\bigcup_{q \in qs} \lambda_q$. As nothing prevents a word from belonging to λ_{q0} and λ_{q1} with $q0 \neq q1$, we take the maximum of the scores obtained for each word of the query. Moreover, we take into account the weights of the words of the query. The score of the words in λ_q are multiplied by $weight(q)$. The detailed score function of the words contained in $\bigcup_{q \in qs} \lambda_q$ is provided below:

$$\forall w \in \bigcup_{q \in qs} \lambda q, \quad score2(w, qs) = \max_{q \in qs} (\ weight(q) \times score(w, q)\ )$$

Let us consider the query sentence $qs =$ 'a canadian biotechnology company offers a cure for lung cancer'. The left hand side of Figure 3 gives an example of the portion of Wordnet that we built around the concepts that compose the meaning of the query. On the right hand side, we provide part of $\bigcup_{q \in qs} \lambda_q$ and the score of these words.

Once the list of words in the semantic neighborhood of the query words is built and scored, the system searches these words inside the resource documents.

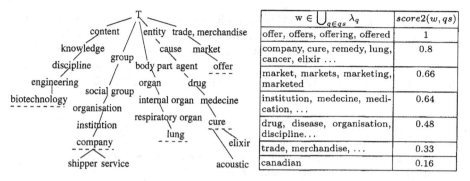

$w \in \bigcup_{q \in qs} \lambda_q$	$score2(w, qs)$
offer, offers, offering, offered	1
company, cure, remedy, lung, cancer, elixir ...	0.8
market, markets, marketing, marketed	0.66
institution, medecine, medi-cation, ...	0.64
drug, disease, organisation, discipline...	0.48
trade, merchandise, ...	0.33
canadian	0.16

Fig. 3. Query sentence concepts extracted from Wordnet and their scores for qs = 'a canadian biotechnology company offers a cure for lung cancer'.

Every word of the documents is compared to the words in $\bigcup_{q \in qs} \lambda_q$. As stated in formula 2 below, a resource word will get a non-null value if it belongs to $\bigcup_{q \in qs} \lambda_q$.

$$\forall\, w \in rs, score3(w, qs) = \begin{cases} score2(w, qs) & \text{if } w \in \bigcup_{q \in qs} \lambda_q \\ 0 & otherwise \end{cases} \quad (2)$$

The last step is to evaluate the relevance of a resource sentence. We propose to sum the scores of the words that compose the sentence:

$$\forall rs, score4(rs, qs) = \sum_{w \in rs} score3(w, qs)$$

In order to normalize to 1 the score function, we divide $score(rs, qs)$ by the score4 of the query sentence. Indeed, this is not enough to normalize the score function because a resource sentence may contain twice or thrice as many words as the query sentence and have a score two or three times the score of the query sentence. Nevertheless, if this situation happens, we bound the score to 1. With this formula, a highly rated resource sentence offers an interesting specialisation of part of the query sentence regardless of the rest of the information contained in the resource sentence.

Definition 2. *The ontological score of a resource sentence is defined by*

$$oScore(rs, qs) = \min\left(\frac{score4(rs, qs)}{score4(qs, qs)}, 1\right) = \min\left(\frac{score4(rs, qs)}{\sum_{q \in qs} weight(q)}, 1\right)$$

In Figure 4, we provide a sample of resource sentences for the query 'canadian biotechnology company offer cure lung cancer' automatically derived from qs='a canadian biotechnology company offers a cure for lung cancer' and their respective ontological scores. In this example, the score of the query sentence is $score4(qs, qs) = \sum_{q \in qs} weight(q) = 0.16+0.8+0.8+1+0.8+0.8+0.8 = 5.16$. The first column of the table contains resource sentences retrieved from the Internet. The scores of the words in $\bigcup_{q \in qs} \lambda_q$ are indicated in parenthesis. The second column gives the ontologicalScore of each sentence as defined in Definition 2.

resource sentence extracted from a URL retrieved by Google	oScore
your cancer(0.8) risk: the source on prevention - the harvard center for cancer(0.8) prevention developed this interactive site to estimate your risk of cancer(0.8) and offer(1.0) personalized tips for better health http://www.searchbeat.com/cancer.htm(ranked 1st by Google[1])	0.66
the company(0.8) is also developing dna plasmid drugs(0.4) for arthritis, psoriasis, and breast/prostate cancer(0.8) http://library.bjmu.edu.cn/calis_med/nk/organ.htm(ranked 2nd by Google)	0.39
qlt phototherapeutics inc., a canadian(0.2) biotechnology(0.8) company(0.8), announced today that the food and drug(0.4) administration (fda) has given marketing(0.75) clearance for its light-activated drug(0.4), photofrin(r) (porfimer sodium) for injection, as a potentially curative(0.8) treatment for certain types of early-stage, microinvasive lung(0.8) cancer(0.8) http://www.slip.net/~mcdavis/whnufeb8.html(ranked 40th by Google)	1.0
potential cancer(0.8) indications include breast, ovarian, acute myeloid leukemia(0.8) , sarcoma(0.8) http://www.jkf.org/news/Aug01.html(ranked 44th by Google)	0.46

Fig. 4. Sentences from documents retrieved by Google and their ontological scores.

Experiments have shown that sentences that are scored above 0.8 are interesting sentences. All sentences above this threshold will be candidate sentences for the whole conceptual matching process. The main drawback of the ontological measure is that it presumes that the relevance of a sentence depends on the relevance of the words that compose the sentence. This is too strong an assumption: a sentence may acquire its relevance from the surrounding context or use irrelevant words to satisfy the query (as in periphrases). Section 6 proposes a broader discussion on the adequacy and the performance of this measure.

3   Extraction of Conceptual Structures

Our approach to extract conceptual graphs uses a set of transformation rules in order to transform syntactic structures into conceptual structures[2]. This section summarises the transformation process that we presented in [6].

The extraction of CGs from a natural language sentence is divided into 3 parts: 1) we use a grammatical parser to exhibit the syntactic structure of the

---

[1] The rankings have been obtained on 01/10/2003.

[2] sesei has been implemented using the Notio API [9].

sentence, 2) we transform this structure into an intermediate form (called grammatical graph) and 3) we apply of set of transformation rules to the grammatical graph in order to get a CG representing the original sentence.

In order to obtain the syntactic structure of the sentence, we use a grammatical parser (called Conexor) developped by a Finish company[1]. This parser [3] is based upon a functional dependency grammar (FDG), directly inspired by Tesnière's model [12]. Conexor transforms a sentence into a syntactic structure called FDG tree which emphasizes the grammatical nature of the syntagmas recognized in a sentence, and also gives their syntactic relations.

Thus, our system transforms a FDG tree into an intermediate formalism that we call Grammatical Graph (GG). A GG is a conceptual graph representing the grammatical knowledge provided by Conexor's output. To realise this step, we simply map 1) each syntagma to a concept representing the syntagma (syntagmatic concept), 2) each grammatical relation between syntagmas to a conceptual relation linking concepts obtained from the previous step, 3) each POS modifier to a new concept linked by an 'attr' relation to the syntagmatic concept. This lossless operation allows us to unify grammatical structures and conceptual structures into a common formalism: conceptual graphs. Consequently, any remaining transformation process relies on graph operations.

The final step of the transformation process consists in applying a set of transformation rules in order to build a conceptual graph representing the sentence. Sesei rules are written in conceptual graphs. Each rule is composed of a premise and a conclusion. The premise is a portion of a grammatical graph and the conclusion contains the conceptual graph associated to the grammatical structure of the premise and certain relations that indicate the link between syntagmatic concepts of the premise and semantic concepts of the conclusion.

The characteristics of the subsystem that extracts the conceptual structures have been presented in details in [6]. The new version of the system uses an augmented set of 76 rules that can handle a broad variety of syntactic phenomena such as verb conjugations, active and passive voices, auxiliaries and verb-chains, possession, interrogative form, adverbial relations, relative clauses, etc. The rules have been optimised in order to deliver reasonable computing times. Although most of the operations involved in the transformation are NP-hard, the system mostly manipulates 'small' graphs (a graph contains at most 80 concepts).

Figure 5 illustrates the distribution of 96 graphs found in the 100 first pages of Google for the query sentence qs="a canadian biotechnology company offers a cure for lung cancer". Each point represents the time needed to transform a grammatical graph that contains a given number of concepts (the number of concepts of a GG is roughly equivalent to the number of syntagmas in the natural language sentence plus ten). Most of the input grammatical graphs contain less than fifty concepts and can be converted into a CG in less than one second.

---

[1] See the URL http://www.Conexoroy.com/products.htm for an on-line demonstration of the analysers.

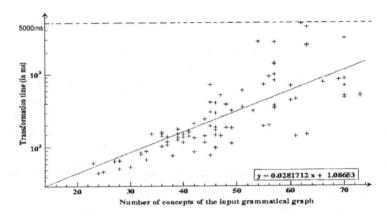

Fig. 5. Distribution of 96 grammatical graphs representing resource sentences and the time to convert them in CG using a Linux based Pentium 4 PC at 1.4Ghz.

4   Comparison of Conceptual Structures

After the extraction process, we obtain a knowledge base that is composed of a concept type hierarchy (extracted from Wordnet), a relation type hierarchy, and a set of conceptual graphs (query and resource graphs) whose node types are contained in the hierarchies. Henceforth, the remaining of the document retrieval process is to sort resource graphs (and thereby the documents that contain them) in order to evaluate how relevant they are to satisfy the query graph. This section provides insights into the metric used for graph ranking.

Many approaches have been proposed to evaluate the similarity of two conceptual graphs. In [7], the authors propose a new algorithm for conceptual graph matching and define a theoretical framework that decomposes the semantic similarity into three elements : structure, surface and theme similarity. Nevertheless, the authors do not propose a clear interest function that can be used to evaluate how relevant a graph is. More recently, [13] have proposed a new CG matching technique but the measure requires the user to identify entry-points in CGs in order to compare them. In our context, this technique is not useable because we do not have any control on the resource graphs and cannot select an entry-point for the matching algorithm.

The semantic distance used by sesei to compare CGs is an adaptation of a distance proposed in [4, 5] which aims at evaluating the amount of shared information between two conceptual graphs. The shared information is obtained by generalising the two graphs. Thus, the more specialised the common generalisation is, the higher it is rated.

This measure has been developped to compare any pair of graphs and was used to build clusters of conceptual structures for text mining [4, 5]. Nevertheless, we adapted this distance because we did not want to evaluate the similarity between two graphs but how compatible a resource graph is with a query graph. Our idea is that a good resource graph satisfies the query graph even if it adds some new information. The evaluation of the common generalisations of the resource graph and the query graph is performed by only considering the information contained in the query graph.

In order to evaluate the semantic distance between two graphs G1 and G2, we first build their common generalisation Γ that contains the information shared by G1 and G2 (G1 is always the query graph in our application and G2 is the resource graph). Let us define two operators: C and R that respectively return the set of concepts and the set of relations of a given graph. Definitions 3 and 4 specify $C(\Gamma)$ and $R(\Gamma)$. According to these definitions, it is possible to find several generalisations for two given graphs. Our application implements an algorithm that generates all the possible generalisations of G1 and G2. The best generalisation is selected after the system has evaluated all the alternatives. Please note that the system records which concepts of G1 and G2 are used to create a concept of Γ so that we can easily retrieve the projections of the concepts of Γ in the original graphs.

Definition 3. *Let Γ be a common generalisation of G1 and G2, π_{G1} and π_{G2} be two projections that map a concept of Γ to its projected concepts in the original graphs G1 and G2. The concepts set $C(\Gamma)$ is defined by*

$$\forall c \in C(\Gamma), type(c) = \begin{cases} type(\pi_{G2}(c)) & \text{if } type(\pi_{G1}(c)) \text{ is } \top \\ leastCommonSubtype(\pi_{G1}(c), \pi_{G2}(c)) & otherwise \end{cases}$$

According to Definition 3, the concepts of Γ are derived from the concepts of G1 and G2. A pair of concepts in G1 and G2 produces a concept of Γ if they have a non-trivial least common super type (e.g. different from $\top$). This rule has one exception: the type of the common concept may be the type of the resource concept when the type of the query concept $(type(\pi_{G1}(c)))$ is $\top$. This exception allows the system to handle questions. Examples are provided in Figure 6.

In Definition 4, the relations of the common generalisation Γ are created from the relations of G1 and G2. There is a relation between two concepts of Γ iff the projections of these concepts are connected by relations of the same type in G1 and G2. This definition entails that the relations shared by two graphs depend on their shared concepts.

Definition 4. *Let Γ be a common generalisation of G1 and G2, its set of relations $R(\Gamma)$ is defined by*

$$\forall r \in R(\Gamma), \forall c, d \in C(\Gamma) \text{ connected by } r,$$
$$\exists r1 \in R(G1)|\ r1 \text{ connects } \pi_{G1}(c) \text{ and } \pi_{G1}(d), type(r) = type(r1)$$
$$and\ \exists r2 \in R(G2)|\ r2 \text{ connects } \pi_{G2}(c) \text{ and } \pi_{G2}(d), type(r) = type(r2)$$

Figure 6 exemplifies the construction of a common generalisation. In this example, the query graph is built from the sentence: 'Who offers a cure for cancer?' and the resource graph is built from: 'a big company will market a sedative'; we build the generalisation [company]<-(agnt)-[market]-(ptnt)->[medecine]. The common generalisation contains the information of the resource graph that satisfies the query graph.

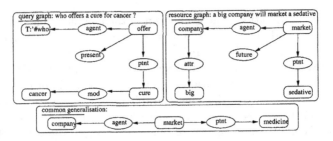

Fig. 6. A query graph and a resource graph and their common graph.

Once we have generated all the common generalisations of G1 and G2, we still need to evaluate the relevance of these generalisations. This evaluation is based on the conceptual similarity and the relational similarity defined below.

The *conceptual similarity* evaluates the similarity between the set of concepts of Γ and G1. It allows to estimate whether or not G2 contains the same concepts as G1. Conceptual similarity is a simple quotient as stated in formula 3. Ideally, the conceptual similarity should take into account the weights of the concepts that compose a query graph in order to emphasize their importance in the meaning of the query. Unfortunately, we didn't have time to implement such a metric. Thus, we consider each of the concepts of the query graph as equally valuable. This is the reason why we merely divide the number of concepts of Γ by the number of concepts of G1.

$$sc = \frac{|\{c \in \Gamma\}|}{|\{c \in G1\}|} \tag{3}$$

The *relational similarity* is a little bit harder to define because it relies on the notions of projection and neighborhood of a subgraph. As mentioned before, the system keeps information about the way it combines the original graphs when building a generalisation. Thus, it is possible to implement a projection process as the reciprocal of the generalisation process. The resulting graph of the projection of Γ in G1 is called $\Pi_{\Gamma G1}$. On the other hand, the neighborhood of a subgraph SG is called $\Omega(SG)$ and is specified in Definition 5

Definition 5. *Let SG be a proper subgraph of a conceptual graph G, then $\Omega(SG)$ is a subgraph of G such that*

$$c \in \mathcal{C}(\Omega(SG)) \iff \begin{cases} c \in \mathcal{C}(SG) \ or \\ \exists r \in \mathcal{R}(G), \exists c' \in \mathcal{C}(SG) | r \ is \ connected \ to \ c \ and \ c' \end{cases}$$

$$r \in \mathcal{R}(\Omega(SG)) \iff \exists c' \in \mathcal{C}(SG) | r \ is \ connected \ to \ c'$$

In order to evaluate the relational similarity, we compare the value of the relations that connect the concepts of Γ and the value of the relations that connect the concepts of the projection of Γ in G1. In this way, the metric verifies that the common concepts of the query graph and the resource graph are used in the same way, using the same relations.

$$sr = \frac{\sum\limits_{r \in \Gamma} weight(r)}{\sum\limits_{r \in \Omega(\Pi_{\Gamma G1})} weight(r)} \tag{4}$$

In Formula 4, $weight(r)$ is the weight of a relation. Unlike the formula proposed by [4] in which the weight of a relation is calculated from the weights of the concepts it connects, we decided to use a set of fixed weights for relations because concepts have no weight in our metric. The weights have been set empirically by studying a reasonable number of graphs. These values seem to give coherent results but we need to fine tune them. Table 7 gives a sample of the weights of the relations that we actually use in sesei.

relation type	agent	ptnt	rcpt	manr	ptim	mod	attr	present	future	link
fixed weight	1.0	1.0	1.0	0.4	0.4	0.4	0.4	0.2	0.2	0.1

Fig. 7. Weights of the relations used to compute the relational similarity.

Finally, the conceptual similarity and the relational similarity are composed into a single value of semantic similarity between the resource graph and the query graph. Experiments led us to keep the settings suggested by [5]:

$$semanticScore(g1, g2) = 0.5 \times sc + 0.5 \times sc \times sr$$

In this formula, the value of sr is weakened by sc in order to represent the fact that the relational similarity is only evaluated between the concepts shared by both graphs. The weights of sc and sr are equal because we bestow the same importance on relations and concepts: a good resource graph will need to contain concepts close to the concepts of the query and the same relations between these concepts. Figure 8 contains a sample set of resource sentences and their semantic scores for the query sentence qs = 'a canadian biotechnology company offers a cure for lung cancer'.

5 Tests and Results

Our application allows to step through each stage of the retrieval process via a GUI. Due to space limitations, we only provide a single screenshot of our application. Figure 9 represents the result pane for the query qs = 'a canadian biotechnology company offers a cure for lung cancer'. The upper part of the picture lists the URLs returned by Google sorted by semantic score for the following query graph:

```
[canadian]<-(attr)-[biotechnology]<-(attr)-[company]<-(agnt)-
-[offer]-(ptnt)->[cure]-(mod)->[cancer]-(attr)->[lung]
```

resource sentence extracted from a URL retrieved by Google	score
your cancer risk: the source on prevention - the harvard center for cancer(X) prevention developed this interactive site to estimate your risk of cancer and offer(X) personalized tips for better health http://www.searchbeat.com/cancer.htm(ranked 1st by Google	0.14
the company(X) is also developing dna plasmid drugs(X) for arthritis, psoriasis, and breast/prostate cancer(X) http://library.bjmu.edu.cn/calis_med/nk/organ.htm(ranked 2nd by Google)	0.21
qlt phototherapeutics inc., a canadian(X) biotechnology(X) company(X), announced today that the food and drug(X) administration (fda) has given marketing(X) clearance for its light-activated drug, photofrin(r) (porfimer sodium) for injection, as a potentially curative(X) treatment for certain types of early-stage, microinvasive lung(X) cancer(X) http://www.slip.net/~mcdavis/whnufeb8.html(ranked 40th by Google)	0.66
potential cancer(X) indications include breast, ovarian, acute myeloid leukemia , sarcoma http://www.jkf.org/news/Aug01.html(ranked 44th by Google)	0.07

Fig. 8. Sentences retrieved by Google and their semantic scores.

When the user selects a URL, sesei reveals the list of the sentences of the selected URL sorted by semantic score (in the lower part of the window). The interface display several informations about the different data structures (URL, sentences, FDGTree, GG, CG, etc.) and the time needed to perform each stage of the retrieval process. An interesting feature is the ability to compare the ranking of each URL according to both Google and sesei. As depicted, the best result for this search was found in the web page http://www.slip.net/~mcdavis/whnufeb8.html(ranked 40th by Google) with a semantic score of 0.66. For this query, sesei filtered the 100 first pages returned by Google in less than 9 minutes on a Linux based 1.4Ghz Pentium 4 system. Four minutes were used to download the content of the pages and to find 96 candidate sentences and four more minutes were dedicated to extracting and comparing the conceptual structures.

The method and the tool that we developped aim at extracting conceptual graphs from natural language documents and at comparing the resulting conceptual structures in order to increase the accuracy of document retrieval. We developped a search engine based on this technique as a proof of concept. Indeed, sesei is not a real search engine but a filter atop of a classic search engine. We use Google as a source of potentially valuable documents and to build test corpuses for sesei's queries. Hence, sesei does not maintain its own index of documents and has to parse every document returned by Google , thus it cannot be compared to a classic search engine in terms of speed.

Although sesei cannot be compared to a classic search engine, our technique offers some interesting features for document retrieval. For instance, sesei can present to the user the portions of a document that are relevant to her/his query and not only the name of the most relevant documents.

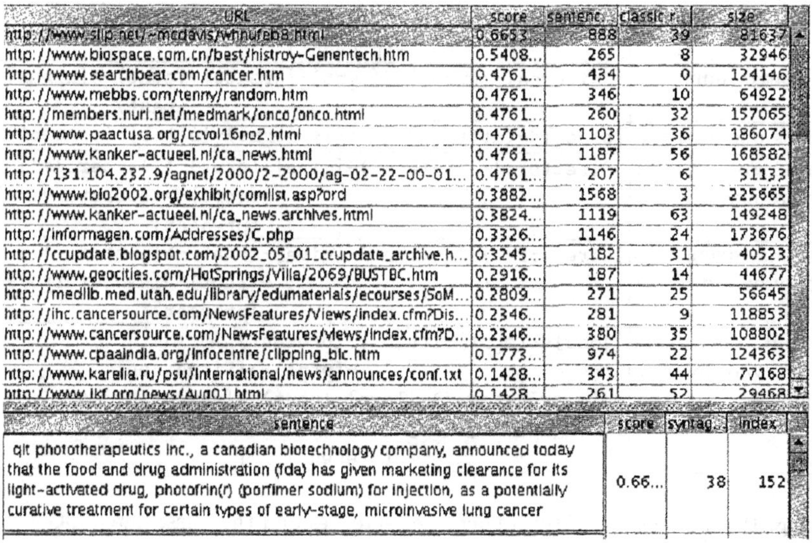

URL	score	sentenc.	classif. r	size
http://www.slip.net/~mcdavis/whnufeba.html	0.6653...	888	39	81637
http://www.biospace.com.cn/best/histroy-Genentech.htm	0.5408...	265	8	32946
http://www.searchbeat.com/cancer.htm	0.4761...	434	0	124146
http://www.mebbs.com/tenny/random.htm	0.4761...	346	10	64922
http://members.nuri.net/medmark/onco/onco.html	0.4761...	260	32	157065
http://www.paactusa.org/ccvol16no2.html	0.4761...	1103	36	186074
http://www.kanker-actueel.nl/ca_news.html	0.4761...	1187	56	168582
http://131.104.232.9/agnet/2000/2-2000/ag-02-22-00-01...	0.4761...	207	6	31133
http://www.bio2002.org/exhibit/comlist.asp?ord	0.3882...	1568	3	225665
http://www.kanker-actueel.nl/ca_news.archives.html	0.3824...	1119	63	149248
http://informagen.com/Addresses/C.php	0.3326...	1146	24	173676
http://ccupdate.blogspot.com/2002_05_01_ccupdate_archive.h...	0.3245...	182	31	40523
http://www.geocities.com/HotSprings/Villa/2069/BUSTBC.htm	0.2916...	187	14	44677
http://medlib.med.utah.edu/library/edumaterials/ecourses/SoM...	0.2809...	271	25	56645
http://ihc.cancersource.com/NewsFeatures/Views/index.cfm?Dis...	0.2346...	281	9	118853
http://www.cancersource.com/NewsFeatures/views/index.cfm?D...	0.2346...	380	35	108802
http://www.cpaaindia.org/infocentre/clipping_blc.htm	0.1773...	974	22	124363
http://www.karelia.ru/psu/International/news/announces/conf.txt	0.1428...	343	44	77168
http://www.ikf.org/news/Aug01.html	0.1428...	261	52	29468

sentence	score	syntag.	index
qlt phototherapeutics inc., a canadian biotechnology company, announced today that the food and drug administration (fda) has given marketing clearance for its light-activated drug, photofrin(r) (porfimer sodium) for injection, as a potentially curative treatment for certain types of early-stage, microinvasive lung cancer	0.66...	38	152

Fig. 9. Screenshot of sesei (results pane).

Another advantage of our document retrieval technique is that is has been easy to transform sesei into a simple question answering system. The application accepts questions like 'who built the pyramids in egypt?' or 'where are the Burtinskaya steppes?'. In this case, the interrogative pronoun is associated to a concept of type T in the query graph (as in Figure 6). Thus, any concept will match this part of the query as long as it is related to concepts which are comparable with the other concepts of the query. More generally, sesei has been easily extended to interpret a broad variety of syntactic structures. These results validate the technique of automatic extraction of conceptual structures through a set of transformation rules. Figure 10 gives a sample of sentences and their representation in CGLF obtained using our application.

Moreover, the system creates a knowledge base as a result of the document retrieval process. This knowledge base contains several conceptual structures related to the topic of the query and defines a concept type hierarchy for this domain. Although a rule-based approach for extracting conceptual structures provides a coarse-grained representation of the semantics of the documents, we believe that this representation can be used as a basis for broader experiments in natural language processing and knowledge representation.

On the other hand, both our technique and our application present some limitations. First, it is difficult to provide a clear evaluation of the metrics used by sesei both for the preselection of candidate sentences (Section 2) and for the comparison of conceptual structures (Section 4) in terms of recall and precision. The corpus that we use (i.e. documents returned by Google) allows us to explore many different kinds of domains and documents but does not contain any 'objective' measure of the relevance of a document. We think that the results

a cat is on a mat	`[cat]<-(agnt)-[be]-(loc)->[mat]` `-(present)`
who created linux ?	`[Universal:'#who']<-(agnt)-[create]-(ptnt)->[linux]` `-(past)`
the black dog eats	`[black]<-(attr)-[dog:'#']<-(agnt)-[eat*b]-(present)`

a director gives a letter to a secretary	`[director:'*']<-(agnt)-[give]-(ptnt)->[letter:'*']`
	                  `-(rcpt)->[secretary:'*']`
	                  `-(present)`

a letter is given to a secretary by a director	`[director:'*']<-(agnt)-[give]-(ptnt)->[letter:'*']`
	                  `-(rcpt)->[secretary:'*']`
	                  `-(present)`

john thinks mary loves a sailor who lives in Dresden	`[john]<-(agnt)-[think]-(present)`
	                  `-(ptnt)->[Situation:`
	    `[mary]<-(agnt)-[love]-(ptnt)->[sailor:'*']<-`
	`-(mod_agnt)-[live]-(loc)->[Dresden]`
	                  `-(present)    ]`

Fig. 10. A sample of natural language sentences and their associated CG.

have been improved by sesei in comparison to a classic search engine like Google but the quantitative evaluation of the performances of sesei has still to be done. Secondly, both metrics assume that a resource sentence contains its meaning in itself and do not consider the context defined by previous sentences. But, in natural language documents, sentences are not autonomous units of meaning. They often contain pronouns and other references to concepts previously defined in the text or even involve knowledge implicitly known by readers and writers.

6   Conclusion

During our reseach, we have developped a system capable of retrieving documents written in natural language using a semantic representation (expressed by CGs) of both resource documents and the query sentence. This experiment shows that the CG community has developped usefull tools and efficient algorithms that can be plugged together and with other projects (like Wordnet) to deliver novel and broader applications.

The framework we propose in this paper is operational and provides a foundation for new promising experiments. We plan to explore this new search space in several ways such as 1) joining resource graphs when they are extracted from sentences close to each other in a resource document, 2) improving the accuracy and granularity of the semantic representation of sentences and texts, and 3) experimenting with query expansion mechanisms to improve the recall of the system, etc.

We hope to release our application and its source code under a GPL licence in a near future, and we will quickly provide a Web demonstration of sesei in order to get some feedback from the research community.

References

1. C. Barrière. *From a Children's First Dictionary to a Lexical Knowledge Base of Conceptual Graphs.* PhD thesis, Université Simon Fraser, 1997.
2. I. Mel'čuk. Leçon inaugurale. Collège de France, Chaire Internatonale, 1997. www.fas.umontreal.ca/ling/olst/FrEng/melcukColldeFr.pdf.
3. T. Järvinen and P. Tapanainen. Towards an implementable dependency grammar, 1998. http://citeseer.nj.nec.com/73472.htm.
4. M. Montes-y Gómez. *Minería de texto empleando la semenjanza entre estructuras semánticas.* PhD thesis, Centro de Investigación en Computación, Instituto Polit'ecnico Nacional,Mexíco, 2002.
5. M. Montes-y Gómez, A. Gelbukh, A. López López, and R Baeza-Yates. Flexible Comparison of Conceptual Graphs. In *Proc. of the 12th International Conference on Database and Expert Systems Applications (DEXA 2001)*, volume N 2113 of *LNCS*, Munich, Germany, 2001. Springer-Verlag. http://cseg.inaoep.mx/~mmontesg/papers/FlexibleComparisonCG-dexa01.pdf.
6. S. Nicolas, B. Moulin, and G. Mineau. Extracting Conceptual Structures from English Texts Using a Lexical Ontology and a Grammatical Parser. In *Sup. Proc. of 10th International Conference on Conceptual Structures, ICCS'02*, 2002. http://s-java.ift.ulaval.ca/~steff/iccs2002-final.pdf.
7. J. Poole and J. A. Campbell. A Novel Algorithm for Matching Conceptual and Related Graphs. In *Proceedings of ICCS 1995*, 95.
8. C. Roux, D. Proux, F. Rechennmann, and L. Julliard. An Ontology Enrichment Method for a Pragmatic Information Extraction System gathering Data on Genetic Interactions. http://citeseer.nj.nec.com/318185.htm.
9. F. Southey and J.G. Linders. Notio - A Java API for Developing CG Tools. In *Proceedings of ICCS '99*, volume 1640 of *LNCS*, pages 262–271. Springer, 1999. http://backtrack.math.uwaterloo.ca/CG/projects/notio/NotioForICCSFinal.ps.zip.
10. J.F. Sowa and E.C. Way. Implementing a semantic interpreter using conceptual graphs. *IBM Journal of Research and Development*, 1986.
11. A. Tajarobi. La reconnaissance automatique des hyponymes. Master's thesis, Département de langues et linguistique, Université Laval, 1998.
12. L. Tesnière. Eléments de syntaxe structurale. Editions Klincksieck, Paris, 1959.
13. J. Zhong, H. Zhu, J. Li, and Y. Yu. Conceptual Graph Matching for Semantic Search. In *Proceedings of ICCS 2002*, volume 2393 of *LNCS*, pages 93–106. Springer, 2002.

A Semantic Search Approach by Graph Matching with Negations and Inferences

Kewei Tu, Jing Lu, Haiping Zhu, Guowei Liu, and Yong Yu

Department of Computer Science and Engineering,
Shanghai JiaoTong University, Shanghai, 200030, P.R.China
tukw@sjtu.edu.cn, robert_lu@263.net, hpzhu@sjtu.edu.cn,
liugw2000@163.net, yyu@mail.sjtu.edu.cn

Abstract. Research on semantic search has become heated these years. In this paper we propose an approach focusing on searching for resources with descriptions. The knowledge representation we employ is based on conceptual graphs and is expressive with negation. We carry out semantic search by graph matching, which can be performed in polynomial time. Before matching we enrich the resource graphs with background knowledge by a deductive graph inference, so as to improve the search performance. The processing of negations and the graph inference method are two important contributions of this paper.

1 Introduction

Today more and more information has been provided in the Internet. One could find almost anything he is interested in. However, more and more time has to be spent in filtering irrelevant information when searching for that something. Complex searching techniques have been employed in today's keyword based search engines. With the goal of further improving search accuracy and efficiency, semantic search based on structured data representation has been intensively studied these years. Many semantic search techniques have been proposed, and some search engines have been developed, including OntoSeek[1], WebKB-2[2], etc.

We have also been devoted to this research area for years, focusing on searching for resources identified by descriptions. The resource or query description is often ad hoc, that is, there are often various descriptions for the same object. And in many cases the description is complicated. So precise search may not perform well. In our approach, we use a restricted but still expressive conceptual graph notation to represent the descriptions and carry out semantic search by graph matching. And before matching we enrich the resource graphs with background knowledge by deductive inference, so as to improve the search performance. We have made a primary study on semantic matching in the past. [3] introduced the matching method whose representation is non-nested with no negation. Although we now choose the clothing domain to demonstrate our approach, the method itself is domain independent and can be applied to other domains.

A. de Moor, W. Lex, and B. Ganter (Eds.): ICCS 2003, LNAI 2746, pp. 378–391, 2003.

The rest of this paper is organized as follows. Sect.2 overviews our whole architecture for semantic search. Sect.3 describes the representation we employ. Sect.4 presents our knowledge enrichment by inference. Sect.5 introduces in detail the semantic matching. Sect.6 compares our work with related work. Sect.7 makes a conclusion in the end.

2   Overview of the Approach

The whole architecture of our approach is shown in Fig.1. Before performing the semantic search, we gather domain resources from the web. Then the descriptions of the resources will be extracted and converted to our representation based on conceptual graphs (using the resource converter in Fig.1). Method to convert various kinds of resources depends on the original representation of the resources. In our previous work, we have developed a system converting resources described in natural language to conceptual graphs by a machine learning approach[4]. With some improvements it can be used here as a kind of converter. After the conversion, the resource graphs will be stored into our resource repository. When the enquiry is entered, it will also be translated into our representation by converters. We use the 'entry' concept of graph, which is introduced in Sect.3, to organize and manage the resource repository and to pilot the query process.

From the view of the query handler module, the input consists of the query graph and the candidate graphs fetched from the resource repository according to the entry of the query graph, and the output is the consensus ranking of the

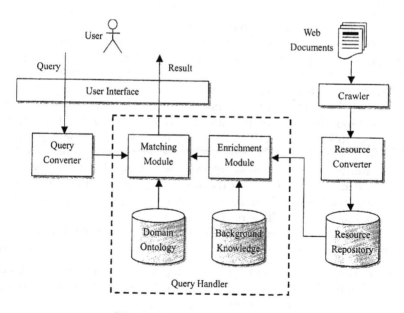

Fig. 1. Overview of the approach

candidates. The ordered answers out of these candidates will be returned to the user.

The query handler module contains two parts: enrichment module and matching module. When a query begins, first the enrichment module will work on the candidate resource graphs by means of inference, so as to gain more information and make the matching module perform better. A background knowledge base will provide the rules needed by the inference. Then the enriched resource graph and the query graph will be matched in the matching module, supported by domain ontology. We now use WordNet[5] as the ontology. In the future we may use more specialized domain ontology instead of it.

The details will be introduced in the following sections, including our representation, the enrichment module and the matching module.

3   Representation

In our approach, we adopt a subset of the classical conceptual graph notation[12]. The subset still holds enough ability to represent negative descriptions as well as assertions.

As presented in [6, 12], the conjunction of two graphs is represented simply by drawing both of them on the same sheet of paper; negation is represented by a negative proposition concept partitioning the negative proposition from the outer context. Since all Boolean operators can be reduced to negation and conjunction, they can be represented in terms of nested negative contexts. And with coreference link, universal quantifier can be represented.

To support conceptual graph matching in our approach, we made the following restrictions on conceptual graph representation:

- As each graph represents a resource object in our approach, we request that every graph has an entry concept, which is the center of the graph designating the represented object and serves as the start position in matching the graph. (See [3] for detailed discussion on the definition and feasibility of entry) The entry of a graph must be in the outermost context, i.e. it can't be in any negative context. Every concept node in the graph should be able to be accessed from the entry through directed relations and undirected coreference links. We mark the entry of a graph when the graph is generated from the description.
- Only coreference links can cross context boundaries, while arcs of relations can't, as demanded in the classical conceptual graph notation. Our restriction is for every context boundary there must be one and only one coreference link that crosses it. This restriction actually guarantees every negative context links to an outer context while avoids some intractable but rare situations such as graphs with coreference link cycles across context boundaries.

These restrictions seem reasonable, as each graph is to describe one object. In our practice, most descriptions can be represented in this way. So we believe our representation is expressive enough for semantic search.

Fig.2 is an example of the representation in our approach.

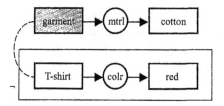

Fig. 2. Conceptual graph for *"a cotton garment that is not a red T-shirt"* (The grey node marks the entry of the graph)

4   Knowledge Enrichment by Inference

In this section, we will introduce the knowledge enrichment phase of our approach. The enrichment is implemented by inference. We will first explain why this phase is needed for semantic search. Then our deductive inference mechanism is formally presented and the enrichment procedure is shown.

4.1   Reasons for the Enrichment

The purpose of knowledge enrichment is to add to the resource conceptual graphs some information that is implicitly contained, so as to make more precise the semantic matching introduced in Sect.5. Properly speaking, there are two main reasons for the enrichment. One is the non-isomorphism phenomenon. The other is inferring new knowledge by using world knowledge.

Non-isomorphism means that one meaning has more than one representation. It is very common in practice. We often have many ways to express the same idea in natural language. When we represent knowledge formally, non-isomorphism still remains. We conclude several familiar kinds of non-isomorphism here:

- **POS.** There are lots of interchangeable uses of different POS forms belonging to one same word. For example, "3 button" and "3 buttoned", "A-line" and "A-lined", etc. In conceptual graph they are interpreted to different concepts with corresponding different relations.
- **Negation.** The meaning of negation can be either implicitly held within one single word or explicitly described using the negative word "not". Take the following as examples: "unlined" and "not lined", "short" and "not long", etc. They are certainly represented differently.
- **Converse Relations.** Expressions appear different when using converse relations. Say, "pocket 1 is to the left of pocket 2" and "pocket 2 is to the right of pocket 1" are differently expressed, but the converse relation pair leads to their same sense.
- **Definition.** A concept in a graph can be substituted by its definition graph without changing the meaning.

The second reason for the enrichment is to infer new knowledge. With world knowledge we can often gain new information from known descriptions. For example, if we know a garment is "stone washed" then it is "prelaundered". New relations can also be introduced. Suppose we get the information that "shirt" is made of "denim" and "denim" is made from "cotton", then we will most probably deduce that this "shirt" is made from "cotton".

4.2   The Deductive Inference

First we define the graph rules, which represent the background knowledge.

Definition 1. *A rule R is a graph: $G_1 \to G_2$, where G_1 and G_2 are two conceptual graphs with coreference link(s) between them. The two concepts linked by a coreference link must be of the same type and must occur in the outermost context. We can replace the coreference link by adding two identical variables as the referents of the two linked concepts. We call G_1 and G_2 the hypothesis and conclusion of the rule respectively.*

The following two graphs are rules representing two examples mentioned in Sect.4.1 ('styl' is the abbr. for 'style', 'matr' for 'matter', and 'mtrl' for 'material'):

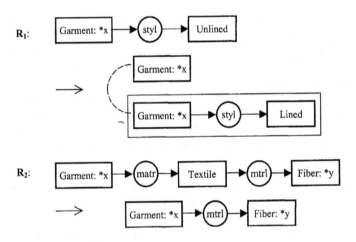

Notice that rules for non-isomorphism should use mutual implication connector instead of implication connector, that is to say, if there is a rule R_1 for non-isomorphism: $G_1 \to G_2$, there should be a rule $R_2 : G_2 \to G_1$.

What does a rule mean? We can represent a rule R in conceptual graph informally with universal quantifier: $\forall x_1, x_2, \ldots, x_n, \neg [G_1 \neg [G_2]]$, where for every $i = 1, 2, \ldots, n$, there are two concepts in G_1 and G_2 respectively, with x_i as the referent and a coreference link between them.

Now we can define the rule of inference.

Definition 2. *Rule of Inference.* *If there is a conceptual graph G and a rule R : $G_1 \to G_2$, and there is a mapping Π from G_1 to G, then we can get a new* *graph. This new graph G' is derived from G and G_2 by inserting G_2 to the same context as $\Pi(G_1)$ and merging each concept c_2 of G_2 with $\Pi(c_1)$. Here c_1 and c_2 denote two concepts in G_1 and G_2 linked by a coreference link, and $\Pi(G_1)$ and $\Pi(c_1)$ respectively denote the image of G_1 and the image concept of c_1 in G. There are some restrictions on the mapping Π from G_1 to G. If in G_1 the dominant concept of c is in an evenly enclosed context, then $\Pi(c)$ in G must be a specialization of c; otherwise if the dominant concept of c is in an oddly enclosed context, then $\Pi(c)$ must be a generalization of c.*

Here is an example illustrating the application of this rule of inference. From the ontology we know 'shirt' is a specialization of 'garment', 'denim' is a specialization of 'textile', and 'cotton' is a specialization of 'fiber'. Therefore if there are the rules R_1 and R_2 shown above and a conceptual graph G (Fig.3a), then we can get a new conceptual graph G'(Fig.3b) by applying the rule of inference twice.

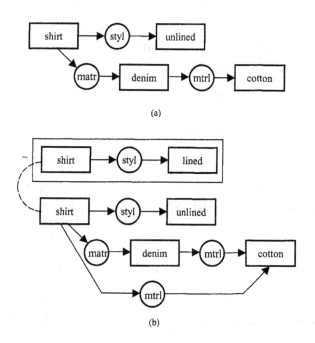

Fig. 3. An example of inference

This rule of inference is sound, or truth-preserving, i.e., if the graph rule and the resource graph are both true, then the result graph is also true. We now give a brief proof by using the first-order inference rules of conceptual graphs [6]. Remember that the rule $R : G_1 \to G_2$ is universally quantified. By universal instantiation, we can link the universally quantified concepts in R to the corresponding ones in G with coreference links. Because G_1 is enclosed at depth 1 when represented in conceptual graph, by erasure rule and insertion rule G_1

can be rewritten to $\Pi(G_1)$. By iteration rule we copy the new form of the rule to the context of $\Pi(G_1)$ in G. Now we have $[\neg [\Pi(G_1) \neg [G_2]] \Pi(G_1)]$ in some context C of G. Then by applying deiteration rule and double negation rule we get $[\Pi(G_1) G_2]$ in C. The coreference links are reserved in the above procedure. So after applying coreference join rule we finally get G' described in Definition 2. All the inference rules used above are sound, so our inference rule is sound too.

On the other hand, this inference is obviously not complete. But the aim of our inference is not to get everything that is true but to make the matching more precise. In fact, we only make the inference for a few steps, as described in the next section. So completeness is not necessary for our approach.

It should be noted that our inference might weaken the description of the resource conceptual graph, instead of restricting it. The simplest case is that, from the resource graph $\neg G_1$ and a rule $G_1 \rightarrow G_2$ with no universally quantified variable, we can get a new graph $\neg [G_1 G_2]$, which is weaker in sense than the resource conceptual graph and can be inferred directly from the resource graph without knowing the background knowledge, i.e., the rule. To be more precise, suppose G_1 is the hypothesis of the rule and G is the resource graph, the result conceptual graph may be weakened if $\Pi(G_1)$ is in an oddly enclosed context of G, and it may be restricted if $\Pi(G_1)$ is in an evenly enclosed context. As shown in the next section, this will affect our use of the deductive inference.

4.3   Knowledge Enrichment Procedure

In our approach, the function of knowledge enrichment is to add background knowledge to the resource graphs. In terms of model theory, this procedure is to restrict the resource representation so as to exclude those models violating the background knowledge represented by the rules. The ideal restricted resource graph is the direct conjunction of the origin graph and the rule, which unfortunately does not meet our representation requirements presented in Sect.3 in some conditions and therefore can't be processed by our matching approach. So we make a compromise between restriction and processability to use the deductive inference introduced in the last section. As we have seen, the resource graph is restricted only if the projection of the rule hypothesis is in an evenly enclosed context of the resource graph. Therefore only in that condition will we do the inference. The entire procedure is described below:

All the rules are stored in a rule base. When a resource conceptual graph needs to be enriched by inference, we will try every rule in the rule base to see whether the hypothesis of the rule and the resource conceptual graph meet the requirements of the inference rule and the forenamed constraint. If so, the deductive approach introduced above will be carried out and new information will be obtained. Before being added to the resource conceptual graph, the new information must be checked so that it won't be a copy of existing information in the resource conceptual graph. We call the last step redundancy detecting.

Such a process will be performed iteratively until no new information could be added or the limit of iteration times is reached. This limit is tunable according to the time and storage limitation. It also prevents the iteration from endlessly going on because of the inference loop, which is caused by such kind of rules that after applying one or more of them serially we will get some non-redundant information which happens to be the hypothesis of the first rule. Notice that such loop can't be prevented by the redundancy-detecting step because the new information gained is actually new.

5   Graph Matching with Negations

This section will explain our matching approach on graphs with negations. We have made a primary study on semantic matching in the past [3]. But with introduction of negations, there are some critical changes in the matching approach. Also there are some other improvements. We will first illustrate how to convert graphs to trees before matching. Then the matching similarity is defined. At last the algorithm is given and evaluated.

5.1   Conversion of the Graphs

Notice that in our approach every graph has an entry which marks the concept described by the whole graph. The entry can't be placed in any negated context and can access every concept of the graph. So we can convert a graph to a tree with the entry as the root. Then matching can be performed on trees instead of graphs.

The conversion is performed by using the entry as the root of the tree and then extending the tree in directed relations and undirected coreference links. Fig.4 is a simple example illustrating the process.

First, notice that the node "button" is duplicated in the resulted tree. This is because there are two possible paths from the entry to the node "button". So the result tree may have more nodes than the original graph. The same node can't be added twice to one path of the tree, so as to break the cycle in the graph.

Second, in the result tree there are two types of links between the concept nodes. One is relation links (solid arrows), which have the same meaning as in a conceptual graph. The other (dashed arrows) is used to represent the coreference links in the original graph. For each group of coreferenced nodes in the graph, there will be a "dummy" node in the result tree, and these nodes will be linked to that dummy node with dashed arrows. This dummy node can be viewed as the individual thing all these coreferenced concept nodes represent. Actually each concept node in the graph without coreference link can also be converted in the tree a concept node with its dummy node linked to it.

Third, the negation symbol on the context "T-shirt colored red" in the original graph is converted to the negation symbol on the node "T-shirt" only. But in fact the negation symbol in the tree means not the negation on the concept node "T-shirt", but the negation on the whole subtree rooted on that node. As in Sect.3 we have made restrictions on negations and coreference links in our graphs, this conversion is always sound.

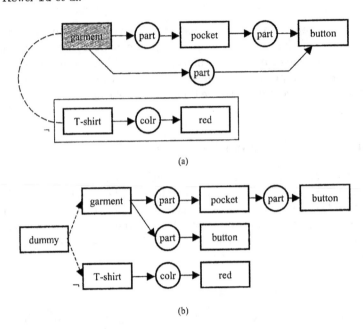

(a)

(b)

Fig. 4. An example of converting a graph (a) into a tree (b) (The grey node of graph (a) marks the entry. 'colr' is the abbr. for 'color'.)

5.2 Similarity Definition

Now we can define the similarity between the query graph and the resource graph. The similarity between graphs is based on the similarity between concepts and relations, which is mainly supported by the domain ontology. As the concept and relation similarity definition used here is the same as in our previous work [3], here we will only focus on the similarity definition between graphs.

In the following discussion, we will use the function $SIM_C(C_Q, C_R)$ to represent the similarity between two concepts, the function $SIM_R(R_Q, R_R)$ to represent the similarity between two relations, and the function $SIM_T(T_Q, T_R)$ to represent the similarity between two subtrees. The subscript Q or R declares the item is from the query or resource graph respectively. The order of parameters in these functions could not be changed, that is, the first parameter must be query item and the second must be resource item. This is because the similarity defined here is not purely tree similarity but the extent to which the resource satisfies the query. The function $W(T_Q)$ represents the weight value of the subtree in the query tree and will be defined later. It represents the importance of the subtree.

The similarity between the query and resource trees or subtrees could be recursively defined as follows, with SIM_C and SIM_R defined in [3]:

Definition 3. $SIM_T(T_Q, T_R) =$

$$
\begin{cases}
\dfrac{\displaystyle\sum_{i=1}^{Num(C_Q)} SIM_T\left(T_Q^i, T_R\right) W\left(T_Q^i\right)}{\displaystyle\sum_{i=1}^{Num(C_Q)} W\left(T_Q^i\right)} & \left(C_Q = dummy\right) \\[6ex]
\dfrac{\displaystyle\sum_{i=1}^{Num(C_R)} SIM_T\left(T_Q, T_R^i\right)}{Num(C_R)} & \left(C_Q \neq dummy, C_R = dummy\right) \\[6ex]
\dfrac{P\left(\displaystyle\operatorname*{Max}_{A(i)} \sum_{i=1}^{Num(C_Q)} SIM_T\left(T_Q^i, T_R^{A(i)}\right) SIM_R\left(R_Q^i, R_R^{A(i)}\right) W\left(T_Q^i\right)\right) + SIM_c\left(C_Q, C_R\right)}{P\left(\displaystyle\sum_{i=1}^{Num(C_Q)} W\left(T_Q^i\right)\right) + 1} & \left(C_Q \neq dummy, C_R \neq dummy\right)
\end{cases}
$$

In the definition, T with subscript represents a tree or subtree, and C with the same subscript represents the root node of T. T^i represents the subtree rooted on the i-th child node of the root of T, and R^i represents the i-th relation of the root of T. $Num(C)$ represents the number of child nodes of the root node C. $A(i)$ is an injection from the subtree labels of the query tree to the subtree labels of the resource tree. If the subtree T^i in the query tree can't find a corresponding subtree in the resource tree, then $A(i)$ is assigned a special value so that the SIM function with $T^{A(i)}$ or $R^{A(i)}$ as the parameter will return 0. P is a discount factor between 0 and 1. The function $W(T_Q)$ represents the weight value of the subtree in the query tree and will be defined later. It represents the importance of the subtree.

The first two formulas in the definition describe how to calculate the similarity while there's dummy node. If the dummy node is the root of the query tree, then the similarity is the weighted average of the similarities of its subtrees with the resource tree. If the dummy node is the root of the resource tree, as there is no weight definition for the resource tree, the result equals to the average of similarities of its subtrees with the query tree. The third formula calculates the similarity between two normal nodes. The main calculation is to find the best possible matching between the query subtrees and the resource subtrees by the Max function. The result similarity is the weighted average of the similarities of subtrees and that of the roots.

In the formula, P is a constant between 0 and 1, which determines how fast the weight of nodes decreases with the depth of the tree. If $P = 1$, all nodes in the query are as important as each other. On the contrary, if $P = 0$, only the entry of the query is considered useful. In our approach, we use $P = 0.5$ by default.

$W(T_Q)$ represents the importance to the user of the description that is represented by the subtree T_Q. If the user describes one of the attributes at length, then that attribute is likely of more importance. So we should add more weight to the subtrees that have more nodes than others. Therefore $W(T_Q)$ is defined as the following:

$$W(T_Q) = \begin{cases} \sum_{i=1}^{Num(C_Q)} W\left(T_Q^i\right) & (C_Q = \text{dummy}) \\ P\left(\sum_{i=1}^{Num(C_Q)} W\left(T_Q^i\right)\right) + 1 & (C_Q \neq \text{dummy}) \end{cases}$$

Users can also modify this function by pointing out the important attributes.

How to calculate the similarity when negations are concerned? Actually, if exactly one of the T_Q and T_R has a negation symbol on the root, the result similarity will be one minus the original similarity of the two trees.

Here is an example showing how the similarity is calculated.

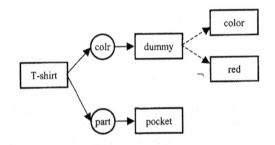

(a) The query tree: a T-shirt which has a pocket and is not red

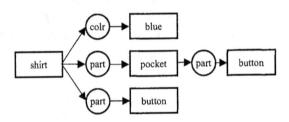

(b) The resource tree: a blue shirt with a buttoned pocket

Fig. 5. An example of mathcing

Before calculating the similarity, we should first calculate the default weight of each subtree in the query tree. According to the formula above, the weights of "color", "blue" and "pocket" are 1, the weight of subtree rooted on "dummy" is 2, and therefore the weight of the whole tree is 2.5.

To calculate the similarity between "T-shirt" and "shirt", the algorithm will first try to match their child nodes. In theory, six similarities (two children in the query match three children in the resource) should be calculated. Because similarity between different relations is zero, in fact only similarities between "dummy" and "blue", "pocket" and "pocket" and "pocket" and "button" are calculated.

The similarity between "dummy" and "blue" is the weighted average of the similarity between "color" and "blue" and the similarity between "not red" and "blue". "Blue" is a kind of "color" and "blue" is in the resource, so the sub-query "color" is satisfied and the similarity is 1. Assume the similarity between "red" and "blue" is 0.2, so the similarity between "not red" and "blue" is 0.8. We now get the similarity between "dummy" and "blue" to be 0.9. On the other side, the similarity between "pocket" and "pocket" equals to 1. As the matching is based on the query, the "button" linked to "pocket" in the resource is ignored. The similarity between "pocket" and "button" is small. So the best matching is "blue" to "dummy" and "pocket" to "pocket".

Finally, although "shirt" is the super type of "T-shirt", it is in the resource. So we assume the similarity between "T-shirt" and "shirt" is 0.7. By the formula we get the final result 0.84.

5.3 Matching Algorithm and Its Evaluation

As described above, our matching method has two steps. The first is to convert the query and resource graphs into trees. The algorithm is similar to depth-first graph traversal. The second is to calculate the similarity of two trees. As the similarity is recursively defined, we use recursion to carry out this step. In matching every pair of nodes, the min cost max flow algorithm is used to get the best match of their child nodes. The evaluation of the time complexity of our algorithm is given below.

The first step is to convert a graph into a tree. In practice, the number of nodes of the generated tree is just a little larger than that of the graph. So the time complexity is linear in the size of the graph in most cases.

The second step is to match two trees. Assume the height of both trees is no larger than h, and there are at most n nodes in one level of a tree. While the min cost max flow algorithm is used for matching, the time complexity is $O\left((a+b)^3 \min(a,b)\right)$, where a and b denote the node number in the two groups for matching. In each level of a tree, we can group the nodes by their parents. Suppose in query tree the number of nodes in each group is $a_1, a_2, \ldots, a_q$ respectively, and in resource tree the number of nodes in each group is $b_1, b_2, \ldots, b_r$ respectively. Then we can calculate the time complexity of the matching in each level of the trees as the following:

$$T = \sum_{i=1}^{q} \sum_{j=1}^{r} \left((a_i + b_j)^3 \min(a_i, b_j)\right)$$

$$= \sum_{i=1}^{q} \sum_{j=1}^{r} \left((a_i^3 + 3a_i^2 b_j + 3a_i b_j^2 + b_j^3) \min(a_i, b_j)\right)$$

$$\leq \sum_{i=1}^{q} \sum_{j=1}^{r} \left((a_i^3 + 3a_i^2 b_j) b_j + (3a_i b_j^2 + b_j^3) a_i\right)$$

$$= \sum_{i=1}^{q} \sum_{j=1}^{r} \left(a_i^3 b_j + 6 a_i^2 b_j^2 + a_i b_j^3 \right)$$

$$< \left(\sum_{i=1}^{q} a_i + \sum_{j=1}^{r} b_j \right)^4$$

$$\leq (2n)^4$$

So, the actual time complexity of the whole algorithm is $O\left(n^4 h\right)$. In practice, most graphs have no more than a few tens of nodes, so we think this time complexity is acceptable.

6 Related Work

Semantic search has been studied for years to improve both recall and precision of information retrieval. Most projects emphasize on selecting from semantic data source the information which perfectly fits the query. In contrast, we use a graph matching approach to carry out imprecise semantic search. Some projects, such as OntoSeek [1], also employ graph matching methods. Our difference from these ones is discussed in the third paragraph. Another feature of our approach is the use of inference to introduce the background knowledge. While many projects have not employed inference, some ones just use limited inference to support query broadening or relaxation, such as *DAML Semantic Search Service* [7]. OntoBroker [8], with SiLRI [9] as its inference engine, employs additional rules expressed in F-logic to provide information about relationships between predicates, therefore it has the capacity to get information by inference which is not stated explicitly in the source structural data. Our approach also employs similar additional rules. Where it differs from OntoBroker is that we query by means of semantic matching. As a result, our approach has to adopt data-driven inference (or forward inference) so that after inference our matching is meaningful.

Sound and complete conceptual graph inference has long been well-established [6, 12]. But we only use a sound and non-complete inference in order to simplify the implementation and make the approach more efficient. The forward chaining inference discussed in [10] is similar to ours. It can be viewed as a simplified version of our inference, for it only deals with simple conceptual graphs which are non-nested with no negation. Therefore its inference is complete. But the expressive power of simple conceptual graphs is too limited, so while reserving the simplicity of the inference, we sacrifice the completeness, which is not important to our method.

Some previoufs work has studied the issue of semantic matching. OntoSeek [1] fetches corresponding nodes first and then checks the arc linkage between them, so as to avoid NP-completeness of such a computation known as Maximum Subgraph Matching. We think such simplification separate matching on nodes from the organization of the graph. So we try to retain graph structure in our similarity definition but confine comparison range to mitigate the computation. Cupid [11] matches on bottom-up traversal and biases matches of schema leaves to perform schema mapping. We share the idea of bottom-up traversal with Cupid, but we prefer matches of the entry. Moreover, like most semantic matching researches, neither of the above projects has the ability to represent and match negation, which is an important feature of our approach.

7   Conclusion and Future Work

In this paper we introduced a semantic search approach by graph matching. Our knowledge representation is based on conceptual graphs with some restrictions, while it is still expressive for semantic search. Before matching we enrich the resource graphs with linguistic and world knowledge by a means based on a formal deductive inference. The semantic matching method calculates the similarity between the query graph and the resource graph in polynomial time. At last the candidate resource graphs are ranked and the best corresponding answers are output. A prototype of our approach is currently under development.

There is still a lot of possible work that remains to be done. Various methods to convert resource and query to our representation could be developed. The rule base used in inference could be automatically built by means of DM methods. Confidence of rules could be introduced to the inference approach. Matching of different levels of the tree structure is also an interesting idea.

References

1. N. Guarino, C. Masolo, and G. Vetere: OntoSeek: Content-Based Access to the Web. IEEE Intelligent Systems, 14(3), pp.70–80.
2. P. A. Martin: General documentation of WebKB-2. Available at http://kvo.itee.uq.edu.au/webkb/WebKB2/doc/generalDoc.html
3. Jiwei Zhong, Haiping Zhu, Jianming Li and Yong Yu: Conceptual Graph Matching for Semantic Search. In Proc. of the 10th Intl. Conf. on Conceptual Structures, (ICCS2002), LNAI 2393, Springer-Verlag, 2002.
4. Lei Zhang and Yong Yu: Learning to Generate CGs from Domain Specific Sentences. In Proc. of the 9th Intl. Conf. on Conceptual Structures, (ICCS2001), LNAI 2120, Springer-Verlag, 2001.
5. George A. Miller: WordNet: An On-line Lexical Database. In the Intl. Journal of Lexicography, Vol.3, No.4, 1990.
6. J. F. Sowa: Conceptual Structures: Information Processing in Mind and Machine. Addison-Wesley. 1984.
7. DAML Semantic Search Service, http://reliant.teknowledge.com/DAML/
8. OntoBroker, http://ontobroker.aifb.uni-karlsruhe.de/index_ob.html
9. S. Decker, D. Brickley, J. Saarela, and J. Angele: A Query and Inference Service for RDF. In: QL'98 - The Query Languages Workshop. 1998.
10. E. Salvat and M.L. Mugnier: Sound and complete forward and backward chainings of graph rules. In Proc. of the 4th Intl. Conf. on Conceptual Structures, ICCS'96, LNAI 1115, Springer-Verlag, 1996.
11. J. Madhavan, P.A. Bernstein, and E. Rahm: Generic Schema Matching with Cupid. In Proc. of the 27th Intl. Conf. on Very Large Databases (VLDB 2001).
12. J. F. Sowa (ed.): Conceptual Graph Standard. Available at http://www.jfsowa.com/cg/cgstand.htm

Using Conceptual Graphs
to Capture Semantics of Agent Communication

Lois W. Harper and Harry S. Delugach

Department of Computer Science
N300 Technology Hall
University of Alabama in Huntsville
Huntsville, AL 35899 USA
{lharper, delugach}@cs.uah.edu

Abstract. Agent communication languages such as KQML and the FIPA ACL serve as metalanguages to define software agent message-passing protocols. These metalanguages are incompatible with each other, preventing intercommunication between agents employing different agent communication languages. The primary hindrance to agent intercommunication is the different underlying semantics of the message passing protocols. Conceptual graphs provide a mechanism to bridge this agent communication barrier by representing the semantics of message-passing protocols in the formal representation of conceptual graphs. Semantic content of the KQML **tell** performative is contrasted with that of the FIPA ACL **inform** performative and represented in conceptual graphs. The intent is that software agents conversant in CGIF may intelligently translate messages between agents employing different agent communication languages.

1 Introduction

A great deal of interest exists in developing enterprise agent-based software applications that will support distributed processes and distributed artificial intelligence. The development of distributed agent-based applications is being pursued both for Internet applications and for other networked environments, such as those found in manufacturing facilities. Three technical challenges that have been identified in supporting these types of agent-based applications are:

- reusability of agent components to support short development lifecycles.
- definition of frameworks that support heterogeneous computing platforms.
- standardization in agent communication languages and protocols.

The third issue, standardization in agent communication languages and protocols, has been studied in light of human communication and Speech Act Theories [1]. Shen, et al [2] state what is needed for intelligent agents to see practical application:

- agents need to be aware of the existence of other agents and of their access mechanisms.
- agents need to know and be capable of communicating their domain ontology.
- agents need to be aware of all available agent services.

A. de Moor, W. Lex, and B. Ganter (Eds.): ICCS 2003, LNAI 2746, pp. 392–404, 2003.

- agents need to know how to monitor other agent tasks for completion success and performance status.
- agents need to know how to specify parameters to customize their service requests.

This list emphasizes that good communication is as critical for successful software agent applications as it is for any successful human endeavor. However, software agents are a new programming paradigm. There are no commonly accepted definitions of what comprises a software agent (mobile or intelligent), nor are software agent systems and agent communication mechanisms standardized. A significant number of different and incompatible agent-based systems will continue to exist until industries can agree upon and apply standards for agent interfaces, agent communication languages and agent computing platforms.

2 Agent Communications and Conceptual Graphs

FIPA ACL and KQML are two Agent Communication Languages (ACLs) that model communication between intelligent software agents after spoken human communication. These models draw from speech act theories, particularly the approach presented by J. R. Searle [3]. Searle's theory considers natural language utterances such as questions, replies and declarations as actions, and also identifies three types of actions associated with an utterance:

- *Locution*, the physical utterance by the speaker
- *Illocution*, the intended meaning of the utterance by the speaker
- *Perlocution*, the action that results from the locution

The intended meaning of a communication act can be termed its *illocutionary force*. The term *performative* identifies the illocutionary force of an illocution and is identified by a verb such as "tell", "convince" or "ask" which is the core meaning of a speech act. The intended meaning of the message passed between two software agents corresponds to illocution in speech act theory. As a result, the messages between intelligent software agents are commonly referred to as *performatives*. The FIPA uses the terms *communicative act* instead of the term *performative*. Since *communicative act* and *performative* are interchangeable, we will use the term *performative* in this paper.

A performative is therefore a fundamental 'unit of communication' between software agents. A performative is a wrapper for some *content* that an agent wants to send or receive. The wrapper indicates who sent the message, what language the content of the message is written in, etc. Agent Communication Languages (ACLs) structure the type and sequence of performatives into communication *protocols* that govern the exchange of messages between software agents. These ACL *protocols* are metalanguages that define both what are acceptable sequences of messages and also the expected outcome of acceptable sequences of messages. Thus a *performative* distinguishes between the type of message and its content, whereas a *protocol* is an expected sequence of messages.

The Foundation for Intelligent Physical Agents (FIPA) began to evaluate specifications for agent technologies in 1995. [12] Their purpose was to promote the success of emerging agent-based applications and services. By making specifications on agent-based technologies available to the global community, the FIPA intended to improve the ability of different agents implementations to operate as societies of agents capable of effective communication acts. [4] The FIPA determined that agent technologies that are grounded in artificial intelligence techniques were sufficiently mature for establishing specifications and FIPA specifications for agent technologies have been developed accordingly [5].

The Defense Advanced Research Project Agency (DARPA) Knowledge Sharing Effort (KSE) initiated research efforts towards knowledge sharing and reuse as early as 1990 [4]. DARPA produced several well-known software languages and utilities that facilitate intelligent agent communications, such as KQML, KIF, Ontolingua and OKBC. KQML was among the first of the ACLs to be developed and used [1], [2]. Researchers who have contributed to the DARPA KSE effort include Tim Finin and Yannis Labrou. A significant side-effect of the separate FIPA and DARPA KSE efforts towards defining and developing agent communication technologies is that the two most commonly known agent communication languages, KQML and the FIPA ACL (hereafter referred to as FIPA), cannot be directly translated from one form to the other [4].

As our modeling language we use conceptual graphs, which are a modeling and reasoning technique based around concepts and relations [6]. A special kind of relation, called an *actor*, is used to denote actions or functional relationships between concepts. With conceptual graphs (CGs), we can model semantics and then use our semantic models to reason about a domain. As an emerging knowledge interchange standard (through CGIF [7]) we can also exchange and re-use our models in other knowledge modeling efforts. The use of CG-based software agents that function as knowledge providers for Internet transactions is already being examined [8]. When conceptual graphs are used to represent the semantics of ACLs, a mechanism is provided that may allow software agents to communicate not only the content of a single message, but also the semantics of each performative and the intended outcome of a sequence of performatives. Since KQML and FIPA cannot be directly translated from one form to the other, their semantic framework must be implicitly represented in the knowledge bases of the software agents designed to be conversant in these languages. Capturing the semantics of these ACLs in CGIF form makes these semantic frameworks explicitly available to other software agents conversant in CGIF. The CGs presented in this paper were drawn using the CharGer Conceptual Graph Editor [13].

The main focus of this paper is to illustrate agent communication semantics in conceptual graphs by using a sample message sent from an agent called A to an agent called B. We then describe its semantics in some detail for each of the agent languages. Our purpose is two-fold: we want to show that conceptual graphs are capable of representing agent communication semantics, and we want to use our representations to illustrate the semantic differences between the KQML and FIPA agent languages.

3 Conceptual Graph Models

This section presents conceptual graphs (CGs) that represent a **tell** KQML performative and an **inform** FIPA performative, and includes the following:

- a description of the **tell** KQML and **inform** FIPA performatives' syntax.
- a simple example proposition that forms the actual content to be communicated, represented in CG form.
- preconditions, postconditions and completion conditions for the **tell** KQML and **inform** FIPA performatives.

The goal of both the **tell** KQML performative and the **inform** FIPA performatives is to convey to some receiving agent that the sending agent believes a particular proposition (contained in the *content* field of a performative) is true. Referring to Table 1, it can be seen that the syntax of the FIPA **inform** performative is identical to KQML's **tell** performative (except for the two different keywords **tell** and **inform**). The syntax of both FIPA and KQML is based on *s*-expressions in LISP; the first element is the performative name and the remaining elements are the keyword/value pairs that make up the performative's arguments. While the syntax of these FIPA and KQML performatives are identical, the semantic frameworks of the two ACLs are substantially different [4], which our example will illustrate. As a result, there is no exact transformation or mapping between the two ACLs, even in this simple example of two syntactically equivalent performatives that appear to support the same goal.

Table 1. KQML **tell** and FIPA **inform** Performatives

KQML tell Performative	FIPA ACL inform Performative
(tell	(inform
: sender agent A	: sender agent A
: receiver agent B	: receiver agent B
: content ("Agent A performs the task negotiate.")	: content ("Agent A performs the task negotiate.")
: language text)	: language text)

The FIPA and KQML ACLs do not impose any constraints on how a proposition is expressed in the content field. For example, the content may be statements in a programming language, a series of binary digits, or a text string as shown here. Consequently, the proposition in the content field may also be expressed in CGIF form.

A query in conceptual graphs is denoted by a graph where one or more concepts are unbound. We show bound and unbound CG representations of the sentence *"Agent A performs the task negotiate."* in Figure 1. The unbound form of this sentence represents a query and is referred to as 'Y' in the KQML specification, but is not referred to by the FIPA ACL specification [9], [5]. The bound form of this sentence represents a proposition declared to be true by the agent sending the sentence. (The bound proposition is referred to as 'X' in both the KQML and FIPA ACL specifications.) In this paper, all references to the content field of this performative refer to the bound and unbound forms of the content sentence of Table 1, represented in CG form in Figure 1.

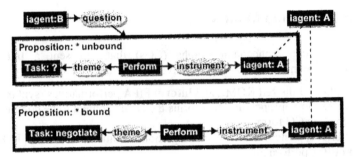

Fig. 1. CG Representation of a Proposition, unbound and bound Forms

We make the following note about the term *agent*. It is customary in CGs to use the relation name *agent* to link an action's concept to the animate-entity concept that performs the action [7]. Unfortunately, this may cause some confusion when the animate–entity itself is a software agent in the sense we are using in this paper. To prevent confusion, intelligent software agents will be denoted in the following CGs by the concept label *iagent*. Also, we will not use the relation *agent* in this paper.

3.1 Illocution of KQML tell and FIPA ACL inform Performatives

The CG representations of the KQML performative **tell** and the FIPA performative **inform** are shown in Figure 2. The left graph is the KQML performative **tell** and the right graph is the FIPA performative **inform**. These graphs are almst identical, as one would expect for performatives that are syntactically equivalent and appear to support the same goal. The goal of both performatives is to convey to the receiving agent (iagent B) that a sending agent (iagent A) believes the proposition in the performative's content field is true.

Looking at the two graphs in Figure 2, we see that they have identical structure and content. A key point of our paper is that their differences are found not in their structure, but in their semantics – their intent as well as the context in which they are used – as we will show in the rest of the paper.

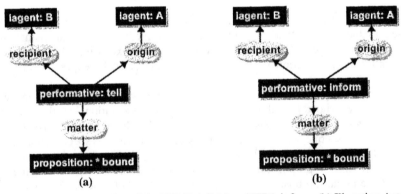

Fig. 2. CG Representation of the KQML **tell** (a) and FIPA **inform** (b) Illocution Act

3.2 Preconditions on Agent A and Agent B

The semantics of KQML is expressed in terms of pre-conditions, post-conditions, and completion conditions for each performative [9]. If the pre- and post-conditions do not hold, a **sorry** or **error** performative is returned and the completion condition cannot occur. [9] A completion condition indicates that the intention, or core meaning, of the performative has been fulfilled.

FIPA expresses communication semantics in terms of a performative's feasibility conditions and its rational effect [5]. Feasibility preconditions are conditions necessary for the sender to perform the illocution act. The rational effect is the expected result(s) of the performative being fulfilled. Thus, in the FIPA specification for the **inform** performative, the feasibility preconditions are specified for the sender and rational effect postconditions are specified for the receiver. In our paper, we use the terms preconditions, postconditions and completion conditions for both the FIPA ACL **inform** performative and the KQML **tell** performative.

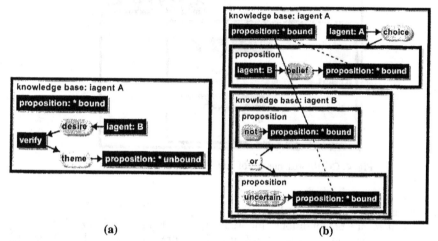

(a)   (b)

Fig. 3. CG Representation of Preconditions on KQML agent A (a) and FIPA agent A (b)

3.2.1 Preconditions on Agent A

The CGs shown in Figure 3 and summarized in Table 2 illustrate that both FIPA and KQML require that agent A holds the bound proposition to be true. However, KQML agent A (Figure 3(a)) also knows that KQML agent B has performed a query with the unbound proposition. The query may have been in the form of the KQML performatives **ask-if**, **ask-all** or **stream-all**. KQML agent A is not required to have prior knowledge of KQML agent B's knowledge base. FIPA agent A (Figure 3(b)) must intend that FIPA agent B comes to believe that the bound proposition is true, but also has two preconditions placed on its knowledge of FIPA agent B's knowledge base. FIPA Agent A must know that FIPA agent B has no knowledge of the bound proposition, or, that FIPA agent B is uncertain of the truth of the bound proposition.

A practical consequence of these differences on the preconditions placed on agent A is that KQML agent A may send the **tell** performative whether or not KQML agent

B has prior knowledge of the bound proposition, whereas FIPA agent A may not send the **inform** performative if FIPA agent A perceives that FIPA agent B has certain knowledge of the bound proposition. [5]

Table 2. Summary of Preconditions on agent A

KQML tell Preconditions (agent A)	FIPA inform Preconditions (agent A)
Agent A holds the bound proposition to be true	Agent A holds the bound proposition to be true
Agent A 'knows' that agent B desires to verify an unbound proposition	Agent B has no knowledge of the bound proposition or is uncertain of the bound proposition

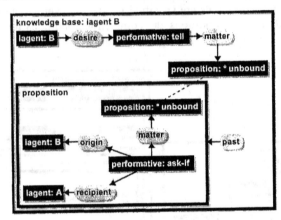

Fig. 4. CG Representation of KQML agent B's Preconditions

3.2.2 Preconditions on Agent B

Although there are preconditions imposed on agent B in KQML [9] there are no preconditions specified for agent B in FIPA [5]. Figure 4 shows that KQML agent B has sent a query concerning the unbound proposition and desires a **tell** performative in return. The **ask-if** performative is represented in Figure 4, although the query sent may also have been one of the KQML **ask-all** or **stream-all** performatives.

In contrast, Figure 5 shows that the FIPA **inform** performative enforces no preconditions on agent B. The FIPA Specification states, "... perhaps agent A has

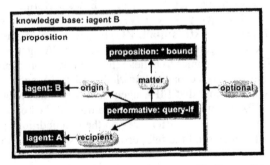

Fig. 5. CG Representation of FIPA agent B's Precondition

been asked," but it does not say that FIPA agent B has to have already placed a prior query with FIPA agent A. [5]

The FIPA specification [5] states that perhaps FIPA agent A has been asked about the proposition. As a result, this is an optional precondition that is not shown in Table 3.

Table 3. Summary of Preconditions on agent B

KQML tell Preconditions (agent B)	FIPA inform Preconditions (agent B)
Agent B has sent an 'ask-if', 'ask-all', or 'stream-all' performative to agent A in the past concerning the unbound proposition	(no mandatory preconditions)
Agent B desires to receive a tell performative concerning the unbound proposition	

3.3 Postconditions for the KQML tell and FIPA ACL inform Performatives

The CGs for the postconditions on agent A make use of the type of node referred to as an *actor*. Actors are used in CGs to represent processes that use their input concepts' referents to change the referents of their output concepts. [10] The actors in these CGs illustrate that the KQML **tell** performative and the FIPA ACL **inform** performative change agent A's view of agent B's knowledge base.

3.3.1 Postconditions on Agent A

A significant difference between the postconditions on agent A for the two ACLs is the effect the performative*s* have on agent A's representation of agent B's knowledge base. Referring to Figure 6 (summarized in Table 4), the viewpoint of KQML agent A (Figure 4(a)) is that the **tell** performative might have changed KQML agent B's unbound proposition to the bound proposition sent to KQML agent B by KQML agent A. The viewpoint of FIPA agent A (Figure 4(a)) is that the **inform** performative has asserted a bound proposition into FIPA agent B's knowledge base.

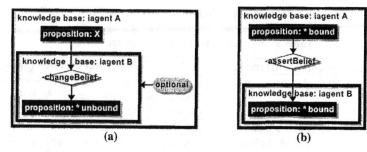

(a)   (b)

Fig. 6. CGs For Postconditions on KQML agent A (**a**) and FIPA agent A (**b**)

Bear in mind that these postconditions are representing agent A's viewpoint of the communication act, i.e., these postconditions reflect agent A's representation of what it knows about agent B's knowledge base.

Table 4. Summary of Postconditions on agent A

KQML tell Postconditions (agent A)	FIPA inform Postconditions (agent A)
Agent A holds the bound proposition to be true	Agent A holds the bound proposition to be true
Agent A's view of agent B's knowledge base shows that agent A has <u>optionally</u> changed the unbound proposition in agent B's knowledge base to the bound proposition.	Agent A's view of agent B's knowledge base shows that agent A has asserted the bound proposition into agent B's knowledge base.

3.3.2 Postconditions on Agent B

The postconditions for agent B are similar between the two ACLS. Figure 7(a) shows that KQML agent B is free to retain the unbound proposition (perhaps while polling a number of agents with the same query), but also knows that KQML agent A believes that the bound proposition is true. Figure 7(b) shows that FIPA agent B knows that FIPA agent A believes that the bound proposition is true, but is not required to accept the bound proposition sent by agent A. [5] states that " whether or not the receiver does, indeed, adopt the proposition will be a function of the receiver's trust in the sincerity and reliability of the sender." This is summarized in Table 5.

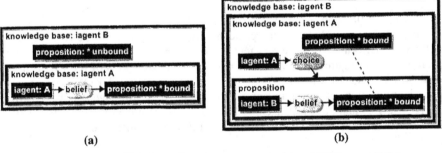

(a)   (b)

Fig. 7. CGs For Postconditions on KQML agent B **(a)** and FIPA agent B **(b)**

Table 5. Summary of Postconditions on agent B

KQML tell Postconditions (agent B)	FIPA inform Postconditions (agent B)
Agent B holds the unbound proposition	Agent B's view of agent A's knowledge base shows that agent A holds the bound proposition to be true.
Agent B's view of agent A's knowledge base shows that agent A believes the bound proposition is true.	Agent B's view of agent A's knowledge base shows that it is agent A's choice that agent B hold the bound proposition to be true.

3.4 Completion Conditions on Agent A for the KQML tell and FIPA ACL inform Performatives

Referring to Figure 8(a), the completion condition applies for KQML agent A as long as KQML agent B does not send a **sorry** or **error** performative. There is no statement as to 'how long' KQML agent A must wait for a **sorry** or **error** performative to be returned. [9] FIPA ACL does not state a completion condition, but does state that the rational effect of the **inform** performative is that FIPA agent B (Figure 8(b)) believes the bound proposition sent by FIPA agent A. However, Section 3.3 noted that FIPA agent B is not required to accept the bound proposition sent by FIPA agent A. Our interpretation is that, whether or not FIPA agent B accepts the bound proposition, it is accepted in FIPA agent A's representation of FIPA agent B's knowledge base.

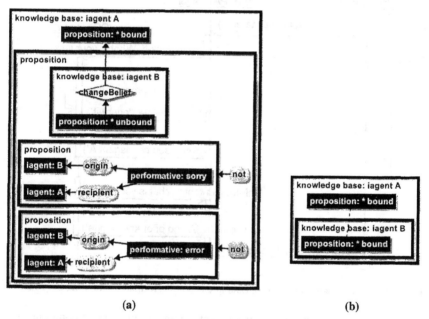

(a)   (b)

Fig. 8. CGs For Completion Conditions on KQML agent B (**a**) and FIPA agent B (**b**)

4 Discussion

Section 3 illustrated the different semantics underlying the KQML **tell** performative and the FIPA ACL **inform** performative. The difference in semantics is substantial although the performatives have the same syntax and goal (both intend to convey to a receiving agent that a sending agent believes a stated proposition is true).

The CGs of these performatives are shown in graphical form, but of course, they may also be represented in Conceptual Graph Interchange Form (CGIF) [7]. As the CGIF standard emerges, more systems will be able to share knowledge bases, so that intelligent software agents that are 'conversant' in CGIF may be written that are capable of bridging the communication gap between different ACLs.

Moore [11] analyzed the semantics of KQML and translated performatives into "more or less equivalent" forms in FLBC, the Formal Language for Business Communication. Moore's goal was not to facilitate communication between software agents conversant in these two ACLs. Rather, Moore states, "... a message for automated electronic communication should more closely reflect its underlying meaning" and proposes FLBC as "exemplar of this approach". The use of CGs to capturing semantics of ACLs meets the need expressed by Moore to represent the underlying semantics of ACLs, but has the significant added advantage of representing both the explicit and implicit semantics of one or more ACLs in CGIF machine-readable form that offers interchange and reasoning capabilities beyond that of an ACL.

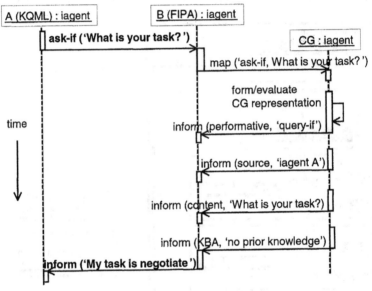

Fig. 9. Sequence Diagram Depicting Example Agent Performative Mapping Sequence

Figure 9 is a UML sequence diagram showing an example agent communication sequence. The purpose of a UML sequence diagram is to show the time-sequence of messages sent between objects. The messages sent between objects are depicted as horizontal lines. Time flows from the top of the diagram downward, although no time scale is implied. The objects at the top of Figure 9 are intelligent software agents.

The sequence diagram in Figure 9 shows an example agent communication scenario. In this example an agent conversant in FIPA receives an **ask-if** performative from an agent conversant in KQML. The content of the **ask-if** performative is the query, "What is your task?" The FIPA agent does not recognize the KQML **ask-if** performative. Since the FIPA agent is intelligent it may choose to ignore the message. However, consider the scenario where the services of a CG agent are available. The FIPA agent may refer the message to the CG agent for translation. This is shown in Figure 9 as a **map** performative. The CG agent evaluates the message, recognizes the performative as being a member of the KQML ACL and returns information that will assist the FIPA agent in responding to the KQML agent's message. In this example, the CG agent informs the agent conversant in FIPA that a similar FIPA ACL performative is query-if, and also informs the FIPA agent that the KQML agent has no prior knowledge of the FIPA agent's task. This is significant, since a practical consequence of the preconditions placed on agents is that a KQML agent may send the **tell** performative whether or not the receiving KQML agent has prior knowledge of a proposition, whereas an FIPA agent may not send the **inform** performative if the sending FIPA agent perceives that the receiving FIPA agent has certain knowledge of a proposition. (Refer to Section 3.2.1) In this example scenario, the agent conversant in FIPA may then return an **inform** performative with the sentence "My task is negotiate."

Such behavior on the part of software agents requires that rules be developed for mapping the performatives in the different ACLs based on the semantic representation of the performatives. While this is not possible from analysis of the syntax of the performatives alone, the conceptual graph representation of the semantic content of a performative may form a basis from which this mapping process occurs.

The net effect of enlisting the mapping services of the CG-based agent is that messages may be sent between agents conversant in different, incompatible agent communication languages. An agent conversant in one ACL may send a performative to an agent conversant in a semantically different and incompatible ACL. The receiving agent may then return an appropriate response. In this scenario, indirect but effective translation of messages between agents conversant in different agent communication languages may occur.

Although this paper examined the semantics of two ACLs (FIPA and KQML), there are also many 'dialects' of KQML, each tailored to a specific application. [2] The same approach may be taken to communicate between agents conversant in different dialects of KQML and also may be useful in facilitating communication between legacy and current versions of the same dialect of an ACL.

Our future work is to examine the other performatives in both FIPA and KQML and represent their semantics in conceptual graph terms. This will give us an opportunity to improve our understanding of the performatives and their corresponding roles in FIPA and KQML ACLs. Our ultimate goal is to find ways that agents using different languages can automatically communicate, even if their underlying semantics differ.

5 Conclusions

In this paper, we have shown how conceptual graphs are capable of representing agent communication semantics, and we illustrated the semantic differences between two performatives used by two of the most commonly known agent communication languages, KQML and FIPA ACL. These performatives were selected because although they are syntactically equivalent and support the same communication goal, they are significantly different in their underlying semantics. The conceptual graphs shown illustrate specific semantic incompatibilities. We also illustrated a scenario of a possible communication sequence between agents conversant in two different ACLs.

Our aim is to overcome semantic incompatibilities by using CG-based translation techniques based on semantic representations in the machine-readable CGIF form.

The conceptual graphs presented in this paper are meant as a proof-of-concept for the approach; we will pursue modeling of other performatives as well. Once we have established semantic models of other performatives used in the FIPA and KQML ACLs, we will be in a position to develop CGIF-conversant agents that can facilitate the transfer of messages (and therefore message content) between agents designed for otherwise-incompatible ACLs.

References

[1] M. N. Huhns and L. M. Stephens, "Multiagent Systems and Societies of Agents," *Multiagent Systems: A Modern Approach to Distributed Artificial Intelligence*, Cambridge: The MIT Press, 1999.

[2] W. Shen, N. Douglas, and J. P. Barthes, *Multi-Agent Systems for Concurrent Intelligent Manufacturing* New York, NY: Taylor & Francis Inc., 2001.

[3] J. R. Searle, *Speech Acts: An Essay in the Philosophy of Language*, Cambridge University Press, Cambridge 1970.

[4] Y. Labrou, T. Finin, and Y. Peng, "Agent Communication Languages: The Current Landscape," *IEEE Intelligent Systems*, vol. 14, pp. 45-52, 1999.

[5] FIPA, "FIPA 97 Specification Part 2: Agent Communication Language," Foundation for Intelligent Physical Agents, October, 1998.

[6] J. F. Sowa, *Conceptual Structures: Information Processing in Mind and Machine*. Reading, Mass.: Addison-Wesley, 1984.

[7] J. F. Sowa, "Conceptual Graph Standard," http://users/bestweb.net/~sowa/cg/cgstandw.htm, 2000.

[8] G. W. Mineau, "A First Step Toward the Knowledge Web: Interoperability Issues Among Conceptual Graph Based Software Agents," in *Conceptual Structures: Integration and Interfaces*, vol. 2393, *Lecture Notes in Artificial Intelligence*, U. Priss, D. Corbett, and G. Angelova, Eds.: Springer-Verlag, 2002, pp. 250-260.

[9] Y. Labrou and T. Finin, "A Semantics Approach for KQML - A General Purpose Communication Language for Software Agents," presented at the Third International Conference on Information and Knowledge Management, 1994.

[10] H. S. Delugach, "Specifying Multiple-Viewed Software Requirements with Conceptual Graphs," *Journal on Systems and Software*, vol. 19, pp. 207-224, 1992.

[11] S. Moore, "KQML & FLBC: Contrasting Agent Communication Languages," *IEEE Proceedings of the 32nd Hawaii International Conference on System Sciences*, 1999.

[12] H. Suguri, "A Standardization Effort for Agent Technologies: The Foundation for Intelligent Physical Agents and Its Activities," *IEEE Proceedings of the 32nd Hawaii International Conference on System Sciences,* 1999.

[13] H. S. Delugach, " CharGer Conceptual Graph Editor," http://www.cs.uah.edu/~delugach/CG, 2003.

Author Index

Lecture Notes in Artificial Intelligence (LNAI)

Lecture Notes in Computer Science